Biopolymers and Biopolymer Blends

Biopolymer and Biopolymer Blends: Fundamentals, Processes, and Emerging Applications showcases the potential of biopolymers as alternative sources to conventional nonbiodegradable petroleum-based polymers. It discusses fundamentals of biopolymers and biopolymer blends from natural and synthetic sources, synthesis, and characterization. It also describes development of desired performance for specific applications in 3D printing and other emerging applications in industry, including packaging, pulp and paper, agriculture, biomedical, and marine.

- Introduces the fundamentals, synthesis, processing, and structural and functional properties of biopolymers and biopolymer blends.
- Explains the fundamental framework of biopolymer blends in 3D printing, featuring current technologies, printing materials, and commercialization of biopolymers in 3D printing.
- Reviews emerging applications, including active food packaging, electronic, antimicrobial, environmental, and more.
- Discusses current challenges and futures prospects.

Providing readers with a detailed overview of the latest advances in the field and a wealth of applications, this work will appeal to researchers in materials science and engineering, biotechnology, and related disciplines.

Emerging Materials and Technologies

Series Editor: Boris I. Kharissov

The *Emerging Materials and Technologies* series is devoted to highlighting publications centered on emerging advanced materials and novel technologies. Attention is paid to those newly discovered or applied materials with potential to solve pressing societal problems and improve quality of life, corresponding to environmental protection, medicine, communications, energy, transportation, advanced manufacturing, and related areas.

The series takes into account that, under present strong demands for energy, material, and cost savings, as well as heavy contamination problems and worldwide pandemic conditions, the area of emerging materials and related scalable technologies is a highly interdisciplinary field, with the need for researchers, professionals, and academics across the spectrum of engineering and technological disciplines. The main objective of this book series is to attract more attention to these materials and technologies and invite conversation among the international R&D community.

Wastewater Treatment with the Fenton Process
Principles and Applications
Dominika Bury, Piotr Marcinowski, Jan Bogacki, Michal Jakubczak, and Agnieszka Jastrzebska

Mechanical Behavior of Advanced Materials:
Modeling and Simulation
Edited by Jia Li and Qihong Fang

Shape Memory Polymer Composites
Characterization and Modeling
Nilesh Tiwari and Kanif M. Markad

Impedance Spectroscopy and its Application in Biological Detection
Edited by Geeta Bhatt, Manoj Bhatt and Shantanu Bhattacharya

Nanofillers for Sustainable Applications
Edited by N.M Nurazzi, E. Bayraktar, M.N.F. Norrrahim, H.A. Aisyah, N. Abdullah, and M.R.M. Asyraf

Chemistry of Dehydrogenation Reactions and its Applications
Edited by Syed Shahabuddin, Rama Gaur and Nandini Mukherjee

Biopolymers and Biopolymer Blends
Fundamentals, Processes, and Emerging Applications
Abdul Khalil H. P. S., Nurul Fazita M. R., and Mohd Nurazzi N.

For more information about this series, please visit: www.routledge.com/Emerging-Materials-and-Technologies/book-series/CRCEMT

Biopolymers and Biopolymer Blends

Fundamentals, Processes, and Emerging Applications

Abdul Khalil H. P. S., Nurul Fazita M. R., and Mohd Nurazzi N.

CRC Press
Taylor & Francis Group
Boca Raton London New York

CRC Press is an imprint of the
Taylor & Francis Group, an **informa** business

First edition published 2024
by CRC Press
2385 Executive Center Drive, Suite 320, Boca Raton, FL 33431

and by CRC Press
4 Park Square, Milton Park, Abingdon, Oxon, OX14 4RN

CRC Press is an imprint of Taylor & Francis Group, LLC

ISBN: 978-1-032-54260-7 (hbk)
ISBN: 978-1-032-54265-2 (pbk)
ISBN: 978-1-003-41604-3 (ebk)

DOI: 10.1201/9781003416043

Typeset in Times
by Apex CoVantage, LLC

Contents

Preface

Biopolymer and Biopolymer Blends: Fundamentals, Processes and Emerging Applications encompasses a retrospective of the classification and the chemistry of biopolymers, including natural biopolymers, synthetic biopolymers, and biodegradable polymers from biomass and marine-based resources. This book aims to provide a comprehensive framework to show the possible utilization of biopolymers in biopolymer blends and bionanocomposites as an alternative source to the conventional non-biodegradable polymers from petroleum-based sources. On top of that, this book describes past and recent developments and applications of biopolymers and their blends in 3D printing and other emerging applications towards the sustainable, intelligent, and high-end applications such as packaging, pulp and paper, agriculture, biomedical, cosmetic, and marine industries. Chapters include several studies and the latest applications in similar fields to show the present and future potential of biopolymers and their blends produced in the advancement of biomaterial technology. Topics range from the various approaches explored and developed for biopolymers and biopolymer blends to their properties in tackling environmental, performance-related improvements, challenges, and factors of the increased use of biocomposites in various industries. The book also discusses the recent developments of biopolymers and biopolymer blends in the processes, applications, and challenges in various high-impact research and applications. The fundamentals of biopolymers and biopolymer blends from natural and synthetic sources and the synthesization process, characterizations, desired performance for specific applications, and current challenges and future prospects for biopolymers and biopolymer blends are also discussed.

This book covers the following topic: (1) biopolymer composites: processing, surface modification, and characteristics; (2) biodegradation and composting of biopolymers; (3) state-of-the-art natural biopolymers for bionanocomposites; (4) biopolymers in 3D printing technology; (5) applications of biopolymer blends and biopolymer-based nanocomposites; (6) starch-based film with essential oils for antimicrobial packaging; (7) chitosan-based chemical sensors: sensing mechanisms and detection capacity; (8) seaweed-based biopolymers for sustainable applications; (9) emergence of biopolymers from sugar palm starch for packaging: characteristics and performance; (10) biopolymers for drug delivery applications: modifications and performance; (11) crosslinking networks of functional biopolymer hydrogels; and (12) current challenges and future prospects of biopolymer blends and biopolymer-based nanocomposites.

This book consolidates various important subjects on biopolymers and biopolymer blends into a single resource, intended for a wide range of readers, including researchers, scientists, and students. It particularly caters to individuals involved in designing with biopolymers and biopolymer blends, exploring alternatives to traditional polymers by transitioning to biopolymers, and aiming to gain a comprehensive understanding of the potential and feasibility of biopolymers in various fields. Additionally, the book contains valuable information that can be beneficial to high school and foundation-level students interested in biopolymers and biopolymer blends.

Authors

Professor Datuk Ts. Dr. Abdul Khalil H.P. Shawkataly, F.I.M.M.M. (U.K.), F.I.A.A.M. (Sweden), known as Abdul Khalil, H.P.S. He completed his Doctoral Degree in the area of Biocomposites at the University of Wales, Bangor, United Kingdom. Currently, he is senior professor at the School of Industrial Technology, Universiti Sains Malaysia (USM), and has served for more than 32 years since 1992. His areas of expertise include biocomposites, biopolymers, and nanocellulose-based materials. Recently, he has been nominated and granted the International Fellow of International Association of Advanced Materials (FIAAM), Sweden, for his contributions in environmental and green nanotechnology in 2021. He was also awarded the 13th National Academic Award in Science, Category Journal Publishing, by the Malaysian Ministry of Higher Education and appointed Academic World Class Professor by the Ministry of Education and Culture, Indonesia, for 2018–2022. He has been listed and awarded as Top Research Scientist Malaysia (TRSM) from 2014 until now. His excellence in research and publication with high numbers of citation have qualified him to be awarded Malaysia's Rising Star Award 2015 (1% Highly Cited Paper in the World–Material Science), Thomson Reuters @ Web of Science, and Malaysia's Research Star Award (MRSA) 2018 in the category High Impact Paper (Natural Sciences)–Elsevier/Scopus, followed by Malaysia's Research Star Award (MRSA) 2019 in High Impact Paper (International Collaboration)–Elsevier/Scopus from the Ministry of Higher Education, Malaysia. At the national and institutional levels, he has many achievements and has been a successful research, publication, administration, and committee head. He has also served as adjunct professor at Universiti Syiah Kuala, Indonesia, and at Universiti Pendidikan Sultan Idris (UPSI), Malaysia. He was on the Advisor Panel of the Malaysian Citation Centre, Ministry of Education; Research Fellow at Institute of Forestry and Forest Product; and on the Standards Malaysia (MS) National Committee for Wood-Based Panels. Also, he was appointed to the editorial board of six ISI/Scopus journals and reviewed more than 200 articles in reputable journals. To date, he has published over 350 scientific articles in WoS/Scopus, with an H-Index of more than 60 and over 16,000 citations. He has also published six academic books, 50 book chapters, and four non-academic books.

Dr. Nurul Fazita, M.R., received bachelor's and master's degree in bioresources and paper and coatings technology at the School of Industrial Technology, Universiti Sains Malaysia. She completed her doctoral degree in the area of biocomposites in 2014 at the University of Auckland, New Zealand. The topic of research for her PhD was *Thermoforming of Composites Made from Bamboo Fabric and Thermoplastic Polymers*. After graduation, she joined the School of Industrial Technology at Universiti Sains Malaysia as a senior lecturer. As a senior lecturer, she taught several courses, such as Statistics with Computer Applications, Fibres and Lignocellulosic Composites, and Basic Bioresources Science and Technology, as well as Bioresources Technology Laboratory. She continues to work and publish in the broader area of biocomposites. She has conducted research for various projects in the areas of thermoforming, biopolymers, natural fiber composites, and 3D printing.

Dr. Mohd Nurazzi N. is a senior lecturer at the School of Industrial Technology, Universiti Sains Malaysia, Penang, Malaysia. Before joining Universiti Sains Malaysia, he was experienced as Post-Doctoral Fellow at the Centre for Defence Foundation Studies, National Defence University of Malaysia, under the Newton Research Grant for the study "Role of Intermolecular Interaction in Conductive Polymer Wrapped MWCNT as Organophosphate Sensing Material Structure". He obtained a diploma in polymer technology from Universiti Teknologi MARA (UiTM) in 2009, a bachelor of science (BSc.) in polymer technology from Universiti Teknologi MARA (UiTM) in 2011, and a master of science (MSc.) from Universiti Teknologi MARA (UiTM) in 2014 under a Ministry of Higher Education Malaysia scholarship. He was awarded a PhD from Universiti Putra Malaysia (UPM) in materials engineering under a Ministry of Higher Education Malaysia scholarship. His main research interests include materials engineering, polymer composites and characterizations, natural fiber composites, and carbon nanotubes for chemical sensors. To date, he has authored and co-authored more than 100 citations indexed in journals, three books, 30 book chapters, 15 conference proceedings/seminars, and three journal special issues as a guest editor on polymer composites, natural fiber composites, and materials science–related subjects.

1 Biopolymer Composites
Processing, Surface Modification and Characteristics

1.1 INTRODUCTION

Since the Renaissance, human civilization has continued through the present. Through the snowball effect, each new intellectual advance leads to further advancements. Countless materials have been innovated, developed and improvised to enhance human lifestyles. Believe it or not, most of the materials are made of polymers, which are a large group of small molecules or monomers combined and linked together in a long chain (Namazi, 2017). Products made from polymers are all around us in the current world. We could be familiar with the top two materials which have been utilized extensively from the Renaissance to the modern age—plastic and rubber. Plastic and rubber have become ubiquitous today. One can spot plastic and rubber material in almost every aspect of our lives, from everyday grocery bags to parts of cars and electronic gadgets. Perhaps we may not even realize that we have used or touched plastic or rubber material in the last minute. In fact, it is not surprising that nearly all fields, including physics, textiles, pharmaceuticals, medicine, molecular biology, biochemistry, construction and mechanical and chemical engineering, have participated in research and development projects related to plastic and rubber. The polymer industry has developed speedily, even faster than copper, steel, aluminum, and other industries (Abdul Khalil et al, 2023). Human life depends much on these materials, as their impact on the present way of life is almost immeasurable.

Diving into the understanding of polymers, they can be divided into two main groups, natural polymers and synthetic polymers. Polymers have been around us in the natural world since the very beginning (e.g., cellulose, starch, polysaccharides, proteins and latex), whereas synthetic polymers, which are usually produced by using carbon atoms provided by petroleum or other fossil fuels, have been exploited since the middle of the 19th century. But what is so special about polymers that they seem to be exceptionally useful to us? The term "polymer" is derived from Greek, where poly means "many" and meres means "parts". Polymers with high molecular weights range between 10,000 and 1,000,000 g/mol or more (Abdul Khalil 2023; Barot et al., 2019) They are made of many small molecules, arrayed in a high number of repeating units called monomers. They are usually bonded together by covalent bonds. When monomers of the same or another type bond together under suitable conditions, a reaction takes place to form a long polymer chain. This is the main

DOI: 10.1201/9781003416043-1

factor that causes polymers to be durable and useful in many fields, which has transformed our modern world.

However, the invention of materials has also come with a price to pay. News regarding the decrease of crude oil or petroleum has been notable as the demand for petroleum-based polymers is increasing. Globally, we consume over 11 billion tons of oil from fossil fuels every year, and crude oil reserves are dropping at the rate of more than 4 billion tons per year (Ali et al., 2017). If this continues, our oil deposits could run dry in 53 years (Kuhns & Shaw, 2018). Lately, news regarding plastic pollution has been alarming for most of us. Previous research estimated that approximately 8 million metric tons (Mt) of macroplastic and 1.5 Mt of microplastic go into the ocean yearly (Ng et al., 2018). If this issue remains unaddressed, the waste could double by 2050 (Lebreton & Andrady, 2019). Countries such as China and India have encountered white pollution due to the extensive usage of plastic in the agricultural sector. Apart from that, the toxins released from the micro-plastics are also something to note, as plastics are resilient to biodegradability. All of these have created concerns for scientists and environmentalists to be more stringent in finding methods, alternatives and new material to mitigate environmental pollution and make our earth a better place to live. One of the new ways is to increase the use of green material and apply green technology instead of using purely synthetic material and technology that would harm the environment.

In the recent years, biopolymers have been emphasized in many sectors, including pharmaceutical, medical, agriculture, building, electronic and automobile, with the ultimate goal to (1) replace existing synthetic polymers for sustainable development, (2) reduce the current struggle of decreasing crude oil and (3) counter the problem of environmental pollution. This is due to the fact that biopolymers are more environmentally friendly compared to crude oil-based polymers. However, biopolymers alone are often not sufficient to stand as an individual material. They need to be reinforced by fillers or combined with other polymers to achieve the desired performance depending on the functionality (Abdul Khalil et al., 2019). This is one of the reasons research on biopolymer blends and bionanocomposites has been surging lately.

Biopolymer blends are materials formed by blending two or more biopolymers, either with a physical or chemical approach. The aim of blending is to improve or tailor properties for certain applications with added value. Nevertheless, there are other motive for blending, too. This includes cost reduction. For instance, poly(lactic) acid (PLA) and starch is one of the most common biopolymer blends, as PLA is able to contribute mechanical and water barrier properties, while starch helps to reduce the material cost. Another case is found in polyhydroxyalkanoates (PHAs), which are versatile yet expensive. Hence, they are often blended with starch to reduce the material cost (Imre & Pukánszky, 2013). A biocomposite is defined as a material made from the combination of two or more distinct constituents to obtain a newly improved material, and a bionanocomposite is defined as biohybrid material composed of biopolymers and an inorganic moiety, showing at least one dimension on the nanometer scale (1 nm = 10^{-9}). However, bionanocomposites are favored over biocomposites, as nanotechnology is growing more popular. They are said to be the materials of the 21st century owing to their uniqueness in applications that is not found in the usual composites. Such unique applications include smart food packaging and tissue engineering in regenerative medicine, electronics and sensors, drug delivery, gene therapy, cosmetics and many others. Almost every sector of industries associated with polymers has begun to venture into nanoscale materials. This motivates materials scientists to further develop and augment bionanocomposites for targeted fields. The two main benefits that researchers pay attention to in the development of bionanocomposites are the remarkable advantages of their synergy with inorganic nanosized solids and their biodegradability. These open the door to possibilities for diverse applications for current and future needs.

1.2 RETROSPECTIVE ON BIOPOLYMERS

Past history has shown that humans have been using natural resources for almost a thousand years. An overview of when humans began to venture into using polymer materials and some of the milestones achieved are reflected in Figure 1.1. As early as 1600 BC, ancient Mesoamericans used a natural material and turned it into balls (Tarkanian & Hosler, 2011). The first attempt at polymer science was by Henri Braconnot in the 1830s. Braconnot, along with Christian Schönbein and a team, synthesized semi-synthetic materials called celluloid and cellulose acetate from the derivatives of the natural polymer cellulose (Thakkar et al., 2020). The term "polymer" was invented by Jöns Jakob Berzelius in 1833 (Suter, 2013).

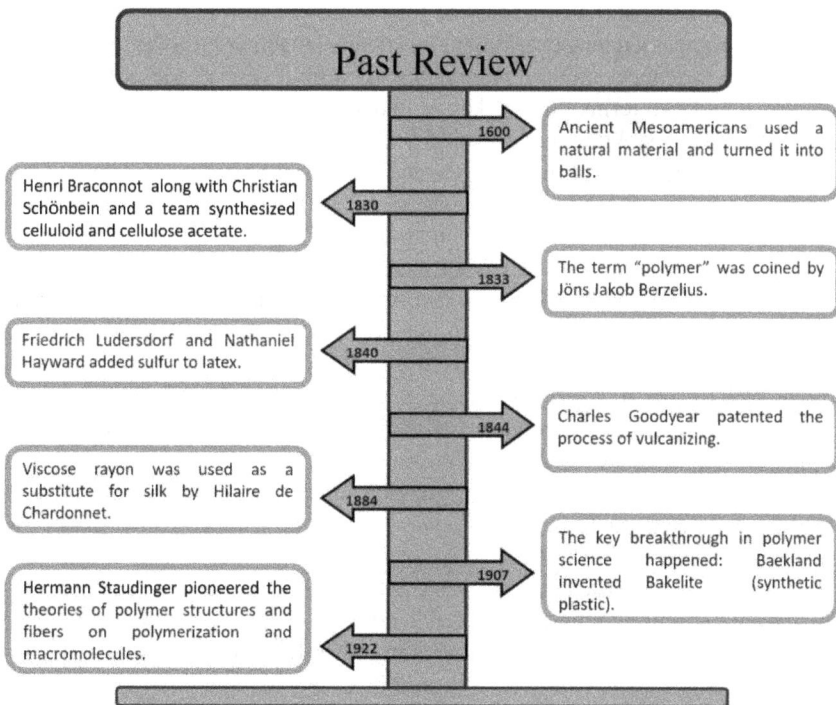

FIGURE 1.1 Review of past polymer science.

In the 1840s, Friedrich Ludersdorf and Nathaniel Hayward added sulfur to latex and discovered that it was able to prevent stickiness. However, it was only in 1844 that Charles Goodyear patented the process of vulcanizing latex with sulfur and heat in the United States, while Thomas Hancock patented a similar process in the United Kingdom. This procedure enhanced natural rubber and prevented it from melting with heat without losing flexibility. This method helped to produce practical products such as waterproofed articles and rubberized materials. Vulcanized rubber made possible the production of tires for bicycles and later for automobiles made by the Goodyear Tire Company (Princi, 2011). In 1884, an artificial fiber plant based on regenerated cellulose, also known as viscose rayon, was used as a substitute for silk by Hilaire de Chardonnet. However, it was extremely flammable (Shabbir & Mohammad, 2017). The key breakthrough in polymer science happened in 1907 when a Belgian-American chemist called Leo Baekland invented Bakelite—a fully synthetic plastic (thermosetting phenol-formaldehyde resin) without any trace of natural molecules. The success of Baekland led many major companies to invest in the work of research and development of new polymers.

Regardless of the advancement of polymer synthesis at that time, the nature of polymers was only known through association theory or aggregate theory by Thomas Graham in 1861, where he hypothesized that polymers were colloids and aggregates of molecules, connecting with each other via unknown intermolecular forces. It was not fully understood until 1922 when Hermann Staudinger (a professor of organic chemistry at the University of Applied Sciences in Zurich) suggested that polymers encompassed long chains of atoms bonded together by covalent bonds. His research was to pioneer modern manipulations of both natural and synthetic polymers. He stated two terms that are key to understanding polymers: polymerization and macromolecules. After only a decade, Staudinger's work was acknowledged and gained acceptance. In 1953, he was awarded the Nobel Prize for his work (Pathak et al., 2014).

However, the commercial polymer industry only surfaced during World War II due to limited resources and supply of natural materials, including silk and rubber. This instigated the increased production of synthetic materials such as nylon and man-made rubber and advanced polymers such as Kevlar and Teflon (Pathak et al., 2014). In 1954, polypropylene was invented by Giulio Natta and then manufactured in 1957 (Sivaram, 2017). As time has passed, these materials have continued to invigorate the growth of the polymer industry and materials from natural polymers to synthetic polymers and now biopolymers.

Demand for biopolymers has been growing rapidly since the last decades, and it is even more noticeable in recent years as the significance of biopolymers has been realized once again. This is because many of us have begun to note that petroleum resources have caused various negative environmental impacts and petroleum resources are not infinite. Fossils and crude oil will eventually run out in the future. Therefore, there is an urgency to finding alternative ways to solve the current issues for a sustainable future. However, what are biopolymers, and why do they attract more attention than fossil fuel-based materials today?

1.3 DEFINITION OF BIOPOLYMERS

The term "biopolymers" has always held ambiguity, as it is often associated with terms such as "biobased" or "biodegradable". According to ASTM D6866-12 and ASTM D7026-13, a biobased polymer is defined as a polymer that comes from renewable sources whereby the carbon content can be determined by the number of carbon traces released from the short CO_2 cycle. Biodegradable polymers, on the other hand, are defined as polymers that are able to biodegrade in the soil via the action of microorganisms and release carbon dioxide and water as their end products. Biodegradable polymers are legally certified by ISO 17088:2012 or ASTM D6400-12 standards. In short, biobased polymers are important for their raw material or resources, while biodegradable polymers are important for their functionality (Niaounakis, 2015a). In order to reduce confusion, biopolymers have been clearly defined based on three major categories (Hassan et al., 2019), as follows.

Biopolymers from renewable sources that are biodegradable

- **Definition:** The first category, biopolymers from renewable sources, can be defined as a kind of polymers composed of repeating monomers that are produced from living beings (Hassan et al., 2019).
- **Source:** This category of biopolymers can be obtained from naturally occurring plants, algae, animals or microorganisms; synthesized chemically from biological starting compounds such as starch, corn and sugar; and produced by fermentative biotechnological processes from microorganisms (Niaounakis, 2015a).
- **Examples:** Examples of biopolymers from this category vary from polysaccharides, protein, lignin and chitosan-based plastics to PLA to PHAs (El-Hadi, 2018; Varma & Gopi, 2021).
- **Further remarks:** These polymeric biomolecules are formed by monomeric units bonding together in covalent bonds. The predominant characteristic that makes biopolymers more environmentally friendly than petroleum-based material is their ability to biodegrade naturally in the environment. Generally, they do not release pollutants but only CO_2 and water in the case of hydrogen, carbon and oxygen compounds (Chen et al., 2016).

Biopolymers from renewable sources that are not biodegradable

- **Definition:** The second category is biopolymers made of biomass or renewable resources that are not biodegradable (Hassan et al., 2019).
- **Source:** The biopolymers under this category are produced from using biomass or renewable sources, such as bioethanol from sugarcane, polyamide 11 from castor oil or specific biopolyesters (Niaounakis, 2015a).
- **Examples:** Examples of existing biopolymers are biopolypropylene (bio-PP), biopolyethylene (bio-PE or green PE) and biopoly(vinyl chloride) (bio-PVC) (Reichert et al., 2020).

- **Further remarks:** These biopolymers are produced using the same method as petroleum-based PEs. The only difference between these biopolymers and petroleum-based PE is the source used. They are said to emit fewer greenhouse gases and be more energy efficient compared to petroleum-based PEs (Reichert et al., 2020).

Biopolymers from non-renewable sources that are biodegradable

- **Definition:** The third category is biopolymers that are produced from certified biodegradable and compostable synthetic crude oil-based sources; some are from biodegradable "aliphatic-aromatic" copolyesters (Hassan et al., 2019).
- **Source:** Can be derived fully from fossil fuels and/or combined with bio-based materials such as polylactic acid, starch and so on (Abdelrazek et al., 2016).
- **Examples:** Examples of these biopolymers are polycaprolactone (PCL), poly(butylenes succinate) (PBS), polybutylene adipate-terephthalate (PBAT) and polyvinyl alcohol (PVOH) (Abdelrazek et al., 2016; Lamnawar et al., 2018; Lins et al., 2015).
- **Further remarks:** The polymer structure contains chemical groups that can be easily broken down by the action of microorganisms. They are ideal for combination with other biodegradable polymers that have high modulus and strength but are very brittle. Some even have water-soluble properties (e.g. PVOH) (Lamnawar et al., 2018).

Rajeshkumar (2021) defined "biopolymers" as polymers that are derived from renewable resources and biological and crude oil-based biodegradable polymers. From the three categories, the first category of biopolymers seems to be the most environmentally friendly, as they are obtained from sources that can be regenerated or grown every year, and they are able to biodegrade via the decomposition action of microorganisms, which eventually returns the material to the environment in a natural way. These are followed by the second category, which is biodegradable even though they are obtained from non-renewable sources. Nevertheless, in recent years, the term "bioplastic" has conventionally been used due to the increase of day-to day usage of plastics. Plastics, like polymers, are a subset of polymers, whereas bioplastics are thermoplastics made from biopolymers from biobased sources such as sugar, seaweed or starch. Biopolymers are a diverse category of materials that include not only bioplastics but also natural polymers like silk, chitosan and fur. For example, PLA is both a bioplastic and a biopolymer.

According to European Bioplastics, a bioplastic is a plastic material that is either biobased, biodegradable or has both properties. To put it another way, 100% biobased plastics can be non-biodegradable, depending on the process applied to produce the bioplastic, while 100% of petroleum-based plastics can be degradable due to their chemical structure, which can be biodegraded by microorganism. Figure 1.2 depicts common types of biopolymers and how they are classified based on biodegradability and biobased content.

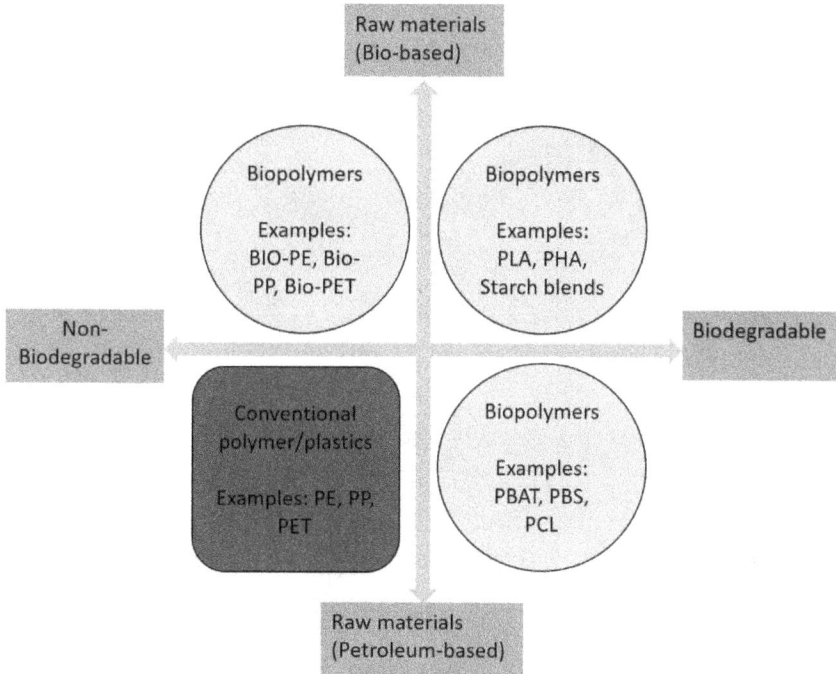

FIGURE 1.2 Biopolymer material coordination chart.

Nevertheless, the preference for using natural biopolymers has increased. The advantages of using biopolymers for the aim of sustainable manufacturing are critical for:

- Resource efficiency
- The resources being cultivated on (at least) an annual basis
- The cascade use principle, as natural biopolymers can first be used for materials and then for energy generation
- Lowering the carbon footprint and greenhouse gas emissions of materials and products, which eventually conserves fossil resources

1.3.1 RESEARCH TRENDS FOR BIOPOLYMERS

In view of the research trends for the past decade, interest in biopolymers has grown, as the number of scientific publications in the field of biopolymers has increased from 2009 to 2023 (over 70,000 publications), as shown in Figure 1.3. Furthermore, biopolymers had the highest number of publications among nine other fields of study from 2009 to 2023, over 10,000, including materials science, chemistry, chemical engineering, chitosan, polymers, nanotechnology, composite material, organic chemistry and nuclear chemistry. The growth of publications in recent years reflects

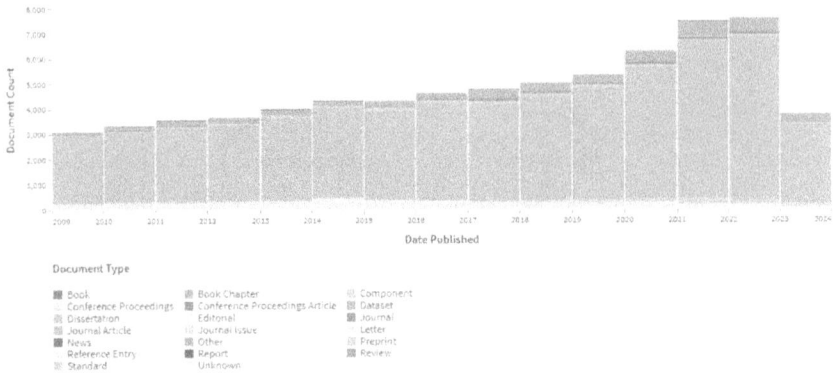

FIGURE 1.3 The trend of publications on biopolymers from 2009 to 2023. (Data extracted from Lens.org.)

the significance of biopolymer innovation, research and development in modeling, processing, synthesizing and production.

1.3.2 MARKET TRENDS FOR BIOPOLYMER-BASED PRODUCTS

Beyond the research field, the global biopolymer market has also been experiencing an increase curve, and it is forecast that the biopolymer market will undergo outstanding growth, with a compound annual growth rate (CAGR) of 21.7% (from USD 10.5 to 27.9 billion) from the year 2020 to 2025, as reported by *Bioplastics and Biopolymers Market*. The data was collected based on the different regions: Asia Pacific (APAC), Europe, North America and the rest of the world (RoW), as indicated in Figure 1.4.

As observed from the trends of research and the global market, it can be deduced that biopolymers have become increasingly important in society in today's world. The growing demand for biopolymer-based materials in the market could be mainly due to several reasons, such as the increasing adoption of green consumerism among consumers as more consumers begin to be aware of the need to reduce the usage of non-renewable resources, the uncertainty about future resources of the petrochemical industry, the inutility of recycling over the years, the ability to use waste as a substance for biopolymer production and the recent use of biopolymers in the medical and pharmaceutical industries (Hojnik et al., 2019). Therefore, it is crucial to emphasize research on biopolymer-based materials to make them functional, practical and economically viable to replace conventional polymers used today.

1.4 BIOPOLYMER BLENDS AND BIONANOCOMPOSITES

In the research field, biopolymer blends and bionanocomposites have been clearly mentioned and studied, particularly in the recent years. In fact, these two terms are

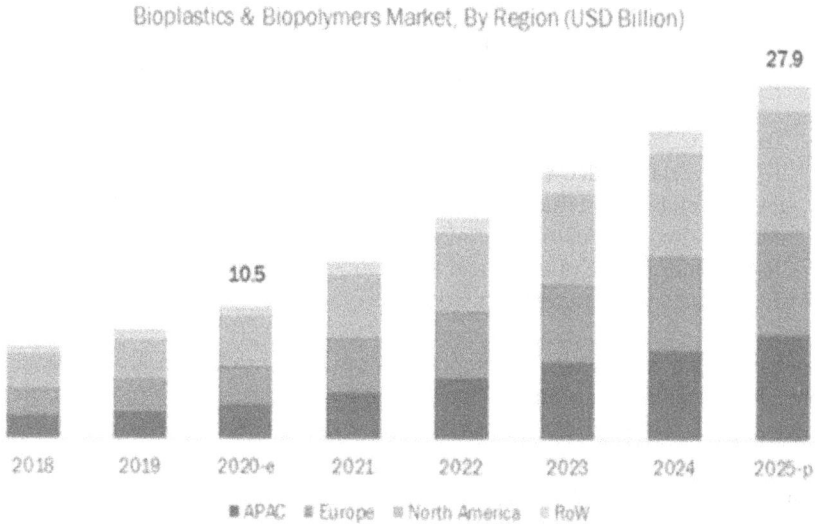

Bioplastics & Biopolymers Market, By Region (USD Billion)

FIGURE 1.4 Bioplastics and biopolymers market by region from 2018 to 2025. (Adapted from *Bioplastics and Biopolymers Market*, April 2020).

sometimes used interchangeably. Although there is a lack of a definite definition for them, biopolymers and bionanocomposites can be briefly introduced as follows (Shaghaleh et al., 2018; Thomas et al., 2013; Vishal, 2017).

- **Biopolymer blends**
  - **Definition:** A material in which two or more types of biopolymers are blended together (Thomas et al., 2013).
  - **Examples of biopolymers:** The common type of blending is blending a biobased polymer with a synthetic polymer (e.g. polyester, starch/PE). Biopolymer blends can be made from a blend between two types of renewable biopolymers (e.g. starch/chitosan, seaweed/starch, etc.) or between a renewable biopolymer and a biodegradable synthetic polymer (e.g. PLA/PBAT, PLA/PCL, etc.)
- **Bionanocomposites**
  - **Definition:** Sometimes known as "nanobiocomposites", "green composites" or "biohybrids". Bionanocomposites are nanocomposites made of a mixture of naturally occurring biopolymers and inorganic moiety. The most prevalent part is the nanostructure of this material.
  - **Examples of bionanocomposites:** They are made of a main matrix from biopolymers, which gives the shape, main function and structural organization, and at the same time are incorporated with nanoparticles, which tunes the structure, value, properties and functionality of the entire composite system. These nanoparticles can serve to improve specific functions of the composite for a particular application.

Regardless of their dissimilarities in the concept and materials used to fabricate them, both biopolymer blends and bionanocomposites share common aims today, which are (Haghighi et al., 2021; Varma & Gopi, 2021):

- To reduce and save the cost of a material.
- To have certain added-value properties of a material such as impact strength, thermal stability, permeability, melt strength, antioxidants, electrical conductivity and antimicrobial properties.
- To widen the scope of usage and application of a material.
- To change and increase degradation rates for better biodegradability.

Biopolymer blends and bionanocomposites have been studied in a wide range of approaches including biomedical, food technology, packaging, electronics, tissue engineering and so on. In fact, many documents and publications have reported on their advantages and potential characteristics to replace non-biodegradable crude oil-based plastics. This is further evaluated through the trend in publications on biopolymers and bionanocomposites in the next subsections.

1.4.1 RESEARCH TRENDS FOR BIOPOLYMER BLENDS AND BIONANOCOMPOSITES

In the past decade, the trend of biopolymer blends has also increased, because demands on biopolymer-based materials and concern for the environment have both increased. To date, research on biopolymer blends for different applications is still actively conducted, as seen in the chronological events listed in Table 1.1.

Biopolymers that are gaining popularity to fabricate biopolymer blends include those directly or indirectly obtained from renewable resources such as PLA, starch, chitosan, alginate, carrageenan and cellulose. Research on bionanocomposites continued to expand when Watzke and Dieschbourg (1994) reported a sol-gel method to fabricate a novel silica-biopolymer nanocomposite using different biopolymers such as gelatin and chitosan. From this study, the system for sol-gel studies under low or microgravitational conditions was developed by comparing the structural features between silica–gelatin bionanocomposites with silica–biopolymer composites of non-gelling biopolymers in microemulsions (e.g., chitosan). Since then, research on bionanocomposites began to emerge widely, and applications in diverse fields can be seen, especially for foods, pharmaceuticals and so on. In the 1990s, research onbionanocomposites inclined to the application of packaging. This could be due to the effort made by researchers to counter the environmental problems caused by conventional plastics.

In a nutshell, natural polymers were once explored by humans a long time ago. However, due to the success of human-made fully synthetic petroleum-based polymers and the rapid commercialization of the materials, humans begin to rely on them, only to realize later that overusage of the materials led to the scarcity of crude oil and deterioration of our environment. Considering the growth of the population, the usage of materials will increase in the years to come. Therefore, it is essential for scientists and industrialists to find more alternatives to curb the problems mentioned by enlarging the resource base, finding ways to use existing raw materials,

TABLE 1.1
Chronological Events for Biopolymer Blends

Year	Remarks	Fields/Applications	Ref.
1977	Starch and ethylene-acrylic acid copolymer	Packaging	(Otey et al., 1977)
1985	Permeability barriers by controlled morphology of polymer blends	Packaging	(Subramanian, 1985)
1989	Starch–gum blends	Food products	(Bielskis et al., n.d.)
1992	Hydrogel–melanin blends	Ocular devices	(Chirila et al., 1992)
1995	Collagen/poly (vinyl alcohol) blends	Biomedical	(Sarti & Scandola, 1995)
1998	Poly (ethylene terephthalate-co-diethylene glycol terephthalate) and polyethylene oxide blends	Implanted biomaterial	(Barcellos et al., 1998)
1999	Polyol-plasticized pullulan–starch blends	Food products	(Biliaderis et al., n.d.)
2000	Microencapsulation of biopolymer blends (gum arabic, mesquite gum and maltodextrin)	Food technology	(Pedroza-Islas et al., 2000)
2001	Starch-based blends	Mulching	(Halley et al., 2001)
2003	Arabic gum, mesquite gum and maltodextrin DE 10 blends	Food products	(Pérez-Alonso et al., 2003)
2006	Caseinate–pullulan bilayers and blends	Food packaging	(Kristo & Biliaderis, 2006)
2008	DNA-conductive polymer blends	Transistors	(Ouchen et al., 2008)
2009	Collagen/biopolymers	Tissue engineering	(Sell et al., 2009)
2010	Lactide and PLA biopolymers	Drinking cups	(Groot & Borén, 2010)
2013	Bis(pyrrolidone-4-carboxylic acid)-based polyamides	Compatibilizer	(Ayadi et al., 2013)
2014	Gelatin–chitosan blend films	Packaging	(Benbettaïeb et al., 2014)
2015	Alginate/keratin blend	Biomedical	(Gupta & Nayak, 2015)
2015	Kappa-carrageenan and cellulose derivatives	Green polymer electrolyte	(Rudhziah et al., 2015)
2015	Poly (butylene-adipate-co-terephtalate)/poly (lactic acid) (PBAT/PLA)	Compatibilizing agents	(Lins et al., 2015)
2016	PCL/polymethyl methacrylate (PMMA) biopolymer blends	Medical applications and some optical systems	(Abdelrazek et al., 2016)

(*Continued*)

TABLE 1.1 (*Continued*)
Chronological Events for Biopolymer Blends

Year	Remarks	Fields/Applications	Ref.
2016	Biomimetic nanofibrillation in two-component biopolymer blends	Transportation and energy-related applications	(Xie et al., 2016)
2016	Pectin and soy protein blends	Meat replacers	(Dekkers, Nikiforidis et al., 2016)
2016	Biopolymer-prebiotic carbohydrate blends	Additive	(Silva et al., 2016)
2016	Protein blends	Food technology	(Dekkers, de Kort et al., 2016)
2017	Blend of organic semiconductors and biopolymers	Printable and flexible phototransistors	(Huang et al., 2017)
2017	Nanostructured cellulose-xyloglucan blends	Water treatment	(Huang et al., 2017)
2017	Agar, gelatin and wax blends	Medicine or biotechnology	(Fuchs et al., 2017)
2017	Plasticized starch/zein blends	Compatibilizing agent	(Favero et al., 2017)
2018	Polylactic acid and polybutyrate adipate terephthalate blends	Film	(Lamnawar et al., 2018)
2018	Polyhydroxybutyrate (PHB) blends	Biomedical	(El-hadi & Abd Elbary, 2018)
2018	Soy protein–agar blends	Wound dressing	(Rivadeneira et al., 2018)
2018	Cellulose, chitosan, starch and gelatin blends	Biosensor	(Elhaes et al., 2018)
2018	Aliphatic-aromatic copolyester and chicken egg white flexible biopolymer blend	Food packaging	(Tiimob et al., 2018)
2018	Chitosan/peg blends	Injectable scaffolds	(Lima et al., 2018)
2018	K-carrageenan and gelatin blends	Tissue engineering	(Tytgat et al., 2018)
2019	Chitosan-based blend hydrogels	Wound healing	(Rasool et al., 2019)
2019	Blends of silk fibroin and collagen	Hair care, cosmetics	(Grabska & Sionkowska, 2019)
2019	Starch–pectin biopolymer blends	Material conservation	(Y & Rao, 2019)
2019	Chitosan-based blend hydrogels	Wound healing	(Rasool et al., 2019)
2019	Chitosan/polyvinyl alcohol (PVA) blends	Burn wounds	(Bano et al., 2019)
2019	Polyhydroxybutyrate and polylactic acid blends	Drug delivery systems	(Harting et al., 2019)

Year	Remarks	Fields/Applications	Ref.
2019	Polylactic acid/poly(D,L-lactic acid) (PDLLA)/ polyhydroxybutyrate blends	Biodegradable optical fibers	(El-Hadi, 2018)
2020	Poly (vinyl alcohol) films enriched with tomato	Active packaging	(Szabo et al., 2020)
2020	Two-step blending in the properties of starch/chitin/ polylactic acid	Biomedical	(Olaiya et al., 2020)
2020	Sodium alginate–assam bora rice starch–based multi-particulate system containing naproxen	Anti-inflammatory drug	(Sarangi et al., 2020)
2020	Biodegradable and antimicrobial polylactic acid–lactic acid oligomer (PLA–OLA) blends containing chitosan-mediated silver nanoparticles with shape memory	Medical	(Sonseca et al., 2020)
2020	Xanthan–curdlan nexus	Edible packaging	(Mohsin et al., 2020)
2020	Molecular modeling analyses for modified biopolymers	HIV protease inhibitors	(Omar et al., 2021)
2020	Biopolymer blends of polyhydroxybutyrate and polylactic acid	Packaging	(Aydemir & Gardner, 2020)
2020	Graphene-based biopolymer TiO_2 electrodes using pyrolysis	Photo-electrocatalysis in water treatment process	(Kaur et al., 2020)
2020	Konjac glucomannan/alginate films enriched with sugarcane vinasse	Mulching	(Santos et al., 2020)
2020	Effect of empty fruit bunches microcrystalline cellulose (MCC) on the thermal, mechanical and morphological properties of biodegradable poly (lactic acid) and polybutylene adipate terephthalate composites	Packaging	(Nor Amira Izzati et al., 2020)
2021	The role of biopolymer-based materials in obstetrics and gynecology applications	Biomedical	(Jummaat et al., 2021)

(Continued)

TABLE 1.1 (*Continued*)
Chronological Events for Biopolymer Blends

Year	Remarks	Fields/Applications	Ref.
2021	Isolation of textile waste cellulose nanofibrillated fiber reinforced in polylactic acid-chitin biodegradable composite for green packaging application	Packaging	(Rizal et al., 2021)
2022	Recent progress in modification strategies of nanocellulose-based aerogels for oil absorption application	Absorbent	(Iskandar et al., 2022)
2022	Insights into the role of biopolymer-based xerogels in biomedical applications	Biomedical	(Abdul Khalil et al., 2022)
2023	Cinnamon-nanoparticle-loaded macroalgal nanocomposite film for antibacterial food packaging applications	Packaging	(Rizal, Abdul Khalil, Abd Hamid et al., 2023)
2023	Coffee waste macro-particle enhancement in biopolymer materials for edible packaging	Packaging	(Rizal, Abdul Khalil, Hamid et al., 2023)

converting waste materials into something useful and producing new materials out of the resources which are abundantly available in nature. Alongside the advancement of technology, the potential of biopolymers can be unleashed, and current biopolymer materials, including biopolymer blends and bionanocomposites, could be used as new-age materials. Nevertheless, how well can these materials perform, what are they made of, how are they fabricated and how sustainable are they? Follow through the chapters to discover food for thought and the science behind these questions.

1.4.2 PROS AND CONS OF BIOPOLYMER BLENDS AND BIONANOCOMPOSITES

Food packaging, medicine, pharmaceuticals and cosmetics all make extensive use of biopolymers. Natural biopolymers especially contain high nutritive value, and they can be used as a food ingredient. Applications such as using chitosan to improve processes such as drug delivery and tissue regeneration, and modern technologies, such as 3D printing, make biopolymers a candidate for a wide range of applications (Diyana et al., 2021). Biopolymer blends and bionanocomposites share common pros and cons, as shown in Figure 1.5.

Although biopolymers show considerable functional properties such as mechanical, thermal and water barriers and could be a good candidate to replace conventional

Pros	Cons
• Biopolymer blends or bionanocomposites are usually sustainable and will help to minimise our reliance on oil. • Lead to more environmentally friendly production of practises and goods. • Due to rising oil prices, product costs are becoming more competitive. • Biopolymer resources are utilised and advocated for recyclable items. • The majority of biopolymer blends and bionanocomposites are biodegradable. • Renewable sources are used to create sustainable biopolymer blends and bionanocomposites.	• Some of the sustainable biopolymer blends or bionanocomposites have limited shelf life (e.g. starch-based films, algae-based films) • Some biopolymer blends and bionanocomposites have a lower performance factor when compared to petroleum-based plastics. • Various biopolymer-based products vary in their environmental impact. • There are concerns that sustainable bioplastics will upset existing recycling methods. • Safety features are questionable.

FIGURE 1.5 Pros and cons of biopolymer blends and bionanocomposites. Reproduced from Diyana et al. (2021).

petroleum-derived plastics in a wide range of applications, they still lack some functionality, especially in durability and shelf life compared to petroleum-based plastics that are used conventionally. In short, biopolymer blends and bionanocomposites have two main advantages over synthetic plastics in terms of practicality: biodegradability and/or compostability and availability from renewable resources. When it comes to commercial interests, the latter appears to be more important, which is blending with non-degradable biopolymers such as bio-polyethylene or bio-polypropylene derived from sugar cane. This is because of the process and final properties, which make them similar to conventional durable plastic. Furthermore, they must be sorted out from conventional plastic when it comes to recycling (Di Bartolo et al., 2021). Another challenge encountered is the use of bionanocomposites with nanofillers in areas such as food packaging. This has been an issue that has raised concerns about nanoparticle migration and the toxicological properties of the associated organomodifiers (Istiqola & Syafiuddin, 2020; Souza et al., 2013). Hence, research is still needed to mitigate these challenges and to improve their properties.

1.5 NATURAL BIOPOLYMERS

Natural biopolymers can be obtained via biological systems from living organisms such as plants, animals and microbes. They are formed naturally during the growth cycle of living organisms. Biopolymers play a crucial role in the growth cycle of an organism mainly to support complex metabolic and cellular activities. The synthesis of biopolymers generally involves enzyme-catalyzed reactions and reactions of chain growth from activated monomers, which are generated inside the cell wall, cytoplasm, organelles, cytoplasmic membrane and the surface of cells and sometimes can be generated through extracellular enzymatic processes. The general biological role of natural biopolymers includes preservation of genetic information; expression of genetic information; catalysis of reactions, energy or other nutrients; protecting against the attack of other cells; sensing of biotic and abiotic factors; storage of carbon; and adhesion to surfaces of other organisms (Aggarwal et al., 2020).

There are in fact pros and cons of biopolymers (Figure 1.6). Natural biopolymers that are generated naturally from natural sources such as plants, animals and microbes are readily degradable by enzymatic action from microorganisms, and the by-products of their biodegradation are water, carbon dioxide CO_2 and some organic matter, which is one of the advantages of being eco-friendly. Moreover, natural biopolymers are abundant in resources, chemically inert, nontoxic and less expensive than synthetic ones; reduce CO_2 emissions; and conserve fossil resources (Gowthaman et al., 2021). However, some disadvantages such as inferior thermal and mechanical properties can be encountered by natural biopolymers (Aggarwal et al., 2020; Babu et al., 2013). Therefore, natural biopolymers are often incorporated with

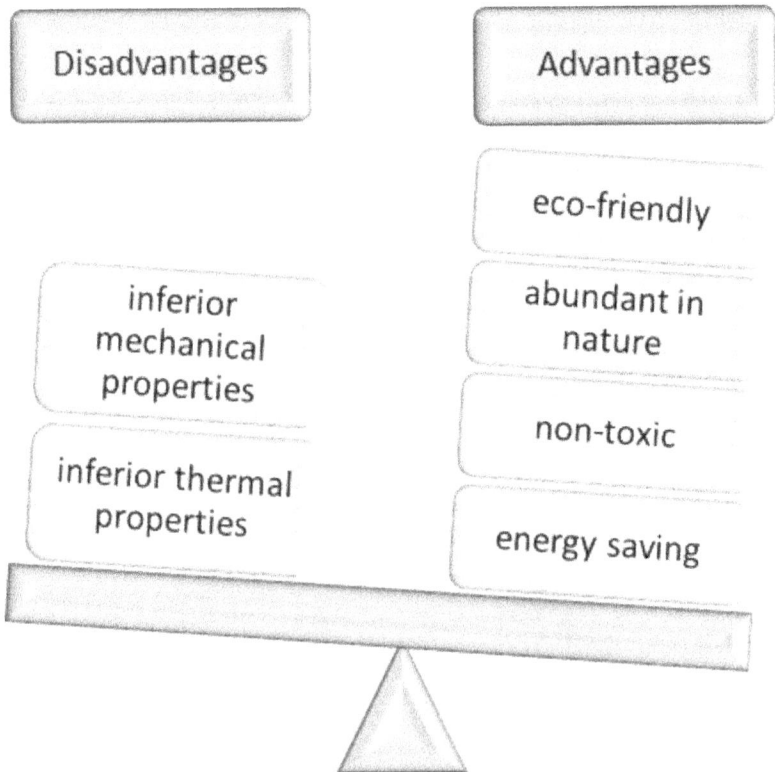

FIGURE 1.6   Advantages and disadvantages of natural biopolymers.

fillers or blended with other polymers to enhance their properties and to extend their application.

1.5.1 PLANT/ALGAE-BASED POLYSACCHARIDES

Plant/algae-based polysaccharides are high molecular weight compounds that bind similar or different monosaccharides together by glyosidic bonds (Ma et al., 2017; Naqash et al., 2017). Plant polysaccharides in particular are usually found in the cell walls of plants. Cellulose, starch, pectin, carrageenan, agar and alginate are among the most abundant biopolymers found on earth that offer possibilities for producing environmentally friendly materials owing to their unique characteristics of being renewable, non-toxic and biodegradable. These promising characteristics enable them to be used in various applications, as displayed in Table 1.2.

Cellulose polysaccharides can be found mainly in the cell wall of the plant, and by content are generally more than 60% of the dry mass of wood (Aggarwal et al., 2020). They are usually insoluble, odorless and rigid in structure, suitable to apply in wound dressing, scaffolds, thickener, wrappers, adhesives, dispersing agents and in drug delivery. However, some fruits such as apples consist of less than 10% cellulose in the cell wall, as they mostly contain pectin polysaccharides (Naqash et al., 2017). Besides the cell wall, pectin can be found in dried citrus peels or apple pomace. The unique feature of pectin is that it has the ability to form gel when it is heated. It is physically coarse to fine, white to yellowish and odorless and has a mucilaginous taste. Previous research has shown its application for antimicrobial action, antiinflammatory and as a thickening and gelling agent for jellies and jams.

On the flip side, starch is not found in cell walls but in roots, bulbs, tubers or seeds of green plants, with the main function as energy storage for the plant. Staple foods like cassava, potato, rice, corn and wheat are rich in starch. Unlike cellulose, starch is soft and flexible in structure, although it has a similar physical appearance to cellulose and is odorless, tasteless and insoluble in water. It has been used in numerous applications such as a thickening agent for puddings; emulsifier, stabilizer and fat replacer in food technology; lubricant in oil boring; and filler in pharmaceutical items. In recent years, polysaccharides such as alginate, agar and carrageenan have been among the promising algae-based polysaccharides that have been of interest in the era of bioplastics compared to traditional methods of utilizing sources from corn and potatoes. This is due to primarily to advantages such as significantly higher productivity, which can be up to 50 times greater than that of conventional startch-based plants; the ability to thrive in a variety of environments; being a non-food resource (in Western countries); complementing terrestrial biomass production; not requiring arable land; and the fact that certain marine plants like seaweeds can flourish in saline water and be cultivated offshore without the need for fertilizers or pesticides (Priyan Shanura Fernando et al., 2019; Usman et al., 2017).

Natural plant polysaccharides are biologically synthetized by plants for energy storage and structural support, whereas, conventionally, plant polysaccharides are extracted through hot water, high-concentration alcohol, alkaline solvent, enzymatic hydrolysis, ultrasonic, microwaves or even gel column chromatography (Ren et al., 2019). Plant proteins can be another alternative source of biopolymers besides plant

TABLE 1.2

The Sources, Physical Features and Applications of Plant/Algae Polysaccharides

Plant/Algae Polysaccharide	Source	Physical Features	Applications	Ref.
Cellulose	Plant tissue, cell wall	Water insoluble, odorless, rigid	Wound dressing, scaffolds, thickener, wrappers, adhesive, dispersing agent, drug delivery, packaging	(Abdul Khalil et al., 2020; Aggarwal et al., 2020)
Starch	Plant seeds, tubers, bulbs, roots	White, tasteless, odorless, insoluble in cold water or alcohol	Thickener, stabilizer, water retention agent, adhesive, excipient, additive, packaging, mulch film	(Syuhada et al., 2018; S. Wang et al., 2015)
Pectin	Plant cell wall, citrus peels, apples pomace	Coarse to fine, white to yellowish, odorless, mucilaginous taste	Antimicrobial action, anti-inflammatory Removes metals, gelling agent, thickening agent, stabilizer, packaging	(Aggarwal et al., 2020; Naqash et al., 2017)
Carrageenan	Red seaweeds	White to slight yellowish, tasteless, odorless, heat-reversible	Thickener, stabilizer, tablet excipient, encapsulating agent for drugs, scaffold material, wound dressing, foliar spray, packaging	(Aggarwal et al., 2020; Cunha & Grenha, 2016)
Agar	Red seaweeds	White, tasteless, odorless, heat-reversible	Laxative, thickener, stabilizer, emulsifier, coagulator, capsules, surgical lubricants, packaging	(Aggarwal et al., 2020; Cunha & Grenha, 2016)
Alginate	Brown seaweeds	White to cream, odorless, tasteless, heat-reversible	Injectable vehicle for tissue engineering, drug delivery, wound dressing; gelling agent; stabilizer of aqueous mixtures, dispersions and emulsions; packaging	(Aggarwal et al., 2020; Cunha & Grenha, 2016)

TABLE 1.3

The Source, Physical Features and Applications of Plant/Algae Proteins

Plant/Algae Proteins	Source	Physical Features	Applications	Ref.
Soy	Crushed or defatted soybean flakes	Sensitive to water, odorless, rigid	Wound dressing, scaffolds, thickener, wrappers, adhesive, dispersing agent, drug delivery	(Aggarwal et al., 2020; Hassan et al., 2019)
Zein	Endosperm of maize	Coarse to fine, white to yellowish, odorless, mucilaginous taste, soluble in aqueous alcohol	Antimicrobial action, anti-inflammatory Remove metals, animal feed	(Aggarwal et al., 2020)
Gluten	Endosperm of wheat	Chalky flavor, stringy mouthfeel	Bread products, imitation meat, stabilizer, food packaging	(Biesiekierski, 2017; Sharma et al., 2017)

polysaccharides. There are many types of plant proteins, and some of those common proteins are listed in Table 1.3. Among them are soy, zein and gluten proteins. Protein biopolymers derived directly from plants have several advantages, including being safe, low-cost, rapidly deployable, natural, renewable and biodegradable, similar to polysaccharides.

Soy protein is one of the most economical proteins that can be obtained from soy beans. It can be extracted from either crushed or defatted soybean flakes that have been removed by dissolving in hexane, followed by dissolution, precipitation and acidifying processes. Soy protein has been a choice of industry for protein isolation due to its high soy protein isolate at approximately 90%, which is free from lipids and carbohydrates. Unfortunately, this limits the application of soy protein, as protein is sensitive to water and hydrophilic in nature. Hence, it is suitable to be used for biomedical applications. Nevertheless, many studies have co-blended soy protein with other biopolymers such as polylactic acid to improve its properties and reduce hydrophilicity.

Zein is the predominant protein found in the endosperm of maize and possesses unique solubility in aqueous alcohol solutions. Zein is considered a waste protein, and it is not an ideal protein for humans due to its poor solubility and imbalanced amino acid profile (Shukla & Cheryan, 2001). However, it can be used as encapsulation for food and drugs. It is also applied in other areas such as textiles, plastics, coatings and adhesives (Garavand et al., 2022).

Gluten can be obtained from the by-products of wheat after washing to remove starch granules and water soluble compounds. Gluten is composed of hundreds of protein monomers or oligopolymers, with two water-insoluble proteins, glutenins and gliadins (Biesiekierski, 2017). Wheat gluten has been used for film forming for edible coating and food products due to its elasticity (Sharma et al., 2017). Due to the restriction on the resources of petroleum, biodegradable materials produced from green plants or marine feedstock are becoming more significant these days. Because of their abundant resources, low cost and good biodegradability, biopolymers from this particular group have been used extensively for diverse applications, be it in the industry or in the research field.

1.5.2 ANIMAL-BASED POLYSACCHARIDES

Animal-based polysaccharides are gaining considerable interest and have been widely utilized, especially in the biomedical, pharmaceutical and cosmetics fields. There are plenty of other types of animal-based polysaccharides. However, this chapter is restricted to chitosan and hyaluronic acid, as they are among the most studied and common types of animal-based polysaccharides. Chitosan, characterized as a functional derivative of chitin, is a linear polysaccharide typically produced through the partial or full deacetylation of chitin using sodium hydroxide (NaOH) generally extracted from shells or crustaceans such as shrimp, lobster and crabs. Usually it appears in a pale, white flaky form. Chitosan is usually obtained through a series of processes from cleaning to demineralizing to deproteinizing and deacetylation (Mohamed et al., 2020). Chitosan is attractive to use in cosmetics, wound dressing, food packaging, water treatment, textiles, pharmaceuticals and scaffolding owing to its non-toxicity, biocompatibility, biodegradability and anti-microbial and anti-inflammatory properties (Table 1.4).

TABLE 1.4
The Source, Physical Features and Applications of Animal Polysaccharides

Animal Polysaccharides	Source	Physical Features	Applications	Ref.
Chitosan	Shellfish and crustacean waste materials	Pale, white and flaky	Cosmetics, wound dressing, paper processing, food packaging, seed coating, plant growth regulator, protein waste recovery, anti-bacterial, flocculating agent, drinking water purification, drug delivery	(Jiménez-Ocampo et al., 2019) (Aggarwal et al., 2020)
Hyaluronic acid	The umbilical cord of newly born children, rooster combs, fermentation broths of streptococcus	Transparent, viscous fluid or white powder	Cosmetics, drug delivery, wound healing, viscosity agent, medication filler, antibacterials	(Fallacara et al., 2018; Graça et al., 2020)

Hyaluronic acid, also known as hyaluronan, is a type of polysaccharide, which is widely found in the extracellular matrix of animals such as such as in rooster crowns, vitreous bodies, umbilical cords, synovial fluid, and skin, although it can be produced synthetically from plants and through fermentation of microorganisms. One of the unique features of hyaluronic acid is the ability to retain water (Graça et al., 2020). Thus, many cosmetics companies use hyaluronic acid in their cosmetic products (Sakulwech et al., 2018). Other applications of hyaluronic acid include preparation of gels for drug delivery, steroid delivery, wound healing, lubricant for osteoarthritic joints, pain relief, enhanced mobility and dermal regeneration (Al-Sibani et al., 2017). Apart from animal-based polysaccharides, animal-based proteins such as collagen, gelatin and keratin have been extensively applied in food and pharmaceutical industries, as shown in Table 1.5. Collagen is the major structure of protein found in animals and consists of 20–30% of the total body proteins (Yoon et al., 2020). It is sourced from fibroblasts and various vertebrates within the body. For example, fibrillar collagen, the most abundant type in vertebrates, assumes a vital role by contributing to the molecular architecture, shape, and mechanical characteristics of tissues. For instance, it enhances tensile strength in the skin and bolsters resistance to traction in ligaments.

TABLE 1.5
The Source, Physical Features and Applications of Animal Proteins

Animal Proteins	Source	Physical Features	Applications	Ref.
Collagen	Invertebrate body walls and cuticles	Hard, fibrous, insoluble protein and molecules form long, thin fibrils	Sutures, dental composites, sausage casings, pore and skin regeneration templates, cosmetics, biodegradable materials, solid support micro-carrier	(Aggarwal et al., 2020; Hassan et al., 2019)
Gelatin	Cattle hide, bones, fish, pig skin, agricultural or non-agricultural	Water-soluble, translucent, flavorless food ingredient, gummy when moist and brittle when dry	Stabilizer, thickener, texturizer, emulsifier, foaming, food wetting agent, pharmaceutical and medical usage, animal feed	(Derkach et al., 2020; Echave et al., 2017; Qiao et al., 2017)
Keratin	Feathers, hair, nails, wool, horn and hooves, stratum corneum and scales	Insoluble in most organic solvents	Absorbents, leather industry, drug delivery system, surgery, food industry, cosmetics, biomedical products, fertilizers, electrode material	(Donato & Mija, 2020)

The molecule is usually rod shaped. Collagen has been applied in tissue-based devices, drug delivery systems, wound healing, nanospheres and hydrogels due to its biodegradable and biocompatible properties, small size, large surface area, high adsorption capacity and ability to disperse in waster.

Gelatin is another substantial animal-based protein that has gained interest in a wide range of applications. It can be produced through denaturation or partial hydrolysis of collagen derived from the skin, tissues and bones of animals. Due to its biodegradability, non-toxicity and cost effectiveness, it has been used widely in forming hydrogels, foaming, emulsifying, food products, animal feed and for pharmaceutical and medical properties. Similar to collagen and gelatin, keratin is also renewable, biodegradable and biocompatible in terms of its characteristics (Donato & Mija, 2020). The primary source of keratin is found in feathers, hair, nails, wool, horn and hooves, stratum corneum and scales. The unique feature of keratin is that it has a high porous network, chemical reactivity and water retention capacity, making it suitable to be utilized as a biosorbent in wastewater treatment, drug delivery systems, surgery, the food industry, cosmetics, biomedical products, fertilizers and electrode material.

1.5.3 MICROORGANISM-BASED POLYSACCHARIDES

Microbial polysaccharides constitute a form of essential commodity that is of growing interest to many industries. Their novel functions, such as consistent chemical and superior physical properties, are advantages of microbial polysaccharides over plant polysaccharides. They can be produced by microorganisms from a variety of sources. Microbial polysaccharides can be obtained from the cell wall or excreted from the cell through extracellular mechanisms and cultivated in bioreactors. Some of the optimization techniques and principles are illustrated in Figure 1.7 to show the scale-up of production of microbial polysaccharides. At the industrial scale, microbial polysaccharides are produced from the exopolysaccharides (EPSs) (i.e. the outermost structure of the microorganism) in large quantity via fermentation processes.

Two main microorganisms that have gained much attention lately in producing microbial polysaccharides are fungi and bacteria (Table 1.6). This chapter covers the two polysaccharides produced by fungi, glucan and pullulan. Levan, dextran, bacterial cellulose and xanthan gum are the other main polysaccharides produced by bacteria that are also covered in this chapter.

Fungi have a cell wall that consists of polysaccharides. Glucan is one of the most abundant microbial polysaccharides found in the cell wall of yeast fungi, and it functions as a regulator of innate immunity. Glucan can be either water soluble or insoluble macromolecules (Yuan et al., 2020). In general, the yeast glucan is part of Japanese food commodities, as it is mostly found in mushrooms (Złotko et al., 2019). It has also been applied in milk yielding, animal feeds, nutritional supplements, plant pests and viral invasion control, coating for surgical instruments, food packaging and cosmetics.

Pullulan can be obtained from ageing stock of the polymorphic fungus *Aureobasidium pullulans*. It can also be synthesized from other microorganisms. Several noted microorganisms are *Aureobasidium spp., Tremella mesenterica,*

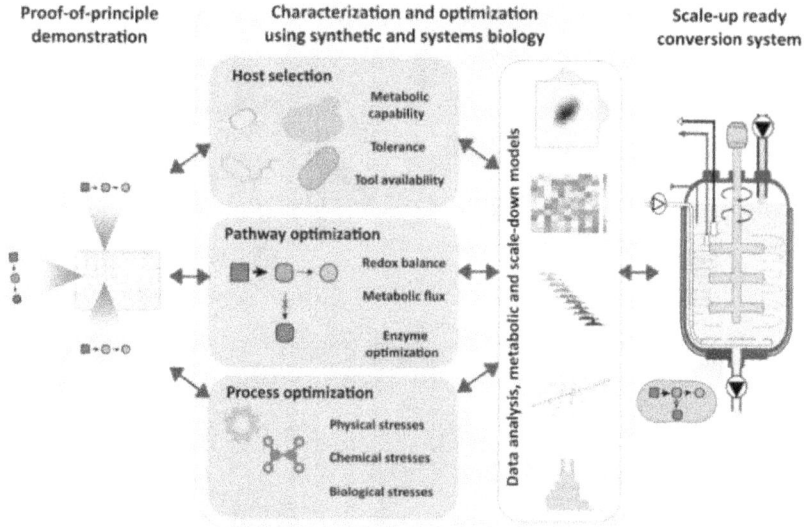

FIGURE 1.7 Microbial polysaccharides obtained at a large scale through cell cultivation in bioreactors. Reproduced from Wehrs et al. (2019).

Cytaria spp. and *Cryphonectria parasitica*, which can be discovered in backwoods soil, ocean water and plant and animal tissues (Singh & Kaur, 2019). Pullulan is highly water soluble and, due to this character by promoting oxygen barrier properties, has high moisture retention capability and is able to prevent fungal growth. These properties enable it to be used in diverse applications such as in food packaging, cosmetics and pharmaceutical industries.

Besides obtaining polysaccharides from fungi, bacteria also play an essential role in producing polysaccharides. The most common polysaccharides produced from bacteria included in this chapter are levan, dextran, bacteria cellulose and xanthan gum. Levan is a class of bacteria polysaccharide excreted from the extracellular homopolysaccharides of d-fructose. Numerous bacteria are able to produce levan, including *Streptococcus salivarius* (i.e. a bacterium of the oral flora), *Lactobacillus sanfranciscensis*, *Bacillus subtilis* and *Bacillus polymyxa*, *Acetobacter xylinum*, *Gluconoacetobacter xylinus*, *Microbacterium levaniformans* and *Zymomonas mobilis* (de Siqueira et al., 2020; Hassan et al., 2019). Levans are a natural adhesive and surfactant, non-viscous and water and oil soluble. They also exhibit immune-stimulating and anti-tumor properties, which are suitable in pharmaceutical and food applications as an emulsifying agent, gelling agent, surface-quality agent, encapsulation, carrier for taste and odor, photographic emulsion, molecular sieve for gel filtration and blood volume extender.

Dextran is a type of exopolysaccharides known to be produced from lactic acid bacteria. Other bacteria such as the *Streptococcus* and *Acetobacter* genera have also been found to produce dextran (Ghimici & Nichifor, 2018). Dextran exhibits a flexible structure attributed to its free rotation of glyosidic bonds (Aggarwal et al.,

TABLE 1.6

The Source, Physical Features and Applications of Microbial Polysaccharides

Microorganism Polysaccharide	Source	Physical Features	Applications	Ref.
Glucan yeast (fungal)	Fungal (*Saccharomyces cerevisiae*) cell wall	Either soluble or insoluble in water	Improve milk yield, animal feeds, nutritional supplements, control plant pests and viral invasions, coating for surgical instruments, food packaging, cosmetics	(Hassan et al., 2019; Yuan et al., 2020)
Pullulan (fungal)	*Aureobasidium pullulans*	White powder dissolute in water; tasteless; odorless; gluey adherent solution; indissoluble in solvents such as ethanol, methanol and acetone	Inhibits fungal growth, low-viscosity filler, stabilizes the quality and texture, binder and stabilizer, protective glaze, stabilizes fatty emulsions, denture adhesive, pharmaceutical coatings	(Aggarwal et al., 2020; Singh & Kaur, 2019)
Levan (bacteria)	*Bacillus subtilis, Bacillus megaterium, Bacillus cereus* and *Bacillus pumilus*	Natural adhesive and surfactant, non-viscous and water and oil soluble	Emulsifying agent helps in preparation, preservation, gelatinization, surface-quality agent, encapsulation, carrier for flavor and odor, photographic emulsion, molecular sieves for gel filtration, blood volume extender	(Aggarwal et al., 2020; de Siqueira et al., 2020; Hassan et al., 2019)

Name	Source/organisms	Properties	Applications	References
Dextran (bacteria)	Dextran sucrase, *Leuconostoc mesenteroides, Saccharomyces cerevisiae, Lactobacillus plantarum* or *Lactobacillus sanfrancisco*	Soluble in water and organic solvents, high viscosity	Solidifying agent, thickening agent, improves surface quality, emulsifier in edible products, soothing, palatable, loaf mass, smoothness, storage life, cryoprotectant, viscosifier, creamy, lower synaeresis, antioxidant for food, water holding capacity, moisture content raised in non-fat mass, functional foods	(Aggarwal et al., 2020; Ghimici & Nichifor, 2018)
Bacterial cellulose (BC) (bacteria)	*Gluconacetobacter, Pseudomonas, Rhizobium, Sarcina, Dickeya* and *Rhodobacter* belong to the *Komagataeibacter* genus	Intact membranes (fiber or pellets form), disassembled bacterial cellulose (BC), and BC nanocrystals (BCNC)	Packaging for foodstuffs, transparent covering, cell divider, permeable, medicine manufacturing industries, water investigation, beauty products, biocompatible, ethyl alcohol manufacturing, conducts electricity, magnetic stuff, human-made blood vessels, scaffold for tissue engineering	(Aggarwal et al., 2020; Gorgieva & Trček, 2019)
Xanthan gum (bacteria)	Plant pathogens such as *Xanthomonas campestris*	Motile, having a single polar flagellum, cream-colored powder soluble in both cold and hot water	Emulsifier and thickening agent texture, viscosity, flavor release, appearance, antiseptic and water control	(Hassan et al., 2019; Nazarzadeh Zare et al., 2019)

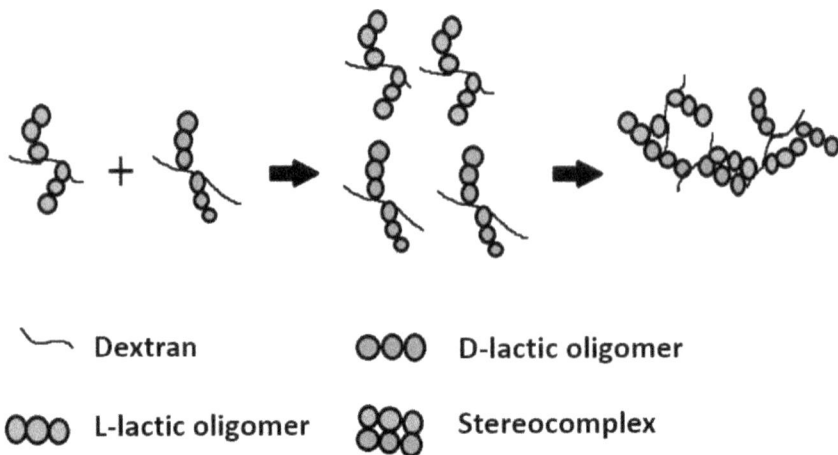

FIGURE 1.8 Gelation of dextran polysaccharide in hydrogel formation.

2020). It is highly soluble in water, biocompatible and biodegradable. Therefore, it possesses the ability to form hydrogen, as illustrated in Figure 1.8, and is normally applied in the pharmaceutical industry as a blood plasma replacement and in the food industry as a thickener for food products, to improve moisture retention and to maintain flavor and appearance.

Bacterial cellulose, also known as microbial cellulose, is another type of exopolysaccharide generated from the bacteria *Komagataeibacter* in a carbon and nitrogen-enriched media. Other bacteria that have been reported to produce bacteria cellulose include *Achromobacter, Alcaligenes, Aerobacter, Agrobacterium, Azotobacter, Gluconacetobacter, Pseudomonas, Rhizobium, Sarcina, Dickeya* and *Rhodobacter* (Rangaswamy et al., 2015). The difference between bacterial cellulose and plant cellulose is that bacterial cellulose does not have the association of lignin, hemicellulose and pectin. Hence, it is much easier to clean and purify compared to plant cellulose. Some of its characteristics are higher water retention, longer drying time and ductility. These makes it preferable to be used in food products as dietary fiber, a thickening agent and packaging; in paper products as high-quality paper; and in pharmaceutical materials for injury and wound dressing and as an immobilizer for *Lactobacillus* cells (Aggarwal et al., 2020; Gorgieva & Trček, 2019).

Xanthan, a bacterial polysaccharide produced from a gram-negative bacterium, *Xanthmonas campestris*, has been extensively studied and is commonly used as a food additive due to its viscosity and stabilizing characteristics (Hassan et al., 2019). Xanthan is found to be stable in high shear pressure, warmth and even acidic environments. It is also known for its biodegradability and non-toxicity (Nazarzadeh Zare et al., 2019). Therefore, it gained favor in the food industry to be used as a food thickener, stabilizer, emulsifier and gelling agent. With the growing demand for polysaccharides in recent years, it becomes imperative to explore alternative sources for polysaccharides rather than relying solely on natural polysaccharides derived from animals, plants or algae. This diversification is essential to ensure food security and supply. Therefore, microbial polysaccharides can serve as an alternative source of polysaccharides since they are more consistent in physical and chemical characteristics compared to natural polysaccharides, and they can be modified to meet targeted needs. Extensive studies have been done to fabricate biopolymer blends or bionanocomposites using microbial polysaccharides through cross-linking with proteins or other biopolymers and even incorporation with fillers to enhance the performance of the material.

1.6   CHEMICAL SYNTHESIS

Synthetic polymers can be synthesized either via chemical or fermentation techniques from renewable resources or fossil fuel. These groups of polymers are classified under synthetics and sub-classified into renewable resources and petroleum-based resources. Among the common biobased polymers synthesized from renewable resources are PLA, polyhydroxybutyrate, bio-polyethylene and bio-polypropylene, whereas biodegradable synthetic polymers that are synthesized from petroleum-based resources include polycaprolactone, polybutyrate adipate terephthalate, polyvinyl alcohol and polybutylene succinate. Synthetic or modified polymers from renewable resources have played a key role in biobased plastic production. Based on their biodegradability, this group of polymers can be further divided into two groups: biodegradable and non-biodegradable polymers. Two common examples of biodegradable polymers are polylactic acid and polyhydroxybutyrate, while two examples of non-biodegradable polymers are bio-polyethylene and bio-polypropylene. On the other hand, certain polymers that are chemically synthesized from petroleum are biodegradable. Some noted examples include polycaprolactone PCL, polybutyrate adipate terephthalate, polyvinyl alcohol and polybutylene succinate. The bonus for this group of polymers is their biodegradability despite exhibiting similar properties to non-biodegradable petroleum-based polymers (Babu et al., 2013; RameshKumar et al., 2020).

1.6.1   RENEWABLE RESOURCES

Synthetic polymers derived from renewable resources are usually acquired either through chemical modification of natural biopolymers such as starch, cellulose or chitin or a two-step process from lignin, starch, plant oil and cellulose. This subsection covers a few promising biobased synthetic polymers that have been use in diverse applications from packaging to pharmaceutical and automobiles (Table 1.8). Among these polymers are PLA, polyhydroxybutyrate, polybutylene succinate, bio-polyethylene and bio-polypropylene.

PLA received great attention in industry and is currently one of the leading biobased polymers used conventionally. PLA is produced from lactic acid, which is naturally present in organic acid via fermentation of sugars from sugar cane, sugar beet or starch. PLA can be synthesized using various established polymerization techniques (Ekiert et al., 2015). These techniques include polycondensation, ring opening polymerization and direct methods such as azeotopic dehydration and enzymatic polymerization. At present, direct polymerization and ring opening polymerization are among the most employed techniques. Several advantages can be found in PLA. These include energy saving in production, where it uses about 25–55% less energy compared to petroleum-based polymer production (Ekiert et al., 2015).

TABLE 1.8

The Source, Physical Features and Applications of Polymers Synthesized from Renewable Resources

Polymers	Source	Physical Features	Applications	Ref.
Polylactic acid	Derived from corn starch, sugarcane	Slight yellowish hue in its natural form, low glass transition temperature (typically between 111 and 145°F)	Food handling and medical implants, packaging, bottles, biodegradable medical devices, electronic devices	(Aggarwal et al., 2020; Ekiert et al., 2015)
Polyhydroxybutyrate	Cells of microorganisms, produced industrially through bacterial fermentation	Stiff and brittle in nature, with low thermal stability and a high degree of crystallinity	Disposable razors, utensils, diapers, containers, sutures, scaffolds, films, paper laminates, bags, containers, automobiles	(McAdam et al., 2020a)
Polybutylene succinate	Derived from polyvinyl acetate from the fermentation of succinic acid	Good oxygen barrier	Agricultural films, drug encapsulation, bags or boxes for both food and cosmetic packaging	(Puchalski et al., 2018)
Bio-polyethylene	Dehydration of bio-ethanol, obtained from glucose (maize, lignocellulose material, wheat, etc.)	Not biodegradable, flexible, durable, printable, transparent, heat resistant, glossy	Packaging films, pouches, boxes, containers, non-disposable carpet, piping	(Siracusa & Blanco, 2020)
Bio-polypropylene	Butylene dehydration of bio-isobutanol obtained from glucose (maize, lignocellulose material, wheat, etc.)	Not biodegradable, flexible, durable, printable, transparent, heat resistant, glossy	Packaging films, bags, containers	(Siracusa & Blanco, 2020)

PLA is biodegradable when it is exposed to the active biological environment. It is compostable and recyclable in the industry, and it is biocompatible. PLA polymers range from amorphous glassy to semi-crystalline to high crystalline, giving it better thermal stability than other biobased polymers such as poly (hydroxyl alkanoate) and poly (ethylene glycol) (PEG) (Arrieta et al., 2017a). These properties made it a promising candidate for food handling and medical implants, packaging, bottles, biodegradable medical devices and electronic devices.

Polyhydroxybutyrate is the by-product of microbial secondary metabolism from microorganisms found in renewable resources (e.g. food waste). This occurs when a microorganism encounters limited nutrients and an unfavorable environment, as shown in Figure 1.9. The most common synthesis approach is through bacterial fermentation, which highly depends on a central carbon metabolite from acetyl-CoA via enzymatic activities (Surendran et al., 2020). In terms of its properties, PHB is stiff and brittle, low in thermal stability yet high in degree of crystallinity. Most PHB exhibits similar properties to petroleum-based polypropylene. PHB is of interest due to its biodegradability when it is exposed to an active biological environment. Owing to its properties, PHB is used as disposable razors, utensils, diapers, containers, sutures, scaffolds, films, paper laminates, bags, containers and parts of automobiles.

PBS is an aliphatic polyester synthesized by polycondensation between succinic acid and 1,4-butanediol (Puchalski et al., 2018). Traditionally, PBS can only be produced from fossil fuel sources. Currently, it can be produced by fermenting succinic acid from renewable feedstock such as sugars, glucose, starch and xylose (Babu et al., 2013). This process has had greater advantages over the chemical process, as it uses renewable feedstock and consumes less energy. Among the famous companies that have developed PBS from renewable feedstock are Mitsubishi Chemical (Japan) and Ajinomoto (Babu et al., 2013). Owing to their high-impact properties, which are similar to those of polypropylene and have the added benefit of being biodegradable, they are applied in a wide range of applications from packaging to agriculture to biomedicine. As mentioned earlier, the starting monomers or base origin of a material are not able to determine the biodegradability of a material; it rather depends on the end-life properties of a material. For instance, bio-polyethylene and bio-polypropylene materials are produced from a starting monomer (i.e. glucose) that can be obtained from renewable resources such as maize, lignocellulose material and wheat. However, the chemical modification and process make them chemically

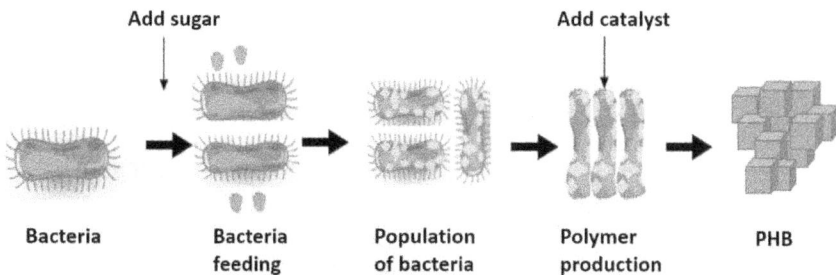

Add sugar   **Add catalyst**

Bacteria   Bacteria   Population   Polymer   PHB
                  feeding   of bacteria   production

FIGURE 1.9   Schematic illustration of production of PHB from microorganisms.

identical to petroleum-based polyethylene and polypropylene. Therefore, they possess properties such as flexibility, durability, printability, transparency, heat resistance and non-biodegradability (Siracusa & Blanco, 2020).

Bio-PP is produced by dehydration of bio-ethanol from glucose. The process starts off with cleaning, shredding or cutting the sugar cane, maize or wheat to obtain the juice, which contains sucrose and is then fermented to obtain ethanol. After the distillation process, the water is removed. Finally, it goes through subsequent polymerization to obtain the ethylene monomer. A pilot scale test of bio-PP was done by a company known as Braskem, but its process has not been revealed (Siracusa & Blanco, 2020). Meanwhile, the process of bio-PP is still under investigation. Generally, bio-PP is obtained through dehydration of bio-isobutanol from glucose, which is obtained from renewable feedstock and goes through polymerization.

1.6.2 PETROLEUM-BASED RESOURCES

Polymers in this group are derived from fossil fuels, such as synthetic aliphatic polyesters, which are obtained from petroleum. In fact, research is still ongoing for some of these polymers to be synthesized from renewable resources to safeguard our fossil fuels. The unique property of the polymers under this group is the ability to biodegrade even though the source is crude oil or fossil-fuel. Polycaprolactone, polybutyrate adipate terephthalate and polyvinyl alcohol are among the common synthetic biodegradable polymers in this group (Table 1.9).

TABLE 1.9

The Source, Physical Features and Applications of Polymers Synthesized from Petroleum-Based Resources

Polymer	Physical Features	Applications	Ref.
Polycaprolactone	Glass transition temperature of about −60°C, good water, oil, solvent and chlorine resistance	Scaffold, drug delivery, implants, dermal fillers, root canal filling, wound dressings, contraceptive devices, fixation devices and tissue engineering	(Espinoza et al., 2020; Shahverdi et al., 2022)
Polybutyrate adipate terephthalate	Flexible and tough, which makes it ideal for combination with other biodegradable polymers that have high modulus and strength but are very brittle	Garbage bags, wrapping films, disposable products (lunch boxes, dishes, cups, etc.,) courier bags, mulch film, shape memory composites, drug delivery	(de Matos Costa et al., 2020; C. Xie et al., 2023)
Polyvinyl alcohol	Crystal clear, water solubility	Papermaking, textiles, a variety of coatings, thermal conductor, cementitious composites	(Aslam et al., 2018; Settier-Ramírez et al., 2020; Q. Wang et al., 2023)

Similar to the properties of PLA yet different in terms of raw material, polycapro-lactone is chemically synthesized by ring-opening polymerization of carprolactone monomer, the raw material for which can be obtained from crude oil (Di Foggia et al., 2010). PCL is an aliphatic polyester that is able to fully biodegrade and is com-postable in an industrial composting system. It is partially crystalline and has a low melting point and a glass transition temperature. This unique property allows it to be used for 3D printing, heat molding and shape memory (Espinoza et al., 2020). The general applications of PCL are scaffolding, drug delivery, dermal fillers, root canal filling and many more in biomedical applications.

Polybutyrate adipate terephthalate is a semi-aromatic copolyester of adipic acid that can be synthesized from 1,4-butanediol and adipic acid and the polymer of dimethyl terephthalate (DMT), the raw material for which is crude oil (de Matos Costa et al., 2020). PBAT consists of aromatic fractions that promote excellent physical properties, while its aliphatic chains exhibit degradation properties, par-ticularly attributed to the cleavage of ester linkages and the interaction between water and their carbonyl groups from benzene rings (de Matos Costa et al., 2020; X. Wang et al., 2019). Thus, PBAT is applied and blended with other thermoplas-tics in disposable products such as garbage bags, disposable cutlery and disposable wrappers.

Polyvinyl alcohol is a synthetic polymer prepared by hydrolysis of polyvinyl ace-tate (Settier-Ramírez et al., 2020). Polyvinyl acetate is made from acetic acid and ethylene. Ethylene is often produced from natural gas or petroleum. PVOH contains a large amount of hydroxyl groups, which are ready to form hydrogen bonding with water, making them hydrophilic and soluble in water (Aslam et al., 2018). Due to this interesting feature, PVOH has benefited a variety of industries, including paper-making, textiles and coatings.

1.7   NATURAL POLYSACCHARIDES

The chemistry of natural polysaccharides, including those from plants, animals and microorganisms, are described in this subsection. These polysaccharides include cellulose, pectin, starch, agar, carrageenan and alginate from plant/algae-based resources; chitosan and hyaluronic acid from animal-based resources; and glucan, pullulan, levan, bacterial cellulose and xanthan gum from microorganisms. This subsection first presents natural polysaccharides derived from plant, followed by animals and microorganisms. Biopolymers produced from natural polysaccharides offer remarkable advantages, such as renewability, biodegradability and biocompat-ibility. Enhanced performance can be attained by blending and incorporating fillers (Torres et al., 2019). Therefore, each section covers not only the basic chemistry of the biopolymers but also examples of biopolymers used for biopolymer blends and bionanocomposites.

1.7.1   CELLULOSE

Cellulose is abundantly found in the cell wall of plants and consists of lignin, hemi-cellulose and cellulose. Each single glucose unit pairs with intermolecular hydrogen

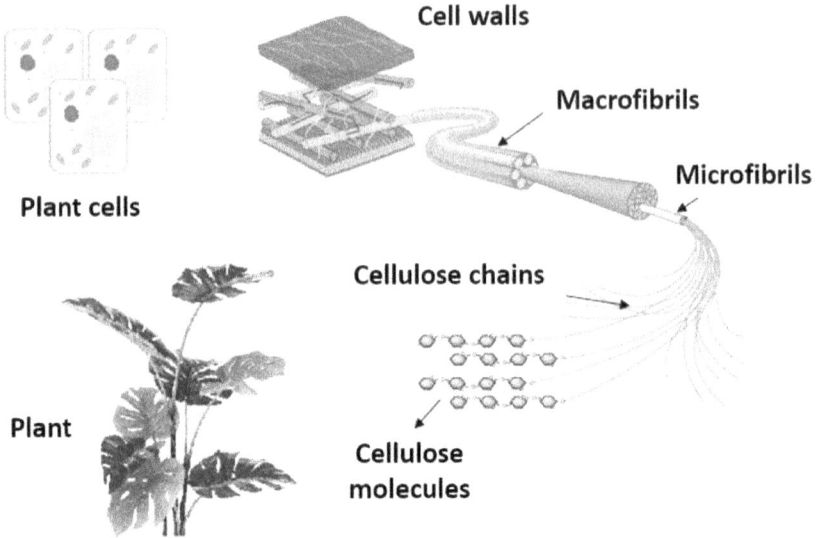

FIGURE 1.10   Schematic structure of plant cellulose microfibrils.

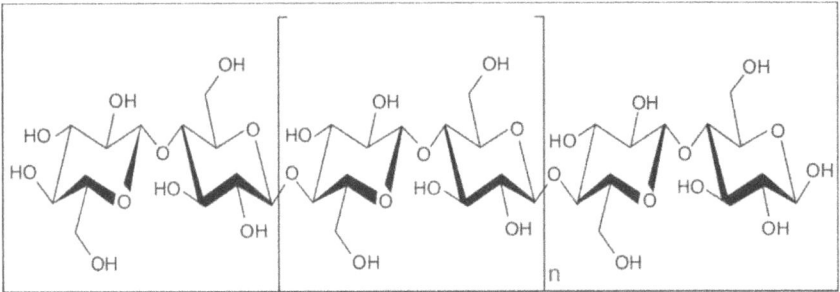

FIGURE 1.11   Chemical structure of cellulose.

bonding to form cellulose microfibrils (Figure 1.10), which contain about 18 cellulose chains (Kubicki et al., 2018).

Cellulose is a polysaccharide with a direct chain of 200 or 300 to over 10,000 β (1→4) associated D-glucose units $(C_6H_{10}O_5)_n$. Cellulose consists of glucose–glucose linkages of 1–4-linked d-glucopyranosyl units arranged in linear chains. The chains form a network of crystalline fibers known as microfibrils, which are approximately 3 nm in diameter (Aggarwal et al., 2020). The chemical structure of cellulose is shown in Figure 1.11.

Basic physical and chemical characteristics of cellulose are density are around 1.5 g/cm^3, melting point within 260–270°C, insoluble in organic solvents; it reacts with halogens, nitric acid and sulfuric acid, and its backbone of sugar forms hydrogen bonds and holds cellulose microfibrils together. The type of bonds that build up cellulose are glycosidic bonds (cellulose is able to form hydrogen bonds with water due to the oxygen atoms in the OH group of cellulose). It also forms inter- and intra-molecular hydrogen bonding between itself and other materials consisting of hydroxyl groups (Aggarwal et al., 2020; Hasan et al., 2018; Wu et al., 2018).

Occurrence: Plants, bacteria, tunicates (sea squirts), wood, algae, etc. (Kargarzadeh et al., 2017).

Biopolymer blends/bionanocomposites: Silver nanoparticles (AgNPs) are used as a bactericidal agent in active food film packaging to produce film with good mechanical and water barrier properties (Hasan et al., 2018). Cellulose–starch hybrid films enhanced mechanical properties attributed to hydrogen bonding between the amylopectin of starch and cellulose used for packaging (Noorbakhsh-Soltani et al., 2018).

1.7.2 STARCH

In leaves and other green tissues, starch is a product of photosynthesis and is temporarily stored in chloroplasts during the day and subsequently broken down during the night. This process is known as "transitory starch" and serves to provide a continuous supply of carbohydrates and energy in the absence of photosynthesis. Many plants, including crop plants like wheat and potatoes, generate starch in their seeds and storage organs such as in their grains and tubers. It is used for germination and sprouting purposes for plants. Starch is a polysaccharide that contains a chain of glucose molecules which are bound together. Depending on the plant, starch is made up of between 20–25% amylose and 75–80% amylopectin (Araújo et al., 2020). It is a semi-crystalline polymer consisting of (1–4) linked α-D glucopyranosyl units (Abdul Khalil et al., 2017). The chemical structure of both amylose and amylopectin is illustrated in Figure 1.12.

The basic chemical formula of a starch molecule is $(C_6H_{10}O_5)_n$. There are two types of polysaccharides found in starch:

1 Amylose: a linear chain of glucose
2 Amylopectin: a highly branched chain of glucose

Basic physical and chemical characteristics of starch are insolubility in cold water, alcohol and other organic solvents; a boiling point of approximately 100°C, similar to cellulose; and the glycosidic bonds are a special type of covalent bond that link from 500 to several hundred thousand glucose monomers together. The linear to branched glucosyl units are composed of hydroxyl groups, which are able to form inter- and intra-molecular hydrogen bonds with other materials that contain hydroxyl groups. The ratio of amylase and amylopectin plays an important role in mechanical properties, as amylase promotes better film properties than amylopectin due to its linear component and consists of water residue, which also contributes to the mechanical properties while reducing glass transition temperature at the same time (Abdul Khalil et al., 2017; Wang et al., 2015).

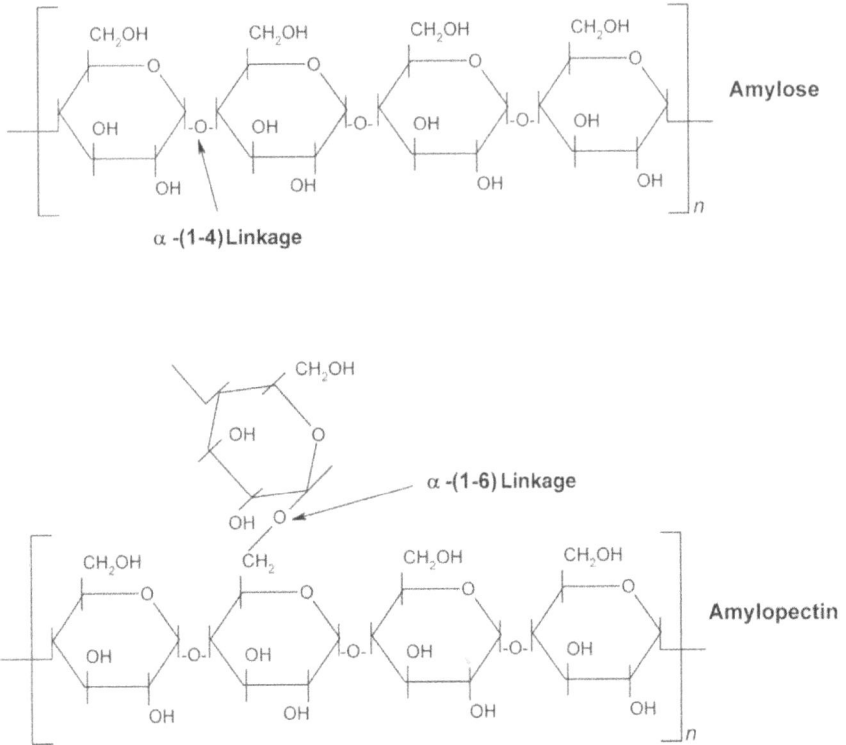

FIGURE 1.12   Chemical structure of starch.

Occurrence: The roots of the cassava plant; the tuber of the potato; the stem pith of sago; and the seeds of corn, wheat and rice (Araújo et al., 2020).

Biopolymer blends/bionanocomposites: Starch is hydrophilic and consists of many OH groups. Therefore, enhancement is needed to expand its application. From biopolymers to biocomposites, enhancement of film properties can be done through incorporation of inorganic fillers such as calcium carbonate, kaolinite, silicate, clay and titanium dioxide (TiO_2) (Abdul Khalil et al., 2017).

1.7.3   PECTIN

Pectin is a structural heteropolysaccharide, which has been used extensively for many years in the food and beverage industry as a thickening agent, a gelling agent and a colloidal stabilizer. It is able to form gel in water under suitable conditions (Naqash et al., 2017).

The distinct chain of pectin consists of esterified D-galacturonic acid in an α-1,4-linked d-chain, which is covalently linked through α-1,2 bonds to methoxy groups in the natural product (Figure 1.13). These uronic acids have carboxyl groups. Pectins are divided into two categories based on their degree of methylation,

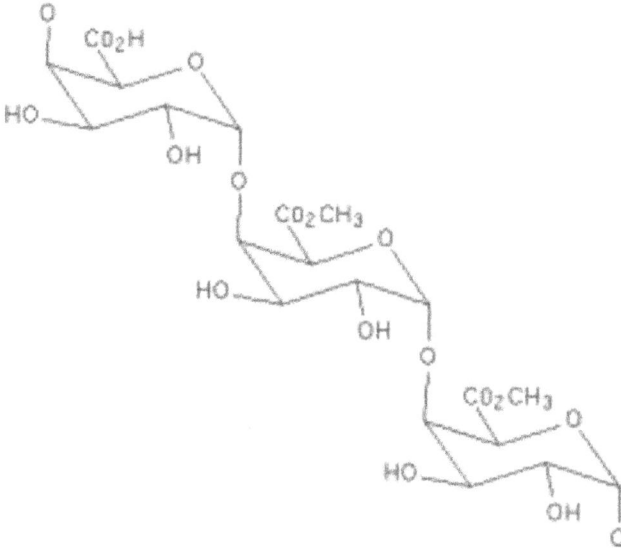

FIGURE 1.13 Chemical structure of pectin.

partially or fully methyl esterified, including low-methoxyl pectins (degree of esterification (DE) < 50%) and high-methoxyl pectins (DE > 50%) (Abdul Khalil et al., 2017).

The basic physical and chemical characteristics of pectin are that protopectin is insoluble in water. However, when the fruit ripens, it is converted to water soluble. The monovalent cation salts of pectinic and pectic acids are usually soluble in water. The ability to form gels and gel formation are caused by hydrogen bonding between free carboxyl groups on the pectin molecules and also between the hydroxyl groups of neighboring molecules. Gel strengths increase upon increasing calcium ion concentration. The melting point of pectin is 140 to 180°C (decompose), and the gelling temperature is in the range of 40 to 100°C. The pH of low-methoxyl pectin ranges from 3 to 3.5, while the pH of high-methoxyl pectin ranges from 2.8–3.6 (Naqash et al., 2017).

Occurrence: Pectin can be derived from protopectin found in the primary cell walls of terrestrial plants and cell walls of higher plants such as citrus peels or apple skin (Hassan et al., 2019).

Biopolymer blends/bionanocomposites: Pectin is generally hydrophilic, which means it is sensitive towards moisture and relative humidity. Hence, blending with other biopolymers such as cellulose, starch or PLA, or incorporation with fillers such as clay and nanocellulose is usually done to solve the problem. Studies have shown that clay improved the tensile strength of a film compared to pure pectin film. Specifically, periodate-oxidized pectin has been combined with chitosan and gelatin to form different structures in biomedical applications (Abdul Khalil et al., 2017).

1.7.4 AGAR

Agar is a water-soluble long-chain algae-based polysaccharide mainly found in red algae and is widely used in the manufacture of gelatin. It also has added advantages to decrease blood sugar concentration and induce an anti-aggregation effect on red blood cells and displays antioxidant, antitumor and antiviral activities (Shahidi & Rahman, 2018).

The chemical formula of agar is $C_{14}H_{24}O_9$. It is a complex mixture of polysaccharides composed of two major fractions, agarose and agaropectin. Agarose is the gelling fraction that forms a neutral linear molecule essentially free of sulfates, the chains of which consist of repeating alternate units of β-1,3-linked-D-galactose and α-1,4-linked 3,6 anhydro-L-galactose units. Agaropectin, the non-gelling fraction, is a sulfated polysaccharide composed of agarose and varying percentages of ester sulfate, D-glucuronic acid and small amounts of pyruvic acid (Abdul Khalil et al., 2017).

The basic physical and chemical characteristics of agar are that it is insoluble in cold water, but it swells considerably, absorbing as much as 20 times its own weight of water. It dissolves readily in boiling water and sets to a firm gel at concentrations as low as 0.50%. Powdered dry agar-agar is soluble in water and other solvents at temperatures between 95 and 100°C. Thermo-reversible gel and agar-agar solution in hot water form a characteristic gel after setting, with a melting point between 85 and 95°C and a gelling point between 32 and 45°C. The viscosity of an agar solution at temperatures above its gelling point is relatively constant at pH ranges from 4.5 to 9.0 and is not greatly affected by ionic strength in pH ranges from 6.0 to 8.0. However, viscosity at constant temperature increases with time once gelation begins. Exposure to high temperature and lower pH may result in lower gel strength Thus, exposing agar-agar solutions to high temperatures above 95°C and pH lower than 6.0 for prolonged periods of time should be avoided (Cotas et al., 2020; Zhang et al., 2017).

Occurrence: From red seaweeds such as *Gracilaria sp., Gracilaria cornea, Gracilaria dominguensis, Gigartina sp.* and *Gelidium sp* (Cotas et al., 2020).

Biopolymer blends/bionanocomposites: A wide range of studies on biopolymer blends and composites have been done, such as agar/starch blend, agar/chitosan blend, soy protein, agar blend, agar/nanocellulose and agar/silver nanoparticles for the use of drug carrier and wound dressing (Abdul Khalil et al., 2017). Agar-graft-PVP and κ-carrageenan (seaweed polysaccharides)-graft-PVP blends in an aqueous medium of pH 7 are capable of forming hydrogels (Prasad et al., 2006). This study demonstrates that graft blends based on seaweed polysaccharides, as described, hold promise for various applications in biomedicine, including tissue engineering, agriculture for water retention, microbiology and pharmaceuticals as hydrogel dressings. These hydrogels can also serve as a sustainable alternative to animal-derived collagen-based materials. An additional study involving agar aerogel incorporating a small-sized zeolitic imidazolate framework and carbon nitride for solar water purification introduces an innovative, cost-effective and convenient solar-triggered solution for sustainable water decontamination using a MOF-based aerogel (Zhang et al., 2017).

1.7.5 CARRAGEENAN

Carrageenan is another type of polysaccharide obtained from red seaweed of the class Rhodophyceae. It appears to be a promising candidate in tissue engineering and regenerative medicine in recent years. The molecular formulae of carrageenan is $C_{23}H_{23}FN_4O_7Zn$. There are three main isomers composed of iota (ι), kappa (κ) and lambda (λ) ι-carrageenans. The difference between these three is the number and position of ester sulfate groups on the repeating galactose units per disaccharide unit. The characteristics of carrageenan are influenced by the sulphate ester group of 3, 6 anhydro-galactose content. The chemical structure of carrageenan is shown in Figure 1.14. It is an anionic sulfated linear polysaccharide formed by a straight chain backbone structure of alternating 1, 3-linked-β-D galactopyranose and 1–4 linked α-D-galactopyranose units. These units occur as the 2- and 4-sulphate or unsulfated, while the 4-linked units occur as the 2-sulphate, 2, 6-disulphate, 3, 6 anhydrid and 3, 6 anhydrid 2-sulphate (Abdul Khalil et al., 2017).

The basic physical and chemical characteristics of carrageenan are that carrageenan is a water-soluble linear sulphated polysaccharide. All carrageenan types are soluble in hot water at temperatures above its gel melting temperature. The normal amplitude of solubility temperature is between 40 and 70°C, depending on the solution concentration and the presence of cations. In cold water, only lambda-carrageenan and the sodium salts of kappa and iota carrageenan are soluble. Potassium and calcium salts from kappa- and iota-type carrageenan are not soluble in cold water but will swell as a function of concentration and type of cations present as well as water temperature and condition of dispersion. Hot aqueous solutions of kappa and iota carrageenan have the ability to form thermo-reversible gels upon cooling.

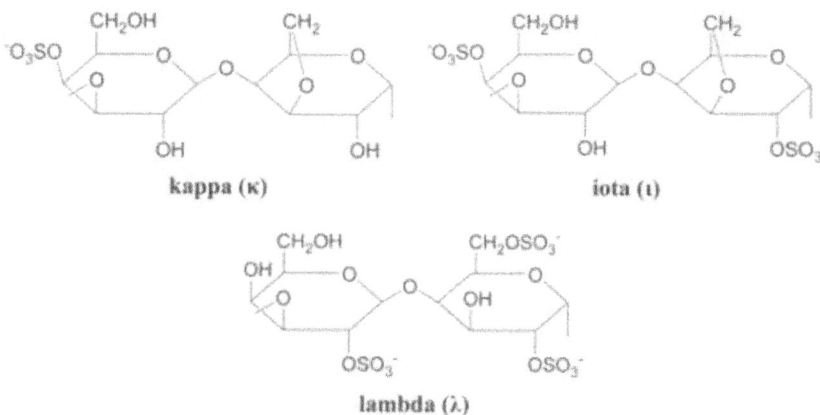

FIGURE 1.14 Chemical structure of kappa, iota and lambda carrageenan.

This phenomenon occurs due to the formation of a double helix structure by the carrageenan polymers. The gel is thermo-reversible. The carrageenan concentration is generally 1.5% by weight of the water solution. Commercial carrageenans are generally available in viscosities ranging from about 5 to 800 cps when measured in 1.5% solutions at 75°C. Carrageenan solutions are quite stable in neutral or alkaline pH (Aggarwal et al., 2020; Seo & Yoo, 2021).

Occurrence: Red algae from *Eucheuma* (*kappaphycus*), *Chondrus*, *Gigartina* and *Hypnea* species (Aggarwal et al., 2020).

Biopolymer blends/bionanocomposites: Carrageenan has been incorporated with fillers and blended with other biopolymers to achieve better physical and mechanical properties. Kappa carrageenan shows a synergy effect with locust beam gum (LBG) in aqueous gel systems. Carrageenan is different from other hydrocolloids, as it is able to interact with milk proteins. Interaction occurs due to the strong electrostatic interaction between the negatively charged ester sulfate groups in the carrageenan molecule with the strong positive charges of the milk casein micella (Seo & Yoo, 2021). Another form of interaction is through links established among ester sulfate groups of carrageenan with carboxylic residues of amino acids that make up the protein. Carrageenan has also been incorporated with mica nanoclays and blended with a pectin matrix to achieve 10% higher mechanical strength compared to the neat film (Abdul Khalil et al., 2017).

1.7.6 ALGINATE

Alginate is another prominent algae-based polysaccharide. It is obtained from brown algae, whose occurrence is mentioned subsequently. The biocompatibility and non-toxicity of alginates make them available for extensive applications, particularly in biomedical fields. The chemical structure of alginate is shown in Figure 1.15. Alginate is mainly composed of uronic acids of M and G blocks, where M blocks refer to 1,4-β-D-mannuronic acid and G blocks refer to 1,4-α-L-guluronic acid, with

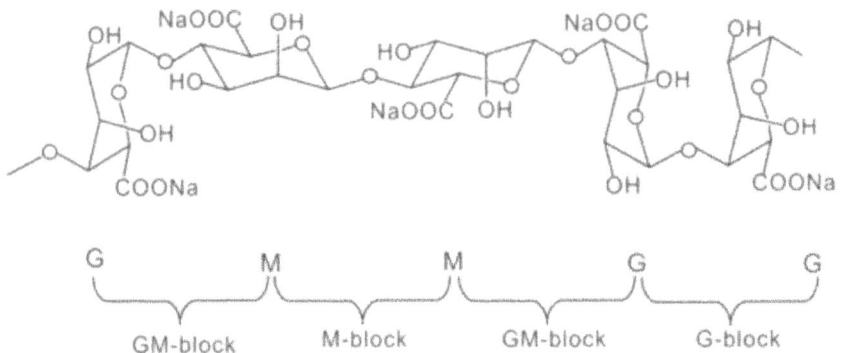

FIGURE 1.15 Chemical structure of alginate.

a homogeneous (poly-G, poly-M) or heterogeneous (GM) block composition, which was proven by partial acid hydrolysis (Wang et al., 2019). Each species of brown seaweed differs in composition and sequences. Such differences usually occur in the ratio of mannuronic and guluronic acid blocks (Abdul Khalil et al., 2017).

The basic physical and chemical characteristics of alginate are its ability to form a uniform and transparent gel when alginic acid interacts with present metal ions such as calcium, sodium or magnesium. It is insoluble in water and organic solvents and thermo-irreversible in room temperature. Alginates have the ability to absorb 200–300 times of their own weight in water to form a viscous gum. The viscosity of alginate solutions increases upon decreasing pH and reaches a maximum around pH 3 to 3.5 due to the protonated alginate backbone that forms hydrogen bonds (Usman et al., 2017; Wang et al., 2019).

Occurrence: Extracted from brown algae (*Phaeophyceae*), including *Laminaria japonica*, *Laminaria digitata*, *Laminaria hyperborea*,
Macrocystis pyrifera and *Ascophyllum nodosum* (Wang et al., 2019).

Biopolymer blends/bionanocomposites: Many studies have been done to
improve the weak mechanical and water barrier properties of the alginate matrix. Some studies have incorporated other polysaccharides such
as PLA, chitosan and hyaluronic acid (HA), as well as soy protein, while
others incorporated minerals as filler in the alginate matrix (Kosik-Kozioł
et al., 2017; Wang et al., 2019; Wongkanya et al., 2017). Such incorporation
includes cellulose nanofibrils and calcium chloride. This process is also
applied in pharmaceuticals as a coating film on drug tablets to control the
release of drug (Volić et al., 2018).

1.7.7 CHITOSAN

Chitosan is a partially deacetylated product of chitin. Chitosan is widely used in the pharmaceutical field, as it contains positive charges that can interact with the negative part of cell membrane. Unlike other polysaccharides that usually contain carbon, hydrogen and oxygen, chitin and chitosan consist of nitrogen (6.89%) (da Silva Alves et al., 2021). The formula of chitosan is $C_6H_{11}O_4N$, and it is a copolymer made up of one amine and two free hydroxyl groups for each monomer. It comprises of β-(1 → 4)-2-acetamido-D-glucose and β-(1 → 4)-2-amino-D-glucose units (da Silva Alves et al., 2021). Chitosan contains two main sugars, glucosamine and N-acetylglucosamine; the proportion for each sugar usually relies on the alkaline treatment (Jiménez-Ocampo et al., 2019).

The basic physical and chemical characteristics are that it is an off-white powder with molecular weight 300–1000 kDa depending on the source of chitin. The degree of acetylation (DA) is 15–25%. Viscosity is 86.4 Pa s (at 20°C, 30 rpm). It has no melting point, as it decomposes upon heating, and chitin and chitosan also contain nitrogen (6.89%), which is the reason for their being of commercial interest. Chitosan is the only positively charged, naturally occurring polysaccharide. Chitosan molecules have both amino and hydroxyl groups. Stable covalent bonds can be formed via various processes including etherification, esterification, and reductive amination.

Chitosan has antibacterial activity and antifungal, mucoadhesive, analgesic and hemostatic properties (Jiménez-Ocampo et al., 2019; Qiao et al., 2017).

> **Occurrence:** Insects, crustaceans, squid, centric diatoms and fungi (Hassan et al., 2019; Qiao et al., 2017).
>
> **Biopolymer blends/bionanocomposites:** Both chitin and chitosan possess reactive hydroxyl and amino groups, but chitosan is usually less crystalline than chitin, which makes chitosan more accessible to reagents and other biopolymers. Several examples of the usage of chitosan for emerging applications include blending it with gelatin and 3-phenylacetic acid for food packaging applications (Liu et al., 2021). Additionally, a chitosan/gelatin/PVA blend was formulated into a hydrogel for wound dressings (Fan et al., 2016), and chitosan/starch/silver nanoparticles were employed to create antimicrobial papers (Jung et al., 2018).

1.7.8 Hyaluronic Acid

Hyaluronic acid is also known as hyaluronan. It receives tremendous attention, particularly in the field of cosmetics and pharmaceuticals. It is an essential component found in the extracellular and pericellular matrixes and the inner cells. It is an unbranched biopolymer, which belongs to heteropolysaccharides. The formula of hyaluronic acid is $C_{14}H_{21}NO_{11}$ n., and the chemical structure is illustrated in Figure 1.16. HA has a linear chain consisting of repeating disaccharide units linked by ß-1,4-glycosidic bonds. Each disaccharide contains N-acetyl-D-glucosamine and D-glucuronic acid connected by ß 1,3-glycosidic bonds.

The basic physical and chemical characteristics are that HA can reach a very high molecular weight, up to 20,000 kDa. HA possesses fundamental physical and chemical characteristics, such as being able to attain an exceptionally high molecular weight, reaching up to 20,000 kDa. Furthermore, HA is notably sensitive to pH,

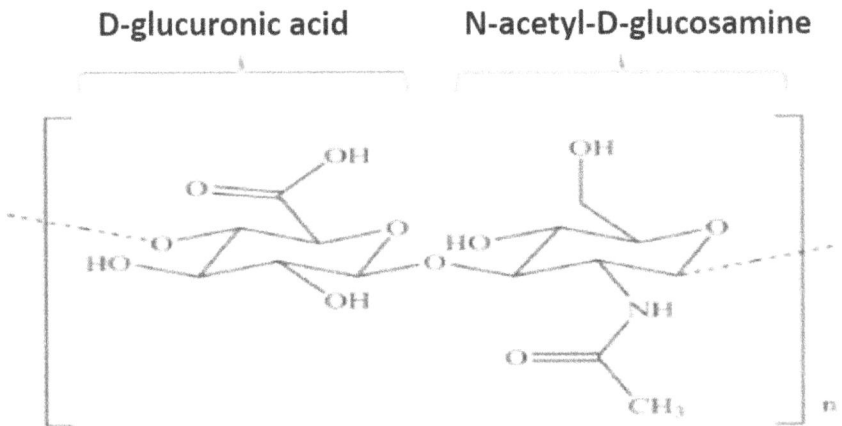

D-glucuronic acid   **N-acetyl-D-glucosamine**

FIGURE 1.16   Chemical structure of HA.

undergoing hydrolysis in both acidic and alkaline environments, and depolymerizes at pH levels exceeding 11 or dropping below 4. HA solution is a non-Newtonian liquid with shear-thinning and viscous behavior. HA has hydrophilic sites such as carboxylic acid (-COOH) or hydroxyl (-OH) groups that readily link with other hydrophilic substrates (Fallacara et al., 2018).

Occurrence: Rooster combs, human umbilical cords, human joint synovial fluid, vitreous humor, human dermis and epidermis, vertebrate tissues and the skin of animals (Fallacara et al., 2018).

Biopolymer blends/bionanocomposites: HA degrades easily and exhibits weak biomedical properties. Hence, there is always a need to improve its performance for end use. Its biocompatibility and free hydroxyl groups allow it to crosslink with other biopolymers such as gelatin and chitosan (Alemdar, 2016). There is also evidence of hyaluronic acid blended with poly (L-lactic acid) for biomedical application (Wang et al., 2017). More recently, chitosan is blended with hyaluronic acid as hydrogel for injectable tissue engineering as well (Lee et al., 2020).

1.7.9 YEAST GLUCAN

Glucan is abundant in the cell walls of fungi. β-glucans are carbohydrates (sugars) that are found in the cell walls of bacteria, fungi, yeasts, algae, lichens and plants such as oats and barley. They are taken as herbal medicines; to prevent and treat cancer, human immunodeficiency virus (HIV) and diabetes; to lower cholesterol; and to increase immune system function. Therefore, glucan is particularly popular in the food and pharmaceutical industries (Yuan et al., 2020). Recently, the application of glucans has been extended to water treatment owing to their effectiveness in heavy metal adsorption (Jiang et al., 2019).

Yeast glucan belongs to the class of β-glucan, and its structure includes two distinct macromolecular components composed of consecutively (1→3)-linked β-D-glucopyranosyl residues, with small numbers of (1→6)-linked branches and a minor component with consecutive (1→6)-linkages and (1→3)-branches (Rahar et al., 2011). Yeast glucan exhibits fundamental physical and chemical characteristics, including a high degree of polymerization (DP > 100). β-glucan is composed of multiple OH groups and β-(1→3) linkages, and its high DP renders it completely insoluble in water. The solubility of yeast glucan increases as the DP decreases. The ratio of soluble to insoluble fractions of β-glucan is significantly influenced by the extraction conditions (Rahar et al., 2011; Yuan et al., 2020).

Occurrence: *Rhynchelytrum repens, Lentinus edodes, Grifola frondosa, Tremella mesenterica, Tremella aurantia, Zea may, Agaricus blazei, Phellinus baummi, Saccharomyces cerevisae* (yeast) and *Agaricus blazei murell* (mushroom) (Rahar et al., 2011; Yuan et al., 2020).

Biopolymer blends/bionanocomposites: The large number of hydroxyl groups in β-glucan allow interaction with other biopolymers. Such evidence is found in brown wheat flour/β-glucan blended with xanthan and guar gum to enhance the food texture, which is absolutely beneficial for frozen dough (Ahmed & Thomas, 2018). Apart from that, the biocompatibility of glucan with multiple OH groups allows them to be easily crosslinked with chitosan and the formation of hydrogels for heavy metal adsorption (Jiang et al., 2019).

1.7.10 PULLULAN

Pullulan is a non-ionic polysaccharide derived from fungus (Figure 1.17). Due to its unique properties such as high solubility and low viscosity, and because it is easily modified and has stability in a wide range of pH values, pullulan has been used in many applications, including as a food additive, blood plasma substitution, flocculant, and adhesive and in film fabrication (Singh & Kaur, 2019). It is now becoming more competitive with natural gums made from marine algae and other plants in terms of cost effectiveness. The chemical formula of pullulan is $(C_6H_{10}O_5)_n$. Pullulan is a linear, non-ionic polysaccharide with a chemical structure composed of maltotriose units: α-$(1 \rightarrow 6)$-linked $(1 \rightarrow 4)$-α-d-triglucosides (Priyadarshi et al., 2021).

The basic physical and chemical characteristics of pullulan are that it is a white powder, odorless and tasteless. It is highly soluble in water and diluted alkali due to a low degree of hydrogen bonds in its crystal form. It has high flexibility of the chain and forms a viscous solution in hot and cold water but does not form a gel. It is less viscous compared to other hydrocolloids, with high stability to sodium chloride and a wide range of pH values from 2 to 11. It is heat resistant and decomposes at high temperatures from 250 to 280°C. It is insoluble in organic solvents, except in dimethylformanide and dimethyl sulfoxide, and the α-1,6-glucosidic linkages in pullulan give structural flexibility (Wang et al., 2019; Priyadarshi et al., 2021).

Occurrence: Fermented from black yeast *Aureobasidium pullulans* and other species of microorganisms, *Tremella mesenterica*, *Teloschistes flavicans*, *Cryphonectria parasitica*, *Cytariaharioti* and different carbon sources from Asian palm kernel, jackfruit seed and rice hull (Priyadarshi et al., 2021; Singh & Kaur, 2019).

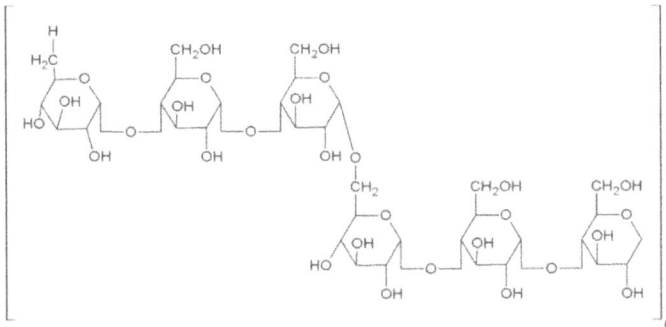

FIGURE 1.17   Chemical structure of pullulan.

Biopolymer blends/bionanocomposites: Due to its unique α-1,6-glucosidic linkages, pullulan is flexible enough to form intra- and inter-molecular bonding with pectin, which eventually enhances water resistance properties (Priyadarshi et al., 2021). This is useful, especially in food packaging applications. Another instance is found in gelatin and pullulan blends incorporated with nanofibers. This composite showed enhanced mechanical properties due to the increase of intermolecular hydrogen bonds and the decrease of intramolecular hydrogen bonds. Because of that, this composite was suggested to be used in tissue engineering (Wang et al., 2019). The modification of pullulan, performed via chemical reactions, blending or incorporation of fillers is usually employed to enhance its uses in as many areas as possible.

1.7.11 DEXTRAN

Dextran is another exopolysaccharide obtained from bacteria in a sucrose-rich media (Ghimici & Nichifor, 2018). It is highly demanded in biomedical, pharmaceutical and tissue engineering applications. This is because it is biocompatible, non-toxic and non-immunogenic (Zheng et al., 2019). Conventionally, it has been used to treat hypovolemia resulting from surgery and other types of bleeding. The formula of dextran is $H(C_6H_{10}O_5)_xOH$, and the chemical structure of dextran is shown in Figure 1.18. It is a hydrophilic polysaccharide composed of α(1,6)-linked glucopyranose units

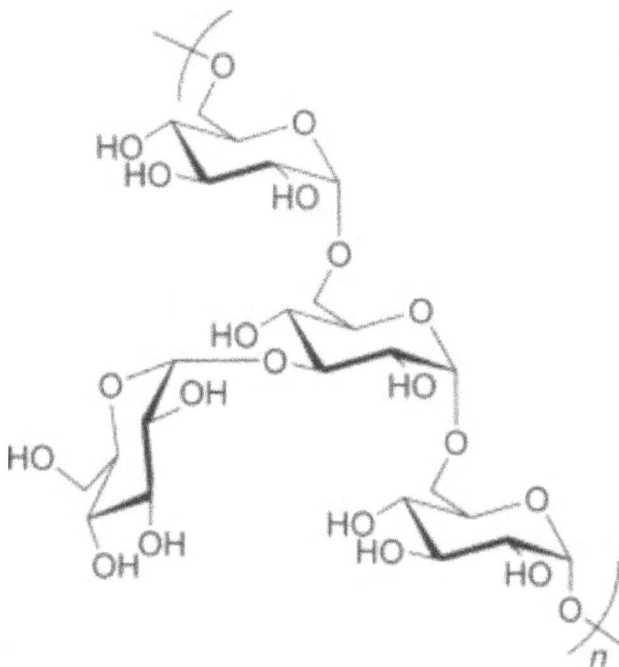

FIGURE 1.18 Chemical structure of dextran.

with α-1 $\rightarrow$ 2, α-1 $\rightarrow$ 3, and α-1 $\rightarrow$ 4 linked side chains attached to the C-3 position of the backbone (Rzayev et al., 2017; Ghimici & Nichifor, 2018).

> **Occurrence:** Produced by bacteria strains from *Leuconostoc, Lactobacillus* and *Streptococcus* (Ghimici & Nichifor, 2018; Hassan et al., 2019).
>
> **Biopolymer blends/bionanocomposites:** Dextran can be easily modified to form a 3D network structure with some of the promising functional groups including tyramine, ethylamine vinyl sulphones, thiols and acrylates (Ghimici & Nichifor, 2018). Dextran is used in biomedical applications owing to its hydroxyl groups, which can be oxidized into aldehydes and function as a crosslinking agent to crosslink with other amino group-containing polymers. One instance of this is the fabrication of a poly (vinyl alcohol)/dextranaldehyde composite hydrogel for wound dressing application (Zheng et al., 2019). It is also a versatile biopolymer to fabricate electrospun nanofiber membranes. These can be found in colloidal nanofiber composites of dextran and folic acid for electro-active platforms (Rzayev et al., 2017). Increasing the fraction of dextran in nanofiber colloids is believed to improve the transmission of folic acid due to the hydrogen bonding linkages.

1.7.12 BACTERIAL CELLULOSE

Bacterial cellulose (BC) is produced via bacteria fermentation in a static culture. Recently, a significant expansion of applications in antibacterial packaging applications in the food and packaging industries and drug delivery systems has been noticeable (Rydz et al., 2018; Treesuppharat et al., 2017). The BC chemical structure is illustrated in Figure 1.19. BC is made up of (1 $\rightarrow$ 4)-*D*-anhydroglucopyranose chains bonded through β-glycosidic linkages. The material's geometry is determined by the intra-molecular and inter-molecular hydrogen-bonding network and hydrophobic and van der Waals interactions, forming parallel chains (Rangaswamy et al., 2015).

The basic physical and chemical characteristics of BC are its high crystallinity (84–89%) and high polymerization degree. It has a higher surface area than the

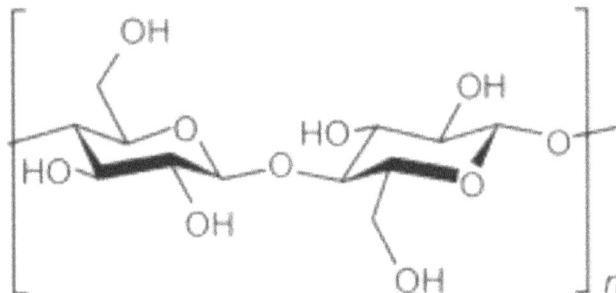

FIGURE 1.19   Chemical structure of bacterial cellulose.

cellulose obtained from plant sources (high aspect ratio of fibers with diameter 20–100 nm), high flexibility (Young's modulus of 15–18 GPa) and high water-holding capacity (over 100 times its own weight) with the specific surface area (37 m^2/g) (Rangaswamy et al., 2015) (Treesuppharat et al., 2017).

Occurrence: Bacterial cellulose is produced extracellularly by gram-negative bacterial cultures of *Gluconacetobacter, Acetobacter, Agrobacterium, Achromobacter, Aerobacter, Sarcina, Azobacter, Rhizobium, Pseudomonas, Salmonella* and *Alcaligenes*. Among the bacteria, *Komagataeibacter xylinus* is the most efficient in producing bacterial cellulose (Rangaswamy et al., 2015).

Biopolymer blends/bionanocomposites: In biomedical applications, bacterial cellulose nanocrystals/regenerated chitin fibers (BCNC/RC) for suture biomaterial, 2,2,6,6-tetramethylpiperidinyloxy (TEMPO)-oxidized BC with Ag nanoparticles, have been developed for wound dressing application (Wu et al., 2018). Another biomedical application for drug delivery was achieved by bacterial cellulose/gelatin hydrogel composites crosslinked with glutaraldehyde. The composite showed excellent dimensional stability caused by the intra- and inter-molecular hydrogen bonding interaction between bacterial cellulose and gelatin (Treesuppharat et al., 2017). For packaging applications, it was observed that BC blended with PCL showed better mechanical properties and biodegradability than neat PCL due to the contribution of high tensile properties of BC due to homogenous distribution of PCL throughout the BC network (Rydz et al., 2018).

1.7.13 XANTHAN GUM

Xanthan gum is an anionic heteropolysaccharide produced through the fermentation of simple sugar by a specific kind of bacteria. It was first discovered in the 1960s and then commercialized in the 1970s. It is a recognized stabilizer, additive and thickening agent used for toothpaste, cream and lotions, as it is known for its biocompatibility, biodegradability and water solubility (Kumar et al., 2017). The chemical formula of xanthan gum is $C_{35}H_{49}O_{29}$, and the chemical structure of xanthan gum shows a long-chain polysaccharide, which consists of -glucose, d-mannose and d-glucuronic acid as building blocks with a high number of trisaccharide side chains on every glucose (Figure 1.20). The beta-D-glucoses are linked $(1 \rightarrow 4)$ to form a backbone similar to cellulose (Sworn, 2021).

The basic physical and chemical characteristics are that xanthan gum easily dissolves at room temperature and is soluble in cold water. It has a high molecular weight (on the order of 1000 kDa) and shows high pseudoplastic flow. It has high viscosity and decreases with shear force applied. The viscosity remains the same from 0 to 100°C and from a pH of 1 to 13. It is highly enzyme resistant and has synergistic interaction with galactomannans (e.g. guar gum, konjac and locust bean gum). It is stable in a wide range of pH, temperature and organic solvents. It consists of a large number of hydroxyl groups that are able to form hydrogen bonding with other biopolymers (Kumar et al., 2017; Sworn, 2021).

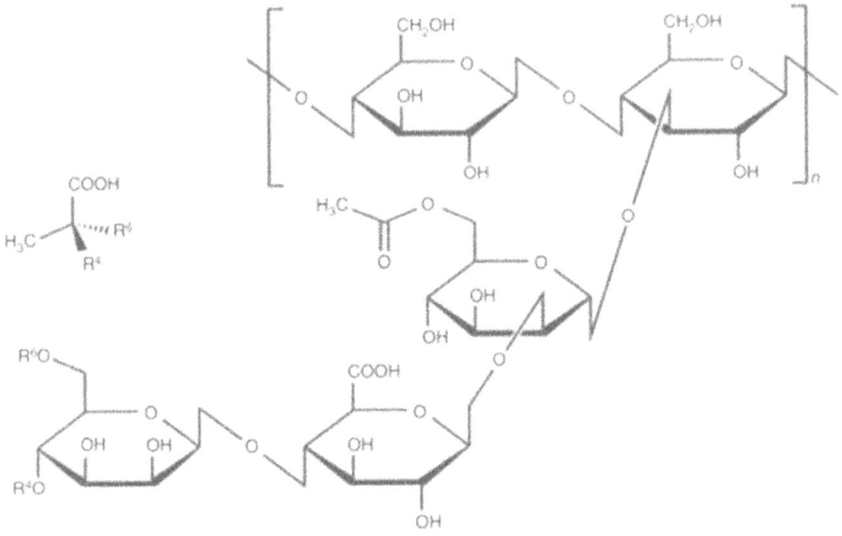

FIGURE 1.20 Chemical structure of xanthan gum.

Occurrence: Produced extracellularly by a pure-culture fermentation process of carbohydrate by the bacterium known as *Xanthomonas campestris* (Aggarwal et al., 2020).

Biopolymer blends/bionanocomposites: Xanthan gum can achieve better physical performances when it is blended with other biopolymers and incorporated with fillers such as alginate, cellulose nanocrystals and halloysite nanotubes, mainly due to the formation of electrostatic attraction and hydrogen bonding that promote uniform dispersion and result in enhanced mechanical and thermal properties (Kumar et al., 2017). A crosslinked hydrogel nanocomposite of xanthan gum for adsorption of crystal violet dye in water purification applications showed the effectiveness of dye adsorption due to the presence of electrostatic and hydrogen bonding between the hydroxyl group of xanthan gum and the polar nitrogen atom within the crystal violet dye (Mittal et al., 2021).

1.8   NATURAL PROTEINS

The chemistry of natural proteins, including those mainly from plants and animals, are detailed in this section, starting with natural plant proteins, which include soy protein, zein protein and gluten protein, and followed by animal proteins, including collagen, gelatin and keratin. Similar to the sequence of description in the section on natural polysaccharides, the chemical structure, chemical characteristics and some notable examples of natural proteins used in biopolymer blends or bionanocomposites are detailed accordingly.

1.8.1 SOY PROTEIN

Soy protein can be found in soybeans. One of the important groups of minor compounds present in soybean that has received tremendous attention is a class of phytoestrogen called isoflavones. Isoflavone compounds were considered non-nutrients, as they did not provide energy and did not even function as vitamins (Dan Ramdath et al., 2017; Yu et al., 2016). However, they play a key role in preventing numerous diseases, so they are often incorporated in health-promoting substances. Daidzein and genistein are the most common isoflavones, whose characteristic chemical structure (the B-ring is linked to the C3 position of the C-ring instead of the C2 position) resembles the structure of estrogens, in particular 17-β estradiol (Yu et al., 2016) (Figure 1.21).

The basic physical and chemical characteristics of soy protein are that the molecular weight of soy protein ranges from 300,000 to 600,000 kDa. The polypeptide chains of soy protein polymers are associated with and entangled in a complicated three-dimensional structure by disulfide and hydrogen bonds. Soy protein is abundant, with a great amount of polar functional groups such as hydroxyl, amino and carboxylic groups (Dan Ramdath et al., 2017; Han et al., 2017).

Occurrence: Soybeans that have been dehulled and defatted (Aggarwal et al., 2020; Hassan et al., 2019).

Biopolymer blends/bionanocomposites: Soy protein has certain drawbacks such as low water resistance and mechanical properties that limit its application. Therefore, soy protein needs to be modified to improve mechanical properties, water resistance and productive life to facilitate its application (Tian et al., 2018). Recently, soy protein resins have been used to fabricate green composites for many applications such as hydrogels, adhesives, plastics, films, coatings and emulsifiers (Tian et al., 2018). In another instance, water resistance and thermal abilities could improve due to the addition of graphene. This improved performance may be due to the hydrogen bonds and π–π interactions between graphene and the soy protein isolate (SPI) matrix, which gives SPI–graphene films wide potential application in drug delivery, packaging and the food industry (Han et al., 2017).

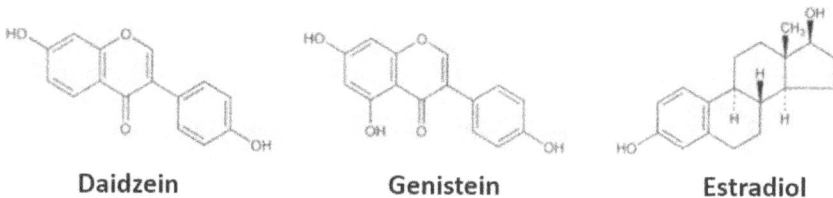

Daidzein   Genistein   Estradiol

FIGURE 1.21 Chemical structure of isoflavones and estradiol.

$$\text{Zein} \quad \left[\begin{array}{c} O \\ \| \\ -HN \quad \overset{}{C} \quad R- \end{array} \right]_m$$

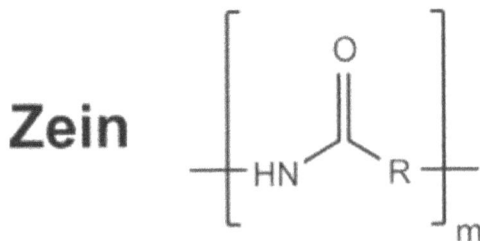

FIGURE 1.22   Chemical structure of zein.

1.8.2   ZEIN

Zein is a plant protein isolated from corn. It has been found exclusively in corn endo-sperm cells (Figure 1.22) (Shukla & Cheryan, 2001). Originally, the primary pro-duction of zein came from the by-products of starch and oil during the wet milling process, and zein protein was usually incorporated into animal feed. Zein was con-sidered a waste protein and did not receive much attention until recent years. Several uses were found for zein before petroleum, including the production of fibers, adhe-sives, buttons, binders and coatings. There is currently a growing interest in zein as a polymeric material, in part due to the perceived negative impact of plastic on solid-waste disposal. Moreover, zein offers many advantages as a raw material to produce plastics, coatings and films (Lorenzo et al., 2018). Zein is a combination of different peptide chains linked by disulfide bonds. These peptides can be classified according to their solubility, charge and molecular size. There are two major fractions of zein, α and β. α-zein is the major protein; it accounts for about 80% of the total prolamin present in corn and is soluble in 60–95% aqueous ethanol. Besides that, α-Zein con-tains less histidine, arginine, proline and methionine than β-zein. On the other hand, β-zein accounts for approximately 10% of the total zein content in corn, is relatively unstable, precipitates and coagulates frequently and has therefore not been a con-stituent of commercial zein preparations.

The molecular weight of zein varies from 22 to 27 kDa. It has an isoelectric pH of 6.228 and is the major storage protein (accounting for 35–65%) in corn. Pure zein is clear, odorless, tasteless, hard, water insoluble and edible. One of the defining char-acteristics of zein is its insolubility in water. This hydrophobicity is attributed to the presence of a high proportion of nonpolar amino acid residues. The poor water solubil-ity and imbalanced amino acid profile make zein a less than ideal protein for human consumption. Zein is soluble only in aqueous ethanol and proteins containing higher concentrations of amino acids, such as glutamic acid, proline, leucine and alanine, and lower amounts of basic and acidic amino acids. Therefore, it is highly hydrophobic.

Occurrence: Maize (corn).

Biopolymer blends/bionanocomposites: A series of reports showed that due to its nontoxicity, biodegradability and good biocompatibility, zein is a prom-ising natural polymer for use as a scaffold in drug delivery systems, wound healing and food packaging (Lorenzo et al., 2018). It has been successfully

blended with other biopolymers such as silk, collagen, chitosan, polycaprolactone and poly(L-lactide) (Elzoghby et al., 2015). Additionally, the ability
of zein and its resins to form tough, glossy, hydrophobic grease-proof coatings and their resistance to microbial attack has been of commercial interest (Sharif et al., 2019). Potential applications of zein include uses in fiber,
adhesives, coatings, ceramics, inks, cosmetics, textiles, chewing gum and
biodegradable plastics (Lorenzo et al., 2018).

1.8.3 GLUTEN

Gluten is a Latin word that means "glue", due to its ability to hold grains like wheat,
barley and rye together (Biesiekierski, 2017). Gluten is a mixture of hundreds of
distinct proteins within the same family, although it is primarily made up of two
different classes of proteins: gliadin, which gives bread the ability to rise during bak-
ing, and glutenin, which is responsible for dough's elasticity (Lorenzo et al., 2018;
Patni et al., 2011). Research has been performed to develop techniques for converting
wheat gluten into more useful products. Plant protein from wheat shows the advantage for usage as films and plastics because of its abundant resources, low cost and
good biodegradability, making it a promising substitute for petroleum-based plastics.
Gluten proteins are termed prolamins. They are proteins attached to starch in the
endosperm, which consists of gliadin (the water-soluble component) and glutenin
(the water-insoluble component); they bind to each other to form a network that supports dough and allows bread to be light and fluffy (Sharma et al., 2017). Amino
acids present in both gliadin and glutenin help the two proteins form hydrogen bonds
with each other (Sharma et al., 2017).

Gluten is a protein complex which is insoluble in water, although there may be
amounts of soluble proteins trapped in the gluten matrix. Despite its insolubility and
its hydrophobic nature, gluten absorbs about twice its dry weight in water to form
the gluten network. It is sticky, extensible and elastic and has high water absorption
capacity and cohesivity (Biesiekierski, 2017; Patni et al., 2011).

Occurrence: Wheat, rye, barley, triticale, spelt, einkorn, emmer and kumut
(Biesiekierski, 2017).

Biopolymer blends/bionanocomposites: Many factors are involved in plasticizer selection, including molecular structures, polarities, required product qualities, properties and costs. Various promising plasticizers that can
be utilized in making wheat gluten–based bioplastic are glycerol, xylan,
dicarboxylic acid, lactic acid, water and octanoic acid. Various crosslinking agents can also be used to improve the properties of the film,
including aldehyde, 1-ethyl-3-(3-dimethylaminopropyl) carbodiimide,
N-hydroxysuccinimide and silica. For instance, additives like xylan or fillers like silica and hydro ethyl cellulose, lubricators like salicyclic acid and
binders like urea and sodium hydroxide can be incorporated into gluten in
various ratios to form biodegradable composite films (Patni et al., 2011;
Sharma et al., 2017). Furthermore, there is evidence of electrospun nanofibers blended with gluten for biomedical applications (Aziz et al., 2019).

1.8.4 COLLAGEN

Collagen refers to a family of proteins that are the primary structural component of connective tissues, such as skin and cartilage. The substance makes up about a third of all the protein within the mammalian body, more than any other type of protein in the body by mass. There are 29 known types of collagen (Yoon et al., 2020). The configuration of this protein greatly affects its role in tissue architecture. Collagen contains many periodically repeating 3-amino-acid sequences containing Gly. The Gly consists of repeating sequences, including about 1,400 amino acid residues. The most common tripeptide unit of collagen is Gly-Pro-Hyp, and these three residues form one helical turn. There are several collagens, but 80–90% of them belong to types I, II and III, for example, type II collagen with a molecular mass of 1461.64 g mol^{-1} and a molecular formula of $C_{65}H_{102}N_{18}O_{21}$ (Figure 1.23). It is often found to group in three molecules and twist to form collagen aggregation about 290 nm long and 1.5 nm in diameter (Vázquez-Portalatín et al., 2016).

The melting point of collagen is in the range of 30 to 50°C, and the boiling point is between 160 and 190°C. It is poorly soluble in water and has a high content of alpha helix structures linked together with covalent bonds. As a structural protein, collagen has excellent biocompatibility and cell adhesion and can promote cell proliferation and differentiation (Liu et al., 2019).

Occurrence: Collagen occurs throughout the body, but especially in the skin, bones and connective tissues (Hassan et al., 2019).

Biopolymer blends/bionanocomposites: Owing to the mechanical structure, collagen has high tensile strength and is a nontoxic, easily absorbable, biodegradable and biocompatible material. Therefore, it has been used for many medical applications such as in treatment for tissue infection, drug delivery systems and gene therapy. Collagen matrices or sponges can be used to treat wounds for tissue regrowth and reinforcement. It is biocompatible and readily crosslinks with other biopolymers such as chitosan and HA, for example, hydroxyapatite/collagen composites, chitosan-collagen composites that have been seeded with cells for tissue-engineered heart

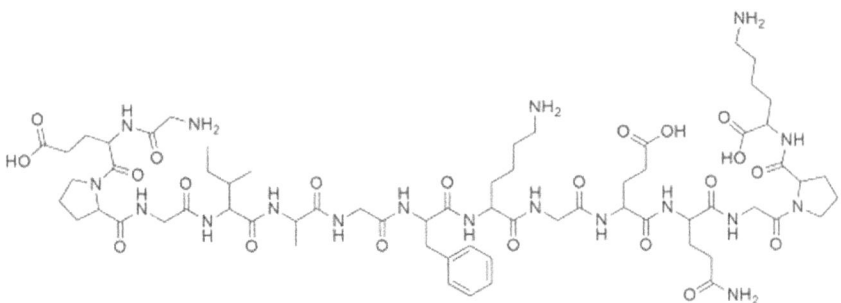

Type II-collagen

FIGURE 1.23   Chemical structure of type II collagen.

valves, and 3D composites derived from blends of chitosan and collagen incorporating HA as crosslinker for wound healing applications (Bu & Li, 2018; Fu et al., 2017; Sionkowska et al., 2016).

1.8.5 GELATIN

Gelatin is an extracellular matrix protein, which allows it to be used in applications such as wound dressings, drug delivery and gene transfection. There are two types of gelatin, Type A and Type B. Type A gelatin is derived by acid hydrolysis of collagen and has 18.5% nitrogen. Type B is derived by alkaline hydrolysis containing 18% nitrogen and no amide groups. Elevated temperatures cause the gelatin to melt and form coils, whereas lower temperatures result in coil to helix transformation (Ninan et al., 2015). Gelatin is a heterogeneous mixture of single- or multi-stranded poly-peptides, each with extended left-handed proline helix conformations and containing between 300 and 4,000 amino acids. The basic chemical structure of gelatin consists of functional groups such as NH_2 and COOH (Bazmandeh et al., 2020). Based on the process used for its manufacture, gelatin is obtained either as Type A or B. Type A gelatin is obtained by acidic treatment of collagen and has an isoelectric point (pI) between 7.0 and 9.0. Type B gelatin, however, is obtained by alkaline hydrolysis of collagen and has a pI between 4.8 and 5.0 (Ninan et al., 2015).

Gelatin is a polyampholyte that gels below 35 to 40°C. The heterogeneous nature of the molecular weight profile of this biopolymer is affected by pH and temperature, which in turn affects the noncovalent interactions and phase behavior of gelatin in solution. Crosslinking and/or hardening or could be as complex as ligand-mediated active targeting at the cellular level and contains many functional groups like NH_2 and COOH, which allow gelatin to be modified using nanoparticles and biomolecules (Bazmandeh et al., 2020; Voron'ko et al., 2016).

Occurrence: Gelatin is a denatured form of collagen obtained by acid or alkaline collagen processing (Derkach et al., 2020).

Biopolymer blends/bionanocomposites: Gelatin polymer is often used in dressing wounds, where it acts as an adhesive. Scaffolds and films with gelatin allow for the scaffolds to hold drugs and other nutrients that can be used to supply a wound for healing (Bazmandeh et al., 2020). However, there are some inherent problems of gelatin-based films, such as poor mechanical and water resistance. These can be overcome by applying recently developed nanocomposite technology. Gelatin-based nanocomposite films prepared with various types of nanofillers, such as nanoclay, organic fillers and nanometals, have exhibited increased film properties, along with other novel properties, such as antimicrobial activity, antioxidant activity and UV-screening properties (Derkach et al., 2020). Subsequently, gelatin-based nanocomposite films blended with certain organic fillers and inorganic fillers such as nanosilver, nanocopper (CuNPs), zinc oxide (ZnO) and titanium dioxide (TiO_2) nanoparticles exhibited strong antimicrobial activity against foodborne pathogenic microorganisms (Ninan et al., 2015).

1.8.6 KERATIN

Keratin is the major structural fibrous protein to form the hair, wool, feathers, nails and horns of many kinds of animals. Keratin can be classified into two groups: soft keratin and hard keratin. It may be present in two conformations, α-helix and β-sheet. Compared to other proteins, keratin-based materials have higher stability and are not degraded by enzymes (Shavandi et al., 2017).

The chemical structure of keratin is shown in Figure 1.24. A keratin protein is defined by a primary structure based on amino acid chains. These chains vary in number and sequence of amino acids, polarity, charge and size. However, similarities exist in their structure independently of the species of animal or function. Small modifications in the keratin's amino acid sequence cause significant property modification, since these sequences determine the whole molecular structure and the nature of the covalent or ionic bonds. The sulfur-containing amino acids, methionine and cysteine, as shown in Figure 1.24, have an even greater influence due to their role in establishing intra- or inter-molecular disulfide bonds (Shavandi et al., 2017).

Keratin is extremely insoluble in water and organic solvents and has high sulfur content and filament-forming proteins. It has a high concentration of cysteine, 7 to 20% of the total amino acid residues, that form inter- and intra-molecular disulfide bonds. Cysteine-rich proteins are endowed by nature with high mechanical strength owing to the large number of disulfide bonds, and the elastic nature of keratin fiber is due to the interplay between α-helix and β-sheet configuration of the protein. The disulfide linkage between cysteine molecules present as intrachain and interchain bonds is responsible for its good stability and lower solubility. Therefore, the dissolution and the extraction of keratin are difficult compared to other natural polymers, such as collagen or starch (Hassan et al., 2019; Ma et al., 2017).

Occurrence: There are two types of keratin, which can be differentiated as α- and β-keratins. The former is widely found in wool, hair, horn, nails, hooves and the outermost layer of the skin, also known as the stratum corneum. β-keratins are the real part of hard avian and reptilian tissues, for example, plumes, hooks, mouths of flying creatures and scales and paws of reptiles (Hassan et al., 2019; Aggarwal et al., 2020).

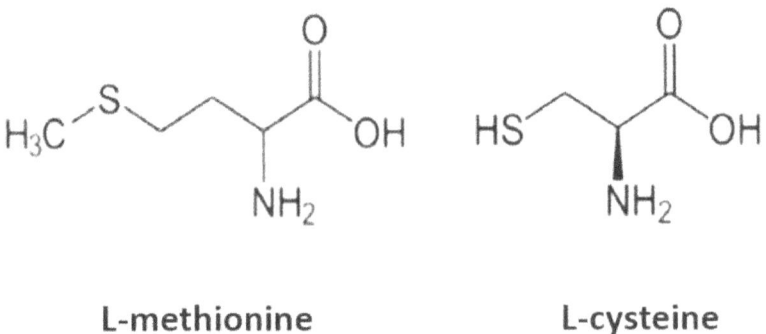

L-methionine   **L-cysteine**

FIGURE 1.24   Chemical structure of keratin.

Biopolymer blends/bionanocomposites: Keratin composites that are prepared without any additives suffer from a brittle structure, which limits
their applications. Therefore, several studies have tried to overcome this
weak structure by reinforcing with additives or blending with natural or
synthetic polymers (Singamneni et al., 2019). Improved mechanical properties were generally achieved. In this regard, there has been increasing
interest in reinforcing the keratin matrix with naturally derived green compounds. As an example, keratin blends with chitosan have been proposed
for wound healing and artificial skin substitutes (Lin et al., 2017). On the
other hand, the interaction between keratin and synthetic polymers has also
been widely and deeply studied. The relationship between poly(ethylene
oxide) (PEO) and keratin-blended films was explored in order to develop
a keratin-based material with improved structural properties (Ma et al.,
2017). The improved structural properties of keratin/PEO blends enable the
development of keratin materials for use as scaffolds for cell growth, wound
dressing and drug-delivery membranes.

1.9 SYNTHETIC POLYMERS FROM RENEWABLE RESOURCES

The chemical structure and characteristics of synthetic polymers from renewable
resources are detailed in this section. Synthetic polymers from renewable resources
covered in this section are PLA, polyhydroxybutyrate, polybutylene succinate, bio-
polyethylene and bio-polypropylene. In addition, several examples of using these
synthetic polymers in biopolymer blends and bionanocomposites are detailed in this
section.

1.9.1 POLYLACTIC ACID

Polylactic acid is a promising alternative to petroleum-based plastics that have been
widely used in packaging, electronic, automotive and biomedical applications. This
is because polylactides can break down into nontoxic products such as water and
carbon dioxide during degradation under an active biological environment and
are biocompatible (Baran & Yildirim Erbil, 2019). Thus, they reduce the amount
of plastic waste. The most common way to produce PLA is through fermentation
from corn or sugarcane feedstock, which is from 100% renewable resources. The
general chemical structure of PLA is shown in Figure 1.25. PLA is made up of lactic acid monomer (2-hydroxypropanoic acid, $HO-CH_3-CH-COOH$), which is the
simplest 2-hydroxycarboxylic acid with a chiral carbon atom. It can exist in two
stereocomplex crystallization of enantiomeric forms: L-lactic and D-lactic. Utilizing
poly (L-lactide) (PLLA) and poly (D-lactide) (PDLA) represents a robust approach
for substantially improving material properties, including stability and biocompatibility. This enhancement arises from the potent intermolecular interactions between
L-lactyl and D-lactyl units, which have proven a pivotal strategy in advancing applications of PLA (Baran & Yildirim Erbil, 2019).

Pure PLA exhibits a melting point of 180°C and a glass transition temperature of
60°C. PLA, on the other hand, is a hydrophobic polymer with poor toughness and

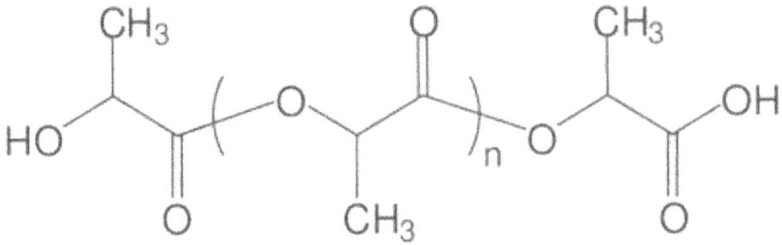

FIGURE 1.25   Chemical structure of PLA.

low thermal and crystallization capabilities and is prone to degradation via hydrolysis and occasionally microbial attack, unlike polysaccharides, which possess more reactive side chain groups. PLA exhibits tensile strength, flexural strength and specific gravity in the ranges of approximately 61 to 66 MPa, 48 to 110 MPa and 1.24 g/cm^3, respectively. This is due to the presence of hydroxyl and acid functional groups in the lactic acid molecule, enabling esterification reactions (Baran & Yildirim Erbil, 2019; Chotiprayon et al., 2020; Coppola et al., 2018).

> **Occurrence:** Produced by polycondensation and/or ring-opening polymerization of lactic acid from the natural feedstock such as corn, potatoes or sugar beets (Baran & Yildirim Erbil, 2019).
>
> **Biopolymer blends/bionanocomposites:** PLA can be blended with other flexible polymers, which can act as plasticizers to improve its mechanical strength and reduce its brittleness. PLA is often blended with starch thermoplastic and coir fibers to increase biodegradability and reduce the cost of the resulting blend (Chotiprayon et al., 2020). The use of chitin as a reinforcing filler provides a promising expansion on the applications of PLA composite film in the biomedical field (Olaiya et al., 2019). This is due to PLA composite film showing a significant increase of glass transition, primarily due to the increase of intermolecular bonds between PLA, chitin and starch. PLA usually exhibits a poor gas barrier and water vapor barrier. Therefore, it is often incorporated with nanoclay or carbon nanotubes to promote better thermal and mechanical properties compared to neat PLA (Coppola et al., 2018).

1.9.2   POLYHYDROXYBUTYRATE

Polyhydroxybutyrate is a member of polyhydroxyalkanoate family, which has been extensively studied for commercial use in packaging, medical and coating materials. The interesting feature of PHB is that it has properties similar to petroleum-derived polypropylene, yet it is biodegradable under aerobic and anaerobic condition and biocompatible (Mohandas et al., 2017). PHB, a homopolymer, contains monomer 3-hydroxybutyric acid (HBA) units. It has a methyl functional group (CH$_3$) and an ester linkage group (-COOR). This chain and functional group play an important role in giving the material thermoplastic, hydrophobic, high crystalline, and brittle characteristics (McAdam et al., 2020b).

PHB is highly crystalline due to its linear chain structure and shorter pendant groups. It has higher gas barrier properties than polyethylene and polypropylene (PP), polyethylene terephthalate (PET) and polyvinylchloride (PVC) is more rigid but less flexible than PP. It is able to degrade to D-3 hydroxybutyric acid *in vivo* together with low toxicity. It is thermally unstable, especially when close to its melting point, 175 to 180°C, and its glass transition is below 5°C (McAdam et al., 2020a).

Occurrence: It is a product of microbial secondary metabolism in the cells of bacteria. Bacteria strains include *Ralstonia eutropha, Alcaligenes* spp., *Azotobacter* spp., *Bacillus* spp., *Nocardia* spp., *Pseudomonas* spp., *Rhizobium* spp. and *Ralstonia eutropha* (McAdam et al., 2020a).

Biopolymer blends/bionanocomposites: PHB possesses high crystallinity. However, its rigidity and low impact resistance limit its application. Therefore, it is usually blended with other polymers to enhance its mechanical properties and processability. One instance is biodegradable blend nanocomposites that are produced from polylactic acid, poly(3-hydroxybutyrate) and cellulose nanocrystals (NCs) with dicumyl peroxide (DCP) as a crosslinking agent for 3D printing. In this case, PHB acts as a nucleating agent to induce PLA crystallization into a more ordered crystalline structure, attributed to its higher crystallinity compared to PLA (Frone et al., 2020). Due to the high crystallinity of PHB, it has been used as a filler for starch thermoplastic (TPS), and it resulted in better thermal stability compared to neat TPS (Florez et al., 2019).

1.9.3 POLYBUTYLENE SUCCINATE

Poly(butylene succinate) is a commercially available aliphatic polyester with many interesting properties, including biodegradability, melt processability and thermal and chemical resistance. Therefore, PBS is a cost-effective alternative to other polymers such as PLA, PBAT and PHB. It can be used either as a matrix polymer or in combination with other polymers such as PLA and PCL. Possible applications include food packaging, mulch film, plant pots, hygiene products, fishing nets and fishing lines (Luyt & Malik, 2018). PBS is an aliphatic polyester, prepared by polycondensation of 1,4-butanediol and succinic acid via chemical synthesis. These monomers can be obtained either from renewable or fossil-based resources (de Matos Costa et al., 2020). The chemical structure of PBS is shown in Figure 1.26. Due to its

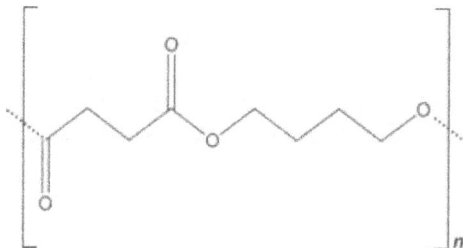

FIGURE 1.26 Chemical structure of PBS.

semi-crystalline nature, thermal stability and melting point lower than that of other commercially available biodegradable polymers, PBS is an excellent candidate for the production of biodegradable films for packaging. It has good barrier properties, as PBS's oxygen permeability (PO) is about four times lower than that of low-density polyethylene (LDPE) and about double that of PLA.

Occurrence: It can be produced from renewable resources such as sugars, glucose, starch and xylose via bacteria fermentation (de Matos Costa et al., 2020).

Biopolymer blends/bionanocomposites: PBS exhibits low thermal stability, poor mechanical properties and slow crystallization rates. In order to enhance its properties, many studies have been done to incorporate inorganic materials or blend with other polymers. Inorganic materials such as clay and zeolite have been used to enhance the physical properties of PBS/ montmorillonite (MMT) nanocomposites (Hwang et al., 2012). Ultimately, PBS ionomers showed improved mechanical physical properties because of ionic interactions. The enzymatic hydrolysis behavior of PBS ionomers was drastically accelerated with increasing ionic content because of reduction in crystallinity and improved hydrophilicity. It was also blended with biodegradable polymers and PCL and incorporated with carbon nanotubes to achieve higher mechanical, thermal and electric conductivity compared to neat PBS (Gumede et al., 2018).

1.9.4 BIO-POLYETHYLENE

Biobased polyethylene falls into the category of polymers from renewable resources, considering the source of starting material (i.e. sugar cane), which can be obtained from renewable resources. It is produced through the fermentation of sugar to obtain bio-ethanol from plant resources. The bio-ethanol serves as the source for production. Bio-PE has similar characteristics to polyethylene derived from crude oil. Hence, the chemical structure, applications and recycling are all identical (Siracusa & Blanco, 2020). This also means that bio-PE is not biodegradable. From polymerization, there are basically two types of bio-PE: bio-low density polyethylene (bio-LDPE) and bio-high density polyethylene (bio-HDPE) (Siracusa & Blanco, 2020). The non-biodegradability of bio-PE is mainly due to the C–C and C–H bonds (σ bonds), whose bond energy is on the order of 300–600 kJ/mol (Martínez-Romo et al., 2015).

Occurrence: Starting materials obtained from renewable resources such as sugar cane, sugar beet and wheat grain (Siracusa & Blanco, 2020).

Biopolymer blends/bionanocomposites: Since bio-PE is not biodegradable, many studies have incorporated biodegradable polymers to improve the biodegradability. This can be done through blending with biodegradable polymers or incorporation of fillers. One example is found in bio-PE/PCL blends (Bezerra et al., 2019). An other instance is binary blends of bio-HDPE with PLA by melt compounding. The results showed that multifunctionalized vegetable oils and peroxides were needed to enhance the miscibility between bio-HDPE and PLA in order to achieve higher mechanical properties for packaging and kitchen utensil applications (Quiles-Carrillo et al., 2019).

1.10 SYNTHETIC POLYMERS FROM PETROLEUM-BASED RESOURCES

This section presents the chemical structure and characteristics of synthetic polymers from petroleum-based resources. In addition, several examples of using these synthetic polymers in biopolymer blends and bionanocomposites are detailed in this section. These include polycaprolactone, polyvinyl alcohol and polybutyrate adipate terephthalate.

1.10.1 Polycaprolactone

Poly(ε-caprolactone) is a biodegradable aliphatic, which belongs to the poly(a-hydroxyl acid) group and has been widely used in biomedical applications. This is mainly due to its biocompatibility and bioresorbability. Furthermore, the FDA has approved the usage of PCL for human implants, as toxicological assays show it is non-mutagenic and innocuous in animals (Espinoza et al., 2020). PCL is a polyester produced by ring-opening polymerization of epsilon-caprolactone, which is commonly derived from fossil carbon (Ali Akbari Ghavimi et al., 2015). Similar to most biodegradable polymers, the main chains of PCL consist of carbon. The chemical structure of PCL is shown in Figure 1.27.

The thermal, physical and mechanical properties of PCL depend on its degree of crystallinity, which varies with molecular weight and reaches a maximum value of 69% in PCL. PCL exhibits high ductility and flexibility due to its low glass transition temperature of around −60°C. It can be easily processed through methods such as melt extrusion, film blowing, injection molding and melt-spinning. PCL demonstrates excellent chain flexibility and can be synthesized in a range of molecular weights. While PCL is inherently hydrophobic, it is soluble in a wide range of aromatic, polar and chlorinated hydrocarbons but remains insoluble in aliphatic hydrocarbons, alcohols and glycols. The molecular weight of 15,000 g mol^{-1} means the material is brittle, and the molecular weight of 40,000 g mol^{-1} indicates that the material is soft and semi-crystalline in nature. PCL is able to degrade by

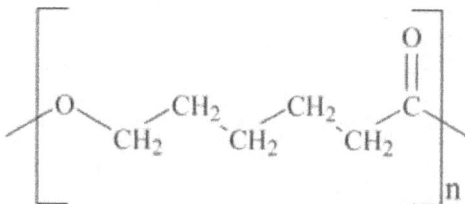

FIGURE 1.27 Chemical structure of PCL.

hydrolysis in physiological conditions (such as in the human body), attributed to its ester linkages, but PCL has a longer degradation time than poly(lactic-co-glycolic) acid (PLGA) and PLA. PCL is thermally more stable than polylactic acid and completely degradable through enzymatic activities (Ali Akbari Ghavimi et al., 2015; Gumede et al., 2018).

   Occurrence: Raw material obtained from petroleum and can be synthesized by polycondensation of 6-hydroxyhexanoic acid as well as ring-opening polymerization of ε-caprolactone using Tin(II) 2-ethylhexanoate [Sn(Oct)$_2$] as a catalyst (Gumede et al., 2018).

   Biopolymer blends/bionanocomposites: PCL undergoes degradation through the hydrolysis of its ester linkages under physiological conditions, making it a highly sought-after material for implantable biomaterial applications. Researchers have explored encapsulating various drugs within PCL beads to achieve controlled release and targeted drug delivery. Additionally, PCL is frequently blended with starch to create cost-effective and biodegradable materials. Furthermore, low molecular weight PCL can be incorporated into other polymers to enhance their resistance to weathering. For example, PCL/PEG-PCL microspheres were added to estradiol, and it was proven to promote efficacy in encapsulation for drugs (Espinoza et al., 2020). On the other hand, PCL is often mixed with starch, as it was reported that PCL and starch mutually benefited whereby the PCL was able to alter the humidity sensitivity of starch, while starch was able to speed up the degradation rate of PCL, a function that is crucial in biomedical applications (Ali Akbari Ghavimi et al., 2015). It is believed that the interaction between PCL and starch is formed through the hydrogen bonding between ester carbonyl of PCL and the hydroxyl group of starch.

1.10.2 Polyvinyl Alcohol

In early years, the principal application of polyvinyl alcohol was in textile sizing. In recent years, PVOH have received much attention in biomedical devices because it is soluble in water, biodegradable and biocompatible and has a low protein absorption property and chemical resistance (Aslam et al., 2018). PVOH results from a hydrolysis reaction, which removes acetate groups from PVAc molecules without disrupting their long-chain structure (Aslam et al., 2018). The unique chemical feature is the presence of an OH group, which enables PVOH to interact with water and biopolymers with OH groups (Ali et al., 2018). The chemical structure of the resulting vinyl alcohol repeating units is as shown in Figure 1.28.

   PVOH is soluble in water but insoluble in organic solvent and only sparsely soluble in ethanol. It shows compatibility with a number of polymers, and it can be easily mixed with various natural materials, which extends the range of its

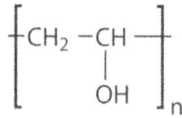

FIGURE 1.28   Chemical structure of PVOH.

applicability. It is biodegradable due to its excellent hydrophilicity, ascribed to the presence of hydroxyl groups on the carbon atoms, and is a semicrystalline polymer with high dielectric strength and high transparency (Aslam et al., 2018; Settier-Ramírez et al., 2020).

Occurrence: A hydrolysis product of polyvinyl acetate (Aslam et al., 2018).

Biopolymer blends/bionanocomposites: PVOH has often been used for blending with other polymers. Blending PVOH with other polymers creates opportunities to improve its processibility, to modify physical properties and to reduce cost. For instance, PVOH is used as sizing agent and is able to provide greater strength to textile yarns, which causes them to be resistant to oils and greases. It is also used in adhesives and emulsifiers to make water-soluble film. Furthermore, it is easily modified to form other polymers. PVA can be made into resins of polyvinyl butyral (PVB) and polyvinyl formal (PVF) with the reaction of butyraldehyde ($CH_3CH_2CH_2CHO$) and formaldehyde (CH_2O) (Aslam et al., 2018). Furthermore, the presence of OH groups serves to form hydrogen bonds with other polymers such as poly(vinylpyrrolidone) (PVP). One example is a blend of PVP and PVP doped with cerium nitrate developed by Ali et al. (2018), which is suitable in optical, electronic and organic semiconductor applications.

1.10.3   POLYBUTYRATE ADIPATE TEREPHTHALATE

Polybutyrate adipate terephthalate has been widely marketed as plastic packaging, shopping bags and garbage bags as an alternative to LDPE. Among the distinguished companies that have developed PBAT-based materials are BASF, Novamont, BIOTECH and KINGFA (Jian et al., 2020). The aromatic portion of PBAT contributes outstanding physical properties, while its aliphatic chains facilitate degradation under various conditions, including soil degradation without requiring temperature control. This makes PBAT a highly promising material for disposable packaging products. The biodegradation of PBAT primarily involves hydrolysis catalyzed by microbial enzymes. During this process, the butylene adipate (BA) units in PBAT, especially the non-crystalline portion, tend to degrade more rapidly compared to the butylene terephthalate (BT) units with a crystalline structure. PBAT is an aliphatic-aromatic co-polyester of adipic acid 1,4-butanediol

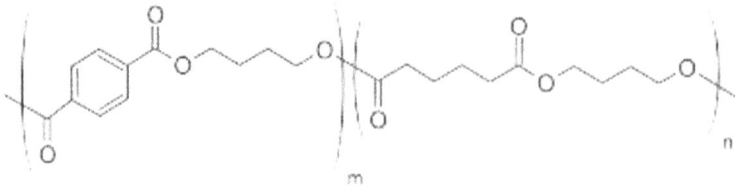

FIGURE 1.29   Chemical structure of PBAT.

and terephthalic acid from dimethyl terephthalate. The chemical structure of the PBAT polymer is shown in Figure 1.29. The real structure of PBAT occurs in a random block (Jian et al., 2020).

PBAT shows biodegradability due to the aliphatic unit in the molecule chain, which is attributed to the cleavage of ester linkages and the ability to react with water, with its carbonyl group located in benzene rings. The mechanical properties of PBAT are more flexible than those of most biodegradable polyesters such as poly (lactic acid) and poly (butylene-co-succinate) and are similar to those of low-density PE. Due to the absence of structural order in PBAT, it has high flexibility and a low elastic modulus, which makes it ideal to be blended with other biodegradable polymers (Fu et al., 2020; Siracusa & Blanco, 2020).

Occurrence: Polycondensation reaction of butanediol, adipic acid and terephthalic acid from the raw material petroleum (Siracusa & Blanco, 2020).

Biopolymer blends/bionanocomposites: The properties of pure PBAT are not sufficient for most applications because it has higher production costs or lower mechanical properties when compared with conventional plastics. The development of a PBAT market will only be possible when production costs decrease or its properties are improved. Therefore, PBAT is often mixed with low-cost material like starch or PLA to decrease the end price and enhance the mechanical properties while maintaining the biodegradability of the composites. The biodegradation behavior of PBAT, PLA and PLA/PBAT blends in freshwater with sediment was affected by the content, aggregation structure, hydrophilia and biodegradation mechanisms of the components, where the crystallinity of PLA/PBAT composites increases with the degradation time, which is a result of the faster biodegradation of amorphous regions by microorganisms and enzymes, while elevated PLA content (25, 50, 75%) in composites leads to a higher increase in the O/C content ratio after degradation. After 24 months of degradation, an increase in the relative peak area proportion of C-O to C=O is observed (Fu et al., 2020; Jian et al., 2020). It is also incorporated with fillers, as done by Shankar and Rhim (2019), where the zinc oxide nanoparticles are reinforced in PLA/PBAT blends to enhance optical and mechanical properties along with antibacterial properties.

1.11 POLYMER BLENDS

Generally, blending is the method of thoroughly combining different materials to create a homogenous product. The terms blending and mixing are often used interchangeably, but there are some distinctions. Blending is typically a gentle method of combining materials, while mixing is usually a more vigorous process. Blending can help to enhance product quality by uniformly coating particles, diffusing liquids and fusing particles together. Therefore, blending is one of the methods that is commonly used for polymer modification in order to produce a new product with the desired properties (Imre & Pukánszky, 2013). Blending polymers is a good way to come up with novel materials with better properties (John, 2015). Altering surface properties such as friction coefficient, fostering adhesion, adding color, improving stability, increasing output and gaining easy-opening features are just a few of the benefits of blending (Morris, 2016). In other words, by correctly selecting the component polymers, the properties of the blends can be manipulated according to their end use (Markovic & Visakh, 2017). In addition, there are some times when two natural polymers will coexist as a blend in nature, such as elastin and collagen in skin, where unique mechanical and structural properties can be demonstrated by such mixes (Sionkowska, 2015).

According to Sperling (2000) Amos et al. (1954) were credited with pioneering the field of polymer blending, and it was further developed a generation after (Aylsworth, 1914). Amos et al. (1954) invented high-impact polystyrene (HIPS). However, Ostromislensky was the first to come up with the concept of rubber-toughened polystyrene in 1927 (Sperling, 2000). Without stirring, Ostromislensky dissolved rubber in a styrene monomer and polymerized it in order to impart impact resistance to the resulting material. With little toughening action, the result was a continuous rubber phase and a discontinuous plastic phase. Subsequently, Amos and his team made a substantial advancement by introducing agitation and shearing to the reaction between rubber and styrene solution. This innovative approach led to the evolution of HIPS as a widely used thermoplastic, primarily produced through continuous bulk processing. Today, HIPS stands as one of the top five most-produced families of thermoplastic polymeric materials in terms of production volume (Bonilla-Cruz et al., 2013). Following this success, the development of acrylonitrile butadiene styrene (ABS) resins, as well as a wide range of other multicomponent polymer materials, occurred almost immediately.

The field of polymer blends has experienced an increasing rate of study over the past three decades due to the interest of researchers in blending methods to produce new materials (Sionkowska, 2011). Because people could make compositions with properties that homopolymers and statistical copolymers could not match, polymer blends and composites became a central part of polymer science and engineering (Sperling, 2000). Greater hardness and effect resistance, higher modulus, higher use temperature, wider temperature range of sound and vibration damping and so on are examples of such properties. In the same vein, Imre and Pukánszky (2013) also reported that blending of polymers is a technology that was developed in the 1970s or even earlier. On the other hand, according to Yu et al. (2006), various starch-polyolefin blends were developed in the 1970s and 1980s.

Blending a biopolymer with another polymer that does not have to be biodegradable is a beneficial way of changing its properties (Niaounakis, 2015b). Also, according a study reported by Imre and Pukánszky (2013), in the development of new polymeric materials, blending two or more polymers offers a great advantage by tailoring the properties of the polymeric materials over a wide range. Because of their theoretical and practical relevance, polymer blends have been extensively researched (Markovic & Visakh, 2017). Blending one biopolymer with another, such as PLA and starch, is the most commonly studied, but thermoplastic phenol formaldehyde resin/poly(e-caprolactone), chitosan/soy protein, PHB/cellulose acetate butyrate and PLA/poly (butylene succinate) are also worth mentioning (Imre & Pukánszky, 2013).

Biopolymers are characterized as successful, long-term alternatives to traditional petrochemical-based products. However, their massive use has been hindered by some unsatisfactory material properties of these polymers, and the processability of nearly all pure environmentally friendly biopolymers is not equivalent to that of commercial thermoplastics (Niaounakis, 2015b). Besides, blending may be a much cheaper and quicker way to obtain the desired properties rather than the copolymerization technique (Tokiwa et al., 2009). A detailed examination of biopolymeric blends and composites, as well as their applications in diverse industrial sectors, is easy to find (Rogovina & Vikhoreva, 2006). Furthermore, blends can also help in the development of new low-cost and high-performance products. These new blends are transforming polymers from their sources into new value-added products (Yu et al., 2006).

Polymer blends that have been produced can be utilized in a variety of applications due to their unique properties (Markovic & Visakh, 2017). Polymer blends are generally categorized as either homogeneous (molecularly miscible) or heterogeneous (immiscible) blends. However, according to Qin (2016), in a polymer blend or mixture, at least two polymers are combined to produce a new material with distinct physical properties, similar to metal alloys. Polymer blends can be categorized into three groups: immiscible, miscible and compatible polymer blends. In immiscible polymer blends, the constituent polymers occur in different stages, with different glass transition temperatures (T_g); in other words, they are blends that exhibit more than two phases. Usually, this kind of blend consists of two T_g values, since the two components of the blend are phase separated. Meanwhile, miscible polymer blends are blends that are frequently made from polymers with comparable chemical structures, resulting in a single-phase polymer blend. One glass transition temperature has been identified. Also, compatible polymer blends are known as immiscible polymer blends, where the physical properties are macroscopically uniform owing to sufficiently strong interactions between the constituent polymers. They are also blends that are useful and have inhomogeneity on a small enough scale that it is not noticeable when in use.

To gain an advantage in processing or performance properties, the base and additive polymer should be compatible or miscible. Therefore, one of the most significant factors affecting the final polymer properties is the miscibility of the blend. The benefits of making miscible blends include single-phase morphology and mechanical property reversibility (Tokiwa et al., 2009). However, in cases where blended polymers exhibit incompatibility, resulting in the separation of the polymer mixture into distinct phases due to differences in molecular weight and viscosity, or when they display a coarsely dispersed structure, imperfections may become apparent on the surface of injection-molded products. In such situations, a surface exfoliation technique is often employed to improve the quality of the molded article (Niaounakis, 2015b). For example, miscible polymer blends include combinations like poly(styrene) (PS)/poly(phenylene oxide) and poly(styrene-acrylonitrile)/poly(methyl methacrylate), while immiscible blends encompass poly(propylene PP)/PS and PP/poly(ethylene).

According to Niaounakis (2015b), there are a few measurements that can be used to determine the miscibility of a blended polymer. A measurement of the optical clarity of a polymer blended film, an appropriate measurement of mechanical characteristics and glass transition temperature (T_g) measurement using differential scanning calorimetry (DSC) are standard experimental procedures for assessing polymer blend miscibility. From an optical properties perspective, films made from polymer blending can be seen clearly, which indicates that the blended polymers are miscible. Otherwise, if the blended polymers are immiscible, the resulting film will look as if it contains foreign matter; that is, it is not optically clear. As indicated by Markovic and Visakh (2017), optical transparency is common in miscible mixes. Similarly, the mechanical properties of a blended polymer, such as tensile strength, are often in an intermediate phase between the blend components.

As has been noted, the state of miscibility of a blended polymer can be determined by a few measurements, one of which involves measurement of the glass transition temperature by using DCS. This measurement is characterized under thermal properties. In addition, a miscible amorphous blend will have a single T_g intermediate between the homopolymer elements. Meanwhile, an incompatible or partly miscible blend would have several T_g values, where T_g will have intermediate values for partially miscible blends, referring to partially miscible stages rich in one of the components. A study reported by Ibrahim and Kadum (2010) shows that good miscibility is where, between two values of pure polymers, there is only one glass transition temperature. In their study, they blended polystyrene and acrylonitrile-butadiene-styrene in different ratios through a single-screw extruder, which is called the melt blending method. In addition, Macknight and Karasz (1989) also stated that a miscible polymer blend is one that meets the thermodynamic requirements of a single-phase system. For instance, if the blend is made up of two components, A and B, a single-phase mixture would form at a constant temperature and pressure.

Also, Morris (2016) stated that the most basic blends can be created by combining ingredients in an extruder, which is used to convert the resin into a film or coating. To achieve the desired properties in more complex blends, specialized screw designs or customized compounding equipment may be needed. Moreover, even when properties can be controlled by inserting specific functional layers within a multilayer film, blending is crucial for some applications, especially packaging. Blending may be used to tailor special properties, such as barrier performance, into a layer. Blending may be required to make the polymer stable enough to extrude or to ensure that the surface properties of the extruded product are satisfactory. Thus, the final blend characteristics will be determined by the thermodynamics and the thermal and rheological properties of the polymers, as well as the flow and stress history, which is process dependent.

As eloquently stated by Aravind et al. (2009), blends of two different polymers are not always miscible or compatible with one another. Meanwhile, according to George et al. (2010), most polymer blends are immiscible to varying degrees, with complex phase morphologies that are dependent on the chemical nature of the constituents and their individual rheological properties, except for a few polymer blends that are thermodynamically miscible. In the same vein John (2015) found that in the case of immiscible blends, phase separation can occur, resulting in inferior compounded product properties. As a result, using graft and block copolymers to control phase morphology is essential. Even though immiscible systems are preferable in certain situations because each component keeps its own properties, miscible blends produce an average of individual properties. They are distinguished by a narrow interphase, weak interfacial adhesion and coarse morphology, all of which contribute to poor final properties (George et al., 2010).

1.12 METHODS OF POLYMER BLENDING

As studied by Patil et al. (2016), polymer blends may be synthesized or processed using a variety of techniques. Each process has advantages and disadvantages. The methods that will be discussed are melt blending, mill mixing and fine powder mixing technology, the solution casting method, freeze drying, latex blending and interpenetrating polymer network technology. It is worth mentioning that the solution casting method is one of the methods that gained interest with most researchers, especially in a film preparation (Patil et al., 2016). In the same vein, Shundo and Ijioto (1966) also used some of the blending methods in their experiments with natural rubber and styrene-butadiene rubber blends, such as solution blending, latex blending, roll blending and Banbury mixer blending, in order to find which type of blending method is more effective in producing uniform blends.

1.12.1 MELT BLENDING

Preparation of polymer blends can be done through melt blending. Melt blending is the most common method of preparing polymer blends in practice, as it offers contamination-free polymer blend preparation processes (Patil et al., 2016). Due to the lack of organic solvents, melt blending is environmentally friendly. Melt blending integrates with today's manufacturing systems like extrusion and injection molding. Melt blending has gained popularity as a result of its potential in industrial applications. Furthermore, according to Rane et al. (2018), for producing a clay/polymer nanocomposite of thermoplastics and elastomeric polymeric matrix, melt blending method is the preferred approach. In addition, Kango et al. (2014) indicated that melt blending is the most convenient and widely used technique for creating polymer hybrids with inorganic nanoparticles, such as semiconductor nanoparticles. Ma et al. (2006) also used melt blending to produce nanocomposites of silane modified zinc oxide, ZnO and polystyrene resin to produce polystyrene resin nanocomposites with antistatic properties. In order to produce polymer nanocomposites through melt blending, inorganic nanoparticles are dispersed into the polymer matrix and undergo an extrusion process (Kango et al., 2014).

In the melt blending process, individual components are processed and melted using specialized devices such as extruders and temperature controllers. Typically, extruders or batch mixers combine the blend components while they are still molten. According to Sadiku and Ogunniran (2013), the best way to disperse one polymer in another immiscible polymer is to use vigorous mechanical stirring in compounders at high temperatures when all components are in a molten state. The melt blending process is done in the presence of an inert gas such as helium or neon. To achieve a uniform mixture of all raw materials, the raw materials are subjected to a specific chamber containing extruders. The temperature is raised to a desirable level, and all additional materials are melted as a result. Process conditions such as mixing duration, operating temperature and pressure, in addition to the composition of the constituents, are vital in achieving desirable blend properties. As eloquently stated by Giles et al. (2005), if the extruder temperature profile is incorrect, the product ingredients are incorrectly formulated, the melt temperature at the end of the extruder is incorrect, the puller at the end of the line is running at the incorrect speed, the cooling bath temperature is incorrect or any other incorrect operating condition occurs, the product that will be produced will not meet customer requirements. An incorrect setting at the start of the process may also result in an unacceptable product at the end of the line after a considerable amount of value has been added. According to Khan et al. (2019), this method is generally thought to be an excellent technique for the preparation of polymers, with the exception that it may appear to be too expensive at times.

1.12.2   Mill Mixing and Fine Powder Mixing Techniques

Another method that is used to produce polymer blends is the mill mixing and fine powder mixing technique. Mixing components are mixed through milling and grinding in this simple, straightforward approach. This is accomplished through the use of various types of milling devices and grinders. The most common methods for achieving powder blending and particle size reduction are ball milling and attritor milling (Mehrotra, 2014). The raw materials are ground to produce the finest powder, which is then blended together until the final state is reached, which is at the micro level. This is to make sure that the mixture is uniformly mixed. If the blend product meets the criteria, it is ready to be used for the next process where additional processing is needed in order to obtain the desired polymer blend products. However, if the powder cannot be reworked, the milling process may begin, or it may be scrapped. Bunbury mixers, also known as Master Mixers, are a widely used design for mechanical mixing of polymer ingredients, making them ideal for polymer blend synthesis (D. H. Killheffer, 1962). This mixer is made up of two counter-rotating multilobed cams in a mixing chamber (Witt, 1984).

1.12.3   Solution Casting Method

The solution casting method is a polymer preparation that is favored by most researchers due to its easy and simple procedures (Khan et al., 2019). Solution casting is a simple and flexible procedure for making polymer films in the laboratory. Simply, solution casting is the process of dispersing a chosen polymer component in a polymer solution and then casting the film using standard coating procedures (Grothe et al., 2012). According to Mathew and Oksman (2014), solution casting is the oldest method of producing plastic films and was invented by Eastman Kodak in the 19th century to produce photographic films. The solution system's primary components are polymer and solvent, but different additives may be added as well (Galiano, 2014). When it comes to preparing the casting solution, the polymer chosen is crucial. In fact, the polymer must be soluble in the chosen solvent at a concentration that is directly related to the final product application. To generate homogeneous composite films, the polymer must have good solubility in the chosen solvent and excellent dispersion in the polymer solution.

Khan et al. (2019) explained this method briefly, where the first step is that the components of the mixture are dissolved in a common solvent and vigorously stirred. Next, the selected solvent is used to dissolve the chosen polymers. It should be observed that solvent selection is critical, as it also plays a significant role in the solution casting procedure. Then, to get a homogeneous solution, the solution mixture is stirred for a certain amount of time. The stirred process can be done by using a homogenizer or magnetic stirrer or can be manually stirred by using a glass rod. A further step is the solution then is cast on a prepared mold. Finally, the end product is collected for further characterization. The procedure has the advantages of allowing the system to blend quickly without consuming a lot of energy, and the drying process of the solution on a surface does not cause any additional mechanical or thermal stress (Siemann, 2005).

1.12.4 FREEZE DRYING

Freeze drying, also known as lyophilization, is another polymer blending method. As eloquently stated by H. Zhang (2018), in the pharmaceutical, biological and agricultural industries, freeze-drying is a common drying method. Meanwhile, according to Ghalia and Dahman (2016), freeze drying is a method of removing residual solvent from a material in order to make a dry powder that can be loaded into a cell easily. When dealing with temperature-sensitive chemicals, this is the preferred drying method because it produces dry powders or cakes that are easy to store. The duration of drying the sample depends on the amount of liquid water present in the sample. In this method, the component polymers are quenched to a very low temperature during freeze-drying, causing the solution to freeze. Then, the solvent is removed by a sublimation process, which means it changes from solid state to a vapor phase. It is worth mentioning that this method has a few benefits, which are that heat-sensitive materials are less likely to degrade when dried at low temperatures, it is friendlier to antioxidant activity than other drying techniques and the physical form of the dry product can be attractive (Santiago & Moreira, 2020).

1.12.5 LATEX BLENDING

Latex is an important term in the polymer industry, and it has a particular meaning. It is used to create a very fine stable dispersion of polymer particles in any aqueous medium (Datta, 2013). Latex blending is thus a novel way for preparation of most common polymer blends and other polymerization methods (Khan et al., 2019). Shundo and Ijioto (1966) stated that latex blending was an effective method in producing a uniform blend rather than other mechanical methods. However, these researchers reported that latex blending methods in preparation of polymer blends are not practical in the manufacturing industry. Latex blending is a dispersed mixture of two or more distinct types of latex components. Both kinds of particles contribute to the properties of the film formed after the dispersion has dried. The properties of hard-soft latex blends have been studied by a number of scientists. Hard latex is made of a polymer with a T_g above room temperature, whereas soft latex is made of a polymer with low T_g.

1.12.6 INTERPENETRATING POLYMER NETWORKS

The other method of preparation polymer blends is an interpenetrating polymer network (IPN). An IPN is polymer consisting of two or more molecularly interlaced networks that are not covalently bonded to each other and cannot be separated until chemical bonds are broken (Alemán et al., 2007). However, an IPN is not defined as a combination of two or more preformed polymer networks. The components in these systems are grown in such a manner that they are connected to each other, but not by any chemical bonds. As a result, special methods are often needed to create such polymer networks, as simple mixing of two or more polymers will not be sufficient to produce an IPN. There are a few types of IPN: semi-interpenetrating polymer networks (SIPNs), simultaneous interpenetrating polymer networks and sequential interpenetrating polymer network (Karak, 2012). A SIPN is created when a monomer polymerizes in the presence of a polymer. Alemán et al. (2007) define a SIPN as a polymer made up of one or more polymer networks that are linear or branched polymers, with at least one of the networks being penetrated on a molecular scale. The polymer portion of the network can be split from the network in this case without affecting any chemical bonds. Meanwhile, a simultaneous interpenetrating polymer network is interpreted as a method for forming component networks at the same time, whereas a sequential interpenetrating polymer network is described as a procedure for forming a second component's network after the first component's network has been established.

1.13 PROPERTIES OF POLYMER BLENDS

Since the development of polymer blends has gained interest from researchers, understanding the performance or properties of the polymer blends is crucial. Each property has its own importance depending on the end use of the final product or application. Here are a few properties that are usually considered when preparing polymer blends: electrical, mechanical, thermal and optical properties. These properties can be controlled by using an appropriate amount of polymer components during polymer blends.

1.13.1 ELECTRICAL PROPERTIES

Every property of a polymer blend depends on the behavior of polymer components incorporated into the blend. The properties of pure polymers can be enhanced by modification or through incorporation of fillers to produce the desired product. The demands of a specific application force these polymeric materials to work under specific conditions (e.g., chemical, mechanical and electrical) (Parameswaranpillai et al., 2014). It is worth mentioning that most researchers have an interest in investigating and reviewing conducting polymers because these types of polymer blends are increasingly being used in various electrical and electronic applications (Ameen et al., 2006). Furthermore, conducting polymers are on the verge of being commercially viable (Xavier, 2014). Ameen et al. (2006) conducted an experiment on polyaniline-polyvinyl chloride (PANI-PVC) blends doped with sulfamic acid (dopant) in aqueous tetrahydrofuran to investigate the temperature dependence of direct current conductivity. To determine the impact of sulfamic acid (dopant) in the temperature range from 300–400K, the DC conductivity of PANI-PVC blended films was tested. It was concluded that the conductivity of PANI-PVC blends increases with the amount of doped PANI.

1.13.2 MECHANICAL PROPERTIES

Mechanical properties are important properties that need to be considered when producing any product, especially products that need extra strength. These characteristics are essential when marketing a product, as customers will evaluate its strength before purchasing it. Abdul Khalil et al. (2019) stated that mechanical properties are evaluated by the stiffness and resistance to load exertion. The main fundamental parameters that are generally calculated to describe the mechanical properties of polymer blends are tensile strength, Young's modulus, hardness and ductility (Khan et al., 2019). To evaluate their products, researchers typically use a variety of parameters and correlate them with mechanical properties (Pukánszky & Tüdõs, 1990). An article published by Hamad et al. (2016) studied the mechanical properties and compatibility of a PLA/PS polymer blend. They synthesized the blend by using a twin screw extruder and molded it using an injection machine. These researchers found that the mechanical properties of the polymer blend, tensile strength and elongation, increased when the PLA content increases. Thus, based on the mechanical property performances, they concluded that the PLA/PS blend has higher compatibility compared to other polyester/PS blends.

1.13.3 THERMAL PROPERTIES

Apart from electrical and mechanical properties, thermal properties are also important properties when preparing a polymer blend. The glass transition temperature T_g of a polymer blend is the parameter that is often considered for thermal industrial application, as it is the most common experimental technique. Also, critical temperature, heat capacity, heat deflection temperature, flammability, solidus, thermal expansion, thermal conductivity and other parameters can be used to compare the thermal properties of various polymer blends (Khan et al., 2019). As mentioned, T_g is the major parameter when comparing thermal properties, as it will indicate whether the polymer blend is miscible or immiscible (Zainal & Chan, 2019). Ibrahim and Kadum (2010) studied the influence of polymer blending on mechanical and thermal properties. The thermal properties are determined by using differential scanning calorimetry. The results from the work show a good indication for enhancing the miscibility of polymer blends, as for the two values of pure polymers, there is only one glass transition temperature.

1.13.4 OPTICAL PROPERTIES

One of the most fundamental and essential properties of polymer blends is their optical properties, which aid in extracting a variety of knowledge about the blends (Khan et al., 2019). After the launch of transparent ABS in the market, optical properties have gotten a lot of attention from the scientific community and industry. Although there has been limited progress in this field, these properties are critical to remember when evaluating the suitability of a blend for a specific application (Xavier, 2014). Transparency is one of optical properties that influence consumer acceptance of the product. This is due to the fact that a high level of transparency results in a strong visual display of items. Transparency is correlated to the interaction of radiation between the substances. Takahashi et al. (2012) studied the optical properties for an immiscible blend of poly (methyl methacrylate) and ethylene–vinyl acetate copolymer (EVA). It was stated that when the difference in refractive index between the two components is small, PMMA/EVA exhibits good transparency at room temperature. Furthermore, Errico et al. (2006) reported that a polymer blend system consisting of EVA-*g*-PMMA and PMMA became opaque at high temperatures and transparent at room temperature, which shows that transparency is dependent on the difference in temperature.

1.14 BLENDING BIOPOLYMERS WITH OTHER BIOPOLYMERS

The combination of biopolymers exhibits properties that differ or have some improvements from those of their individual polymer counterparts, making them a preference among most researchers. According to Bonilla et al. (2014), films produced by blending polymers typically have different physical and mechanical properties compared to films made of individual components. Therefore, the combination of two different natural biopolymers is favored by most researchers because it is biodegradable, low in cost and widely available in nature. Researchers focused on renewable resources to create biodegradable materials in order to protect the environment and reduce emissions caused by oil-derived plastics (Mohanty et al., 2000).

1.15 BLENDING OF NATURAL BIOPOLYMERS WITH OTHER NATURAL BIOPOLYMERS

Chitosan, starch and alginate are some examples of natural biopolymers. Chitosan is mostly incompatible with commercial polymers, but it has a good outcome when blended with other biopolymers in a variety of application such as food packaging, biomedical, conductive material and metal complexation resin (Heeres et al., 2013). Chitosan has many excellent characteristics such as biodegradability, biocompatibility, bioactive, nontoxicity, antibacterial properties, gel-forming properties, hydrophilicity and selectively permeability (Rajeswari et al., 2020). Besides chitosan, starch is also a commonly used biopolymer in many applications. According to European Bioplastics (2020), statistics show that approximately 18.7% of starch blends were produced in 2020. Low cost and wide availability make starch a popular biopolymer in many applications (Heeres et al., 2013). Alginate is a hydrophilic polysaccharide commonly found in the cell walls of brown algae that when hydrated forms a viscous gum (Pawar & Edgar, 2012). Alginate is a new type of industrial food coating that binds divalent metal ions to form a gel (Kulig et al., 2016).

Food packaging is one of the many industrial applications for biodegradable materials. To improve food safety, chitosan was chosen because it has antimicrobial properties and can provide edible protective coating (Dutta et al., 2009). The other biopolymer used is starch to reduce water absorption and increase mechanical properties (Alix et al., 2013). Good film-forming capacities in chitosan and starch contribute to the formation of the composite film (Talón et al., 2017). According to Tripathi et al. (2008) chitosan–starch has been formulated, resulting in a potential food packaging application by the preparation of ferulic acid incorporated with a starch–chitosan blend film. Another study shows a chitosan–rice starch blend film was prepared by Brodnjak and Todorova (2018) using chitosan, rice starch and other materials, that is, malic acid and glycerol, with ultrasonic treatment.

Several properties of chitosan–rice starch blend films were investigated for comparison with chitosan and rice starch film. The moisture content was increased due to the presence of rice starch, which can produce a highly crosslinked system and therefore prevent water molecules from penetrating the composite film. Water vapor permeability (WVP) also decreased because of the addition of glycerol. The tensile strength and elongation at breaks were increased due to acoustic activation in an ultrasonic bath. Also, the mechanical resistance was improved and caused homogeneity of the surface. In overall, a chitosan blend with rice starch improved the characteristics of the composite film.

Furthermore, biopolymers are useful in tissue engineering to produce a biodegradable scaffold that acts as a temporary skeleton for accommodating and stimulating new tissue growth (Li et al., 2005). Before biodegradable scaffolds were introduced, bioactive ceramics and polymers were developed as a tissue engineering scaffold, and the drawbacks are that they are inherently brittle and have low biodegradation rates (Li et al., 2005). To overcome these problems, biopolymers were used because they have high biodegradation rates and good mechanical properties. The biopolymer used is chitosan and alginate. Chitosan is known to have good biocompatibility, biodegradability, low toxicity and the ability to be fabricated in different shapes, while alginate is commonly used in bone tissue engineering because it is biocompatible, hydrophilic and biodegradable under normal physiological conditions (Wang et al., 2017).

As well as in tissue engineering, the combination of chitosan and alginate has various applications due to its properties of polyelectrolyte complex fibers, beads, nanolayered polyethylene terephthalate (PET) films, drug-loaded membranes and multilayers (Kulig et al., 2016). Wang et al. (2017) proposed a preparation to blend chitosan with alginate, and the product was called a chitosan–alginate hybrid. The combination between these two polymers resulted strong ionic bonding between the amino group of chitosan and carboxyl group of alginates. The scaffold produced had a highly porous and interconnected pore structure, making it suitable for cell attachment, proliferation and tissue growth and allowing nutrients to pass through and metabolites to exchange (Wang et al., 2007). Moreover, a chitosan–alginate hybrid is compatible with proliferation of olfactory ensheathing cells (OECs) and neural stem cells (NSCs), which are cells for regrowth and self-renewal. Overall, a chitosan–alginate blend is very suitable for use as a scaffold in tissue engineering.

Many researchers are interested in the combination of cellulose and starch. Both cellulose and starch are natural biopolymers from plants and have similar chemical structures. Biodegradable properties, low price and abundant availability of starch and cellulose are the reasons many experiments with cellulose–starch blends have been performed. The unique characteristics of the product of blending cellulose and starch can be used in many applications, such as food packaging, biomedical and environmental (Shang et al., 2019). From Arvanitoyannis and Kassaveti (2009), various types of starch–cellulose blends show excellent mechanical properties, thermal properties, water vapor transmission rate and gas permeability. Generally, blending a cellulose derivative or starch derivative has wide application compared to natural cellulose and starch, but a natural cellulose–starch blend shows promising characteristics with numerous applications.

Miyamoto et al. (2009) investigated the structure and properties of cellulose–starch blend films regenerated from aqueous sodium hydroxide (NaOH) solution. CEKICEL, a cellulose–corn starch blend, was produced as a food material from aqueous NaOH solutions on a commercial scale for the first time. However, limited information is available regarding its properties and structure (Hisano et al., 1991). Miyamoto et al. (2009) published a paper on the structure of a cellulose–corn starch blend film made from a cellulose–corn starch aq. NaOH solution with varied proportions of polymer followed by coagulation of sulfuric acid. The structure and properties were evaluated by using scanning electron microscopy (SEM), wide-angle x-ray diffraction (WAXD), dynamic viscoelastic measurement and other equipment, and the result was that the average pore size of cellulose–starch blend films increased from 1 to 6 m as the starch content increased. These pores existed separately from one another, which may have contributed to the blend films' high water and oil absorbency. Another result was that the crystalline regions of the blends were incompatible, while in the amorphous regions, cellulose and starch were miscible to some degree. Overall, Miyamoto et al. (2009) concluded that the structure and properties of the cellulose–starch blend were scientifically important, and it was expected to be widely used in food material.

Another study on cellulose–starch was conducted by Shang et al. (2019) using cellulose–starch hybrid films plasticized by aqueous $ZnCl_2$ solution. The objective of that paper was to learn how cellulose and starch interact in $ZnCl_2$ solutions as two biopolymers with identical structures and how these interactions affect the structure and properties of the cellulose–starch blend hybrid materials. The experiment was carried out by preparing cellulose–starch hybrid films plasticized by an aqueous $ZnCl_2$ solution, varying the starch/cellulose ratios. The structure and mechanical properties of the film were inspected by using SEM, X-ray diffraction (XRD) and Fourier-transform infrared. The result shows that the mechanical properties, that is, tensile strength and elongation of the cellulose–starch hybrid, were improved, while the crystallinity of starch was reduced. Shang et al. (2019) also notices that in cellulose-rich hybrid films, although chemical interactions were not affected, a higher starch content decreased the material properties.

Besides starch-cellulose blends, there is also a compatible blend between starch and microcrystalline cellulose. MCC is a naturally occurring substance made from partly depolymerized and purified cellulose (Trache et al., 2016). Excipient is a substance that serves as the medium for a drug or other active substance, and it is important in industrial pharmacy. Co-processed excipients combine two or more excipients using a suitable manufacturing process that will not change the chemical structure in the excipient (Patel & Gohel, 2016). Co-processed excipients help to solve the drawbacks that can arise by using general-grade excipients (Mamatha et al., 2017). Limwong et al. (2004) blended rice starch (RS) and MCC to produce a co-processed excipient for direct compression. The method used was a spray-drying technique. Spray-drying is a process that uses a hot gas to rapidly dry a liquid or slurry into a dry powder. The spray-drying technique can overcome conventional emulsion stability issues during storage and distribution of dry products without using harmful organic solvents (Li et al., 2019). Limwong et al. (2004) carried out an experiment with a combination of RS and MCC by varying the RS:MCC ratio. They concluded that an RS and MCC ratio of 7:3 was the best combination that exhibited the result of the co-processed excipient with high compressibility, good flowability and self-disintegration.

Builders et al. (2010) also published that blending starch and MCC can produce excipient. This time, they blended MCC and maize starch (Mst), resulting in a novel multifunctional excipient MCC-Mst. A novel excipient means an excipient that is being used in a drug product for the first time. A multifunctional excipient defined as an excipient that provides added functionalities to the formulation or that has several functions in the formulation (Bhor et al., 2014). A multifunctional excipient provide several benefits, including lower production costs and fewer steps in the manufacturing process (Nachaegari & Bansal, 2004). The method used by Builders et al. (2010) to produce the multifunctional excipient MCC-Mst was known as compatibilized polymer blending by mixing colloidal dispersions of MCC and chemically gelatinized Mst at controlled temperatures. The result shows that in terms of direct compression capacity and disintegration performance, the MCC–Mst composites were comparable to MCC and Mst. The multifunctional MCC-Mst combination demonstrated excellent performance in terms of disintegration efficiency and loading capacity when incorporated into oral tablet formulations designed for the rapid release of active pharmaceutical ingredients (APIs) using the direct compression method.

Pharmaceutical excipients are substances that are added to a pharmaceutical dosage form such as a capsule to help with the production process; to protect, sustain, or enhance stability; or to improve bioavailability or patient acceptability (Haywood & Glass, 2011). Hence, capsules also play an important role in pharmaceuticals for drug delivery. A capsule is an edible package made of gelatine or another suitable material that is packed with medicine to create a unit of dosage (Hadi et al., 2013). Gelatine is a biopolymer that is commonly used in a number of industries due to its biocompatibility and biodegradability. Gelatine also has properties such as high water solubility, non-toxicity, high mechanical strength, elasticity in dry state and moisturizing by binding a significant volume of water. These excellent properties mean it is used in the pharmaceutical, photography and cosmetic industries (Das et al., 2017).

Many scientists have tried to obtain a suitable capsule that can be used in drug delivery. Zhang et al. (2013) published a paper on developing capsule materials by using gelatine–starch blends. They blended gelatine with corn starch, and poly (ethylene glycol) was used as a plasticizer and compatibilizer by varying the gelatine/starch ratio. The result shows that the blend with gelatine/starch in a ratio of 50:50 exhibited a good film and capsule. The properties of the blend consist of high viscosity. The transparency and toughness of the blends is increased due to the existence of PEG. The mechanical properties of the capsule were also investigated and showed an increase in tensile strength and decrease in elongation. The capsule seemed to be more rigid and brittle.

The blend of starch and gelatine is not only for capsules; it can also be used as an alternative to plastics in agriculture, as reported by Rosseto et al. (2019). Plastics used in agriculture are largely derived from synthetic petrochemical polymers. These products need a proper waste management system at the end of their lives, since they can pollute the soil and create high-cost environmental impacts for manufacturers (Blanco et al., 2018). To overcome this problem, many researchers have come out with biodegradable plastic. Biodegradable plastic is often made from low-cost, renewable resources that are abundant in nature, such as starch, gelatine, cellulose and other biopolymers (Cazón et al., 2017). There are many challenges to produce biodegradable plastic; for example, it is very costly compared to conventional plastic, and in term of properties, commodity plastic shows better properties (Imre & Pukánszky, 2013).

Hybridization of starch and gelatin has garnered significant attention in various research studies due to their advantages, including biodegradability and wide availability among consumers, which in turn leads to reduced raw material expenses (Rosseto et al., 2019). The process employed to combine starch and gelatin is known as extrusion. Extrusion is a mechanical procedure involving the feeding of polymers in various forms, such as granules, powders, flakes and pellets, through an extruder screw (Rosato, 1998). Rosseto et al. (2019) showed many methods of combination for starch and gelatine that can produce biodegradable plastics, such as using starch with the addition of hydrocolloid to protect granules from shearing during the manufacturing process; retain moisture; and reduce the blend's syneresis, water solubility and water absorption. Also, the addition of maize starch to gelatine films can increases the thickness, transparency and mechanical strength of the films and also enhance their structure, all of which increases the films' applicability. The authors also stated that when blending cassava starch and gelatine, higher gelatine concentrations increased the water vapor permeability, and mechanical strength values decreased the opacity of the film.

In agricultural production, rapeseed was coated with a biodegradable liquid film made of oxidized maize starch and gelatine from leather scraps. The film was found to increase the rate of rapeseed survival and yield and is suitable for agricultural production. The film has strong water absorption and retention properties, allowing plants to thrive in the most basic conditions. In addition, the ability of agricultural mulch films made from native and phosphorylated corn starch, with and without chitosan surface functionalization, was assessed. Rosseto et al. (2019) concluded that the properties demonstrated by the various compositions studied may allow biodegradable polymer films to be used in agriculture, particularly in mulching, because mulching plastics are normally replaced every crop cycle, whereas biodegradable ones may mix with the soil.

1.16 BLENDING OF SYNTHETIC BIOPOLYMERS WITH OTHER SYNTHETIC BIOPOLYMERS

Blending synthetic biopolymers with other synthetic biopolymers has also gained much attention from researchers and industry. This is because the use of biopolymers in blends can reduce cost and be environmentally friendly due to their properties. Nanda et al. (2011) studied the effect of process engineering on PLA/PHBV blend properties. The biopolymer blend was prepared by the melt blending technique, as this technique is an efficient method for polymer blending. In this study, it was reported that the addition of PLA to PHBV significantly increased the mechanical properties, the tensile strength and Young's modulus of the blend. The increase of strength and modulus of the blend may be due to the properties of PLA, which is high in strength and modulus, as has been reported in other literature (Huda et al., 2006; Richards et al., 2008). Another increase was found in elongation at breaks of the blend with the increase in PLA content. However, the addition of PHBV content to the PHBV/PLA blend was shown to lower the crystallization temperature of PLA. Furthermore, dynamic mechanical analysis showed that the storage modulus was improved with the addition of PLA into the blend system. The authors concluded that this polymer blend can replace petroleum-based polymers due to its eco-friendly properties. Other recent studies on PLA/PHBV blends for various applications have also been reported in other literature (Ferreira & Duek, 2005).

Jiang et al. (2006) proposed a study of a biodegradable blend of polylactide with poly (butylene adipate-co-terephthalate). Combining PLA with PBAT is a safe way to enhance PLA's properties while preserving its biodegradability. PBAT is a polymer that is flexible and fully biodegradable (Chiu et al., 2013; Pietrosanto et al., 2020). The PLA and PBAT blend was made by the melt blending technique, as both polymers were fed into a twin-screw extruder. The authors found that the mechanical properties (tensile strength and modulus) of the polymer blend decreased with the addition of PBAT content. Nonetheless, elongation at breaks and toughness of the blend gradually increased. This is due to the lower tensile strength and modulus of PBAT compared to PLA. A study of miscibility of the blend was done through differential scanning calorimetry analysis. The result showed that the blend was not miscible, as a two-phase system with two glass transitions was obtained even with the addition of PBAT content into the blend. This indicates that there is poor intermolecular interaction between PLA and PBAT. Another result obtained from this study is that the blend viscosity and melt elasticity increased with an increase in PBAT concentration, as PBAT has higher elasticity and viscosity compared to

PLA. Other literature has proposed the effect of compatibilizers on the properties of PLA/PBAT biopolymer blends (Kumar et al., 2010; Al-Itry et al., 2012; Al-Itry et al., 2014).

There is much literature on the investigation of blends of PLA with polycaprolactone biopolymers. PCL is a biodegradable (Takayama & Todo, 2006), biocompatible and eco-friendly polymer that is prepared by ring-opening polymerization of ε-caprolactone with the aid of a catalyst. PCL is primarily used in medical applications (Mofokeng & Luyt, 2015). Recently, PCL has been used in numerous applications such as packaging, agricultural film and manufacturing of disposable material (Malinowski, 2016). One study is from Matta et al. (2014), where the authors discuss the characterization of PLA/PCL blends with various concentrations of PCL. In the study, they produced blends by using a Hakee Rheomix, where the mixer blended the composition of polymeric materials in same processing treatment. From the experiment, it was found that the tensile strength and modulus of the blend decreased due to the formation of PCL spherulites acting as stress concentrators in the PLA matrix. Meanwhile, the viscosity of the blend increased with an increasing PCL concentration, as PCL has higher shear viscosity than PLA. Moreover, it was investigated that the PLA/PCL blend is not miscible, as phase separation in the blend was observed. Similar results of immiscibility of PLA/PCL were reported in Simoes et al. (2009), Gardella et al. (2014) and Urquijo et al. (2015).

Another study of synthetic biopolymer blends is from Yang et al. (2019) and detailed the production of PLA blended with poly (butylene succinate adipate) (PBSA) films containing active compounds to improve the properties of food packaging and its shelf life. Essential oil is often used as active ingredient in polymeric blends for designing active food packaging systems (Alboofetileh et al., 2014; Ma et al., 2017; Chen et al., 2019). The addition of active compounds into polymer materials is one of the steps taken to preserve food freshness as well as to maintain the food quality. PBSA is a copolyester that is synthesized by polycondensation of 1,4-butanediol in the presence of adipic acids (Ahn et al., 2001). In this research, the authors created innovative active films using the extrusion-casting technique. The experimental findings revealed enhancements in two significant mechanical properties: tensile strength and elongation at break. Furthermore, the subsequent evaluation, involving the release rate of active compounds from the film into a food simulant and the film's antioxidant efficiency, also yielded favorable outcomes. As mentioned, one of the main goals is to investigate the shelf life of food by using active films. Thus, the goal was achieved, as the shelf life of the food was extended while preserving the food quality. This is because the antibacterial and antioxidant properties of the PLA/PBSA films were improved when the active compound was released into the food. From this experiment, it was concluded that this biodegradable active film can replace non-biodegradable film in food products, particularly in food packaging.

Poly(hydroxybutyrate) (PHB) is a biodegradable biopolymer that also has been used in various types of polymer blending. PHB has been produced by microorganisms under certain conditions (Bharti & G, 2016; McAdam et al., 2020b). PHB is a polymer with a highly crystalline structure, where its crystallinity is above 50%. PHB has been used widely in medical applications, such as in tissue engineering as scaffolds and in drug carriers (Williams et al., 1999; Zinn et al., 2001; Phillip et al., 2007), and also in packaging applications (Rosa et al., 2004). Arrieta et al. (2017b) reviewed a polymer blend of PLA with PHB for food packaging through the melt blending technique. It was stated that the addition of PHB improved the performance of the pure PLA. To improve the processability of the PLA/PHB blend, plasticizer is often added. Moreover, the incorporation of PHB into the blend as a nucleating agent for PLA increased the PLA crystallinity, thus improving the barrier properties of the PLA/PHB blend. Recent developments of PLA/PHB formulations with improved properties are promising to replace petroleum-based polymers that have been used in food packaging.

1.17 BLENDING OF NATURAL BIOPOLYMERS WITH SYNTHETIC BIOPOLYMERS

Biodegradable synthetic polymers, in general, have many benefits over natural polymers. They are also biopolymers that, due to their biocompatibility and biodegradability, have the potential to replace petroleum-based polymers in a variety of applications, especially in the biomedical field, as a result of their modification and also can be fabricated in a variety of shapes (Tian et al., 2012). For instance, Tănase et al. (2015) reported blending polyvinyl alcohol and starch in their paper. PVA is a synthetic biopolymer that is biodegradable, biocompatible and soluble in water due to the presence of polar alcohol groups, which can form hydrogen bonds with water (Kanatt et al., 2012). Even though PVA is biodegradable, many researchers make an effort to improve its biodegradability by blending it with other polymers. Meanwhile, starch is a natural biopolymer that is frequently used in non-food applications, primarily in paper making and textiles (Vilaseca et al., 2007), as it can be easily found and is abundant. More recently, starch has been used as raw material in producing biodegradable film. Despite its advantages, starch-based materials are known to have some drawbacks, such as lower mechanical properties, poor processibility and extreme sensitive to water (Niranjana Prabhu & Prashantha, 2018).

A study by Tănase et al. (2015) in producing biopolymer blends was proposed. As mentioned, the researchers used PVA and starch for their polymeric blends, such as for food industry and packaging application. In this research, they used PVA as a thermoplastic matrix and starch as a biodegradation agent. They also used glycerol as plasticizer to increase the flexibility behavior of starch and PVA since starch is not fully compatible with PVA. The final product of the blending was a film and was characterized in a few areas, such as optical and permeability properties, which are the main properties in packaging. For optical properties, it showed that films with starch exhibit low light transmission compared to PVA/glycerol films, which means that films with starch incorporation have high opacity.

Thus, this shows that the PVA/starch films have excellent barrier properties, which is the most important category of properties for food packaging, as it can protect food from UV lights. Next is permeability, which also a dominant property for food packaging. The permeability of the films is increased along with an increase in starch. It was stated that the permeability properties are affected by the sample's crystallinity, where the crystallinity decreases when the permeability increases.

From this paper, it can be concluded that these biopolymer blends have good physical properties, which makes them suitable for packaging application.

In the same vein, Bonilla et al. (2014) also produced a biodegradable film based on PVA and chitosan for food packaging products. Yu et al. (2018) stated that PVA and chitosan, which are biocompatible and have strong filming formation, have received a lot of attention as biodegradable polymers because of their environmental benefits. According to Chiellini et al. (2003), PVA has been widely used in the preparation of polymeric blends and composites with other biopolymers. Meanwhile, chitosan films have a lot of potential as active packaging materials because of their low oxygen permeability and antimicrobial activity (Kanatt et al., 2012). Theoretically, chitosan is miscible with PVA, as they can form hydrogen bonds due to the presence of hydroxyl and amine groups in chitosan (Chen et al., 2008). As for that, by using the polymer blending process, PVA was blended with chitosan to improve the physical properties of chitosan-based materials (Chen et al., 2007).

In a paper by Bonilla et al. (2014), it was stated that the addition of chitosan in PVA film reduced film stretchability, which makes it resistant to fracture compared to pure biopolymer films. This is due to the formation of inter-chain bonds that reinforce cohesion between the polymer networks but limit the polymer chain slippage. Furthermore, according to the researchers, the addition of chitosan into PVA films reduced the UV light transmission of the blended film. Lower transmission of UV light can give a blended film good barrier properties, which can protect the food inside the packaging from lipid oxidation induced by UV light.

On the other hand, Tang et al. (2012) discussed biopolymer blending based on starch and polylactic acid. PLA is a biodegradable synthetic polymers, which is the most common polymer that has been used in recent development of polymer blending (Gunatillake & Adhikari, 2003). Among the families of synthetic polymers, PLA is often used to blend with starch to increase the performance of the product, such as to increase the biodegradability and reduce the cost (Tang et al., 2012). The first polymer blend system based on PLA and starch was patented by Ajioka et al. (1995). They produced polymer blend films by using a sealed mixer at the first step and then using a hot-pressed procedure, resulting in a smooth and translucent film.

In recent years, several studies have been done based on this kind of polymer blending. Jun (2000) studied the performance of PLA/starch blending with a combination of reactive agents during the melt extrusion process. Diisocyanates such as toluene diisocyanate (TDI), 1,6-diisocyanatohexane (DIH) and 4,4′-methylenebis (phenylisocyanate) (MDI) have been used as crosslinked agents. The agents increased the tensile strength of the polymer blend. The use of DHI crosslinking agents gave a higher tensile strength compared to other agents. This is because the DHI agent has a long chain structure and is more flexible that other crosslinking agents. However, the elongation at breaks of the blends did not show a significant increase compared to tensile strength even with the addition of reactive agents. This shows that the plasticizing properties of the polymer blend cannot be improved. Jang et al. (2007) characterized the thermal properties and morphology of a PLA/starch compatibilized blend. In the preparation of PLA/starch blends, reactive compatibilizers such as maleic anhydride (MA) and maleated thermoplastic starch (MATPS) are used to enhance interfacial adhesion. They found that MA is a good compatibilizer for PLA/starch blend systems, while MATPS is not. MA has shown good interfacial morphology when it has been incorporated into a blend system. As for the thermal properties, the crystallinity of PLA/starch blends improved with MA as compatibilizer. The increase of crystallinity is due to the effects of plasticization by MA.

Other natural and synthetic biodegradable polymer blends have gained much less recognition than starch-based biodegradable polymer blends. Claro et al. (2016) studied a comparison of PLA/chitosan with a PLA/cellulose acetate blend for packaging applications. Their goal was to produce biodegradable films with improved properties without the addition of an additive or plasticizer. It may perhaps be observed without straying too far afield from the main focus that cellulose acetate is also a biopolymer that is synthesized from an esterification reaction of cellulose with acetic acid and anhydrous acetate by using sulfuric acid as a catalyst (Edgar et al., 2001). Per its name, cellulose acetate is a cellulose derivative that has antimicrobial properties. From their study, the authors found that PLA/chitosan film had the greatest increase in mechanical properties, where the elongation at breaks of the blended film increased compared to pure PLA film. This shows that PLA/chitosan film is the best option to be used in packaging applications compared to PLA/cellulose acetate film. On the other hand, the PLA/cellulose acetate blend did not show improvement in mechanical properties; the tensile strength and elastic modulus resulted in the lowest values. Despite this drawback, the PLA/cellulose acetate blend still showed an increase in elongation at breaks. Based on the result, the authors proved that PLA/chitosan and PLA/cellulose acetate blends do not need any additives or plasticizers in their blends for use in packaging applications.

Another study on biopolymer blends was by Sionkowska (2003), who prepared a collagen/PVP film by using the solution casting method. The work was done to investigate the interaction between collagen and PVP in the blend as well as to find the miscibility state of those blends. It was found that collagen and PVP have good interaction with each other. This can be seen by the result obtained from viscometry, DSC and fourier transform infrared spectroscopy (FTIR). Based on viscometric data, it was shown that the collagen/PVP blend is miscible, as the criteria for miscibility were detected by viscometry. Meanwhile, a DSC test is the common method to determine the miscibility of polymer blends. The main parameter of miscibility is the glass transition temperature (T_g). DSC showed that the glass transition temperatures for the blends and single components were distinct. On the other hand, the FTIR test also showed that collagen and PVP had good interaction, as FTIR could detect the intermolecular interaction between those two polymers. From the test, it was found that the position of amide band A and amide I band of collagen moved to a higher frequency after blending with PVP. This shows that formation of hydrogen bonding occurred between collagen and PVP. The author has also investigated other collagen-based polymer blending in her studies (Sionkowska et al., 2009).

1.18   SURFACE MODIFICATION OF BLENDING

From previous topic, it can be seen that many types of biopolymer blends can be tailored for specific applications. The examples show that the result of compatible blending between two polymers can produce a new material with desired characteristics. However, there are some biopolymer blends that are immiscible with each other, meaning another step must be taken, known as surface modification (Imre & Pukánszky, 2013). By performing surface modification, the blends between two biopolymers become compatible. Surface modification is making many blends compatible, and it increases the advantage of biopolymer blends that can be used in various applications. As matter of fact, surface modification can not only be used in polymers, but it has been used in many ways due to polymer surface properties limiting their intended wide range of applications (Nemani et al., 2018). Fabbri and Messori (2017) stated that the most popular goal of surface modification is to modify the outermost layer of a polymer by introducing functional groups onto the surface to improve certain properties such as barrier properties, adhesion and wettability. The surface of the polymer means the frontier between two different media, characterized by a certain thickness that reflects a gradient of properties (Mincheva & Raquez, 2019).

Generally, surface modification methods have been classified based on the characteristic original material being modified, such as physical, chemical and biological characteristics (Fabbri & Messori, 2017). The easiest, low-cost, scalable way of surface modification is using physical methods. It provides more robust and abrasion-resistant polymer surfaces. It is also eco-friendly because no chemicals are needed in this process (Ozdemir et al., 1999). Besides physical methods, there are chemical methods that use chemicals to modify the surface of a component. Unlike physical methods, chemical methods can enhance the properties of a component without altering the surface roughness. Most chemical surface treatment methods use wet procedures in which the polymer is dipped, coated or sprayed with a chemical to improve its surface properties (Mitra & Saha, 2013). Recent advancements in the utilization of biopolymers highlight the necessity for the introduction of a new category of polymer modification techniques, specifically those rooted in biological approaches. Biological methods involve the use of proteins, peptides, ligands, receptors and other fundamental biomolecules, while additional compounds like drugs and lipids can be incorporated onto the material's surface. These methods encompass physical adsorption, self-crosslinking and chemical conjugation. When it comes to blending, there are several popular strategies for surface modification in polymer blends, including copolymerization, grafting, trans-esterification and the utilization of reactive coupling agents.

1.18.1 Copolymerization

Copolymerization is an effective method used for tailoring polymers. Copolymerization involves polymerizing two separate monomers simultaneously, with the intention of combining both of the structures into a single polymer chain. This greatly broadens and diversifies the range and variety of properties of copolymer molecules, allowing for the incorporation of desirable properties from various monomer units (Scott & Penlidis, 2017). A study by Graebling and Bataille (1994) proposed an experiment on polypropylene/poly(hydroxybutyrate) blends with the presence of PP-g-PHB copolymer. The copolymer of this blend was prepared based

on a three-step procedure, alchoholysis of PHB, grafting of styrene and maleic anhydride on PP and the reaction between the polymers obtained. The morphology of
a PP-PHB blend changes significantly when just the copolymer is present, as the
PHB polymers tend to disappear into the PP matrix. Also, for mechanical properties, tensile strength showed an increase in yield stress but a decrease in break
properties with the presence of the copolymer. Another study by Guo et al. (2019)
reported production of a poly (1,8-octanediol citrate) (POC)/poly(ε-caprolactone)
(PCL) composite elastomer via an in-situ copolymerization blending technique. A
completely transparent elastomer was produced as a result of partial incorporation
of PCL into POC networks. The incorporation of the PCL into the process showed
a positive result where the thermosetting time was decreased, thus producing a non-
sticky elastomer.

1.18.2 GRAFTING

Polymer grafting is a technique that involves the covalent attachment of monomers to a polymer chain(Koshy et al., 2016). This method encompasses two distinct
approaches: grafting-to and grafting-from. Grafting-to involves the attachment of an
end-functionalized polymer chain to a solid substrate, whereas grafting-from entails
initiating the grafting reaction through surface polymerization (Minko, 2008). An
example of biopolymer grafting is in the surface modification of biodegradable electrospun nanofiber scaffolds conducted by Park et al. (2007). Electrospun nanofiber
scaffolds are widely used in tissue engineering to deliver transplanted cells to target
areas and provide mechanical support from physiological loads. They provide a good
microenvironment for cell adhesion, proliferation and differentiation due to their
properties such as tunable porosity, high surface-to-volume ratio and ease of surface
functionalization (Mo et al., 2018). To make a biodegradable electrospun nanofiber scaffold, a synthetic biopolymer was used consisting of PGA, PLLA and PLA.
The method used was grafting-to in situ polymerization of acrylic acid (AA). This
method added a carboxylic functional group to the scaffold, making it hydrophilic
and cell compatible. The result of this surface modification through the grafting
method significantly improved fibroblast adhesion and proliferation on the surface-
modified scaffold.

1.18.3 TRANSESTERIFICATION

Transesterification has been successfully used to produce polymers and blends with
better properties. A study by Lee et al. (2001) discussed the transesterification reaction of barium sulfate ($BaSO_4$) on a poly(butylene terephthalate)/poly(ethylene terephthalate) blend system. In this study, a titanate coupling agent (TCA), which is
organic particles, was introduced into $BaSO_4$ (inorganic particle), which acted as
a surface modifier of $BaSO_4$ in order to improve the interfacial adhesion inorganic
filler and organic polymer matrix. When PBT and PET are blended, a block copolymer is formed from the mixture. It is difficult to separate the block copolymer
state through transesterification; thus, the end result is a random copolymer with
a decrease in molecular weight. It has been concluded that $BaSO_4$ suppresses the

transesterification reaction during the melt blending process, as the melting endotherm peak decreases with an increase in PBT concentration. The result from the experiment shows that the melt blend of PBT/PET with no $BaSO_4$ showed outstanding transesterification reactions with a decrease in mechanical properties compared to PBT/PET with surface modification $BaSO_4$ or modified $BaSO_4$, which showed a slight transesterification reaction but an increase in mechanical properties (flexural strength). Thus, the surface modification was shown to improve the polymer blend properties.

1.18.4 REACTIVE COUPLING AGENT

The use of reactive coupling agents is also one of the ways to improve polymer properties and blends. Jariyasakoolroj and Chirachanchai (2014) proposed a study on a polymer blend of PLA with silane modified starch. Silane is one of coupling agents that function to modify starch via the hydrolyzable and organofunctional reactive sides in order to form covalent bond with PLA. A series of trimethoxy silane coupling agents such as 3-glycidoxypropyl trimethoxysilane (GPMS), 3-aminopropyl trimethoxy silane (APMS) and 3-chloropropyl trimethoxysilane (CPMS) were used to coupleg with starch, which then formed GP-starch, AP-starch and CP-starch, respectively. The surface modification of starch by organofunctional silane coupling agents has successfully been traced based on the FTIR results obtained. Based on the result, it was stated that the coupling of silane on starch might happen via silanol linkages. The nuclear magnetic resonance (NMR) method used to study the structure of starch modified by silane agents gave information about bond formation between the components. Compared to other silane modified starch, only CP-starch formed covalent bonds with PLA during blending, thus providing compatibility between PLA and starch. The modified starch (CP-starch) also played a role as a nucleating agent, where it accelerated the crystallization rate in PLA, and this was proved by the increasing rate in degree of crystallinity of the PLA/CP-starch blend. The mechanical properties of tensile strength and elongation at breaks of the film produced from the reactive blend also increased compared to pure PLA and other PLA/silane-modified starch films.

1.19 CONCLUSIONS

From the classification and chemistry of polymers, it can be concluded that there are variety of polymers being produced either biologically or synthetically via chemical modification. Nevertheless, the preference for polymers used in bio-plastics always emphasizes two characteristics, biodegradability and biobased origins. The main motivation behind the use of either biobased polymers or biodegradable polymers is always to reduce the carbon footprint, greenhouse emissions, waste pollution and exploitation of fossil fuel. Apart from that, the chemical characteristics of each polymer have been studied extensively, as they assist in new or novel material production. This can be seen from the examples given for biopolymer blends and bionanocomposites whereby two or more polymers can be mixed or incorporated with fillers to achieve better performance for targeted usage.

REFERENCES

Abdelrazek, E. M., Hezma, A. M., El-khodary, A., & Elzayat, A. M. (2016). Spectroscopic studies and thermal properties of PCL/PMMA biopolymer blend. *Egyptian Journal of Basic and Applied Sciences*, *3*(1). https://doi.org/10.1016/j.ejbas.2015.06.001

Abdul Khalil, H. P. S., Yahya, E. B., Tajarudin, H. A., Balakrishnan, V., & Nasution, H. (2022). Insights into the role of biopolymer-based xerogels in biomedical applications. *Gels, 8*(6). https://doi.org/10.3390/gels8060334

Abdul Khalil, H. P. S., Yahya E. B., Jummaat, F., Adnan, A. S., Olaiya, N. G., Rizal, S., Abdullah, C. K., Pasquini, D., & Thomas. S. (2023). Biopolymers based aerogels: A review on revolutionary solutions for smart therapeutics delivery. *Progress in Materials Science*, *131*, 101014.

Abdul Khalil, H. P. S., Adnan, A. S., Yahya, E. B., Olaiya, N. G., Safrida, S., Hossain, M. S., Balakrishnan, V., Gopakumar, D. A., Abdullah, C. K., Oyekanmi, A. A., & Pasquini, D. (2020). A review on plant cellulose nanofibre-based aerogels for biomedical applications. *Polymers*, *12*(8). https://doi.org/10.3390/polym12081759

Abdul Khalil, H. P. S., Chong, E. W. N., Owolabi, F. A. T., Asniza, M., Tye, Y. Y., Rizal, S., Nurul Fazita, M. R., Mohamad Haafiz, M. K., Nurmiati, Z., & Paridah, M. T. (2019). Enhancement of basic properties of polysaccharide-based composites with organic and inorganic fillers: A review. *Journal of Applied Polymer Science*, *136*(12). https://doi.org/10.1002/app.47251

Abdul Khalil, H. P. S., Tye, Y. Y., Saurabh, C. K., Leh, C. P., Lai, T. K., Chong, E. W. N., Nurul Fazita, M. R., Hafiidz, J. M., Banerjee, A., & Syakir, M. I. (2017). Biodegradable polymer films from seaweed polysaccharides: A review on cellulose as a reinforcement material. *Express Polymer Letters*, *11*(4). https://doi.org/10.3144/expresspolymlett.2017.26

Aggarwal, J., Sharma, S., Kamyab, H., & Kumar, A. (2020). The realm of biopolymers and their usage: An overview. *Journal of Environmental Treatment Techniques*, *8*(3).

Ahmed, J., & Thomas, L. (2018). Effect of xanthan and guar gum on the pasting, stickiness and extensional properties of brown wheat flour/β-glucan composite doughs. *LWT—Food Science and Technology*, *87*. https://doi.org/10.1016/j.lwt.2017.09.017

Ahn, B. D., Kim, S. H., Kim, Y. H., & Yang, J. S. (2001). Synthesis and characterization of the biodegradable copolymers from succinic acid and adipic acid with 1,4-butanediol. *Journal of Applied Polymer Science*, *82*(11), 2808–2826. https://doi.org/10.1002/app.2135

Ajioka, M., Enomoto, K., Suzuki, K., & Yamaguchi, A. (1995). Basic properties of polylactic acid produced by the direct condensation polymerization of lactic acid. *Bulletin of the Chemical Society of Japan*, *68*(8), 2125–2131.

Alboofetileh, M., Rezaei, M., Hosseini, H., & Abdollahi, M. (2014). Antimicrobial activity of alginate/clay nanocomposite films enriched with essential oils against three common foodborne pathogens. *Food Control*, *36*(1), 1–7. https://doi.org/10.1016/j.foodcont.2013.07.037

Alemán, J., Chadwick, A. V., He, J., Hess, M., Horie, K., Jones, R. G., Kratochvíl, P., Meisel, I., Mita, I., Moad, G., Penczek, S., & Stepto, R. F. T. (2007). Definitions of terms relating to the structure and processing of sols, gels, networks, and inorganic-organic hybrid materials (IUPAC recommendations 2007). *Pure and Applied Chemistry*, *79*(10), 1801–1829. https://doi.org/10.1351/pac200779101801

Alemdar, N. (2016). Fabrication of a novel bone ash-reinforced gelatin/alginate/hyaluronic acid composite film for controlled drug delivery. *Carbohydrate Polymers*, *151*. https://doi.org/10.1016/j.carbpol.2016.06.033

Ali, F. M., Kershi, R. M., Sayed, M. A., & AbouDeif, Y. M. (2018). Evaluation of structural and optical properties of Ce3+ ions doped (PVA/PVP) composite films for new organic semiconductors. *Physica B: Condensed Matter, 538*. https://doi.org/10.1016/j.physb.2018.03.031

Ali, M., Sultana, R., Tahir, S., Watson, I. A., & Saleem, M. (2017). Prospects of microalgal biodiesel production in Pakistan: A review. *Renewable and Sustainable Energy Reviews, 80*. https://doi.org/10.1016/j.rser.2017.08.062

Ali Akbari Ghavimi, S., Ebrahimzadeh, M. H., Solati-Hashjin, M., & Abu Osman, N. A. (2015). Polycaprolactone/starch composite: Fabrication, structure, properties, and applications. *Journal of Biomedical Materials Research—Part A, 103*(7). https://doi.org/10.1002/jbm.a.35371

Al-Itry, R., Lamnawar, K., & Maazouz, A. (2012). Improvement of thermal stability, rheological and mechanical properties of PLA, PBAT and their blends by reactive extrusion with functionalized epoxy. *Polymer Degradation and Stability, 97*(10), 1898–1914. https://doi.org/10.1016/j.polymdegradstab.2012.06.028

Al-Itry, R., Lamnawar, K., & Maazouz, A. (2014). Rheological, morphological, and interfacial properties of compatibilized PLA/PBAT blends. *Rheologica Acta, 53*(7), 501–517. https://doi.org/10.1007/s00397-014-0774-2

Alix, S., Mahieu, A., Terrie, C., Soulestin, J., Gerault, E., Feuilloley, M. G. J., Gattin, R., Edon, V., Ait-Younes, T., & Leblanc, N. (2013). Active pseudo-multilayered films from polycaprolactone and starch based matrix for food-packaging applications. *European Polymer Journal, 49*(6), 1234–1242. https://doi.org/10.1016/j.eurpolymj.2013.03.016

Al-Sibani, M., Al-Harrasi, A., & Neubert, R. H. H. (2017). Effect of hyaluronic acid initial concentration on cross-linking efficiency of hyaluronic acid—Based hydrogels used in biomedical and cosmetic applications. *Pharmazie, 72*(2). https://doi.org/10.1691/ph.2017.6133

Ameen, S., Ali, V., Zulfequar, M., Mazharul Haq, M., & Husain, M. (2006). Synthesis and characterization of polyaniline-polyvinyl chloride blends doped with sulfamic acid in aqueous tetrahydrofuran. *Central European Journal of Chemistry, 4*(4), 565–577. https://doi.org/10.2478/s11532-006-0044-y

Amos, J. L., McCurdy, J. L., & McIntire, O. R. (1954). Method of making linear interpolymers of monovinyl aromatic compounds and a natural or synthetic rubber. *US Patent 2694692*. https://doi.org/10.1145/178951.178972

Araújo, R. G., Rodríguez-Jasso, R. M., Ruiz, H. A., Govea-Salas, M., Rosas-Flores, W., Aguilar-González, M. A., Pintado, M. E., Lopez-Badillo, C., Luevanos, C., & Aguilar, C. N. (2020). Hydrothermal-microwave processing for starch extraction from Mexican avocado seeds: Operational conditions and characterization. *Processes, 8*(7). https://doi.org/10.3390/pr8070759

Aravind, I., Eichhorn, K., Komber, H., Jehnichen, D., Zafeiropoulos, N. E., Ahn, K. H., Grohens, Y., Stamm, M., & Thomas, S. (2009). A study on reaction-induced miscibility of poly (trimethylene terephthalate)/polycarbonate blends. *The Journal of Physical Chemistry B, 113*(6), 1569–1578.

Arrieta, M. P., Samper, M. D., Aldas, M., & López, J. (2017a). On the use of PLA-PHB blends for sustainable food packaging applications. *Materials, 10*(9). https://doi.org/10.3390/ma10091008

Arrieta, M. P., Samper, M. D., Aldas, M., & López, J. (2017b). On the use of PLA-PHB blends for sustainable food packaging applications. *Materials, 10*(9), 1–26. https://doi.org/10.3390/ma10091008

Arvanitoyannis, I. S., & Kassaveti, A. (2009). Starch–cellulose blends. In *Biodegradable polymer blends and composites from renewable resources* (pp. 17–53). John Wiley & Sons. https://doi.org/10.1002/9780470391501.ch2

Aslam, M., Kalyar, M. A., & Raza, Z. A. (2018). Polyvinyl alcohol: A review of research status and use of polyvinyl alcohol based nanocomposites. *Polymer Engineering and Science, 58*(12). https://doi.org/10.1002/pen.24855

Ayadi, F., Mamzed, S., Portella, C., & Dole, P. (2013). Synthesis of bis(pyrrolidone-4-carboxylic acid)-based polyamides derived from renewable itaconic acid—Application as a compatibilizer in biopolymer blends. *Polymer Journal, 45*(7). https://doi.org/10.1038/pj.2012.206

Aydemir, D., & Gardner, D. J. (2020). Biopolymer blends of polyhydroxybutyrate and polylactic acid reinforced with cellulose nanofibrils. *Carbohydrate Polymers, 250.* https://doi.org/10.1016/j.carbpol.2020.116867

Aylsworth, J. W. (1914). Plastic composition. *US Patent 1111284.* https://doi.org/10.1145/178951.178972

Aziz, S., Hosseinzadeh, L., Arkan, E., & Azandaryani, A. H. (2019). Preparation of electrospun nanofibers based on wheat gluten containing azathioprine for biomedical application. *International Journal of Polymeric Materials and Polymeric Biomaterials, 68*(11). https://doi.org/10.1080/00914037.2018.1482464

Babu, R. P., O'Connor, K., & Seeram, R. (2013). Current progress on bio-based polymers and their future trends. *Progress in Biomaterials, 2*(1). https://doi.org/10.1186/2194-0517-2-8

Bano, I., Arshad, M., Yasin, T., Ghauri, M. A. (2019). Preparation, characterization and evaluation of glycerol plasticized chitosan/PVA blends for burn wounds. *International Journal of Biological Macromolecules, 124.*

Baran, E. H., & Yildirim Erbil, H. (2019). Surface modification of 3d printed pla objects by fused deposition modeling: A review. *Colloids and Interfaces, 3*(2). https://doi.org/10.3390/colloids3020043

Barcellos, I. O., Carobrez, S. G., Pires, A. T. N., & Alvarez-Silva, M. (1998). In vivo and in vitro responses to poly(ethylene terephthalate-co-diethylene glycol terephthalate) and polyethylene oxide blends. *Biomaterials, 19*(22). https://doi.org/10.1016/S0142-9612(98)00119-7

Barot, A. A., Panchal, T. M., Patel, A., & Patel, C. M. (2019). Polyester the workhorse of polymers: A review from synthesis to recycling. *Archives of Applied Science Research, 11*(2).

Bazmandeh, A. Z., Mirzaei, E., Fadaie, M., Shirian, S., & Ghasemi, Y. (2020). Dual spinneret electrospun nanofibrous/gel structure of chitosan-gelatin/chitosan-hyaluronic acid as a wound dressing: In-vitro and in-vivo studies. *International Journal of Biological Macromolecules, 162.* https://doi.org/10.1016/j.ijbiomac.2020.06.181

Benbettaïeb, N., Kurek, M., Bornaz, S., & Debeaufort, F. (2014). Barrier, structural and mechanical properties of bovine gelatin-chitosan blend films related to biopolymer interactions. *Journal of the Science of Food and Agriculture, 94*(12). https://doi.org/10.1002/jsfa.6570

Bezerra, E. B., de França, D. C., de Souza Morais, D. D., dos Santo Silva, I. D., Siqueira, D. D., Araújo, E. M., & Wellen, R. M. R. (2019). Compatibility and characterization of Bio-PE/PCL blends. *Polimeros, 29*(2). https://doi.org/10.1590/0104-1428.02518

Bharti, S. N., & G, S. (2016). Need for bioplastics and role of biopolymer PHB: A short review. *Journal of Petroleum & Environmental Biotechnology, 7*(272), 1–3. https://doi.org/10.4172/2157-7463.1000272

Bhor, N. J., Bhusare, S. E., & Kare, P. T. (2014). Multifunctional excipients: The smart excipients. *International Journal of Pure and Applied Bioscience, 2*(5), 144–148.

Bielskis, E., & Leo, A., JK Zeien—US Patent 4, 859,484, & 1989, undefined. (n.d.). *Processed starch-gum blends.* Google Patents.

Biesiekierski, J. R. (2017). What is gluten? *Journal of Gastroenterology and Hepatology (Australia), 32.* https://doi.org/10.1111/jgh.13703

Biliaderis, C., Lazaridou, A., Polymers, I. A.-C., & 1999, undefined. (n.d.). *Glass transition and physical properties of polyol-plasticised pullulan—starch blends at low moisture.* Elsevier.

Blanco, I., Loisi, R. V., Sica, C., Schettini, E., & Vox, G. (2018). Agricultural plastic waste mapping using GIS. A case study in Italy. *Resources, Conservation and Recycling, 137,* 229–242. https://doi.org/10.1016/j.resconrec.2018.06.008

Bonilla, J., Fortunati, E., Atarés, L., Chiralt, A., & Kenny, J. M. (2014). Physical, structural and antimicrobial properties of poly vinyl alcohol-chitosan biodegradable films. *Food Hydrocolloids, 35,* 463–470. https://doi.org/10.1016/j.foodhyd.2013.07.002

Bonilla-Cruz, J., Dehonor, M., Saldivar-Guerra, E., & Gonzalez-Montiel, A. (2013). Polymer modification: Functionalization and grafting. In *Handbook of polymer synthesis, characterization, and processing* (pp. 205–223). https://doi.org/10.1002/9781118480793.ch10

Brodnjak, U. V., & Todorova, D. (2018). *Chitosan and rice starch films as packaging materials* (pp. 275–280). https://doi.org/10.24867/grid-2018-p34

Bu, H., & Li, G. (2018). Comparative investigation of hydroxyapatite/collagen composites prepared by CaCl2 addition at different time points in collagen self-assembly process. *Journal of Materials Science, 53*(9), 6313–6324. https://doi.org/10.1007/s10853-018-2027-8

Builders, P. F., Bonaventure, A. M., Tiwalade, A., Okpako, L. C., & Attama, A. A. (2010). Novel multifunctional pharmaceutical excipients derived from microcrystalline cellulose–starch microparticulate composites prepared by compatibilized reactive polymer blending. *International Journal of Pharmaceutics, 388*(1–2), 159–167. https://doi.org/10.1016/j.ijpharm.2009.12.056

Cazón, P., Velazquez, G., Ramírez, J. A., & Vázquez, M. (2017). Polysaccharide-based films and coatings for food packaging: A review. *Food Hydrocolloids, 68,* 136–148. https://doi.org/10.1016/j.foodhyd.2016.09.009

Chen, C. H., Wang, F. Y., Mao, C. F., Liao, W. T., & Hsieh, C. D. (2008). Studies of chitosan: II. Preparation and characterization of chitosan/poly(vinyl alcohol)/gelatin ternary blend films. *International Journal of Biological Macromolecules, 43*(1), 37–42. https://doi.org/10.1016/j.ijbiomac.2007.09.005

Chen, C. H., Wang, F. Y., Mao, C. F., & Yang, C. H. (2007). Studies of chitosan. I. Preparation and characterization of chitosan/poly(vinyl alcohol) blend films. *Journal of Applied Polymer Science, 116*(5), 1086–1092. https://doi.org/10.1002/app

Chen, H., Li, L., Ma, Y., Mcdonald, T. P., & Wang, Y. (2019). Development of active packaging film containing bioactive components encapsulated in β-cyclodextrin and its application. *Food Hydrocolloids, 90*(September 2018), 360–366. https://doi.org/10.1016/j.foodhyd.2018.12.043

Chen, W. Y., Suzuki, T., & Lackner, M. (2016). Handbook of climate change mitigation and adaptation. In *Handbook of climate change mitigation and adaptation* (2nd ed., Vols. 1–4). https://doi.org/10.1007/978-3-319-14409-2

Chiellini, E., Corti, A., D'Antone, S., & Solaro, R. (2003). Biodegradation of poly (vinyl alcohol) based materials. *Progress in Polymer Science (Oxford), 28*(6). https://doi.org/10.1016/S0079-6700(02)00149-1

Chirila, T. V., Cooper, R. L., Constable, I. J., & Horne, R. (1992). Radiation-absorbing hydrogel—melanin blends for ocular devices. *Journal of Applied Polymer Science, 44*(4). https://doi.org/10.1002/app.1992.070440405

Chiu, H. T., Huang, S. Y., Chen, Y. F., Kuo, M. T., Chiang, T. Y., Chang, C. Y., & Wang, Y. H. (2013). Heat treatment effects on the mechanical properties and morphologies of poly (lactic acid)/poly (butylene adipate-co-terephthalate) blends. *International Journal of Polymer Science, 2013*, 1–11. https://doi.org/10.1155/2013/951696

Chotiprayon, P., Chaisawad, B., & Yoksan, R. (2020). Thermoplastic cassava starch/ poly(lactic acid) blend reinforced with coir fibres. *International Journal of Biological Macromolecules, 156*. https://doi.org/10.1016/j.ijbiomac.2020.04.121

Claro, P. I. C., Neto, A. R. S., Bibbo, A. C. C., Mattoso, L. H. C., Bastos, M. S. R., & Marconcini, J. M. (2016). Biodegradable blends with potential use in packaging: A comparison of PLA/Chitosan and PLA/Cellulose Acetate Films. *Journal of Polymers and the Environment, 24*(4), 363–371. https://doi.org/10.1007/s10924-016-0785-4

Coppola, B., Cappetti, N., Maio, L. Di, Scarfato, P., & Incarnato, L. (2018). 3D printing of PLA/clay nanocomposites: Influence of printing temperature on printed samples properties. *Materials, 11*(10). https://doi.org/10.3390/ma11101947

Cotas, J., Leandro, A., Pacheco, D., Gonçalves, A. M. M., & Pereira, L. (2020). A comprehensive review of the nutraceutical and therapeutic applications of red seaweeds (Rhodophyta). *Life, 10*(3). https://doi.org/10.3390/life10030019

Cunha, L., & Grenha, A. (2016). Sulfated seaweed polysaccharides as multifunctional materials in drug delivery applications. *Marine Drugs, 14*(3). https://doi.org/10.3390/md14030042

Dan Ramdath, D., Padhi, E. M. T., Sarfaraz, S., Renwick, S., & Duncan, A. M. (2017). Beyond the cholesterol-lowering effect of soy protein: A review of the effects of dietary soy and its constituents on risk factors for cardiovascular disease. *Nutrients, 9*(4). https://doi.org/10.3390/nu9040324

Das, M. P., P. R., S., Prasad, K., J. V., V., & M., R. (2017). Extraction and characterization of gelatin: A functional biopolymer. *International Journal of Pharmacy and Pharmaceutical Sciences, 9*(9), 239–242. https://doi.org/10.22159/ijpps.2017v9i9.17618

da Silva Alves, D. C., Healy, B., Pinto, L. A. de A., Cadaval, T. R. S. A., & Breslin, C. B. (2021). Recent developments in chitosan-based adsorbents for the removal of pollutants from aqueous environments. *Molecules (Basel, Switzerland, 26*(3). https://doi.org/10.3390/molecules26030594

Datta, S. (2013). Elastomer blends. In *The science and technology of rubber* (4th ed.). Elsevier Inc. https://doi.org/10.1016/B978-0-12-394584-6.00012-1

Dekkers, B. L., de Kort, D. W., Grabowska, K. J., Tian, B., Van As, H., & van der Goot, A. J. (2016). A combined rheology and time domain NMR approach for determining water distributions in protein blends. *Food Hydrocolloids, 60*. https://doi.org/10.1016/j.foodhyd.2016.04.020

Dekkers, B. L., Nikiforidis, C. V., & van der Goot, A. J. (2016). Shear-induced fibrous structure formation from a pectin/SPI blend. *Innovative Food Science and Emerging Technologies, 36*. https://doi.org/10.1016/j.ifset.2016.07.003

de Matos Costa, A. R., Crocitti, A., de Carvalho, L. H., Carroccio, S. C., Cerruti, P., & Santagata, G. (2020). Properties of biodegradable films based on poly(Butylene succinate) (pbs) and poly(butylene adipate-co-terephthalate) (pbat) blends. *Polymers, 12*(10). https://doi.org/10.3390/polym12102317

Derkach, S. R., Kuchina, Y. A., Kolotova, D. S., & Voron'ko, N. G. (2020). Polyelectrolyte polysaccharide-gelatin complexes: Rheology and structure. *Polymers, 12*(2). https://doi.org/10.3390/polym12020266

de Siqueira, E. C., Rebouças, J. de S., Pinheiro, I. O., & Formiga, F. R. (2020). Levan-based nanostructured systems: An overview. *International Journal of Pharmaceutics, 580*. https://doi.org/10.1016/j.ijpharm.2020.119242

Di Bartolo, A., Infurna, G., & Dintcheva, N. T. (2021). A review of bioplastics and their adoption in the circular economy. *Polymers*, *13*(8). https://doi.org/10.3390/polym13081229

Di Foggia, M., Corda, U., Plescia, E., Taddei, P., & Torreggiani, A. (2010). Effects of sterilisation by high-energy radiation on biomedical poly-(ε-caprolactone)/hydroxyapatite composites. *Journal of Materials Science: Materials in Medicine*, *21*(6). https://doi.org/10.1007/s10856-010-4046-0

Diyana, Z. N., Jumaidin, R., Selamat, M. Z., Ghazali, I., Julmohammad, N., Huda, N., & Ilyas, R. A. (2021). Physical properties of thermoplastic starch derived from natural resources and its blends: A review. *Polymers*, *13*(9). https://doi.org/10.3390/polym13091396

Donato, R. K., & Mija, A. (2020). Keratin associations with synthetic, biosynthetic and natural polymers: An extensive review. *Polymers*, *12*(1). https://doi.org/10.3390/polym12010032

Dutta, P. K., Tripathi, S., Mehrotra, G. K., & Dutta, J. (2009). Perspectives for chitosan based antimicrobial films in food applications. *Food Chemistry*, *114*(4), 1173–1182. https://doi.org/10.1016/j.foodchem.2008.11.047

Echave, M. C., Burgo, L. S., Pedraz, J. L., & Orive, G. (2017). Gelatin as biomaterial for tissue engineering. *Current Pharmaceutical Design*, *23*(24). https://doi.org/10.2174/0929867324666170511123101

Edgar, K. J., Buchanan, C. M., Debenham, J. S., Rundquist, P. A., Seiler, B. D., Shelton, M. C., & Tindall, D. (2001). Advances in cellulose ester performance and applicaton.pdf. crdownload. *Progress in Polymer Science*, *26*, 1605–1688.

Ekiert, M., Mlyniec, A., & Uhl, T. (2015). The influence of degradation on the viscosity and molecular mass of poly(lactide acid) biopolymer. *Diagnostyka*, *16*(4).

El-Hadi, A. M. (2018). Miscibility of crystalline/amorphous/crystalline biopolymer blends from PLLA/PDLLA/PHB with additives. *Polymer—Plastics Technology and Engineering*, *58*(1). https://doi.org/10.1080/03602559.2018.1455863

El-Hadi, A. M., & Abd Elbary, A. M. (2018). Design of the electrically conductive PHB blends for biomedical applications. *Journal of Materials Science: Materials in Electronics*, *29*(19). https://doi.org/10.1007/s10854-018-9743-3

Elhaes, H., Saleh, N. A., & Ibrahim, M. A. (2018). Molecular modeling applications of some bio-polymers blends as biosensor. *Sensor Letters*, *16*(7). https://doi.org/10.1166/sl.2018.3974

Elzoghby, A. O., Elgohary, M. M., & Kamel, N. M. (2015). Implications of protein- and peptide-based nanoparticles as potential vehicles for anticancer drugs. In *Advances in protein chemistry and structural biology* (Vol. 98). https://doi.org/10.1016/bs.apcsb.2014.12.002

Errico, M. E., Greco, R., Laurienzo, P., Malinconico, M., & Viscardo, D. (2006). Acrylate/EVA reactive blends and semi-IPN: Chemical, chemical-physical, and thermo-optical characterization. *Journal of Applied Polymer Science*, *99*(6), 2926–2935. https://doi.org/10.1002/app.22788

Espinoza, S. M., Patil, H. I., San Martin Martinez, E., Casañas Pimentel, R., & Ige, P. P. (2020). Poly-ε-caprolactone (PCL), a promising polymer for pharmaceutical and biomedical applications: Focus on nanomedicine in cancer. *International Journal of Polymeric Materials and Polymeric Biomaterials*, *69*(2). https://doi.org/10.1080/00914037.2018.1539990

Fabbri, P., & Messori, M. (2017). Surface modification of polymers: Chemical, physical, and biological routes. In *Modification of polymer properties*. Elsevier Inc. https://doi.org/10.1016/B978-0-323-44353-1.00005-1

Fallacara, A., Baldini, E., Manfredini, S., & Vertuani, S. (2018). Hyaluronic acid in the third millennium. *Polymers*, *10*(7). https://doi.org/10.3390/polym10070701

Fan, L., Yang, H., Yang, J., Peng, M., & Hu, J. (2016). Preparation and characterization of chitosan/gelatin/PVA hydrogel for wound dressings. *Carbohydrate Polymers, 146*. https://doi.org/10.1016/j.carbpol.2016.03.002

Favero, J., Belhabib, S., Guessasma, S., Decaen, P., Reguerre, A. L., Lourdin, D., & Leroy, E. (2017). On the representative elementary size concept to evaluate the compatibilisation of a plasticised biopolymer blend. *Carbohydrate Polymers, 172*. https://doi.org/10.1016/j.carbpol.2017.05.018

Ferreira, B. M. P., & Duek, E. A. R. (2005). Pins composed of poly (L-lactic acid)/poly (3-hydroxybutyrate-co-hydroxyvalerate) PLLA/PHBV blends : Degradation in vitro. *Journal of Applied Biomaterials & Biomechanics, 3*(1), 50–60.

Florez, J. P., Fazeli, M., & Simão, R. A. (2019). Preparation and characterization of thermoplastic starch composite reinforced by plasma-treated poly (hydroxybutyrate) PHB. *International Journal of Biological Macromolecules, 123*. https://doi.org/10.1016/j.ijbiomac.2018.11.070

Frone, A. N., Batalu, D., Chiulan, I., Oprea, M., Gabor, A. R., Nicolae, C. A., Raditoiu, V., Trusca, R., & Panaitescu, D. M. (2020). Morpho-structural, thermal and mechanical properties of PLA/PHB/Cellulose biodegradable nanocomposites obtained by compression molding, extrusion, and 3d printing. *Nanomaterials, 10*(1). https://doi.org/10.3390/nano10010051

Fu, J. H., Zhao, M., Lin, Y. R., Tian, X. D., Wang, Y. D., Wang, Z. X., & Wang, L. X. (2017). Degradable chitosan-collagen composites seeded with cells as tissue engineered heart valves. *Heart Lung and Circulation, 26*(1). https://doi.org/10.1016/j.hlc.2016.05.116

Fu, Y., Wu, G., Bian, X., Zeng, J., & Weng, Y. (2020). Biodegradation behavior of poly(butylene adipate-co-terephthalate) (PBAT), poly(lactic acid) (PLA), and their blend in freshwater with sediment. *Molecules, 25*(17). https://doi.org/10.3390/molecules25173946

Fuchs, S., Hartmann, J., Mazur, P., Reschke, V., Siemens, H., Wehlage, D., & Ehrmann, A. (2017). Electrospinning of biopolymers and biopolymer blends. *Journal of Chemical and Pharmaceutical Sciences, 2017*(Special 10).

Galiano, F. (2014). Casting solution. In *Encyclopedia of membranes* (Issue 2009, p. 1). https://doi.org/10.1007/978-3-642-40872-4

Garavand, F., Khodaei, D., Mahmud, N., Islam, J., Khan, I., Jafarzadeh, S., Tahergorabi, R., & Cacciotti, I. (2022). Recent progress in using zein nanoparticles-loaded nanocomposites for food packaging applications. *Critical Reviews in Food Science and Nutrition*, 1–21. https://doi.org/10.1080/10408398.2022.2133080

Gardella, L., Calabrese, M., & Monticelli, O. (2014). PLA maleation: An easy and effective method to modify the properties of PLA/PCL immiscible blends. *Colloid and Polymer Science, 292*(9), 2391–2398. https://doi.org/10.1007/s00396-014-3328-3

George, S., Kumari, P., & Panicker, U. G. (2010). Influence of static and dynamic crosslinking techniques on the transport properties of ethylene propylene diene monomer rubber/poly (ethylene-co-vinyl acetate) blends. *Journal of Polymer Research, 17*(2), 161–169. https://doi.org/10.1007/s10965-009-9302-y

Ghalia, M. A., & Dahman, Y. (2016). Advanced nanobiomaterials in tissue engineering: Synthesis, properties, and applications. In *Nanobiomaterials in soft tissue engineering: Applications of nanobiomaterials*. Elsevier Inc. https://doi.org/10.1016/B978-0-323-42865-1.00006-4

Ghimici, L., & Nichifor, M. (2018). Dextran derivatives application as flocculants. *Carbohydrate Polymers, 190*. https://doi.org/10.1016/j.carbpol.2018.02.075

Giles, H. F., Wagner, J. R., & Mount, E. M. (2005). Extrusion process. In *Extrusion* (pp. 1–8). https://doi.org/10.1016/B978-0-8155-1473-2.50002-7

Gorgieva, S., & Trček, J. (2019). Bacterial cellulose: Production, modification and perspectives in biomedical applications. *Nanomaterials, 9*(10). https://doi.org/10.3390/nano9101352

Gowthaman, N. S. K., Lim, H. N., Sreeraj, T. R., Amalraj, A., & Gopi, S. (2021). Advantages of biopolymers over synthetic polymers. In *Biopolymers and their industrial applications* (pp. 351–372). Elsevier. https://doi.org/10.1016/b978-0-12-819240-5.00015-8

Grabska, S., & Sionkowska, A. (2019). Biopolymer films based on the blends of silk fibroin and collagen for applications in hair care cosmetics. *Science and Technology of Polymers and Advanced Materials.* https://doi.org/10.1201/9780429425301-30

Graça, M. F. P., Miguel, S. P., Cabral, C. S. D., & Correia, I. J. (2020). Hyaluronic acid—Based wound dressings: A review. *Carbohydrate Polymers, 241.* https://doi.org/10.1016/j.carbpol.2020.116364

Graebling, D., & Bataille, P. (1994). Polypropylene/polyhydroxybutyrate blends: Preparation of a grafted copolymer and its use as surface-active agent. *Polymer-Plastics Technology and Engineering, 33*(3), 341–356. https://doi.org/10.1080/03602559408013097

Groot, W. J., & Borén, T. (2010). Life cycle assessment of the manufacture of lactide and PLA biopolymers from sugarcane in Thailand. *International Journal of Life Cycle Assessment, 15*(9). https://doi.org/10.1007/s11367-010-0225-y

Grothe, J., Kaskel, S., & Leuteritz, A. (2012). Nanocomposites and hybrid materials. In *Polymer science: A comprehensive reference, 10 volume set* (Vol. 8, pp. 177–209). Elsevier B.V. https://doi.org/10.1016/B978-0-444-53349-4.00206-5

Gumede, T. P., Luyt, A. S., & Müller, A. J. (2018). Review on PCL, PBS, AND PCL/PBS blends containing carbon nanotubes. *Express Polymer Letters, 12*(6). https://doi.org/10.3144/expresspolymlett.2018.43

Gunatillake, P. A., & Adhikari, R. (2003). Biodegradable synthetic polymers for tissue engineering. *European Cells and Materials, 5*, 1–16. https://doi.org/10.22203/eCM.v005a01

Guo, Y., Liang, K., & Ji, Y. (2019). New degradable composite elastomers of POC/PCL fabricated via in-situ copolymerization blending strategy. *European Polymer Journal, 110*(November 2018), 337–343. https://doi.org/10.1016/j.eurpolymj.2018.11.048

Gupta, P., & Nayak, K. K. (2015). Compatibility study of alginate/keratin blend for biopolymer development. *Journal of Applied Biomaterials and Functional Materials, 13*(4). https://doi.org/10.5301/jabfm.5000242

Hadi, M. A., Raghavendra Rao, N. G., Srinivasa Rao, A., Shiva, G., & Akram, J. W. (2013). Impact of capsules as a carrier for multiple unit drug delivery and the importance of HPMC capsules. *Research Journal of Pharmacy and Technology, 6*(1), 34–43.

Haghighi, H., Gullo, M., La China, S., Pfeifer, F., Siesler, H. W., Licciardello, F., & Pulvirenti, A. (2021). Characterization of bio-nanocomposite films based on gelatin/polyvinyl alcohol blend reinforced with bacterial cellulose nanowhiskers for food packaging applications. *Food Hydrocolloids, 113.* https://doi.org/10.1016/j.foodhyd.2020.106454

Halley, P., Rutgers, R., Coombs, S., Kettels, J., Gralton, J., Christie, G., Jenkins, M., Beh, H., Griffin, K., Jayasekara, R., & Lonergan, G. (2001). Developing biodegradable mulch films from starch-based polymers. *Starch/Staerke, 53*(8). https://doi.org/10.1002/1521-379X(200108)53:8<362::AID-STAR362>3.0.CO;2-J

Hamad, K., Kaseem, M., Deri, F., & Ko, Y. G. (2016). Mechanical properties and compatibility of polylactic acid/polystyrene polymer blend. *Materials Letters, 164*, 409–412. https://doi.org/10.1016/j.matlet.2015.11.029

Han, Y., Li, K., Chen, H., & Li, J. (2017). Properties of soy protein isolate biopolymer film modified by graphene. *Polymers, 9*(8). https://doi.org/10.3390/polym9080312

Harting, R., Johnston, K., & Petersen, S. (2019). Correlating in vitro degradation and drug release kinetics of biopolymer-based drug delivery systems. *International Journal of Biobased Plastics, 1*(1). https://doi.org/10.1080/24759651.2018.1563358

Hasan, A., Waibhaw, G., Saxena, V., & Pandey, L. M. (2018). Nano-biocomposite scaffolds of chitosan, carboxymethyl cellulose and silver nanoparticle modified cellulose nanowhiskers for bone tissue engineering applications. *International Journal of Biological Macromolecules*, *111*. https://doi.org/10.1016/j.ijbiomac.2018.01.089

Hassan, M. E., Bai, J., & Dou, D. Q. (2019). Biopolymers—definition, classification and applications. *Egyptian Journal of Chemistry*, *62*(9). https://doi.org/10.21608/EJCHEM.2019.6967.1580

Haywood, A., & Glass, B. D. (2011). Pharmaceutical excipients—where do we begin? *Australian Prescriber*, *34*(4), 112–114. https://doi.org/10.18773/austprescr.2011.060

Heeres, H. J., Maastrigt, F. van, & Picchioni, F. (2013). Polymeric blends with biopolymers. In *Handbook of biopolymer-based materials: From blends and composites to gels and complex networks*. Wiley-VCH (pp. 143–172).

Hisano, J., Goto, A., & Okajima, K. (1991). Edible body and process for preparation thereof. *US Patent 4994285*.

Hojnik, J., Ruzzier, M., & Ruzzier, M. K. (2019). Transition towards sustainability: Adoption of eco-products among consumers. *Sustainability (Switzerland)*, *11*(16). https://doi.org/10.3390/su11164308

Huang, J., Du, J., Cevher, Z., Ren, Y., Wu, X., & Chu, Y. (2017). Printable and flexible phototransistors based on blend of organic semiconductor and biopolymer. *Advanced Functional Materials*, *27*(9). https://doi.org/10.1002/adfm.201604163

Huda, M. S., Drzal, L. T., Mohanty, A. K., & Misra, M. (2006). Chopped glass and recycled newspaper as reinforcement fibers in injection molded poly(lactic acid) (PLA) composites: A comparative study. *Composites Science and Technology*, *66*(11–12), 1813–1824. https://doi.org/10.1016/j.compscitech.2005.10.015

Hwang, S. Y., Yoo, E. S., & Im, S. S. (2012). The synthesis of copolymers, blends and composites based on poly(butylene succinate). *Polymer Journal*, *44*(12). https://doi.org/10.1038/pj.2012.157

Ibrahim, B. A., & Kadum, K. M. (2010). Influence of polymer blending on mechanical and thermal properties. *Modern Applied Science*, *4*(9), 157–161. https://doi.org/10.5539/mas.v4n9p157

Imre, B., & Pukánszky, B. (2013). Compatibilization in bio-based and biodegradable polymer blends. *European Polymer Journal*, *49*(6). https://doi.org/10.1016/j.eurpolymj.2013.01.019

Iskandar, M. A., Yahya, E. B., Abdul Khalil, H. P. S., Rahman, A. A., & Ismail, M. A. (2022). Recent progress in modification strategies of nanocellulose-based aerogels for oil absorption application. *Polymers*, *14*(5). https://doi.org/10.3390/polym14050849

Istiqola, A., & Syafiuddin, A. (2020). A review of silver nanoparticles in food packaging technologies: Regulation, methods, properties, migration, and future challenges. *Journal of the Chinese Chemical Society*, *67*(11). https://doi.org/10.1002/jccs.202000179

Jang, W. Y., Shin, B. Y., Lee, T. J., & Narayan, R. (2007). Thermal properties and morphology of biodegradable PLA/starch compatibilized blends. *Journal of Industrial and Engineering Chemistry*, *13*(3), 457–464.

Jariyasakoolroj, P., & Chirachanchai, S. (2014). Silane modified starch for compatible reactive blend with poly(lactic acid). *Carbohydrate Polymers*, *106*(1), 255–263. https://doi.org/10.1016/j.carbpol.2014.02.018

Jian, J., Xiangbin, Z., & Xianbo, H. (2020). An overview on synthesis, properties and applications of poly(butylene-adipate-co-terephthalate)—PBAT. *Advanced Industrial and Engineering Polymer Research*, *3*(1). https://doi.org/10.1016/j.aiepr.2020.01.001

Jiang, C., Wang, X., Wang, G., Hao, C., Li, X., & Li, T. (2019). Adsorption performance of a polysaccharide composite hydrogel based on crosslinked glucan/chitosan for heavy metal ions. *Composites Part B: Engineering*, *169*. https://doi.org/10.1016/j.compositesb.2019.03.082

Jiang, L., Wolcott, M. P., & Zhang, J. (2006). Study of biodegradable polylactide/poly(butylene adipate-co-terephthalate) blends. *Biomacromolecules*, *7*(1), 199–207. https://doi.org/10.1021/bm050581q

Jiménez-Ocampo, R., Valencia-Salazar, S., Pinzón-Díaz, C. E., Herrera-Torres, E., Aguilar-Pérez, C. F., Arango, J., & Ku-Vera, J. C. (2019). The role of chitosan as a possible agent for enteric methane mitigation in ruminants. *Animals*, *9*(11). https://doi.org/10.3390/ani9110942

John, M. J. (2015). Biopolymer blends based on polylactic acid and polyhydroxy butyrate-co-valerate: Effect of clay on mechanical and thermal properties. *Polymer Composites*, *36*(11), 2042–2050. https://doi.org/10.1002/pc

Jummaat, F., Yahya, E. B., Abdul Khalil, H. P. S., Adnan, A. S., Alqadhi, A. M., Abdullah, C. K., A. K., A. S., Olaiya, N. G., & Abdat, M. (2021). The role of biopolymer-based materials in obstetrics and gynecology applications: A review. *Polymers*, *13*(4), 633. https://doi.org/10.3390/polym13040633

Jun, C. L. (2000). Reactive blending of biodegradable polymers: PLA and starch. *Journal of Polymers and the Environment*, *8*(1), 33–37. https://doi.org/10.1023/A:1010172112118

Jung, J., Kasi, G., & Seo, J. (2018). Development of functional antimicrobial papers using chitosan/starch-silver nanoparticles. *International Journal of Biological Macromolecules*, *112*. https://doi.org/10.1016/j.ijbiomac.2018.01.155

Kamarudin, S. H., Rayung, M., Abu, F., Ahmad, S. B., Fadil, F., Karim, A. A., Norizan, M. N., Sarifuddin, N., Mat Desa, M. S. Z., Mohd Basri, M. S., & Abdullah, L. C. (2022). A review on antimicrobial packaging from biodegradable polymer composites. *Polymers*, *14*(1), 174.

Kanatt, S. R., Rao, M. S., Chawla, S. P., & Sharma, A. (2012). Active chitosan-polyvinyl alcohol films with natural extracts. *Food Hydrocolloids*, *29*(2), 290–297. https://doi.org/10.1016/j.foodhyd.2012.03.005

Kango, S., Kalia, S., Thakur, P., & Kumari, B. (2014). Semiconductor—polymer hybrid materials. *Advance in Polymer Science*, *267*, 283–312. https://doi.org/10.1007/12

Karak, N. (2012). Fundamentals of polymers. In *Vegetable oil-based polymers* (pp. 1–30). https://doi.org/10.1533/9780857097149.1

Kargarzadeh, H., Ahmad, I., Thomas, S., & Dufresne, A. (2017). *Handbook of nanocellulose and cellulose nanocomposites*. John Wiley & Sons.

Kaur, P., Frindy, S., Park, Y., Sillanpää, M., & Imteaz, M. A. (2020). Synthesis of graphene-based biopolymer tio2 electrodes using pyrolytic direct deposition method and its catalytic performance. *Catalysts*, *10*(9). https://doi.org/10.3390/catal10091050

Khan, I., Mansha, M., & Jafar Mazumder, M. A. (2019). Polymer blends. In *Functional polymers* (pp. 513–549). https://doi.org/10.1007/978-3-319-95987-0

Killheffer, D. H. (1962). *Banbury the master mixer*. Palmerton Publishing.

Koshy, T. M., Gowda, D. V, Tom, S., Karunakar, G., Srivastava, A., & Moin, A. (2016). Polymer grafting: An overview. *American Journal of Pharmaceutical Research*, *6*(2), 1–13.

Kosik-Kozioł, A., Costantini, M., Bolek, T., Szöke, K., Barbetta, A., Brinchmann, J., & Święszkowski, W. (2017). PLA short sub-micron fiber reinforcement of 3D bioprinted alginate constructs for cartilage regeneration. *Biofabrication*, *9*(4). https://doi.org/10.1088/1758-5090/aa90d7

Kristo, E., & Biliaderis, C. G. (2006). Water sorption and thermo-mechanical properties of water/sorbitol-plasticized composite biopolymer films: Caseinate-pullulan bilayers and blends. *Food Hydrocolloids*, *20*(7). https://doi.org/10.1016/j.foodhyd.2005.11.008

Kubicki, J. D., Yang, H., Sawada, D., O'Neill, H., Oehme, D., & Cosgrove, D. (2018). The shape of native plant cellulose microfibrils. *Scientific Reports*, *8*(1). https://doi.org/10.1038/s41598-018-32211-w

Kuhns, R. J., & Shaw, G. H. (2018). Peak oil and petroleum energy resources. In *Navigating the energy maze* (pp. 53–63). Springer International Publishing. https://doi.org/10.1007/978-3-319-22783-2_7

Kulig, D., Zimoch-Korzycka, A., Jarmoluk, A., & Marycz, K. (2016). Study on alginate-chitosan complex formed with different polymers ratio. *Polymers, 8*(5), 1–17. https://doi.org/10.3390/polym8050167

Kumar, A., Rao, K. M., & Han, S. S. (2017). Development of sodium alginate-xanthan gum based nanocomposite scaffolds reinforced with cellulose nanocrystals and halloysite nanotubes. *Polymer Testing, 63*. https://doi.org/10.1016/j.polymertesting.2017.08.030

Kumar, M., Mohanty, S., Nayak, S. K., & Rahail Parvaiz, M. (2010). Effect of glycidyl methacrylate (GMA) on the thermal, mechanical and morphological property of biodegradable PLA/PBAT blend and its nanocomposites. *Bioresource Technology, 101*(21), 8406–8415. https://doi.org/10.1016/j.biortech.2010.05.075

Lamnawar, K., Maazouz, A., Cabrera, G., & Al-Itry, R. (2018). Interfacial tension properties in biopolymer blends: From deformed drop retraction method (DDRM) to shear and elongation rheology-application to blown film extrusion. *International Polymer Processing, 33*(3). https://doi.org/10.3139/217.3614

Lebreton, L., & Andrady, A. (2019). Future scenarios of global plastic waste generation and disposal. *Palgrave Communications, 5*(1). https://doi.org/10.1057/s41599-018-0212-7

Lee, E. J., Kang, E., Kang, S. W., & Huh, K. M. (2020). Thermo-irreversible glycol chitosan/hyaluronic acid blend hydrogel for injectable tissue engineering. *Carbohydrate Polymers, 244*. https://doi.org/10.1016/j.carbpol.2020.116432

Lee, S.-S., Kim, J., Park, M., Lim, S., & Choe, C. R. (2001). Transesterification reaction of the BaSO4-Filled PBT/poly(ethylene terephthalate) blend. *Journal of Polymer Science Part B: Polymer Physics, 39*(21), 2589–2597. https://doi.org/10.1002/polb.0000

Li, L., Hui Zhou, C., & Ping Xu, Z. (2019). Self-nanoemulsifying drug-delivery system and solidified self-nanoemulsifying drug-delivery system. *Nanocarriers for Drug Delivery*, 421–449. https://doi.org/10.1016/b978-0-12-814033-8.00014-x

Li, Z., Ramay, H. R., Hauch, K. D., Xiao, D., & Zhang, M. (2005). Chitosan-alginate hybrid scaffolds for bone tissue engineering. *Biomaterials, 26*(18), 3919–3928. https://doi.org/10.1016/j.biomaterials.2004.09.062

Lima, D. B., Almeida, R. D., Pasquali, M., Borges, S. P., Fook, M. L., & Lisboa, H. M. (2018). Physical characterization and modeling of chitosan/peg blends for injectable scaffolds. *Carbohydrate Polymers, 189*. https://doi.org/10.1016/j.carbpol.2018.02.045

Limwong, V., Sutanthavidul, N., & Kulvanich, P. (2004). Spherical composite particles of rice starch and microcrystalline cellulose: A new coprocessed excipient for direct compression. *Aaps Pharmscitech, 5*(2), 40–49. https://doi.org/10.1208/pt050230

Lin, Y. H., Huang, K. W., Chen, S. Y., Cheng, N. C., & Yu, J. (2017). Keratin/chitosan UV-crosslinked composites promote the osteogenic differentiation of human adipose derived stem cells. *Journal of Materials Chemistry B, 5*(24). https://doi.org/10.1039/c7tb00188f

Lins, L. C., Livi, S., Duchet-Rumeau, J., & Gérard, J.-F. (2015). Phosphonium ionic liquids as new compatibilizing agents of biopolymer blends composed of poly(butylene-adipate-co-terephtalate)/poly(lactic acid) (PBAT/PLA). *RSC Advances, 5*(73). https://doi.org/10.1039/c5ra10241c

Liu, X., Zheng, C., Luo, X., Wang, X., & Jiang, H. (2019). Recent advances of collagen-based biomaterials: Multi-hierarchical structure, modification and biomedical applications. *Materials Science and Engineering C, 99*. https://doi.org/10.1016/j.msec.2019.02.070

Liu, Y., Wang, D., Sun, Z., Liu, F., Du, L., & Wang, D. (2021). Preparation and characterization of gelatin/chitosan/3-phenylacetic acid food-packaging nanofiber antibacterial films by electrospinning. *International Journal of Biological Macromolecules, 169.* https://doi.org/10.1016/j.ijbiomac.2020.12.046

Lorenzo, G., Sosa, M., & Califano, A. (2018). Alternative proteins and pseudocereals in the development of gluten-free pasta. In *Alternative and replacement foods* (Vol. 17, pp. 433–458). Elsevier. https://doi.org/10.1016/B978-0-12-811446-9.00015-0

Luyt, A. S., & Malik, S. S. (2018). Can biodegradable plastics solve plastic solid waste accumulation? In *Plastics to energy: Fuel, chemicals, and sustainability implications.* https://doi.org/10.1016/B978-0-12-813140-4.00016-9

Ma, C. C. M., Chen, Y. J., & Kuan, H. C. (2006). Polystyrene nanocomposite materials—Preparation, mechanical, electrical and thermal properties, and morphology. *Journal of Applied Polymer Science, 100*(1), 508–515. https://doi.org/10.1002/app.23221

Ma, H., Shen, J., Cao, J., Wang, D., Yue, B., Mao, Z., Wu, W., & Zhang, H. (2017). Fabrication of wool keratin/polyethylene oxide nano-membrane from wool fabric waste. *Journal of Cleaner Production, 161.* https://doi.org/10.1016/j.jclepro.2017.05.121

Ma, Y., Li, L., & Wang, Y. (2017). Development of antimicrobial active film containing CINnamaldehyde and its application to snakehead (Ophiocephalus argus) fish. *Journal of Food Process Engineering, 40*(5), 1–8. https://doi.org/10.1111/jfpe.12554

Macknight, W. J., & Karasz, F. E. (1989). Polymer blends. In *Comprehensive polymer science and supplements* (pp. 111–130). https://doi.org/10.1002/masy.19880150115

Malinowski, R. (2016). Mechanical properties of PLA/PCL blends crosslinked by electron beam and TAIC additive. *Chemical Physics Letters, 662,* 91–96. https://doi.org/10.1016/j.cplett.2016.09.022

Mamatha, B., Srilatha, D., Sivanarayani, C. H., Kumar Desu, P., & Venkateswara Rao, P. (2017). Co-processed excipients: An overview. *World Journal of Pharmaceutical Research, 6*(15), 224–237. https://doi.org/10.20959/wjpr201715-10078

Markovic, G., & Visakh, P. M. (2017). Polymer blends: State of art. In *Recent developments in polymer macro, micro and nano blends: Preparation and characterisation* (pp. 1–15). Elsevier Ltd. https://doi.org/10.1016/B978-0-08-100408-1.00001-7

Martínez-Romo, A., González-Mota, R., Soto-Bernal, J. J., & Rosales-Candelas, I. (2015). Investigating the degradability of HDPE, LDPE, PE-BIO, and PE-OXO films under UV-B radiation. *Journal of Spectroscopy, 2015.* https://doi.org/10.1155/2015/586514

Mathew, A. P., & Oksman, K. (2014). Processing of bionanocomposites: Solution casting. In *Handbook of green materials* (pp. 35–52). https://doi.org/10.1142/9789814566469_0018

Matta, A. K., Rao, R. U., Suman, K. N. S., & Rambabu, V. (2014). Preparation and characterization of biodegradable PLA/PCL polymeric blends. *Procedia Materials Science, 6,* 1266–1270. https://doi.org/10.1016/j.mspro.2014.07.201

McAdam, B., Fournet, M. B., McDonald, P., & Mojicevic, M. (2020a). Production of polyhydroxybutyrate (PHB) and factors impacting its chemical and mechanical characteristics. *Polymers, 12*(12). https://doi.org/10.3390/polym12122908

McAdam, B., Fournet, M. B., McDonald, P., & Mojicevic, M. (2020b). Production of polyhydroxybutyrate (PHB) and factors impacting its chemical and mechanical characteristics. *Polymers, 12*(12), 1–20. https://doi.org/10.3390/polym12122908

Mehrotra, P. K. (2014). Powder processing and green shaping. In *Comprehensive hard materials* (Vol. 1). Elsevier Ltd. https://doi.org/10.1016/B978-0-08-096527-7.00007-6

Mincheva, R., & Raquez, J.-M. (2019). The surface of polymers. In *Surface modification of polymers: Methods and applications* (pp. 1–30). https://doi.org/10.1002/9783527819249.ch1

Minko, S. (2008). Grafting on solid surfaces:"grafting to" and "grafting from" methods. In *Polymer surfaces and interfaces* (pp. 215–234). https://doi.org/https://doi.org/10.1007/978-3-540-73865-7_11

Mitra, S. K., & Saha, A. A. (2013). Encyclopedia of microfluidics and nanofluidics. In *Encyclopedia of microfluidics and nanofluidics*. https://doi.org/10.1007/978-3-642-27758-0

Mittal, H., Al Alili, A., Morajkar, P. P., & Alhassan, S. M. (2021). Graphene oxide cross-linked hydrogel nanocomposites of xanthan gum for the adsorption of crystal violet dye. *Journal of Molecular Liquids*, *323*. https://doi.org/10.1016/j.molliq.2020.115034

Miyamoto, H., Yamane, C., Seguchi, M., & Okajima, K. (2009). Structure and properties of cellulose–starch blend films regenerated from aqueous sodium hydroxide solution. *Food Science and Technology Research*, *15*(4), 403–412. https://doi.org/10.3136/fstr.15.403

Mo, X., Sun, B., Wu, T., & Li, D. (2018). Electrospun nanofibers for tissue engineering. In *Electrospinning: Nanofabrication and applications* (pp. 719–734). Elsevier Inc. https://doi.org/10.1016/B978-0-323-51270-1.00024-8

Mofokeng, J. P., & Luyt, A. S. (2015). Dynamic mechanical properties of PLA/PHBV, PLA/PCL, PHBV/PCL blends and their nanocomposites with TiO2 as nanofiller. *Thermochimica Acta*, *613*, 41–53. https://doi.org/10.1016/j.tca.2015.05.019

Mohamed, S. A. A., El-Sakhawy, M., & El-Sakhawy, M. A. M. (2020). Polysaccharides, protein and lipid-based natural edible films in food packaging: A review. *Carbohydrate Polymers*, *238*. https://doi.org/10.1016/j.carbpol.2020.116178

Mohandas, S. P., Balan, L., Lekshmi, N., Cubelio, S. S., Philip, R., & Bright Singh, I. S. (2017). Production and characterization of polyhydroxybutyrate from Vibrio harveyi MCCB 284 utilizing glycerol as carbon source. *Journal of Applied Microbiology*, *122*(3). https://doi.org/10.1111/jam.13359

Mohanty, A. K., Misra, M., & Hinrichsen, G. (2000). Biofibres, biodegradable polymers and biocomposites: An overview. *Macromolecular Materials and Engineering*, *276–277*, 1–24. https://doi.org/10.1002/(SICI)1439-2054(20000301)276:1<1::AID-MAME1>3.0.CO;2-W

Mohsin, A., Zaman, W. Q., Guo, M., Ahmed, W., Khan, I. M., Niazi, S., Rehman, A., Hang, H., & Zhuang, Y. (2020). Xanthan-Curdlan nexus for synthesizing edible food packaging films. *International Journal of Biological Macromolecules*, *162*. https://doi.org/10.1016/j.ijbiomac.2020.06.008

Morris, B. A. (2016). Polymer blending for packaging applications. In *Multilayer flexible packaging* (2nd ed., pp. 173–204). Elsevier Inc. https://doi.org/10.1016/B978-0-323-37100-1.00013-2

Nachaegari, S. K., & Bansal, A. K. (2004). Coprocessed excipients for solid dosage forms. *Pharmaceutical Technology*, *28*(1), 52–64.

Namazi, H. (2017). Polymers in our daily life. *BioImpacts*, *7*(2), 73–74. https://doi.org/10.15171/bi.2017.09

Nanda, M. R., Misra, M., & Mohanty, A. K. (2011). The effects of process engineering on the performance of PLA and PHBV blends. *Macromolecular Materials and Engineering*, *296*(8), 719–728. https://doi.org/10.1002/mame.201000417

Naqash, F., Masoodi, F. A., Rather, S. A., Wani, S. M., & Gani, A. (2017). Emerging concepts in the nutraceutical and functional properties of pectin: A review. *Carbohydrate Polymers*, *168*. https://doi.org/10.1016/j.carbpol.2017.03.058

Nazarzadeh Zare, E., Makvandi, P., Borzacchiello, A., Tay, F. R., Ashtari, B., & Padil, V. T. V. (2019). Antimicrobial gum bio-based nanocomposites and their industrial and biomedical applications. *Chemical Communications*, *55*(99). https://doi.org/10.1039/c9cc08207g

Nemani, S. K., Annavarapu, R. K., Mohammadian, B., Raiyan, A., Heil, J., Haque, M. A., Abdelaal, A., & Sojoudi, H. (2018). Surface modification of polymers: Methods and applications. *Advanced Materials Interfaces*, *5*(24), 1–26. https://doi.org/10.1002/admi.201801247

Ng, E. L., Huerta Lwanga, E., Eldridge, S. M., Johnston, P., Hu, H. W., Geissen, V., & Chen, D. (2018). An overview of microplastic and nanoplastic pollution in agroecosystems. *Science of the Total Environment, 627.* https://doi.org/10.1016/j.scitotenv.2018.01.341

Niaounakis, M. (2015a). Biopolymers: Applications and trends. In *Biopolymers: Applications and trends.* https://doi.org/10.1016/c2014-0-00936-7

Niaounakis, M. (2015b). Blending. In *Biopolymers: Processing and products* (pp. 117–185). https://doi.org/10.1016/B978-0-323-26698-7.00003-9

Ninan, N., Muthiah, M., Park, I. K., Wong, T. W., Thomas, S., & Grohens, Y. (2015). Natural polymer/inorganic material based hybrid scaffolds for skin wound healing. *Polymer Reviews, 55*(3). https://doi.org/10.1080/15583724.2015.1019135

Niranjana Prabhu, T., & Prashantha, K. (2018). A review on present status and future challenges of starch based polymer films and their composites in food packaging applications. *Polymer Composites, 39*(7), 2499–2522. https://doi.org/10.1002/pc.24236

Noorbakhsh-Soltani, S. M., Zerafat, M. M., & Sabbaghi, S. (2018). A comparative study of gelatin and starch-based nano-composite films modified by nano-cellulose and chitosan for food packaging applications. *Carbohydrate Polymers, 189.* https://doi.org/10.1016/j.carbpol.2018.02.012

Nor Amira Izzati, A., John, W. C., Nurul Fazita, M. R., Najieha, N., Azniwati, A. A., & Abdul Khalil, H. P. S. (2020). Effect of empty fruit bunches microcrystalline cellulose (MCC) on the thermal, mechanical and morphological properties of biodegradable poly (lactic acid) (PLA) and polybutylene adipate terephthalate (PBAT) composites. *Materials Research Express, 7*(1), 015336. https://doi.org/10.1088/2053-1591/ab6889

Olaiya, N. G., Nuryawan, A., Oke, P. K., Abdul Khalil, H. P. S. , Rizal, S., Mogaji, P. B., Sadiku, E. R., Suprakas, S. R., Farayibi, P. K., Ojijo, V., & Paridah, M. T. (2020). The role of two-step blending in the properties of starch/chitin/polylactic acid biodegradable composites for biomedical applications. *Polymers, 12*(3). https://doi.org/10.3390/polym12030592

Olaiya, N. G., Surya, I., Oke, P. K., Rizal, S., Sadiku, E. R., Ray, S. S., Farayibi, P. K., Hossain, M. S., & Abdul Khalil, H. P. S. (2019). Properties and characterization of a PLA-chitin-starch biodegradable polymer composite. *Polymers, 11*(10). https://doi.org/10.3390/polym11101656

Omar, A., Ezzat, H., Elhaes, H., & Ibrahim, M. A. (2021). Molecular modeling analyses for modified biopolymers. *Biointerface Research in Applied Chemistry, 11*(1). https://doi.org/10.33263/BRIAC111.78477859

Otey, F. H., Westhoff, R. P., & Russell, C. R. (1977). Biodegradable films from starch and ethylene-acrylic acid copolymer. *Industrial and Engineering Chemistry Product Research and Development, 16*(4). https://doi.org/10.1021/i360064a009

Ouchen, F., Kim, S. N., Hay, M., Zate, H., Subramanyam, G., Grote, J. G., Bartsch, C. M., & Naik, R. R. (2008). DNA-conductive polymer blends for applications in biopolymer-based field effect transistors (FETs). *Nanobiosystems: Processing, Characterization, and Applications, 7040.* https://doi.org/10.1117/12.801358

Ozdemir, M., Yurteri, C. U., & Sadikoglu, H. (1999). Physical polymer surface modification methods and applications in food packaging polymers. *Critical Reviews in Food Science and Nutrition, 39*(5), 457–477. https://doi.org/10.1080/10408699991279240

Parameswaranpillai, J., Thomas, S., & Grohens, Y. (2014). Polymer blends: State of the art, new challenges, and opportunities. In *Characterization of polymer blends: Miscibility, morphology and interfaces* (pp. 1–6). https://doi.org/10.1002/9783527645602.ch01

Park, K., Ju, Y. M., Son, J. S., Ahn, K. D., & Han, D. K. (2007). Surface modification of biodegradable electrospun nanofiber scaffolds and their interaction with fibroblasts. *Journal of Biomaterials Science, Polymer Edition, 18*(4), 369–382. https://doi.org/10.1163/156856207780424997

Patel, H., & Gohel, M. (2016). A review on development of multifunctional co-processed excipient. *Journal of Critical Reviews*, *3*(2), 48–54.

Pathak, S., Sneha, C., & Mathew, B. B. (2014). Bioplastics: Its timeline based scenario & challenges. *Journal of Polymer and Biopolymer Physics Chemistry*, *2*(4).

Patil, H., Tiwari, R. V., & Repka, M. A. (2016). Hot-melt extrusion: From theory to application in pharmaceutical formulation. *AAPS PharmSciTech*, *17*(1), 20–42. https://doi.org/10.1208/s12249-015-0360-7

Patni, N., Yadava, P., Agarwal, A., & Maroo, V. (2011). Study on wheat gluten biopolymer: A novel way to eradicate plastic waste. *Indian Journal of Applied Research*, *3*(8), 253–255. https://doi.org/10.15373/2249555x/aug2013/81

Pawar, S. N., & Edgar, K. J. (2012). Alginate derivatization: A review of chemistry, properties and applications. *Biomaterials*, *33*(11), 3279–3305. https://doi.org/10.1016/j.biomaterials.2012.01.007

Pedroza-Islas, R., Alvarez-Ramírez, J., & Vernon-Carter, E. J. (2000). Using biopolymer blends for shrimp feedstuff microencapsulation II: Dissolution and floatability kinetics as selection criteria. *Food Research International*, *33*(2). https://doi.org/10.1016/S0963-9969(00)00015-6

Pérez-Alonso, C., Báez-González, J. G., Beristain, C. I., Vernon-Carter, E. J., & Vizcarra-Mendoza, M. G. (2003). Estimation of the activation energy of carbohydrate polymers blends as selection criteria for their use as wall material for spray-dried microcapsules. *Carbohydrate Polymers*, *53*(2). https://doi.org/10.1016/S0144-8617(03)00052-3

Phillip, S., Keshavarz, T., & Roy, I. (2007). Polyhydroxyalkanoates: Biodegradable polymers with a range of applications. *Journal of Chemical Technology & Biotechnology*, *82*(3), 233–247. https://doi.org/10.1002/jctb

Pietrosanto, A., Scarfato, P., Di Maio, L., Nobile, M. R., & Incarnato, L. (2020). Evaluation of the suitability of poly(lactide)/poly(butylene-adipate-co-terephthalate) blown films for chilled and frozen food packaging applications. *Polymers*, *12*(4), 1–18. https://doi.org/10.3390/POLYM12040804

Prasad, K., Mehta, G., Meena, R., & Siddhanta, A. K. (2006). Hydrogel-forming agar-graft-PVP and κ-carrageenangraft-PVP Blends: Rapid synthesis and characterization. *Journal of Applied Polymer Science*, *102*(4). https://doi.org/10.1002/app.24145

Princi, E. (2011). *Handbook of polymers in paper conservation*. Smithers Rapra.

Priyadarshi, R., Kim, S. M., & Rhim, J. W. (2021). Pectin/pullulan blend films for food packaging: Effect of blending ratio. *Food Chemistry*, *347*. https://doi.org/10.1016/j.foodchem.2021.129022

Priyan Shanura Fernando, I., Kim, K. N., Kim, D., & Jeon, Y. J. (2019). Algal polysaccharides: Potential bioactive substances for cosmeceutical applications. *Critical Reviews in Biotechnology*, *39*(1). https://doi.org/10.1080/07388551.2018.1503995

Puchalski, M., Szparaga, G., Biela, T., Gutowska, A., Sztajnowski, S., & Krucińska, I. (2018). Molecular and supramolecular changes in polybutylene succinate (PBS) and polybutylene succinate adipate (PBSA) copolymer during degradation in various environmental conditions. *Polymers*, *10*(3). https://doi.org/10.3390/polym10030251

Pukánszky, B., & Tüdõs, F. (1990). Miscibility and mechanical properties of polymer blends. *Makromolekulare Chemie. Macromolecular Symposia*, *38*(1), 221–231.

Qiao, C., Ma, X., Zhang, J., & Yao, J. (2017). Molecular interactions in gelatin/chitosan composite films. *Food Chemistry*, *235*. https://doi.org/10.1016/j.foodchem.2017.05.045

Qin, Y. (2016). Applications of advanced technologies in the development of functional medical textile materials. *Medical Textile Materials*, 55–70. https://doi.org/10.1016/b978-0-08-100618-4.00005-4

Quiles-Carrillo, L., Montanes, N., Jorda-Vilaplana, A., Balart, R., & Torres-Giner, S. (2019). A comparative study on the effect of different reactive compatibilizers on injection-molded pieces of bio-based high-density polyethylene/polylactide blends. *Journal of Applied Polymer Science, 136*(16). https://doi.org/10.1002/app.47396

Rahar, S., Swami, G., Nagpal, N., Nagpal, M., & Singh, G. (2011). Preparation, characterization, and biological properties of β-glucans. *Journal of Advanced Pharmaceutical Technology & Research, 2*(2). https://doi.org/10.4103/2231-4040.82953

Rajeshkumar, L. (2021). Biodegradable polymer blends and composites from renewable resources. In *Biodegradable polymers, blends and composites.* https://doi.org/10.1016/B978-0-12-823791-5.00015-6

Rajeswari, A., Gopi, S., Jackcina Stobel Christy, E., Jayaraj, K., & Pius, A. (2020). Current research on the blends of chitosan as new biomaterials. In *Handbook of chitin and chitosan.* INC. https://doi.org/10.1016/b978-0-12-817970-3.00009-2

RameshKumar, S., Shaiju, P., O'Connor, K. E., & P, R. B. (2020). Bio-based and biodegradable polymers—State-of-the-art, challenges and emerging trends. *Current Opinion in Green and Sustainable Chemistry, 21.* https://doi.org/10.1016/j.cogsc.2019.12.005

Rane, A. V., Kanny, K., Abitha, V. K., & Thomas, S. (2018). Methods for synthesis of nanoparticles and fabrication of nanocomposites. In *Synthesis of inorganic nanomaterials.* Elsevier Ltd. https://doi.org/10.1016/b978-0-08-101975-7.00005-1

Rangaswamy, B. E., Vanitha, K. P., & Hungund, B. S. (2015). Microbial cellulose production from bacteria isolated from rotten fruit. *International Journal of Polymer Science, 2015.* https://doi.org/10.1155/2015/280784

Rasool, A., Ata, S., & Islam, A. (2019). Stimuli responsive biopolymer (chitosan) based blend hydrogels for wound healing application. *Carbohydrate Polymers, 203.* https://doi.org/10.1016/j.carbpol.2018.09.083

Reichert, C. L., Bugnicourt, E., Coltelli, M. B., Cinelli, P., Lazzeri, A., Canesi, I., Braca, F., Martínez, B. M., Alonso, R., Agostinis, L., Verstichel, S., Six, L., De Mets, S., Gómez, E. C., Ißbrücker, C., Geerinck, R., Nettleton, D. F., Campos, I., Sauter, E., . . . Schmid, M. (2020). Bio-based packaging: Materials, modifications, industrial applications and sustainability. *Polymers, 12*(7). https://doi.org/10.3390/polym12071558

Ren, Y., Bai, Y., Zhang, Z., Cai, W., & Del Rio Flores, A. (2019). The preparation and structure analysis methods of natural polysaccharides of plants and fungi: A review of recent development. *Molecules, 24*(17). https://doi.org/10.3390/molecules24173122

Richards, E., Rizvi, R., Chow, A., & Naguib, H. (2008). Biodegradable composite foams of PLA and PHBV using subcritical CO 2. *Journal of Polymers and the Environment, 16*(4), 258–266. https://doi.org/10.1007/s10924-008-0110-y

Rivadeneira, J., Audisio, M. C., & Gorustovich, A. (2018). Films based on soy protein-agar blends for wound dressing: Effect of different biopolymer proportions on the drug release rate and the physical and antibacterial properties of the films. *Journal of Biomaterials Applications, 32*(9). https://doi.org/10.1177/0885328218756653

Rizal, S., Abdul Khalil, H. P. S., Abd Hamid, S., Yahya, E. B., Ikramullah, I., Kurniawan, R., & Hazwan, C. M. (2023). Cinnamon-nanoparticle-loaded macroalgal nanocomposite film for antibacterial food packaging applications. *Nanomaterials, 13*(3), 560. https://doi.org/10.3390/nano13030560

Rizal, S., Abdul Khalil, H. P. S., Hamid, S. A., Ikramullah, I., Kurniawan, R., Hazwan, C. M., Muksin, U., Aprilia, S., & Alfatah, T. (2023). Coffee waste macro-particle enhancement in biopolymer materials for edible packaging. *Polymers, 15*(2). https://doi.org/10.3390/polym15020365

Rizal, S., Olaiya, F. G., Saharudin, N. I., Abdullah, C. K., N. G., O., Mohamad Haafiz, M. K., Yahya, E. B., Sabaruddin, F. A., Ikramullah, & Abdul Khalil H. P. S. (2021). Isolation of textile waste cellulose nanofibrillated fibre reinforced in polylactic acid-chitin biodegradable composite for green packaging application. *Polymers, 13*(3), 325. https://doi.org/10.3390/polym13030325

Rogovina, S. Z., & Vikhoreva, G. A. (2006). Polysaccharide-based polymer blends: Methods of their production. *Glycoconjugate Journal*, *23*(7–8), 611–618. https://doi.org/10.1007/s10719-006-8768-7

Rosa, D. S., Lotto, N. T., Lopes, D. R., & Guedes, C. G. F. (2004). The use of roughness for evaluating the biodegradation of poly-β-(hydroxybutyrate) and poly-β-(hydroxybutyrate-co-β-valerate). *Polymer Testing*, *23*(1), 3–8. https://doi.org/10.1016/S0142-9418(03)00042-4

Rosato, D. V. (1998). The complete extrusion process. In *Extruding plastics* (pp. 1–53). https://doi.org/10.1007/978-1-4615-5793-7_1

Rosseto, M., Krein, D. C., Balbé, P., & Dettmer, A. (2019). Starch-gelatin film as an alternative to the use of plastics in agriculture: A review. *Journal of the Science of Food and Agriculture*, *99*(15), 6671–6679. https://doi.org/10.1002/jsfa.9944

Rudhziah, S., Rani, M. S. A., Ahmad, A., Mohamed, N. S., & Kaddami, H. (2015). Potential of blend of kappa-carrageenan and cellulose derivatives for green polymer electrolyte application. *Industrial Crops and Products*, *72*. https://doi.org/10.1016/j.indcrop.2014.12.051

Rydz, J., Musiol, M., Zawidlak-Wegrzyńska, B., & Sikorska, W. (2018). Present and future of biodegradable polymers for food packaging applications. *Biopolymers for Food Design*. https://doi.org/10.1016/B978-0-12-811449-0.00014-1

Rzayev, Z. M. O., Bunyatova, U., & Şimşek, M. (2017). Multifunctional colloidal nanofiber composites including dextran and folic acid as electro-active platforms. *Carbohydrate Polymers*, *166*. https://doi.org/10.1016/j.carbpol.2017.02.100

Sadiku, E. R., & Ogunniran, E. S. (2013). Compatibilization as a tool for nanostructure formation. In *Nanostructured polymer blends*. Elsevier Inc. https://doi.org/10.1016/B978-1-4557-3159-6.00004-3

Sakulwech, S., Lourith, N., Ruktanonchai, U., & Kanlayavattanakul, M. (2018). Preparation and characterization of nanoparticles from quaternized cyclodextrin-grafted chitosan associated with hyaluronic acid for cosmetics. *Asian Journal of Pharmaceutical Sciences*, *13*(5). https://doi.org/10.1016/j.ajps.2018.05.006

Santiago, A., & Moreira, R. (2020). Drying of edible seaweeds. In *Sustainable seaweed technologies* (pp. 131–154). Elsevier Inc. https://doi.org/10.1016/b978-0-12-817943-7.00004-4

Santos, N. L., Ragazzo, G. de O., Cerri, B. C., Soares, M. R., Kieckbusch, T. G., & da Silva, M. A. (2020). Physicochemical properties of konjac glucomannan/alginate films enriched with sugarcane vinasse intended for mulching applications. *International Journal of Biological Macromolecules*, *165*. https://doi.org/10.1016/j.ijbiomac.2020.10.049

Sarangi, M. K., Rao, M. E. B., Parcha, V., & Upadhyay, A. (2020). Tailoring of colon targeting with sodium alginate-Assam bora rice starch based multi particulate system containing Naproxen. *Starch/Staerke*, *72*(7–8). https://doi.org/10.1002/star.201900307

Sarti, B., & Scandola, M. (1995). Viscoelastic and thermal properties of collagen/poly(vinyl alcohol) blends. *Biomaterials*, *16*(10). https://doi.org/10.1016/0142-9612(95)99641-X

Scott, A. J., & Penlidis, A. (2017). Copolymerization. In *Reference module in chemistry, molecular sciences and chemical engineering* (pp. 1–11). https://doi.org/10.1016/B978-0-12-409547-2.13901-0

Sell, S. A., McClure, M. J., Garg, K., Wolfe, P. S., & Bowlin, G. L. (2009). Electrospinning of collagen/biopolymers for regenerative medicine and cardiovascular tissue engineering. *Advanced Drug Delivery Reviews*, *61*(12). https://doi.org/10.1016/j.addr.2009.07.012

Seo, C. W., & Yoo, B. (2021). Effect of κ-carrageenan/milk protein interaction on rheology and microstructure in dairy emulsion systems with different milk protein types and κ-carrageenan concentrations. *Journal of Food Processing and Preservation*, *45*(1). https://doi.org/10.1111/jfpp.15038

Settier-Ramírez, L., López-Carballo, G., Gavara, R., & Hernández-Muñoz, P. (2020). PVOH/ protein blend films embedded with lactic acid bacteria and their antilisterial activity in pasteurized milk. *International Journal of Food Microbiology, 322.* https://doi. org/10.1016/j.ijfoodmicro.2020.108545

Shabbir, M., & Mohammad, F. (2017). Sustainable production of regenerated cellulosic fibres. In *Sustainable fibres and textiles.* https://doi.org/10.1016/B978-0-08-102041-8. 00007-X

Shaghaleh, H., Xu, X., & Wang, S. (2018). Current progress in production of biopolymeric materials based on cellulose, cellulose nanofibers, and cellulose derivatives. *RSC Advances, 8*(2). https://doi.org/10.1039/c7ra11157f

Shahidi, F., & Rahman, Md. J. (2018). Bioactives in seaweeds, algae, and fungi and their role in health promotion. *Journal of Food Bioactives, 2.* https://doi.org/10.31665/ jfb.2018.2141

Shahverdi, M., Seifi, S., Akbari, A., Mohammadi, K., Shamloo, A., & Movahhedy, M. R. (2022). Melt electrowriting of PLA, PCL, and composite PLA/PCL scaffolds for tissue engineering application. *Scientific Reports, 12*(1). https://doi.org/10.1038/ s41598-022-24275-6

Shang, X., Jiang, H., Wang, Q., Liu, P., & Xie, F. (2019). Cellulose–starch hybrid films plasticized by aqueous ZnCl 2 solution. *International Journal of Molecular Sciences, 20*(3), 1–17. https://doi.org/10.3390/ijms20030474

Shankar, S., & Rhim, J. W. (2019). Effect of types of zinc oxide nanoparticles on structural, mechanical and antibacterial properties of poly(lactide)/poly(butylene adipate-co-terephthalate) composite films. *Food Packaging and Shelf Life, 21.* https://doi.org/10.1016/j. fpsl.2019.100327

Sharif, N., Fabra, M. J., & López-Rubio, A. (2019). Nanostructures of zein for encapsulation of food ingredients. In *Biopolymer nanostructures for food encapsulation purposes* (pp. 217–245). Elsevier. https://doi.org/10.1016/B978-0-12-815663-6.00009-4

Sharma, N., Khatkar, B. S., Kaushik, R., Sharma, P., & Sharma, R. (2017). Isolation and development of wheat based gluten edible film and its physicochemical properties. *International Food Research Journal, 24*(1).

Shavandi, A., Silva, T. H., Bekhit, A. A., & Bekhit, A. E. D. A. (2017). Keratin: Dissolution, extraction and biomedical application. *Biomaterials Science, 5*(9). https://doi. org/10.1039/c7bm00411g

Shukla, R., & Cheryan, M. (2001). Zein: The industrial protein from corn. *Industrial Crops and Products, 13*(3). https://doi.org/10.1016/S0926-6690(00)00064-9

Shundo, A., & Ijioto, A. (1966). Polymer blends: Blending methods for. *Journal of Applied Polymer Science, 10*(6), 939–953.

Siemann, U. (2005). Solvent cast technology—A versatile tool for thin film production. *Progress in Colloid and Polymer Science, 130*(June), 1–14. https://doi.org/10.1007/ b107336

Silva, E. K., Zabot, G. L., Cazarin, C. B. B., Maróstica, M. R., & Meireles, M. A. A. (2016). Biopolymer-prebiotic carbohydrate blends and their effects on the retention of bioactive compounds and maintenance of antioxidant activity. *Carbohydrate Polymers, 144.* https://doi.org/10.1016/j.carbpol.2016.02.045

Simoes, C. L., Viana, J. C., & Cunha, A. M. (2009). Mechanical properties of poly(e-caprolactone) and poly(lactic acid) blends. *Journal of Applied Polymer Science, 112*(5), 345–352. https://doi.org/10.1002/app

Singamneni, S., Velu, R., Behera, M. P., Scott, S., Brorens, P., Harland, D., & Gerrard, J. (2019). Selective laser sintering responses of keratin-based bio-polymer composites. *Materials and Design, 183.* https://doi.org/10.1016/j.matdes.2019.108087

Singh, R. S., & Kaur, N. (2019). Understanding response surface optimization of medium composition for pullulan production from de-oiled rice bran by Aureobasidium pullulans. *Food Science and Biotechnology*, 28(5). https://doi.org/10.1007/s10068-019-00585-w

Sionkowska, A. (2003). Interaction of collagen and poly(vinyl pyrrolidone) in blends. *European Polymer Journal*, 39(11), 2135–2140. https://doi.org/10.1016/S0014-3057(03)00161-7

Sionkowska, A. (2011). Current research on the blends of natural and synthetic polymers as new biomaterials: Review. *Progress in Polymer Science (Oxford)*, 36(9), 1254–1276. https://doi.org/10.1016/j.progpolymsci.2011.05.003

Sionkowska, A. (2015). The potential of polymers from natural sources as components of the blends for biomedical and cosmetic applications. *Pure and Applied Chemistry*, 87(11–12), 1075–1084. https://doi.org/10.1515/pac-2015-0105

Sionkowska, A., Kaczmarek, B., Lewandowska, K., Grabska, S., Pokrywczyńska, M., Kloskowski, T., & Drewa, T. (2016). 3D composites based on the blends of chitosan and collagen with the addition of hyaluronic acid. *International Journal of Biological Macromolecules*, 89. https://doi.org/10.1016/j.ijbiomac.2016.04.085

Sionkowska, A., Skopinska-Wisniewska, J., & Wisniewski, M. (2009). Collagen-synthetic polymer interactions in solution and in thin films. *Journal of Molecular Liquids*, 145(3), 135–138. https://doi.org/10.1016/j.molliq.2008.06.005

Siracusa, V., & Blanco, I. (2020). Bio-polyethylene (Bio-PE), Bio-polypropylene (Bio-PP) and Bio-poly(ethylene terephthalate) (Bio-PET): Recent developments in bio-based polymers analogous to petroleum-derived ones for packaging and engineering applications. *Polymers*, 12(8). https://doi.org/10.3390/APP10155029

Sivaram, S. (2017). Giulio Natta and the origins of stereoregular polymers. *Resonance*, 22(11). https://doi.org/10.1007/s12045-017-0568-9

Sonseca, A., Madani, S., Muñoz-Bonilla, A., Fernández-García, M., Peponi, L., Leonés, A., Rodríguez, G., Echeverría, C., & López, D. (2020). Biodegradable and antimicrobial pla—ola blends containing chitosan-mediated silver nanoparticles with shape memory properties for potential medical applications. *Nanomaterials*, 10(6). https://doi.org/10.3390/nano10061065

Souza, P. M. S., Morales, A. R., Marin-Morales, M. A., & Mei, L. H. I. (2013). PLA and montmorilonite nanocomposites: Properties, biodegradation and potential toxicity. *Journal of Polymers and the Environment*, 21(3). https://doi.org/10.1007/s10924-013-0577-z

Sperling, L. H. (2000). History and development of polymer blends and ipns. In *Applied polymer science: 21st century* (Issue 2, pp. 343–354). American Chemical Society Division of Polymeric Materials: Science and Engineering. https://doi.org/10.1016/b978-008043417-9/50020-9

Subramanian, P. M. (1985). Permeability barriers by controlled morphology of polymer blends. *Polymer Engineering & Science*, 25(8). https://doi.org/10.1002/pen.760250810

Surendran, A., Lakshmanan, M., Chee, J. Y., Sulaiman, A. M., Thuoc, D. Van, & Sudesh, K. (2020). Can polyhydroxyalkanoates be produced efficiently from waste plant and animal oils? *Frontiers in Bioengineering and Biotechnology*, 8. https://doi.org/10.3389/fbioe.2020.00169

Suter, U. W. (2013). Why was the macromolecular hypothesis such a big deal? *Advances in Polymer Science*, 261. https://doi.org/10.1007/12_2013_251

Sworn, G. (2021). Xanthan gum. In *Handbook of hydrocolloids* (pp. 833–853). Elsevier. https://doi.org/10.1016/B978-0-12-820104-6.00004-8

Syuhada, N., Yazid, M., Abdullah, N., Muhammad, N., & Matias-Peralta, H. M. (2018). Application of starch and starch-based products in food industry. *Journal of Science and Technology*, 10(2), 144–174. https://doi.org/10.30880/jst.2018.10.02.023

Szabo, K., Teleky, B. E., Mitrea, L., Călinoiu, L. F., Martău, G. A., Simon, E., Varvara, R. A., & Vodnar, D. C. (2020). Active packaging-poly (vinyl alcohol) films enriched with tomato by-products extract. *Coatings, 10*(2). https://doi.org/10.3390/coatings10020141

Takahashi, S., Okada, H., Nobukawa, S., & Yamaguchi, M. (2012). Optical properties of polymer blends composed of poly(methyl methacrylate) and ethylene-vinyl acetate copolymer. *European Polymer Journal, 48*(5), 974–980. https://doi.org/10.1016/j. eurpolymj.2012.02.009

Takayama, T., & Todo, M. (2006). Improvement of impact fracture properties of PLA/PCL polymer blend due to LTI addition. *Journal of Materials Science, 41*(15), 4989–4992. https://doi.org/10.1007/s10853-006-0137-1

Talón, E., Trifkovic, K. T., Nedovic, V. A., Bugarski, B. M., Vargas, M., Chiralt, A., & González-Martínez, C. (2017). Antioxidant edible films based on chitosan and starch containing polyphenols from thyme extracts. *Carbohydrate Polymers, 157*, 1153–1161. https://doi.org/10.1016/j.carbpol.2016.10.080

Tănase, E. E., Popa, M. E., Râpă, M., & Popa, O. (2015). Preparation and characterization of biopolymer blends based on polyvinyl alcohol and starch. *Romanian Biotechnological Letters, 20*(2), 10306–10315.

Tang, X. Z., Kumar, P., Alavi, S., & Sandeep, K. P. (2012). Recent advances in biopolymers and biopolymer-based nanocomposites for food packaging materials. *Critical Reviews in Food Science and Nutrition, 52*(5), 426–442. https://doi.org/10.1080/10408398. 2010.500508

Tarkanian, M. J., & Hosler, D. (2011). America's first polymer scientists: Rubber processing, use and transport in mesoamerica. *Latin American Antiquity, 22*(4). https://doi. org/10.7183/1045-6635.22.4.469

Thakkar, R., Thakkar, R., Pillai, A., Ashour, E. A., & Repka, M. A. (2020). Systematic screening of pharmaceutical polymers for hot melt extrusion processing: A comprehensive review. *International Journal of Pharmaceutics, 576*. https://doi.org/10.1016/j. ijpharm.2019.118989

Thomas, S., Durand, D., Chassenieux, C., & Jyotishkumar, P. (2013). Handbook of biopolymer-based materials: From blends and composites to gels and complex networks. In *Handbook of biopolymer-based materials: From blends and composites to gels and complex networks.* https://doi.org/10.1002/9783527652457

Tian, H., Guo, G., Fu, X., Yao, Y., Yuan, L., & Xiang, A. (2018). Fabrication, properties and applications of soy-protein-based materials: A review. *International Journal of Biological Macromolecules, 120.* https://doi.org/10.1016/j.ijbiomac.2018.08.110

Tian, H., Tang, Z., Zhuang, X., Chen, X., & Jing, X. (2012). Biodegradable synthetic polymers: Preparation, functionalization and biomedical application. *Progress in Polymer Science (Oxford), 37*(2), 237–280. https://doi.org/10.1016/j.progpolymsci. 2011.06.004

Tiimob, B. J., Rangari, V. K., Mwinyelle, G., Abdela, W., Evans, P. G., Abbott, N., Samuel, T., & Jeelani, S. (2018). Tough aliphatic-aromatic copolyester and chicken egg white flexible biopolymer blend with bacteriostatic effects. *Food Packaging and Shelf Life, 15.* https://doi.org/10.1016/j.fpsl.2018.01.001

Tokiwa, Y., Calabia, B. P., Ugwu, C. U., & Aiba, S. (2009). Biodegradability of plastics. *International Journal of Molecular Sciences, 10*(9), 3722–3742. https://doi.org/10.3390/ ijms10093722

Torres, F. G., Troncoso, O. P., Pisani, A., Gatto, F., & Bardi, G. (2019). Natural polysaccharide nanomaterials: An overview of their immunological properties. *International Journal of Molecular Sciences, 20*(20). https://doi.org/10.3390/ijms20205092

Trache, D., Hussin, M. H., Hui Chuin, C. T., Sabar, S., Fazita, M. R. N., Taiwo, O. F. A., Hassan, T. M., & Haafiz, M. K. M. (2016). Microcrystalline cellulose: Isolation, characterization and bio-composites application: A review. *International Journal of Biological Macromolecules*, *93*, 789–804. https://doi.org/10.1016/j.ijbiomac.2016.09.056

Treesuppharat, W., Rojanapanthu, P., Siangsanoh, C., Manuspiya, H., & Ummartyotin, S. (2017). Synthesis and characterization of bacterial cellulose and gelatin-based hydrogel composites for drug-delivery systems. *Biotechnology Reports*, *15*. https://doi.org/10.1016/j.btre.2017.07.002

Tripathi, S., Mehrotra, G. K., & Dutta, P. K. (2008). Chitosan based antimicrobial films for food packaging applications. *E-Polymers*, *December 2015*. https://doi.org/10.1515/epoly.2008.8.1.1082

Tytgat, L., Vagenende, M., Declercq, H., Martins, J. C., Thienpont, H., Ottevaere, H., Dubruel, P., & Van Vlierberghe, S. (2018). Synergistic effect of κ-carrageenan and gelatin blends towards adipose tissue engineering. *Carbohydrate Polymers*, *189*. https://doi.org/10.1016/j.carbpol.2018.02.002

Urquijo, J., Guerrica-Echevarría, G., & Eguiazábal, J. I. (2015). Melt processed PLA/PCL blends: Effect of processing method on phase structure, morphology, and mechanical properties. *Journal of Applied Polymer Science*, *132*(41), 1–9. https://doi.org/10.1002/app.42641

Usman, A., Khalid, S., Usman, A., Hussain, Z., & Wang, Y. (2017). Algal polysaccharides, novel application, and outlook. In *Algae based polymers, blends, and composites: Chemistry, biotechnology and materials science*. https://doi.org/10.1016/B978-0-12-812360-7.00005-7

Varma, K., & Gopi, S. (2021). Biopolymers and their role in medicinal and pharmaceutical applications. In *Biopolymers and their industrial applications*. https://doi.org/10.1016/b978-0-12-819240-5.00007-9

Vázquez-Portalatĺn, N., Kilmer, C. E., Panitch, A., & Liu, J. C. (2016). Characterization of collagen type I and II blended hydrogels for articular cartilage tissue engineering. *Biomacromolecules*, *17*(10). https://doi.org/10.1021/acs.biomac.6b00684

Vilaseca, F., Mendez, J. A., Pèlach, A., Llop, M., Cañigueral, N., Gironès, J., Turon, X., & Mutjé, P. (2007). Composite materials derived from biodegradable starch polymer and jute strands. *Process Biochemistry*, *42*(3), 329–334. https://doi.org/10.1016/j.procbio.2006.09.004

Vishal, P. (2017). Bionanocomposite: A review. *Austin Journal of Nanomedicine & Nanotechnology*, *5*, 1–3.

Volić, M., Pajić-Lijaković, I., Djordjević, V., Knežević-Jugović, Z., Pećinar, I., Stevanović-Dajić, Z., Veljović, Đ., Hadnadjev, M., & Bugarski, B. (2018). Alginate/soy protein system for essential oil encapsulation with intestinal delivery. *Carbohydrate Polymers*, *200*. https://doi.org/10.1016/j.carbpol.2018.07.033

Voron'ko, N. G., Derkach, S. R., Kuchina, Y. A., & Sokolan, N. I. (2016). The chitosan-gelatin (bio)polyelectrolyte complexes formation in an acidic medium. *Carbohydrate Polymers*, *138*. https://doi.org/10.1016/j.carbpol.2015.11.059

Wang, A., Ao, Q., Wei, Y., Gong, K., Liu, X., Zhao, N., Gong, Y., & Zhang, X. (2007). Physical properties and biocompatibility of a porous chitosan-based fiber-reinforced conduit for nerve regeneration. *Biotechnology Letters*, *29*(11), 1697–1702. https://doi.org/10.1007/s10529-007-9460-0

Wang, B., Wan, Y., Zheng, Y., Lee, X., Liu, T., Yu, Z., Huang, J., Ok, Y. S., Chen, J., & Gao, B. (2019). Alginate-based composites for environmental applications: A critical review. *Critical Reviews in Environmental Science and Technology*, *49*(4). https://doi.org/10.1080/10643389.2018.1547621

Wang, G., Wang, X., & Huang, L. (2017). Feasibility of chitosan-alginate (Chi-Alg) hydrogel used as scaffold for neural tissue engineering: A pilot study in vitro. *Biotechnology and Biotechnological Equipment, 31*(4), 766–773. https://doi.org/10.1080/13102818. 2017.1332493

Wang, Q., Xiong, Z., Zhang, J., Fang, Z., Lai, M., & Ho, J. (2023). Impact of polyvinyl alcohol fiber on the full life-cycle shrinkage of cementitious composite. *Journal of Building Engineering, 63*. https://doi.org/10.1016/j.jobe.2022.105463

Wang, S., Guan, S., Wang, J., Liu, H., Liu, T., Ma, X., & Cui, Z. (2017). Fabrication and characterization of conductive poly (3,4-ethylenedioxythiophene) doped with hyaluronic acid/poly (L-lactic acid) composite film for biomedical application. *Journal of Bioscience and Bioengineering, 123*(1), 116–125. https://doi.org/10.1016/j.jbiosc.2016.07.010

Wang, S., Li, C., Copeland, L., Niu, Q., & Wang, S. (2015). Starch retrogradation: A comprehensive review. *Comprehensive Reviews in Food Science and Food Safety, 14*(5). https://doi.org/10.1111/1541-4337.12143

Wang, X., Peng, S., Chen, H., Yu, X., & Zhao, X. (2019). Mechanical properties, rheological behaviors, and phase morphologies of high-toughness PLA/PBAT blends by in-situ reactive compatibilization. *Composites Part B: Engineering, 173*. https://doi.org/10.1016/j.compositesb.2019.107028

Wang, Y., Guo, Z., Qian, Y., Zhang, Z., Lyu, L., Wang, Y., & Ye, F. (2019). Study on the electrospinning of gelatin/pullulan composite nanofibers. *Polymers, 11*(9). https://doi.org/10.3390/polym11091424

Watzke, H. J., & Dieschbourg, C. (1994). Novel silica-biopolymer nanocomposites: The silica sol-gel process in biopolymer organogels. *Advances in Colloid and Interface Science, 50*(C). https://doi.org/10.1016/0001-8686(94)80021-9

Wehrs, M., Tanjore, D., Eng, T., Lievense, J., Pray, T. R., & Mukhopadhyay, A. (2019). Engineering robust production microbes for large-scale cultivation. *Trends in Microbiology, 27*(6). https://doi.org/10.1016/j.tim.2019.01.006

Williams, S. F., Martin, D. P., Horowitz, D. M., & Peoples, O. P. (1999). PHA applications: Addressing the price performance issue I. Tissue engineering. *International Journal of Biological Macromolecules, 25*(1–3), 111–121. https://doi.org/10.1016/S0141-8130(99)00022-7

Witt, W. (1984). Polymer blends. *Kunststoffe—German Plastics, 74*(10), 35–37.

Wongkanya, R., Chuysinuan, P., Pengsuk, C., Techasakul, S., Lirdprapamongkol, K., Svasti, J., & Nooeaid, P. (2017). Electrospinning of alginate/soy protein isolated nanofibers and their release characteristics for biomedical applications. *Journal of Science: Advanced Materials and Devices, 2*(3). https://doi.org/10.1016/j.jsamd.2017.05.010

Wu, H., Williams, G. R., Wu, J., Wu, J., Niu, S., Li, H., Wang, H., & Zhu, L. (2018). Regenerated chitin fibers reinforced with bacterial cellulose nanocrystals as suture biomaterials. *Carbohydrate Polymers, 180*. https://doi.org/10.1016/j.carbpol.2017.10.022

Xavier, S. F. (2014). Properties and performance of polymer blends. In *Polymer blends handbook* (pp. 1031–1201). https://doi.org/10.1007/978-94-007-6064-6

Xie, C., Xiong, Q., Wei, Y., Li, X., Hu, J., He, M., Wei, S., Yu, J., Cheng, S., Ahmad, M., Liu, Y., Luo, S., Zeng, X., Yu, J., & Luo, H. (2023). Fabrication of biodegradable hollow microsphere composites made of polybutylene adipate co-terephthalate/polyvinylpyrrolidone for drug delivery and sustained release. *Materials Today Bio, 20*, 100628. https://doi.org/10.1016/j.mtbio.2023.100628

Xie, L., Xu, H., Li, L. Bin, Hsiao, B. S., Zhong, G. J., & Li, Z. M. (2016). Biomimetic nanofibrillation in two-component biopolymer blends with structural analogs to spider silk. *Scientific Reports, 6*. https://doi.org/10.1038/srep34572

Y, S., & Rao, P. (2019). Material conservation and surface coating enhancement with starch-pectin biopolymer blend: A way towards green. *Surfaces and Interfaces*, *16*. https://doi.org/10.1016/j.surfin.2019.04.011

Yang, C., Tang, H., Wang, Y., Liu, Y., Wang, J., Shi, W., & Li, L. (2019). Development of PLA-PBSA based biodegradable active film and its application to salmon slices. *Food Packaging and Shelf Life*, *22*, 1–9. https://doi.org/10.1016/j.fpsl.2019.100393

Yoon, D., Cho, Y. S., Joo, S. Y., Seo, C. H., & Cho, Y. S. (2020). A clinical trial with a novel collagen dermal substitute for wound healing in burn patients. *Biomaterials Science*, *8*(3). https://doi.org/10.1039/c9bm01209e

Yu, J., Bi, X., Yu, B., & Chen, D. (2016). Isoflavones: Anti-inflammatory benefit and possible caveats. *Nutrients*, *8*(6). https://doi.org/10.3390/nu8060361

Yu, L., Dean, K., & Li, L. (2006). Polymer blends and composites from renewable resources. *Progress in Polymer Science (Oxford)*, *31*(6), 576–602. https://doi.org/10.1016/j.progpolymsci.2006.03.002

Yu, Z., Li, B., Chu, J., & Zhang, P. (2018). Silica in situ enhanced PVA/chitosan biodegradable films for food packages. *Carbohydrate Polymers*, *184*(October 2017), 214–220. https://doi.org/10.1016/j.carbpol.2017.12.043

Yuan, H., Lan, P., He, Y., Li, C., & Ma, X. (2020). Effect of the modifications on the physicochemical and biological properties of β-glucan-a critical review. *Molecules*, *25*(1). https://doi.org/10.3390/molecules25010057

Zainal, N. F. A., & Chan, C. H. (2019). Crystallization and melting behavior of compatibilized polymer blends. In *Compatibilization of polymer blends: Micro and nano scale phase morphologies, interphase characterization, and properties* (pp. 391–433). Elsevier Inc. https://doi.org/10.1016/B978-0-12-816006-0.00014-1

Zhang, H. (2018). Introduction to freeze-drying and ice templating. In *Ice templating and freeze-drying for porous materials and their applications* (pp. 1–27). https://doi.org/10.1002/9783527807390.ch1

Zhang, N., Liu, H., Yu, L., Liu, X., Zhang, L., Chen, L., & Shanks, R. (2013). Developing gelatin-starch blends for use as capsule materials. *Carbohydrate Polymers*, *92*(1), 455–461. https://doi.org/10.1016/j.carbpol.2012.09.048

Zhang, W., Shi, S., Zhu, W., Huang, L., Yang, C., Li, S., Liu, X., Wang, R., Hu, N., Suo, Y., Li, Z., & Wang, J. (2017). Agar aerogel containing small-sized zeolitic imidazolate framework loaded carbon nitride: A solar-triggered regenerable decontaminant for convenient and enhanced water purification. *ACS Sustainable Chemistry and Engineering*, *5*(10). https://doi.org/10.1021/acssuschemeng.7b02376

Zheng, C., Liu, C., Chen, H., Wang, N., Liu, X., Sun, G., & Qiao, W. (2019). Effective wound dressing based on Poly (vinyl alcohol)/Dextran-aldehyde composite hydrogel. *International Journal of Biological Macromolecules*, *132*. https://doi.org/10.1016/j.ijbiomac.2019.04.038

Zinn, M., Witholt, B., & Egli, T. (2001). Occurrence, synthesis and medical application of bacterial polyhydroxyalkanoate. *Advanced Drug Delivery Reviews*, *53*(1), 5–21. https://doi.org/10.1016/S0169-409X(01)00218-6

Złotko, K., Wiater, A., Waśko, A., Pleszczyńska, M., Paduch, R., Jaroszuk-Ściseł, J., & Bieganowski, A. (2019). A report on fungal (1→3)-α-D-glucans: Properties, functions and application. *Molecules*, *24*(21). https://doi.org/10.3390/molecules24213972

2 Biodegradation and Compostable Biopolymers

2.1 BIODEGRADATION AND COMPOSTABLES IN GENERAL

Plastic film is made from a petroleum-based material that is nonbiodegradable due to the fact that it is made from oil. Thompson et al. (2009) stated that around 4% of global oil supply is used as feedstock for the production of plastics, with a comparable amount of energy consumed in the process. Plastics make up only 10% of all recycled garbage, but they make up a significantly larger proportion of the litter that accumulates along shorelines (Barnes et al., 2009). It could take years for non-biodegradable materials to degrade. As a result, it has become a problem that will pollute our environment and fill landfills with non-biodegradable waste. Non-biodegradable materials can clog streams, destroy marine organisms, and pollute a river and its surrounding areas. A broad variety of oil-based polymers are currently used in packaging applications, according to a journal report by (Kamarudin et al. 2022). They are almost all non-biodegradable, and some of them are impossible to recycle or reuse since they are complicated composites in which the contamination levels are different.

Plastic bags create a lot of waste, and they are the most widely used material in people's lives. They are mostly made of polyethylene (PE), which is not biodegradable and takes a long time to break down. As a result, a large amount of garbage ends up in landfills. As the planet moves toward a more environmentally sustainable economy, many scientists have worked to create materials that are more environmentally friendly. Nevertheless, it is not an easy move because it necessitates a lengthy procedure and high manufacturing costs. According to environmental issues emerging in recent years, biodegradable polymers are the latest manufacturing components to be used as raw materials in manufacturing biobased materials. Biodegradable polymers can be prepared through a few methods such as modification, chemical synthesis, microbiological synthesis, enzymatic synthesis, and chemo-enzymatic synthesis (Zeng et al., 2016). The use of biodegradable polymers will help to ensure sustainability and reduce the environmental impact of oil-based polymer disposal (Abdul Khalil, 2023). Many people want to lead the way in reducing the negative impact on the environment. With the advent of bioplastic or biodegradable polymers as a replacement for oil-based or non-biodegradable polymers, it is possible to reduce the negative impact on the environment while still serving as a blueprint for other industries to manufacture environmentally friendly materials.

Biodegradable polymers can withstand deterioration during their usage while also being biodegradable at the end of their useful life (Surya et al. 2022). A biodegradable

DOI: 10.1201/9781003416043-2

polymer, according to Leja and Lewandowicz (2010), is a polymer that degrades by microorganism metabolism. As a result, a biodegradable polymer can break down into carbon dioxide, water, and biomass (Wittaya, 2009). The conversion process of biopolymers into gases is called mineralization. When all biodegradable products or biomass have been absorbed and all carbon has been converted to carbon dioxide, mineralization is complete. Activated sludges, for instance, are used to process sewage flows in a wastewater treatment system in order to biotransform the organic compounds to complete their mineralization (Poznyak et al., 2019). Since the carbon dioxide emitted is already part of the biological carbon cycle, it does not lead to an increase in greenhouse gases (Song et al., 2009). According to Imre and Pukánszky (2013), such polymers are degradable if their end products result in a decrease in molecular weight compounds due to chain scission in the backbone, and degradation is completed. Biodegradable polymers can be made from a variety of sources, including wood and microorganisms (Vieira et al., 2011).

Degradation is one of the disposal methods for polymers. Degradation is a type of decomposition that comes to a halt when polymers are fragmented by heat, moisture, sunlight, or enzymes, resulting in weakening of the polymers' chains (Mohee et al., 2008; Song et al., 2009; Rujnić-Sokele & Pilipović, 2017; Ryan et al., 2018). Meanwhile, biodegradation is a term used to describe specifically engineered biodegradable polymers that undergo complete mineralization (Kyrikou & Briassoulis, 2007). As the number of biopolymers generated has grown, so has interest in polymer biodegradation. The most challenging aspect of making biodegradable polymers is refining their chemical, physical, and mechanical properties, as well as their biodegradability (Leja & Lewandowicz, 2010). Many polymers classified as "biodegradable" are simply "bioerodable", "photodegradable", or just partly biodegradable. Biodegradation is a form of decomposition that involves biological activity (Zeng et al., 2016). In the same vein, Azevedo et al. (2008) stated that the progressive destruction of a substrate mediated by complex biological activity is also referred to as biodegradation. Meanwhile, according to terminology used in environmental engineering, degradation refers to the deterioration or breakdown of materials that happens when microorganisms use an organic substrate as a source of carbon and energy (Poznyak et al., 2019). For most chemicals released into the atmosphere, biodegradation is expected to be the primary process of destruction. The degradation and assimilation of polymers by living microorganisms to create degradation products is via this mechanism. The microorganisms that cause biodegradation are bacteria and fungi (Gautam et al., 2007), which include yeasts and molds.

Biodegradation is known to occur in three processes: biodeterioration, biofragmentation, and assimilation, without neglecting the abiotic influences where the process can stop at each stage (Lucas et al., 2008). Biodeterioration is when microbial communities, other decomposer organisms, and abiotic factors all work together to break down biodegradable materials into tiny fractions. Deterioration is a form of surface deterioration that changes a material's mechanical, physical, and chemical properties (Hueck, 2001; Walsh, 2001). The composition and properties of polymer materials influence microbial growth. Environmental conditions such as humidity, temperature, and emissions in the atmosphere are also significant factors to consider (Lugauskas et al., 2003). Microbial deterioration can occur in several ways: physical by adhering to material surfaces; chemical through the development of microorganisms into materials; and enzymatic, which involves enzymes such as lipases, ureases, and proteases. Also, biofragmentation is a lytic process where the bonds within polymers are cleaved or broken down, thus generating oligomers and monomers (Lugauskas et al., 2003). The last stage of biodegradation is assimilation, which is described as a mechanism of microbial cells integrating atoms from particles of polymeric materials (A. Glaser, 2019).

Biodegradation can occur either in aerobic or anaerobic conditions (Rizzarelli & Carroccio, 2014). Under aerobic conditions, where oxygen is present and carbon dioxide is produced, and in anaerobic conditions, where nitrate, sulfate, or another compound is present and methane is produced, biodegradation may take place (Grima et al., 2000; Kyrikou & Briassoulis, 2007; Poznyak et al., 2019). Aerobic biodegradation is defined as the destruction of microbial materials by microorganisms upon the presence of oxygen as final electron acceptor. In anaerobic biodegradation systems, degradation occurs when the compound is fully consumed by microorganisms, resulting in methane emissions. As mentioned, oxygen is the electron acceptor in aerobic metabolism. Microbial species readily adapt and attain high densities if biodegradation follows this trend of metabolism. As a consequence, the rate of biodegradation is easily restricted by oxygen availability rather than the intrinsic microbial capacity to degrade the polymer or other contaminants. In aerobic conditions, biodegradation is measured by the percentage of carbon content converted to carbon dioxide. In the interval, anaerobic biodegradation, which is in the absence of oxygen, uses nitrate or sulfate as the last electron acceptor. The rate of degradation under this system is limited by the reaction rate of active microorganisms, which results in slow adaption and thus requires months or years to degrade.

Some biopolymers are biodegradable, and some are compostable. In this part, the composting of biopolymers will be discussed. According to Rudnik (2008), compostable polymers were first introduced in the 1980s. In short, compostable biopolymers are polymers that undergo a composting process and will break down by 90% within six months or less. To be a compostable polymer, it should meet the following criteria: no visually distinguishable polymer traces and no ecotoxicity, which means there is no toxic material produced during the composting process (Sadeghi & Mahsa, 2015). Natural organic materials made from starches produced from rice, potato, tapioca, or other plants and vegetable matter have recently been mixed with biodegradable polymers to produce compostable products that can further reduce the environmental effect of carbon footprints. Rudnik (2008) states that compostable polymers can be prepared via blending methods, biotechnological techniques such as fermentation and extraction, and also through polymerization. The leading method is the blending procedure, as this method does not require much cost and energy.

Starch-based polymeric blends are some of the extensively compostable biopolymer blends that are commonly being researched. For instance, Novamont, which is an innovation company, prepared a few polymeric blends based on starch by blending the starch with other biopolymers in the presence of a plasticizer or water by using an extruder (Bastioli, 1998). Meanwhile, an example of a compostable biopolymer produced via the biotechnological route is poly(hydroxybutyrate-co-hydroxy valerate) (PHBV), and one from polymerization is polylactic acid (PLA). Biocompostable materials provide an environmentally friendly alternative to other viable options and can help minimize social and economic inequality, reduce the effect of our consumption on the atmosphere, and create possibilities for making a cleaner and more prosperous world (Sadeghi & Mahsa, 2015). To recap, a material must meet the following conditions to be labelled "compostable": mineralization process, disintegration into a composting device, and completion of biodegradation during composting system end use.

Compostable biopolymers may be classified based on their origin or method of preparation. Compostable biopolymers are made from renewable and petrochemical materials (Sadeghi & Mahsa, 2015). Compostable biopolymers from renewable resources are cellulose, chitosan, proteins, PLA, and poly(3-hydroxybutyrate) (PHB). Plants and microorganisms both generate cellulose, which is a ubiquitous and plentiful biopolymer (Mohamad Haafiz et al., 2013). Many agricultural by-products, such as sugarcane, sorghum bagasse, corn stalks, rye, wheat, oats, and rice straws, can be used to make cellulose pulp (Rudnik, 2008). On the other hands, chitosan is the most common polysaccharide after cellulose. Chitosan is a high molecular weight biopolymer and a deacetylated derivative of chitin that can be found in marine crustacean shells and fungus cell walls (Di Martino et al., 2005). The next compostable biopolymer is proteins, which are random copolymers of amino acids (Rudnik, 2008). Protein also can be classified based on its origin, plants and animals. Soy and potato are examples of protein from plants, while collagen and silk are examples from animal origin. Moreover, PLA is the only renewable resource thermoplastic polymer manufactured at a wide scale, over 140,000 tons per year among the available biopolymers (Mukherjee & Kao, 2011). PLA has been used widely in preparation of biopolymer blends and also in the development of biopolymeric materials. Next is PHB, which is a type of polyhydroxyalkanoate (PHA) (Bonartsev et al., 2011). PHB is synthesized by cells under growth-limiting conditions, which occur where the carbon supply is abundant but nitrogen, phosphorus, magnesium, oxygen, or sulfur is in a limiting concentration (Vishnuvardhan Reddy et al., 2009).

2.2 METHODS OF BIODEGRADATION AND COMPOSTING OF BIOPOLYMERS

There are several different methods of biodegradation that are mostly used in order to determine the biodegradation of biopolymer products: soil burial (Xu & Hanna, 2005; Parvin et al., 2010; Mittal et al., 2016; Rapisarda et al., 2019) and microbiological (Benedict et al., 1983) and enzymatic methods (Nobes et al., 1998). Because of its resemblance to real waste disposal conditions, soil burial is a common and normal procedure for the biodegradation of biopolymers. Azahari et al. (2011) worked on biodegradation of polyvinyl alcohol (PVOH)/corn starch (CS) blend films by burying the films in soil and compost. The biodegradation was determined by enzymatic degradation rate, weight loss of samples in the soil burial test, and tensile strength of the films. The results for enzymatic degradation showed that the degradation rate of films increases with an increase in starch content. This was due to the presence of excess hydroxyl groups that absorbed more enzymatic solution. A higher weight loss was observed in blended film with a high content of corn starch compared to pure

PVOH film. This is attributed to the high biodegradability of corn starch compared to PVOH. For the tensile properties, the result showed a decreasing trend in tensile strength and elongation at break when the corn starch content in the films increased, which is due to the amorphous nature of starch. According to this report, the increase in starch content increased the degradation of biopolymer films.

Microorganisms are involved in the deterioration of both synthetic and natural polymers (Gu, 2003). Macromolecular chains are dissolved in the presence of microorganisms such as bacteria and fungi, which initiates the biodegradation process. The molecular concentrations, molecular weights, and the presence of specific microorganisms on the surfaces of polymer materials all influence microbial degradation. Since polymer-degrading microorganisms are also responsible for the degradation of polymeric products, many researchers have done studies based on microbial degradation. Ruiz-Dueñas and Martínez (2009) studied the microbial degradation of lignin in nature. Further, the molecular architecture of the lignin polymer, in which various non-phenolic phenylpropanoid units form a complex three-dimensional network connected by a variety of ether and carbon–carbon bonds, makes it highly resistant to chemical and biological degradation. Extracellular haemperoxidases working synergistically with peroxide-generating oxidases have formed a novel technique for lignin degradation based on unspecific one-electron oxidation of the benzenic rings in the different lignin substructures by ligninolytic microbes. Similar reports based on microbial biodegradation of biopolymer and biopolymer blends can also be found in other works (Fields et al., 1974; Benedict et al., 1983; Siracusa, 2019).

The growing use of polymers in industry has resulted in the environmentally sensitive problem of waste disposal. Enzymes play an important part in polymer biodegradation (Banerjee et al., 2014). Tokiwa and Calabia (2006) explained the biodegradation of polylactide (PLA) in their study. The hydrolysis of aliphatic polyesters is a two-step enzymatic degradation process. The first step is for the enzyme to bind to the substrate's surface through its surface-binding domain, and the second step is for the ester bond to be hydrolyzed. Kikkawa et al. (2002) performed a study on enzymatic degradation on poly(L-lactide) (PLLA) film. In their study, it was concluded that the rate of degradation in the free amorphous region of the film was faster than in the restricted amorphous region. Other literature has also discussed the role of enzymes in biodegradation of biopolymers (Nakamura et al., 2001) and biopolymer mixtures (Ishigaki et al., 1999; Vikman et al., 1999; Spiridon et al., 2008). Tsuji and Ishizaka (2001) used enzyme to study the mechanisms of PLLA/poly(ε-caprolactone) (PCL) blend degradation. They prepared porous biodegradable PCL films by removing PLLA from blended films using proteinase K as a hydrolysis enzyme for PLLA. The results showed that the enzymatic hydrolysis rate of blended films increased, which is suggested to be due to the hydrolysis of PLLA that is catalyzed by proteinase K at interfaces of PLLA- and PCL-rich phases and also at the film surfaces.

Composting is an option for waste treatment where material undergoes recovery and produces a useful product (Song et al., 2009). Based on the definition from ASTM D6400, composting is defined as "a managed process that controls the biological decomposition and transformation of biodegradable materials into a humus-like substance called compost" (Rudnik, 2008). In other words, the composting process has the capability to convert biodegradable materials into valuable soil amendment materials (compost). Compost is a flexible commodity made from the composting and biodegradation of organic waste that is generated in commercial or domestic environments (Sadeghi & Mahsa, 2015). Compost is an organic matter that is rich in nutrients, which can improve soil quality, help plant development, enhance water-holding ability, and minimize the use of chemical fertilizers. Compost also has other advantages such as acting as a soil conditioner, organic fertilizer, and natural pesticide. Adding compost to soil can improve the soil structure by reducing bulk density and improve porosity, which then results in better plant growth. Also, adding compost to soil can modify and stabilize the soil pH, depending on the soil and compost pH. Furthermore, another capability of compost is to control soil erosion to reduce topsoil loss. As mentioned, compost is capable of holding water; thus it can remain on the soil surface to reduce the amount of rain that falls onto the soil.

Organic matter is broken down by microorganisms such as bacteria and fungi during the composting process, resulting in the production of carbon dioxide, heat, water, and compost. There are three type of composting methods: in-vessel methods, aerated static pile methods, and windrow methods (Sadeghi & Mahsa, 2015). For the in-vessel method, significant volumes of waste may be processed in a small amount of space compared to other composting methods (Rudnik, 2019). Also, in this method, the organic material is composted in a silo, drum, concrete-lined trench, batch tub, or other similar equipment. The environmental conditions are well controlled and regulated, and the substance is aerated by being mechanically mixed and agitated. For instance, Ghorpade et al. (2001) conducted a composting process in at laboratory scale by using extruded PLA sheets combined with yard waste. This experiment was to examine the composting capability of PLA in combination with garden waste. In the experiment, the mixture of PLA and garden waste was mixed in a composting vessel and stayed for four weeks. The result showed that the yard waste compost/PLA mixture with 30% PLA concentration lowered the compost pH. Moreover, it was also observed that the PLA molecular weight was reduced as an effect of composting. The study demonstrated that the addition of PLA concentrations below 30% to yard waste compost is an efficient composting method.

In the aerated static pile method, organic waste is mixed into a large pile and then aerated by drawing air out of the pile or allowing air to flow through the pile (Rudnik, 2019). According to the U.S. Environmental Protection Agency (EPA), composting in an aerated static pile generates compost very easily within three to six months. This method can handle a largely homogeneous blend of organic waste and compostable municipal solid waste. However, the aerated static pile method is ineffective for composting animal waste or grease from the food processing industry. In a study reported by Colón et al. (2013), the composting process involving compostable diapers combined with the organic fraction of municipal solid waste shows the mixture underwent composting in a static, forced-aerated composting reactor over a period of 41 days. It was reported that the composting process and the compost produced was not affected by the presence of compostable diapers, as no pathogenic microorganisms were found during the full-scale composting experiment. The study's significant point is that the composting of an organic fraction of municipal solid waste with compostable diapers may be a novel way to transform this waste into high-quality compost.

The windrow composting method involves forming compostable materials into rows of long piles known as windrows (Sadeghi & Mahsa, 2015). In this method, the materials are aerated by turning the pile either by manually or mechanically on a daily basis. This pile will produce enough heat to keep temperatures stable. It is small enough to cause oxygen to flow to the center of the windrow. The composted component is ready for assembly and shipment to end users when it has reached the desired stage of decomposition (Arvanitoyannis, 2013). Itävaara et al. (1997) proposed a study on biodegradable polymer windrow composting. It was observed that several environmental factors influence the degradation of biodegradable samples, including temperature, pH, oxygen concentration, and water content. The experiment also did a trial on plant growth in which they used barley and radish seed to see the effect of degradation products from biodegradable samples on them. The result showed that the compost made from a polymer sample did not show any toxicity in plant growth.

2.3 BIODEGRADABILITY OF BIOPOLYMERS

Biodegradable biopolymers are biopolymers that can be degraded by naturally occurring microorganisms such as bacteria, fungi, and algae into natural elements (biomass), carbon dioxide, and water (Wojtowicz, 2010). Biodegradable biopolymers are very eco-friendly because they are made from sustainable natural resources, and in the end, they will degrade into material that is harmless to the environment. Using biodegradable biopolymers conserves nature and produces less pollution; thus, it will make the earth move toward a healthy environment. Many researchers nowadays are focused on producing biodegradable biopolymers to replace conventional non-biodegradable polymers. Undoubtedly, non-biodegradable polymers have advantages in term of technical properties and low cost, making them the most favorable alternative for many industries such as food packaging and biomedicine (Bassas-Galià, 2017; Babooram, 2020). For example, polystyrene was commercialized in large volumes for food packaging in the late 1940s (Robertson, 2019). Although it has many advantages, using non-biodegradable synthetic polymers for food packaging is not a wise choice because they are made from petroleum-based resources, which are non-renewable energy and will take a thousand years to degrade, which can cause trouble in the current ecosystem (Kumaran et al., 2020). In biomedicine, non-biodegradable scaffolds were used in many applications such as defect repair, tissue healing, and cell transplantation. Using non-biodegradable scaffolds required an extra operation to remove it compared to biodegradable scaffolds that can degrade without requiring any surgical removal (Stewart et al., 2018; Mo et al., 2018).

As stated earlier, the use of biopolymers helps to maintain sustainability and to reduce waste, but not all biopolymers are biodegradable. Polythioester (PTE) is an example of a non-biodegradable biopolymer, which is contrary to the idea of all natural-based polymers being biodegradable (Steinbüchel, 2005). Although PTE is synthesized by microbial fermentation, PTE is non-biodegradable by microorganism (Elbanna et al., 2004). Besides PTE, other biobased polymers such as bio-polyethene and bio-polypropylene are also non-biodegradable biopolymers, and these polymers are widely used in bioplastic applications (Andreeßen & Steinbüchel, 2019). Some bioplastics need to be non-biodegradable to produce more durable material. Non-biodegradable bioplastics might not reduce waste, but manufacturing using biobased polymers can surely reduce greenhouse gas (GHG) emission and depletion of fossil resources (Allison & Bassett, 2015).

Polymer are theoretically biodegradable, which means that they can be degraded by microbes under the right conditions, but many degrade at such slow rates that they are classified as nonbiodegradable (Patel et al., 2011). According to the EN 13432 standard, a material or product is biodegradable if it can undergo a specific degradation process caused by biological activity and can be measured by a standardized test method under specific environmental conditions within a specified period (Wojtowicz, 2010). In other words, a biodegradable polymer must fully decompose and break down into natural elements within a short period of time after disposal.

The assessment of biodegradability of biopolymers can differ depending on the following factors. The factors that influence the degradation of polymers are dependent on the physiochemical structural and environmental conditions and microbial population where the polymers are exposed (Folino et al., 2020). The basic properties that cause the degradation and biodegradation of polymers are their physical and chemical structures. Biodegradation rates differ among biopolymers due to differences in the structural and physicochemical properties of their surfaces, which allow for stronger or weaker microorganism attack on the surface (Volova et al., 2010; Bátori et al., 2018). The physicochemical structure of biopolymers encompasses aspects such as molecular structure, polymer chain length, crystallinity, and polymer composition. In particular, the chain length of the biopolymer influences its biodegradation, where longer polymer chains are generally more prone to degradation.

However, the crystallinity of the polymer is also a significant biodegradation parameter, as the amorphous parts of the polymer are simpler to degrade than the crystalline parts (Massardier-Nageotte et al., 2006). In a study by Woolnough et al. (2008), it was stated that a higher degree of crystallinity leads to a slower rate of biodegradation, with the amorphous phase degrading first. They mentioned that biopolymer poly(3-hydroxyoctanoate) (PHO), which is an example of MCL-PHA, has lower crystallinity than poly [(3-hydroxybutyrate)-co-(3-hydroxyvalerate)] [P(HB-co-HV)]. Thus, P(HB-co-HV) was found to degrade faster than PHB in fresh water as well as in soil. As mentioned, polymer formulation also affects the biodegradation of biopolymers. Since many microorganisms are needed to target the different functions of the polymer, the more complicated the formula, the less degradable it is. For instance, polymers with rings seem to be more resistant to degradation.

Biopolymers have the potential to biodegrade in natural habitats, which is a beneficial property. One of the most important aspects of their creation is the need to prevent the accumulation of petroleum-based wastes, some of which are not biodegradable in the ecosystem, especially in the oceans. Biodegradability, as previously noted, is dependent on the chosen environment and can vary from one environment to another. Thus, the rate of biodegradability of biopolymers is affected by the environmental conditions, as well as the type of biopolymers used. Karamanlioglu et al. (2017) did a review on environmental factors involved in PLA biopolymer degradation and also the degradation of PLA in various environments. Since biotic and abiotic influences coexist in nature, environmental degradation refers to the whole degradation process of given contents. Humidity, temperature, and catalytic species (pH and the presence of enzymes or microorganisms) are examples of environmental conditions that affect biodegradation. For instance, PLA degrades more quickly in alkaline conditions because hydroxide ions catalyze the cleavage of ester groups during hydrolysis. As a result, a high concentration of hydroxide ions in alkaline media accelerates PLA degradation (Cam et al., 1995; Tsuji & Ikada, 1998).

Different environments such as soil and compost also influence the biodegradation of biopolymers. Karamanlioglu et al. (2017) reported that PLA degradation in soil is much slower than in compost medium, owing to the environmental conditions present in compost, which has higher moisture content and temperature range and thus encourages PLA hydrolysis and assimilation by thermophilic microorganisms. Ohkita and Lee (2006) did a study on biodegradability of PLA and PLA films. In the study, they observed the biodegradation of pure PLA and PLA film buried in soil. After six weeks, pure PLA showed only a little degradation. Meanwhile, PLA film buried in soil for 120 days at 25°C showed no degradation as measured by weight loss of the film. Regardless, scanning electron microscopy (SEM) detected actinomycete threads on the PLA film, but there were no signs of degradation on the PLA film surface.

Beside the physical and chemical structure and environmental conditions the biopolymers are exposed to, microbial communities are also one of the factors in biodegradation of biopolymers. Biodegradation is caused by more than 90 different microorganisms in diverse environments (Emadian et al., 2017; Thakur et al., 2018). For example, PHAs can be directly degraded by microorganisms such as bacteria and fungi (Jendrossek et al., 1996; Folino et al., 2020). Since water is an essential requirement for microorganism growth and reproduction, water and moisture may play an important role in the biodegradation of biopolymers (Trivedi et al., 2016). As moisture levels rise, so does microbial activity, and biopolymers biodegrade at a faster rate. Commonly, polymers with shorter chains, more amorphous parts, and simpler formulas are more vulnerable to microorganism biodegradation. The influence of fungi and bacteria is one major cause, emphasizing the importance of microbial communities in biodegradation.

Many studies have focused on role of bacteria and fungi in biopolymer biodegradation. For PHA compounds, a few bacteria such as *Bacillus, Pseudomonas*, and *Streptomyces* were reported as PHA-degrading microorganisms (Jendrossek et al., 1996). Itävaara et al. (2002) conducted a study on biodegradation of polylactides by soil bacteria in aerobic (aquatic) conditions. Polylactic acid degraded up to 90% after 55 to 60 days, and poly-L-lactic acid degraded up to 50% after 40 to 45 days, according to the researchers. Since such microorganisms have a small enzyme population in the absence of oxygen, oxidation rates are lower in anaerobic conditions (Thakur et al., 2018). On the other hand, *Tritirachium album* was the first fungus to be discovered capable of degrading PLA in the scientific literature (Jarerat & Tokiwa, 2001). Trivedi et al. (2016) did a review on the effect of microbes on biopolymers. It was stated that biopolymer PCL can be degraded by the fungi *Penicillium* and *Aspergillus*. In the review, it is mentioned that PCL can be degraded by *Aspergillus* strain ST-01, which was isolated from soil and incubated at 50°C for six days.

2.4   COMPOSTING OF BIOPOLYMERS

Compostable biopolymers are similar to biodegradable biopolymers, with the same goal of reducing waste on earth. However, compostable biopolymers degrade at compost sites with specific conditions. Compared to biodegradation, the composting process will yield excellent results where material completely degrades much faster into carbon dioxide, water, inorganic compounds, and biomass with no distinguishable or toxic residue (Kale et al., 2007). The outcome of the composting process can be utilized as bio-fertilizer to replace chemical fertilizer. Chemical fertilizer is responsible for greenhouse effects, environmental pollution, and the death of soil organisms and marine inhabitants (Ayilara et al., 2020). These results cause farmers to focusing on producing bio-fertilizers by either industrial composting and natural composting, also known as home composting or backyard composting. Generally, the industrial composting process is much faster compared to natural composting because biodegradation is optimized by controlling the conditions such as shredding and controlling the temperature and oxygen level.

Natural composting can be done at home without the specific conditions of industrial composting. This is because natural composting can compost material in the home environment, room temperature, and natural microbial community. Typically, natural composting consists of organic material such as food waste, wood, yard trimmings, and so on (Reyes-Torres et al., 2018). With the emergence of biopolymer products, they also can be included as a substance for natural composting. However, not all biodegradable biopolymers are suitable to undergo natural composting due to factors such as taking long time to compost naturally at ambient temperatures (Song et al., 2009). It is suggested to compost only products with certification. Examples of certification bodies that offer a home compostable certification program are TÜV Austria (formerly known as AIB Vinçotte) and DIN Certco (Endres & Siebert-Raths, 2011). The process of natural composting can occur in a small pile or preferably in a composting bin. The composting process may take up to two years, but with turning, it can be reduced to three to six months. Turning usually done manually, or it can be done with the help of an advanced composting bin equipped with rotating drums (Kale et al., 2007). Ventilation is also important to ensure the composting material undergoes aerobic processes to produce a stable, non-toxic, pathogen-free, and plant nutrient–rich product (Ermolaev et al., 2014).

The purpose of industrial composting is to convert biodegradable waste of biological origin into stable, sanitized products for agricultural use. It is commercial composting at a large scale with certain conditions to help increase the biodegradation rate (Kale et al., 2007). Some conditions that mean industrial composting can degrade faster at a large scale are high temperatures, around 50–70°C; high moisture; and the oxygen content in aerobic process. These controlled conditions allow increasing microbial activity, and the degradation process becomes faster (Endres & Siebert-Raths, 2011). Materials that can be use for industrial composting are the same as materials for natural composting, such as mechanical pulp, food waste, and home-compostable certificate biopolymers. However, industrial composting can compost what natural composting cannot, for example, PLA, which takes a long time to degrade in natural composting (Rujnić-Sokele & Pilipović, 2017). Like home composting, industrial composting also has certification bodies that offer certification for industrial composting. Some example organizations are DIN Certco, TÜV Austria, and the Australasian Bioplastics Association (ABA). Those organizations refer to the EN 13432 standard, which is needed to pass tests related to the biodegradability, disintegrability, ecotoxicity, and quality of compost of the material (Briassoulis et al., 2010). There are three main phases in industrial composting: the mesophilic phase, then the thermophilic phase, and finally the cooling or maturation phase (Palaniveloo et al., 2020).

2.5   ENVIRONMENTAL IMPACT OF BIOPOLYMERS

Synthetic polymers such as PP, PE, and HDPE are extensively employed in food packaging and food containers due to their durability, supply, and cost-effectiveness. However, the disposal of these packaging materials has become a solid waste management challenge and a significant contributor to environmental pollution (Kamarudin et al., 2022). Also, durability and the undesirable accumulation of synthetic polymers are the main threats to the environment through contaminating natural resources such as water quality and soil fertility (Pathak & Navneet, 2017). To overcome these drawbacks, biobased and biodegradable polymeric materials may be among the most appropriate alternatives for certain applications due to concerns about the global climate and the growing difficulties in handling solid wastes (Sudesh & Iwata, 2008).

In recent years, the utilization of biopolymers has shown a marked increase in interest in a variety of industries such as packaging, agriculture, medicine, and others (Shamsuddin et al., 2018). Various biopolymers, biosynthetic polymers, chemosynthetic polymers, their blends, and composites have been studied for packaging applications. Commonly, biopolymers are mostly derived from renewable resources and can be classified into three groups: polymers made from a renewable resource/biomass, such as agro-polymers from agricultural resources; polymers from microbial products or animal origin, such as polyhydroxyalkanoates, which may be useful in medical and pharmaceutical practices; and chemically synthesized biodegradable polymers obtained from petrochemical resources or modified from natural polymers, as presented in Figure 2.1 (Kumaran et al., 2020).

Typical biopolymers such as polylactic acid and starches have made significant inroads in the packaging industry due to their advantages such as good strength and flexibility, non-toxicity, oxygen impermeability, strong moisture resistance, storage stability over a large temperature range, and low cost for both the raw materials and the processing technology. This shows the characteristics of biopolymers meet the criteria required for food packaging (Sudesh & Iwata, 2008; Buffum et al., 2015).

The development of these biopolymers has several potential benefits over petroleum-based polymers, such as cost effectiveness, environmental friendliness, and user-friendly materials. Mostly, the uses of biodegradable materials are intended to lead towards long-term sustainability and a reduction in the environmental impacts of oil-based polymer disposal (Song et al., 2009). According to universal researchers, the production of biopolymers does not pose any significant threat to human beings, plants, and animals. Instead, biopolymers can lead to a safe and healthy environment

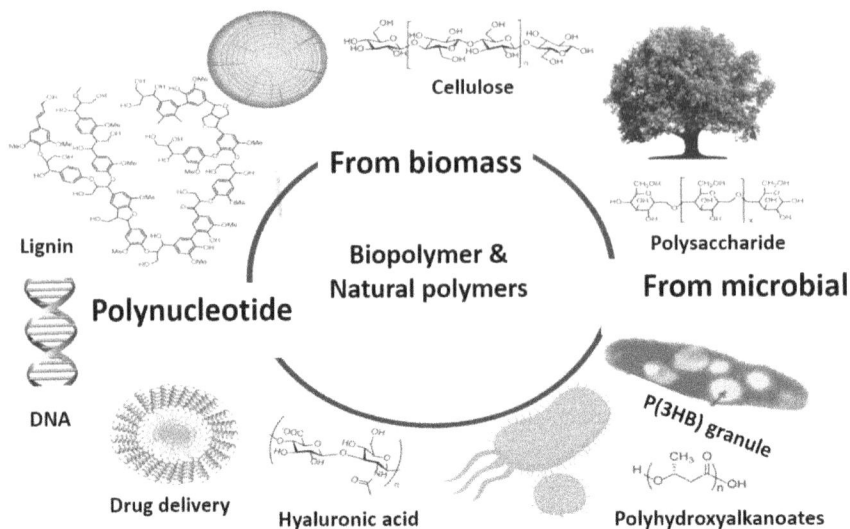

FIGURE 2.1   Structures of the three major sources of biopolymers depending upon origin.

for universal sustainable development (Shamsuddin et al., 2018). Furthermore, the impact created on the environment by biopolymers, especially for CO_2 emissions, is lower compared to synthetic polymers. This is because they could create a "carbon neutral" life cycle, in which the net amount of carbon dioxide emitted into the atmosphere remains relatively constant (Sudesh & Iwata, 2008; Balart et al., 2020). However, the biggest obstacle to using biopolymers as a substitute for petroleum-based polymers is their thermal instability at low temperatures, as well as their low mechanical strength (Sadasivuni et al., 2020).

Other than that, biopolymers have the ability to degrade within a short lifespan compared to synthetic polymers (Shamsuddin et al., 2018). The most widely accepted ASTM definition of biodegradable polymers states that the degradation of these polymers occurs through the activity of naturally occurring microorganisms, such as bacteria, fungi, and algae. This natural decomposition process results in the production of products such as carbon dioxide (CO_2) and water (Mohanty et al., 2000). In fact, bacterial and fungal organisms are the most common biological agents in nature, and they can degrade both natural and synthetic polymers in different ways (Pathak & Navneet, 2017).

During the bioremediation process where microorganisms are employed to remediate contaminated environments, essential nutrients and certain chemicals are provided to support their growth and development. This enables these microorganisms to effectively eliminate pollutants present in contaminated areas (Nandal et al., 2015). Studies in the literature have reported the use of biopolymers such as polypeptides and polysaccharides as critical factors in the processes of bioremediation, recovery of polluted environments, and remediation of heavy metals and petroleum derivatives through natural biopolymers, since they present a high affinity with biological systems (Bedor et al., 2020). As a result, all attempts should be made to procure products from natural sources that have a high rate of biodegradation in the ecosystem, occupying positions and displacing conventional plastics in order to restore the environment that has been harmed so far by the indiscriminate use of synthetic polymers and prevent further deterioration (Kamarudin et al., 2022).

2.6 CONCLUSIONS

Conventional synthetic polymers are challenging to recycle due to their highly heterogeneous nature, which leads to environmental pollution. The environmental impacts associated with plastic waste management techniques cannot be ignored. In contrast, bioplastics offer a viable solution to reduce dependence on conventional petrochemical-based synthetic plastics. Renewable resources like agricultural crop residues and woody biomasses can be used as raw materials to recover the building blocks of bioplastics and biopolymers. Furthermore, bioplastics are biocompatible and biodegradable and possess mechanical properties equal to or better than petrochemical-based plastics. Microbe-derived biopolymers show promise in various ways, but their production still needs to be cost effective. Currently, biodegradable packaging is used for food items that do not require high impermeability to oxygen and water vapor and have a short storage period, such as fresh green groceries and fruits, or for long-storage products like dumplings and fries that do not demand excessive impermeable properties. However, the range of available films offers various properties that can be tailored for packaging other food products with stricter requirements. Recent advancements in biotechnology and material engineering have positively impacted the production of biopolymers and the understanding of their biodegradation characteristics.

Further research and development in recycling techniques, large-scale production, cost reduction, durability, sustainability, greenhouse gas emissions, and optimized biodegradation are needed for bioplastics. Inedible plant residues, microbial biomass, and derivatives containing cellulose fibers and lignin biopolymers could be suitable alternatives for bioplastic manufacturing. Bioplastics are considered environmentally friendly products in terms of sustainability and environmental risk assessment. Therefore, government support and social awareness are crucial to facilitate a shift from petrochemical products to biobased products.

REFERENCES

Abdul Khalil, H. P. S., Yahya, E. B., Jummaat, F., Adnan, A. S., Olaiya, N. G., Rizal, S., Abdullah, C. K., Pasquini, D., & Thomas, S. (2023). Biopolymers based aerogels: A review on revolutionary solutions for smart therapeutics delivery. *Progress in Materials Science*, *131*, 101014.

Allison, E. H., & Bassett, H. R. (2015). Climate change in the oceans: Human impacts and responses. *Science*, *350*(6262), 778–782. https://doi.org/10.1126/science.aac8721

Andreeßen, C., & Steinbüchel, A. (2019). Recent developments in non-biodegradable biopolymers: Precursors, production processes, and future perspectives. *Applied Microbiology and Biotechnology*, *103*(1), 143–157. https://doi.org/10.1007/s00253-018-9483-6

Arvanitoyannis, I. S. (2013). Waste management for polymers in food packaging industries. In *Plastic films in food packaging* (1st ed.). Elsevier Inc. https://doi.org/10.1016/B978-1-4557-3112-1.00014-4

Ayilara, M. S., Olanrewaju, O. S., & Babalola, O. O. (2020). Waste management through composting: Challenges and potentials. *Sustainability, 12*(11), 1–23.

Azahari, N. A., Othman, N., & Ismail, H. (2011). Biodegradation studies of polyvinyl alcohol/corn starch blend films in solid and solution media. *Journal of Physical Science, 22*(2), 15–31.

Azevedo, H. S., Santos, T. C., & Reis, R. L. (2008). Controlling the degradation of natural polymers for biomedical applications. In *Natural-based polymers for biomedical applications* (pp. 106–128). https://doi.org/10.1533/9781845694814.1.106

Babooram, K. (2020). Brief overview of polymer science. In *Polymer science and nanotechnology* (pp. 3–12). https://doi.org/10.1016/b978-0-12-816806-6.00001-7

Balart, R., Montanes, N., Dominici, F., Boronat, T., & Torres-Giner, S. (2020). Environmentally friendly polymers and polymer composites. *Materials, 13*(21), 1–6. https://doi.org/10.3390/ma13214892

Banerjee, A., Chatterjee, K., & Madras, G. (2014). Enzymatic degradation of polymers: A brief review. *Materials Science and Technology (United Kingdom), 30*(5), 567–573. https://doi.org/10.1179/1743284713Y.0000000503

Barnes, D. K. A., Galgani, F., Thompson, R. C., & Barlaz, M. (2009). Accumulation and fragmentation of plastic debris in global environments. *Philosophical Transactions of the Royal Society B: Biological Sciences, 364*(1526), 1985–1998. https://doi.org/10.1098/rstb.2008.0205

Bassas-Galià, M. (2017). Rediscovering biopolymers. In *Handbook of hydrocarbon and lipid microbiology* (pp. 529–550). https://doi.org/10.1007/978-3-319-50436-0

Bastioli, C. (1998). Properties and applications of Mater-Bi starch-based materials. *Polymer Degradation and Stability, 59*(1–3), 263–272.

Bátori, V., Åkesson, D., Zamani, A., Taherzadeh, M. J., & Sárvári Horváth, I. (2018). Anaerobic degradation of bioplastics: A review. *Waste Management, 80*, 406–413. https://doi.org/10.1016/j.wasman.2018.09.040

Bedor, P. B. A., Caetano, R. M. J., De Souza, F. G., & Leite, S. G. F. (2020). Advances and perspectives in the use of polymers in the environmental area: A specific case of PBS in bioremediation. *Polimeros, 30*(2), 1–10. https://doi.org/10.1590/0104-1428.02220

Benedict, C. V, Cook, W. J., Jarrett, P., Cameron, J. a, Huangt, S. J., & Belli, J. P. (1983). Fungal degradation of polycaprolactones. *Journal of Applied Polymer Science, 28*(1), 327–334.

Bonartsev, A. P., Bonartseva, G. A., Shaitan, K. V., & Kirpichnikov, M. P. (2011). Poly(3-hydroxybutyrate) and poly(3-hydroxybutyrate)-based biopolymer systems. *Biochemistry (Moscow) Supplement Series B: Biomedical Chemistry, 5*(1), 10–21. https://doi.org/10.1134/S1990750811010045

Briassoulis, D., Dejean, C., & Picuno, P. (2010). Critical review of norms and standards for biodegradable agricultural plastics Part II : Composting. *Journal of Polymers and the Environment, 18*, 364–383. https://doi.org/10.1007/s10924-010-0222-z

Buffum, K., Pacheco, H., & Shivkumar, S. (2015). Environmental effects on the properties of biopolymer service-ware products. *Polymer—Plastics Technology and Engineering, 54*(5), 506–514. https://doi.org/10.1080/03602559.2014.935410

Cam, D., Hyon, S. H., & Ikada, Y. (1995). Degradation of high molecular weight poly(l-lactide) in alkaline medium. *Biomaterials, 16*(11), 833–843. https://doi.org/10.1016/0142-9612(95)94144-A

Colón, J., Mestre-Montserrat, M., Puig-Ventosa, I., & Sánchez, A. (2013). Performance of compostable baby used diapers in the composting process with the organic fraction of municipal solid waste. *Waste Management, 33*(5), 1097–1103. https://doi.org/10.1016/j.wasman.2013.01.018

Di Martino, A., Sittinger, M., & Risbud, M. V. (2005). Chitosan: A versatile biopolymer for orthopaedic tissue-engineering. *Biomaterials, 26*(30), 5983–5990. https://doi.org/10.1016/j.biomaterials.2005.03.016

Elbanna, K., Lütke-Eversloh, T., Jendrossek, D., Luftmann, H., & Steinbüchel, A. (2004). Studies on the biodegradability of polythioester copolymers and homopolymers by polyhydroxyalkanoate (PHA)-degrading bacteria and PHA depolymerases. *Archives of Microbiology, 182*(2–3), 212–225. https://doi.org/10.1007/s00203-004-0715-z

Emadian, S. M., Onay, T. T., & Demirel, B. (2017). Biodegradation of bioplastics in natural environments. *Waste Management, 59*, 526–536. https://doi.org/10.1016/j.wasman.2016.10.006

Endres, H.-J., & Siebert-Raths, A. (2011). End-of-life options for biopolymers. In *Engineering biopolymers* (pp. 225–243). https://doi.org/10.3139/9783446430020.fm

Ermolaev, E., Sundberg, C., Pell, M., & Jönsson, H. (2014). Bioresource technology greenhouse gas emissions from home composting in practice. *Bioresource Technology, 151*, 174–182. https://doi.org/10.1016/j.biortech.2013.10.049

Fields, R. D., Rodriguez, F., & Finn, R. K. (1974). Microbial degradation of polyesters: Polycaprolactone degraded by P. pullulans. *Journal of Applied Polymer Science, 18*(12), 3571–3579. https://doi.org/10.1002/app.1974.070181207

Folino, A., Karageorgiou, A., Calabrò, P. S., & Komilis, D. (2020). Biodegradation of wasted bioplastics in natural and industrial environments: A review. *Sustainability, 12*(15), 1–37. https://doi.org/10.3390/su12156030

Gautam, R., Bassi, A. S., & Yanful, E. K. (2007). A review of biodegradation of synthetic plastic and foams. *Applied Biochemistry And Biotechnology, 141*(2), 85–108.

Ghorpade, V. M., Gennadios, A., & Hanna, M. A. (2001). Laboratory composting of extruded poly(lactic acid) sheets. *Bioresource Technology, 76*(1), 57–61. https://doi.org/10.1016/S0960-8524(00)00077-8

Glaser, J. A. (2019). Biological degradation of polymers in the environment. In *Plastics in the environment* (pp. 73–94). https://doi.org/10.5772/intechopen.85124

Grima, S., Bellon-Maurel, V., Feuilloley, P., & Silvestre, F. (2000). Aerobic biodegradation of polymers in solid-state conditions: A review of environmental and physicochemical parameter settings in laboratory simulations. *Journal of Polymers and the Environment, 8*(4), 183–195. https://doi.org/10.1023/A:1015297727244

Gu, J. D. (2003). Microbiological deterioration and degradation of synthetic polymeric materials: Recent research advances. *International Biodeterioration and Biodegradation, 52*(2), 69–91. https://doi.org/10.1016/S0964-8305(02)00177-4

Hueck, H. J. (2001). The biodeterioration of materials: An appraisal. *International Biodeterioration and Biodegradation.* https://doi.org/10.1016/S0964-8305(01)00061-0

Imre, B., & Pukánszky, B. (2013). Compatibilization in bio-based and biodegradable polymer blends. *European Polymer Journal, 49*(6), 1215–1233. https://doi.org/10.1016/j.eurpolymj.2013.01.019

Ishigaki, T., Kawagoshi, Y., Ike, M., & Fujita, M. (1999). Biodegradation of a polyvinyl alcohol-starch blend plastic film. *World Journal of Microbiology and Biotechnology, 15*(3), 321–327. https://doi.org/10.1023/A:1008919218289

Itävaara, M., Karjomaa, S., & Selin, J. F. (2002). Biodegradation of polylactide in aerobic and anaerobic thermophilic conditions. *Chemosphere, 46*(6), 879–885. https://doi.org/10.1016/S0045-6535(01)00163-1

Itävaara, M., Vikman, M., & Venelampi, O. (1997). Windrow composting of biodegradable packaging materials. *Compost Science and Utilization*, *5*(2), 84–92. https://doi.org/10. 1080/1065657X.1997.10701877

Jarerat, A., & Tokiwa, Y. (2001). Degradation of poly(L-lactide) by a fungus. *Macromolecular Bioscience*, *1*(4), 136–140. https://doi.org/10.1002/1616-5195(20010601)1:4<136::AID-MABI136>3.0.CO;2-3

Jendrossek, D., Schirmer, A., & Schlegel, H. G. (1996). Biodegradation of polyhydroxyalkanoic acids. *Applied Microbiology and Biotechnology*, *46*(5–6), 451–463. https://doi.org/10.1007/s002530050844

Kale, G., Kijchavengkul, T., Auras, R., Rubino, M., Selke, S. E., & Singh, S. P. (2007). Compostability of bioplastic packaging materials: An overview. *Macromolecular Bioscience*, *7*(3), 255–277. https://doi.org/10.1002/mabi.200600168

Kamarudin, S. H., Rayung, M., Abu, F., Ahmad, S. B., Fadil, F., Karim, A. A., Norizan, M. N., Sarifuddin, N., Mat Desa, M. S. Z., Mohd Basri, M. S., & Samsudin, H. (2022). A review on antimicrobial packaging from biodegradable polymer composites. *Polymers*, *14*(1), 174. https://doi.org/10.3390/polym14010174

Karamanlioglu, M., Preziosi, R., & Robson, G. D. (2017). Abiotic and biotic environmental degradation of the bioplastic polymer poly(lactic acid): A review. *Polymer Degradation and Stability*, *137*, 122–130. https://doi.org/10.1016/j.polymdegradstab.2017.01.009

Kikkawa, Y., Abe, H., Iwata, T., Inoue, Y., & Doi, Y. (2002). Crystallization, stability, and enzymatic degradation of poly(L-lactide) thin film. *Biomacromolecules*, *3*(2), 350–356. https://doi.org/10.1021/bm015623z

Kumaran, S. K., Chopra, M., Oh, E., & Choi, H.-J. (2020). Biopolymers and natural polymers. In *Polymer science and nanotechnology* (pp. 245–256). https://doi.org/10.1016/b978-0-12-816806-6.00011-x

Kyrikou, I., & Briassoulis, D. (2007). Biodegradation of agricultural plastic films: A critical review. *Journal of Polymers and the Environment*, *15*(2), 125–150. https://doi.org/10.1007/s10924-007-0053-8

Leja, K., & Lewandowicz, G. (2010). Polymer biodegradation and biodegradable polymers: A review. *Polish Journal of Environmental Studies*, *19*(2), 255–266.

Lucas, N., Bienaime, C., Belloy, C., Queneudec, M., Silvestre, F., & Nava-Saucedo, J. E. (2008). Polymer biodegradation: Mechanisms and estimation techniques: A review. *Chemosphere*, *73*(4), 429–442. https://doi.org/10.1016/j.chemosphere.2008.06.064

Luckachan, G. E., & Pillai, C. K. S. (2011). Biodegradable polymers: A review on recent trends and emerging perspectives. *Journal of Polymers and the Environment*, *19*(3), 637–676. https://doi.org/10.1007/s10924-011-0317-1

Lugauskas, A., Levinskaite, L., & Peĉiulyte, D. (2003). Micromycetes as deterioration agents of polymeric materials. *International Biodeterioration and Biodegradation*, *52*(4), 233–242. https://doi.org/10.1016/S0964-8305(03)00110-0

Massardier-Nageotte, V., Pestre, C., Cruard-Pradet, T., & Bayard, R. (2006). Aerobic and anaerobic biodegradability of polymer films and physico-chemical characterization. *Polymer Degradation and Stability*, *91*(3), 620–627. https://doi.org/10.1016/j.polymdegradstab.2005.02.029

Mittal, A., Garg, S., Kohli, D., Maiti, M., Jana, A. K., & Bajpai, S. (2016). Effect of cross linking of PVA/starch and reinforcement of modified barley husk on the properties of composite films. *Carbohydrate Polymers*, *151*, 926–938. https://doi.org/10.1016/j.carbpol.2016.06.037

Mo, X., Sun, B., Wu, T., & Li, D. (2018). Electrospun nanofibers for tissue engineering. In B. D. Yu, X. Wang, & Jianyong (Eds.), *Electrospinning: Nanofabrication and applications* (pp. 719–734). Elsevier Inc. https://doi.org/10.1016/B978-0-323-51270-1.00024-8

Mohamad Haafiz, M. K., Eichhorn, S. J., Hassan, A., & Jawaid, M. (2013). Isolation and characterization of microcrystalline cellulose from oil palm biomass residue. *Carbohydrate Polymers, 93*(2), 628–634. https://doi.org/10.1016/j.carbpol.2013.01.035

Mohanty, A. K., Misra, M., & Hinrichsen, G. (2000). Biofibres, biodegradable polymers and biocomposites: An overview. *Macromolecular Materials and Engineering, 276–277*, 1–24. https://doi.org/10.1002/(SICI)1439-2054(20000301)276:1<1::AID-MAME1>3.0.CO;2-W

Mohee, R., Unmar, G. D., Mudhoo, A., & Khadoo, P. (2008). Biodegradability of biodegradable/degradable plastic materials under aerobic and anaerobic conditions. *Waste Management, 28*(9), 1624–1629. https://doi.org/10.1016/j.wasman.2007.07.003

Mukherjee, T., & Kao, N. (2011). PLA based biopolymer reinforced with natural fibre: A review. *Journal of Polymers and the Environment, 19*(3), 714–725. https://doi.org/10.1007/s10924-011-0320-6

Nakamura, K., Tomita, T., Abe, N., & Kamio, Y. (2001). Purification and characterization of an extracellular poly(L-lactic acid) depolymerase from a soil isolate, amycolatopsis sp. strain K104–1. *Applied and Environmental Microbiology, 67*(1), 345–353. https://doi.org/10.1128/AEM.67.1.345-353.2001

Nandal, M., Solanki, P., Rastogi, M., & Hooda, R. (2015). Bioremediation: A sustainable tool for environmental management of oily sludge. *Nature Environment and Pollution Technology, 14*(1), 181–190.

Nobes, G. A. R., Marchessault, R. H., Briese, B. H., & Jendrossek, D. (1998). Microscopic visualization of the enzymatic degradation of poly(3HB-co-3HV) and poly(3HV) single crystals by PHA depolymerases from Pseudomonas leimoignei. *Journal of Environmental Polymer Degradation, 6*(2), 99–107. https://doi.org/10.1023/A:1022858206228

Ohkita, T., & Lee, S. H. (2006). Thermal degradation and biodegradability of poly (lactic acid)/corn starch biocomposites. *Journal of Applied Polymer Science, 100*(4), 3009–3017. https://doi.org/10.1002/app.23425

Palaniveloo, K., Amran, M. A., Norhashim, N. A., Mohamad-Fauzi, N., Peng-Hui, F., Hui-Wen, L., Kai-Lin, Y., Jiale, L., Chian-Yee, M. G., Jing-Yi, L., Gunasekaran, B., & Razak, S. A. (2020). Food waste composting and microbial community structure profiling. *Processes, 8*(6), 723–753. https://doi.org/10.3390/pr8060723

Parvin, F., Rahman, M. A., Islam, J. M. M., Khan, M. A., & Saadat, A. H. M. (2010). Preparation and characterization of starch/PVA blend for biodegradable packaging material. *Advanced Materials Research, 123–125*, 351–354. https://doi.org/10.4028/www.scientific.net/AMR.123-125.351

Patel, P. N., Jyoti Sen, D., Parmar, K. G., Nakum, A. N., Patel, M. N., Patel, P. R., & Patel, V. R. (2011). Biodegradable polymers: An ecofriendly approach in Newer Millenium. *Asian Journal of Biomedical and Pharmaceutical Sciences, 1.*

Pathak, V. M., & Navneet. (2017). Review on the current status of polymer degradation: A microbial approach. *Bioresources and Bioprocessing, 4*(1). https://doi.org/10.1186/s40643-017-0145-9

Poznyak, T. I., Oria, I., & Poznyak, A. S. (2019). Biodegradation. In *Ozonation and biodegradation in environmental engineering* (pp. 353–388). https://doi.org/10.1016/B978-0-12-812847-3.00023-8

Rapisarda, M., La Mantia, F. P., Ceraulo, M., Mistretta, M. C., Giuffrè, C., Pellegrino, R., Valenti, G., & Rizzarelli, P. (2019). Photo-oxidative and soil burial degradation of irrigation tubes based on biodegradable polymer blends. *Polymers, 11*(9), 1–13. https://doi.org/10.3390/polym11091489

Reyes-torres, M., Oviedo-ocaña, E. R., Dominguez, I., Komilis, D., & Sánchez, A. (2018). A systematic review on the composting of green waste: Feedstock quality and optimization strategies. *Waste Management*, *77*, 486–499. https://doi.org/10.1016/j. wasman.2018.04.037

Rizzarelli, P., & Carroccio, S. (2014). Modern mass spectrometry in the characterization and degradation of biodegradable polymers. *Analytica Chimica Acta*, *808*, 18–43. https:// doi.org/10.1016/j.aca.2013.11.001

Robertson, G. L. (2019). History of food packaging. In *Reference module in food science* (pp. 1–49). Elsevier. https://doi.org/10.1016/b978-0-08-100596-5.22535-3

Rudnik, E. (2008). Compostable polymer materials—definitions, structures and methods of preparation. In *Compostable polymer materials* (pp. 11–48). https://doi.org/10.1016/ b978-008045371-2.50004-4

Rudnik, E. (2019). Composting methods and legislation. In *Compostable polymer materials* (pp. 127–161). https://doi.org/10.1016/b978-0-08-099438-3.00005-7

Ruiz-Dueñas, F. J., & Martínez, Á. T. (2009). Microbial degradation of lignin: How a bulky recalcitrant polymer is efficiently recycled in nature and how we can take advantage of this. *Microbial Biotechnology*, *2*(2), 164–177. https://doi.org/10.1111/j. 1751-7915.2008.00078.x

Rujnić-Sokele, M., & Pilipović, A. (2017). Challenges and opportunities of biodegradable plastics: A mini review. *Waste Management and Research*, *35*(2), 132–140. https://doi. org/10.1177/0734242X16683272

Ryan, C. A., Billington, S. L., & Criddle, C. S. (2018). Biocomposite fiber-matrix treatments that enhance in-service performance can also accelerate end-of-life fragmentation and anaerobic biodegradation to methane. *Journal of Polymers and the Environment*, *26*(4), 1715–1726. https://doi.org/10.1007/s10924-017-1068-4

Sadasivuni, K. K., Saha, P., Adhikari, J., Deshmukh, K., Ahamed, M. B., & Cabibihan, J. J. (2020). Recent advances in mechanical properties of biopolymer composites: A review. *Polymer Composites*, *41*(1), 32–59. https://doi.org/10.1002/pc.25356

Sadeghi, G. M. M., & Mahsa, S. (2015). Compostable polymers and nanocomposites—A big chance for planet earth. In *Recycling materials based on environmentally friendly techniques* (pp. 63–104). https://doi.org/10.5772/59398

Shamsuddin, I. M., N, S., M, A., & MK, A. (2018). Biodegradable polymers for sustainable environmental and economic development. *MOJ Bioorganic & Organic Chemistry*, *2*(4), 192–194. https://doi.org/10.15406/mojboc.2018.02.00080

Siracusa, V. (2019). Microbial degradation of synthetic biopolymers waste. *Polymers*, *11*(6), 1–18. https://doi.org/10.3390/polym11061066

Song, J. H., Murphy, R. J., Narayan, R., & Davies, G. B. H. (2009). Biodegradable and compostable alternatives to conventional plastics. *Philosophical Transactions of the Royal Society B: Biological Sciences*, *364*(1526), 2127–2139. https://doi.org/10.1098/ rstb.2008.0289

Spiridon, I., Popescu, M. C., Bodârlău, R., & Vasile, C. (2008). Enzymatic degradation of some nanocomposites of poly(vinyl alcohol) with starch. *Polymer Degradation and Stability*, *93*(10), 1884–1890. https://doi.org/10.1016/j.polymdegradstab.2008.07.017

Steinbüchel, A. (2005). Non-biodegradable biopolymers from renewable resources: Perspectives and impacts. *Current Opinion in Biotechnology*, *16*(6), 607–613. https:// doi.org/10.1016/j.copbio.2005.10.011

Stewart, S. A., Domínguez-Robles, J., Donnelly, R. F., & Larrañeta, E. (2018). Implantable polymeric drug delivery devices: Classification, manufacture, materials, and clinical applications. *Polymers*, *10*(12), 1–24. https://doi.org/10.3390/polym10121379

Sudesh, K., & Iwata, T. (2008). Sustainability of biobased and biodegradable plastics. *Clean— Soil, Air, Water*, *36*(5–6), 433–442. https://doi.org/10.1002/clen.200700183

Surya, I., Hazwan, C. M., Abdul Khalil, H. P. S., Yahya, E. B., Suriani, A. B., Danish, M., & Mohamed, A. (2022). Hydrophobicity and biodegradability of silane-treated nanocellulose in biopolymer for high-grade packaging applications. *Polymers, 14*(19), 4147.

Thakur, S., Chaudhary, J., Sharma, B., Verma, A., Tamulevicius, S., & Thakur, V. K. (2018). Sustainability of bioplastics: Opportunities and challenges. *Current Opinion in Green and Sustainable Chemistry, 13*, 68–75. https://doi.org/10.1016/j.cogsc.2018.04.013

Thompson, R. C., Moore, C. J., Vom Saal, F. S., & Swan, S. H. (2009). Plastics, the environment and human health: Current consensus and future trends. *Philosophical Transactions of the Royal Society of London. Series B, Biological Sciences, 364*, 2153–2166. https://doi.org/10.1098/rstb.2009.0053

Tokiwa, Y., & Calabia, B. P. (2006). Biodegradability and biodegradation of poly(lactide). *Applied Microbiology and Biotechnology, 72*(2), 244–251. https://doi.org/10.1007/s00253-006-0488-1

Trivedi, P., Hasan, A., Akhtar, S., Siddiqui, M. H., Sayeed, U., Kalim, M., & Khan, A. (2016). Role of microbes in degradation of synthetic plastics and manufacture of bioplastics. *Journal of Chemical and Pharmaceutical Research, 8*(3), 211–216. www.jocpr.com

Tsuji, H., & Ikada, Y. (1998). Properties and morphology of poly(L-lactide). II. Hydrolysis in alkaline solution. *Journal of Polymer Science, Part A: Polymer Chemistry, 36*(1), 59–66. https://doi.org/10.1002/(SICI)1099-0518(19980115)36:1<59::AID-POLA9>3.0.CO;2-X

Tsuji, H., & Ishizaka, T. (2001). Preparation of porous poly(ε-caprolactone) films from blends by selective enzymatic removal of poly(L-lactide). *Macromolecular Bioscience, 1*(2), 59–65. https://doi.org/10.1002/1616-5195(20010301)1:2<59::aid-mabi59>3.0.co;2-6

Vieira, M. G. A., Da Silva, M. A., Dos Santos, L. O., & Beppu, M. M. (2011). Natural-based plasticizers and biopolymer films: A review. *European Polymer Journal, 47*(3), 254–263. https://doi.org/10.1016/j.eurpolymj.2010.12.011

Vikman, M., Hulleman, S. H. D., Van Der Zee, M., Myllärinen, P., & Feil, H. (1999). Morphology and enzymatic degradation of thermoplastic starch-polycaprolactone blends. *Journal of Applied Polymer Science, 74*(11), 2594–2604. https://doi.org/10.1002/(SICI)1097-4628(19991209)74:11<2594::AID-APP5>3.0.CO;2-R

Vishnuvardhan Reddy, S., Thirumala, M., & Mahmood, S. K. (2009). Production of PHB and P (3HB-co-3HV) biopolymers by Bacillus megaterium strain OU303A isolated from municipal sewage sludge. *World Journal of Microbiology and Biotechnology, 25*(3), 391–397. https://doi.org/10.1007/s11274-008-9903-3

Volova, T. G., Boyandin, A. N., Vasiliev, A. D., Karpov, V. A., Prudnikova, S. V., Mishukova, O. V., Boyarskikh, U. A., Filipenko, M. L., Rudnev, V. P., Xuân, B. B., Dũng, V. V., . . . & Gitelson, I. I. (2010). Biodegradation of polyhydroxyalkanoates (PHAs) in tropical coastal waters and identification of PHA-degrading bacteria. *Polymer Degradation and Stability, 95*(12), 2350–2359. https://doi.org/10.1016/j.polymdegradstab.2010.08.023

Walsh, J. H. (2001). Ecological considerations of biodeterioration. *International Biodeterioration and Biodegradation.* https://doi.org/10.1016/S0964-8305(01)00063-4

Wittaya, T. (2009). Microcomposites of rice starch film reinforced with microcrystalline cellulose from palm pressed fiber. *International Food Research Journal, 16*(4), 493–500.

Wojtowicz, A. (2010). Biodegradability and compostability of biopolymers. In *Thermoplastic starch: A green material for various industries* (pp. 55–76). https://doi.org/10.1002/9783527628216.ch3

Woolnough, C. A., Charlton, T., Yee, L. H., Sarris, M., & Foster, L. J. R. (2008). Surface changes in polyhydroxyalkanoate films during biodegradation and biofouling. *Polymer International, 57*(9), 1042–1051. https://doi.org/10.1002/pi

Xu, Y., & Hanna, M. A. (2005). Preparation and properties of biodegradable foams from starch acetate and poly(tetramethylene adipate-co-terephthalate). *Carbohydrate Polymers*, *59*(4), 521–529. https://doi.org/10.1016/j.carbpol.2004.11.007

Zeng, S. H., Duan, P. P., Shen, M. X., Xue, Y. J., & Wang, Z. Y. (2016). Preparation and degradation mechanisms of biodegradable polymer: A review. In *IOP conference series: Materials science and engineering* (Vol. 137, p. 012003). https://doi.org/10.1088/1757-899X/137/1/012003

3 State-of-the-Art Natural Biopolymers for Bionanocomposites

3.1 BIONANOCOMPOSITES

Before understanding why bionanocomposites play an important role in today's research and industry fields, it is crucial to lay out the fundamental context of bionanocomposites. A bionanocomposite system is basically formed from a combination of two or more materials with different properties (Zabihi et al., 2018), as shown in Figure 3.1. The different properties of the material have the ability to complement each other, resulting in enhanced physical, mechanical, and thermal properties.

Bionanocomposite is commonly used to describe nanocomposites that incorporate a biologically derived polymer (biopolymer) in conjunction with an inorganic component, with at least one dimension at the nanoscale to exhibit significant performance on a material. Back in the early 1990s, Toyota pioneered the use of nanocomposites. During their research, they made a breakthrough by demonstrating that montmorillonite (MMT) could be exfoliated into individual nanoparticles. This breakthrough substantially improved the dimensional stability, water resistance, and gas barrier properties of nylon-6. Consequently, this discovery sparked widespread interest and initiated numerous research endeavors in the field of nanocomposites (Hussain et al., 2006). However, bionanocomposites made of biobased materials or biopolymers have been separated out to be an individual class from nanocomposites. This is due to the difference in preparation process, properties, and functionalities found between nanocomposites and bionanocomposites (M. H. Mousa et al., 2016).

Bionanocomposites have gained tremendous attention in interdisciplinary areas, including biology, materials science, and nanotechnology. The curve of research is progressing, as shown in Chapter 1, due to the rise of environmental concerns. These materials are often referred to as natural nanocomposites, green nanocomposites, and biobased nanocomposites, owing to their reliance on constituents derived from biological sources (Sen, 2020). Jeevanandam et al. (2018) defined them as materials that contain a constituent or constituents of biological origin and nanoparticles with at least one dimension in the range of 1–100 nm. The biological origin refers to natural biopolymers such as polysaccharides, proteins, and natural gums derived from plants, algae, or animals (Jeevanandam et al., 2018).

In the present day, numerous bionanocomposites have been meticulously developed through extensive research. Among the various bionanocomposites, which encompass films, hydrogels, and aerogels, Table 3.1 provides a comprehensive overview of their respective advantages and disadvantages in comparison to conventional

DOI: 10.1201/9781003416043-3

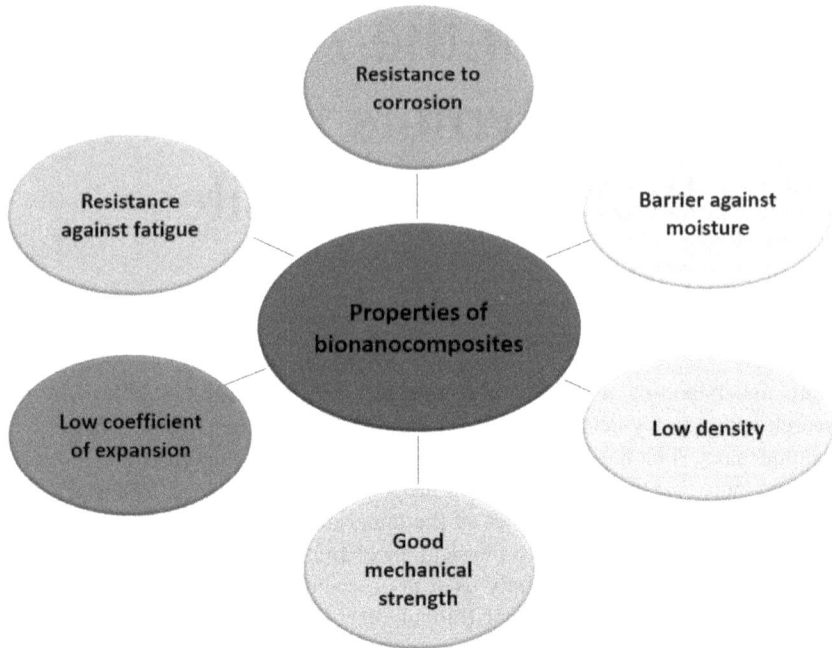

FIGURE 3.1 Six main properties of bionanocomposites, which result in improvement of physical, mechanical, and thermal properties.

TABLE 3.1

Advantages and Disadvantages of Bionanocomposites over Nanocomposites

Advantages	Disadvantages
• Biodegradable	• Lower mechanical strength
• Biocompatible	• Lower thermal stability
• Renewable resources	• Higher moisture absorption
• Abundant	• Preformed polymers
• Reduced packaging volume, weight, and waste	
• Some are edible	
• Function as carriers for antimicrobial and antioxidant agents	
• Microencapsulation and controlled release of active ingredients	

---

nanocomposites (Borgonovo & Apelian, 2011; Thakur et al., 2017; Gunputh & Le, 2020). Nonetheless, the enhancement properties of bionanocomposites depend on the characteristics of the biopolymers, the stoichiometric ratio of the constituent materials, the crosslinking among the constituent materials, and the biocompatibility between the biopolymer matrix and the reinforcement filler or particles. The interaction between fillers at nanometer-scale particles acts as a bridge in the biopolymer matrix (Cader Mhd Haniffa et al., 2016).

3.2 CONSTITUENTS OF BIONANOCOMPOSITES

Bionanocomposites are characterized by one or more discontinuous phase distributed in one continuous phase, whereby the continuous phase refers to the matrix, while the dispersed or discontinuous phase refers to the reinforcement particles or fillers (Siqueira et al., 2010; Abdul Khalil et al., 2019). The discontinuous phase has at least one dimension, roughly 10^{-9} m (Figure 3.2).

Examples of biopolymer matrices used to develop bionanocomposites can be from natural biopolymers that are fully biodegradable, biopolymers that are partially biodegradable thermoplastic polymers, thermoset polymers, or fully biodegradable petroleum-based biodegradable polymer matrices, as mentioned in Chapter 2. However, natural biopolymers and synthetic biodegradable polymers incorporated with nano-scale fillers/particles play a major role in the fabrication of bionanocomposites, as they determine the end result of the biodegradability properties (Othman, 2014). The matrix/continuous phase is made up of natural biopolymers, synthetic biodegradable polymers, or a combination of both materials. Each of these materials inherently possesses different characteristics, including molecular arrangement, active functional groups, bonding nature, thermal behavior, and solubility (George et al., 2020; Zabihi et al., 2018). The matrix phase is like the bulk material that contains particles or fillers dispersed in it. It acts like a glue cementing the particles or fillers together, as illustrated in Figure 3.3. Apart from the continuous phase, fillers play an equally important role to enhance the performance of the end product. This is because the fillers are able to change the characteristics of the matrix they are applied to and support the matrix by reducing shrinkage; increasing thermal tolerance; providing a reinforcement effect; increasing strength, especially tensile,

Matrix (continuous phase)

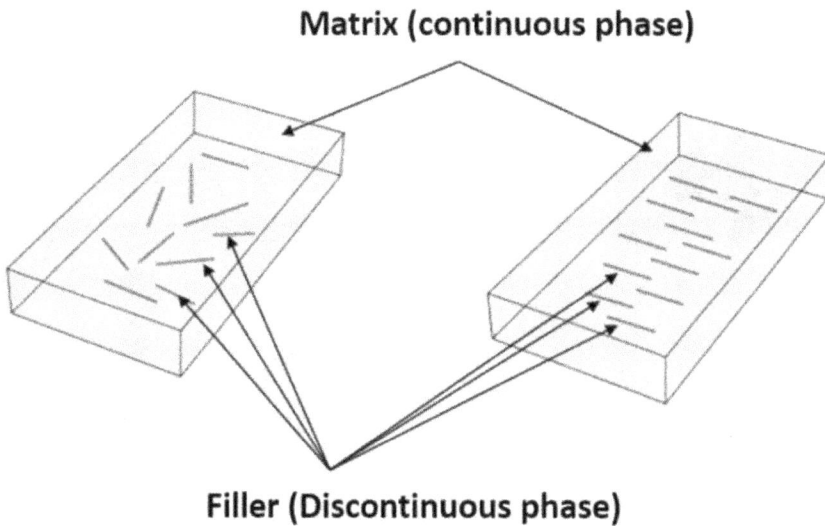

Filler (Discontinuous phase)

FIGURE 3.2 Phases of bionanocomposites between the dispersion of filler and matrix.

and resistance to tears and compression; enhancing the exposure to solvents; and compounding cost reduction (Palem et al., 2018; Rajak et al., 2019).

Fillers and reinforcements differ in terms of features and specifications. Fillers can be divided into two main groups, reinforcing fillers (e.g. carbon black, silica, and fibers) and inert fillers (e.g. clay, calcium carbonate, etc.) (Fahim et al., 2018; Gutiérrez et al., 2019; Gojayev et al., 2020). Carbon fiber reinforcement is a non-woven, carbon fiber usually used to reduce cracking and extend life in bionano-composites (Thakur et al., 2017; Jeevanandam et al., 2018). Different polymer reinforcement fillers/inert fillers are used to enhance the properties of the matrix. Some fillers can be targeted to improve specific properties such as brightness, density, abrasion, fineness, and oil absorption capability (Zare et al., 2017; Mishra et al., 2019), for instance, fiber reinforcement for concrete to enhance properties such as low shrinkage, good thermal expansion, substantial modulus of elasticity, high tensile strength, improved fatigue, and impact resistance (Müller et al., 2017; Fu et al., 2019; Abdul Khalil et al., 2019). Among the many fillers, the common types of filler structures applied in bionanocomposites are spherical, rod-like, and layered structures, as shown in Figure 3.3.

When designing bionanocomposites, there are many methods and techniques to synthesize bionanocomposites (Sanusi et al., 2021). Nevertheless, a few pointers

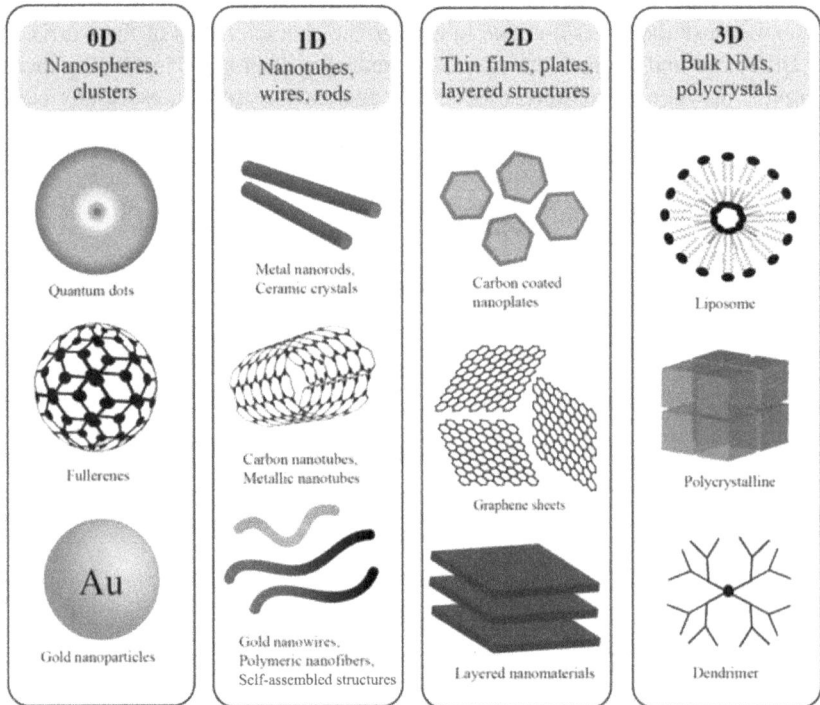

FIGURE 3.3 Common types of filler structures applied in bionanocomposites, including nanomaterials (NMs) with different dimensionalities. Reproduced from Poh et al. (2018).

must be considered prior to fabrication to prevent unnecessary wastage and effort (Dantas de Oliveira and Augusto Gonçalves Beatrice, 2019; Dong et al., 2018; Rajak et al., 2019):

- Determine the proportion of matrix and dispersed phase based on the intended use of the composite.
  - **Reasons:** Excessive fillers can give rigidity to the bionanocomposites. Thus, there is a balance to be struck that normally requires optimization.
- Determine the size and shape of the fillers of the dispersed phase.
  - **Reasons:** This is because the size and shape of the particles/fillers can directly impact the reinforcement effect on the matrix. Usually, smaller particles provide more surface area for contact with the matrix, while longer fibers exhibit better reinforcement.
- Ensure compatibility between the matrix and the particles.
  - **Reasons:** The interface between the matrix and fillers controls the overall performance of the bionanocomposite. The strength of a composite depends not only on the properties of the matrix but on how well it incorporates and interacts with the particles and fibers of the dispersed phase.

Bionanocomposites have been applied in various industries, such as aerospace, food, biomedical, tissue engineering, paint, packaging, and glass coating (Thakur et al., 2017). Nonetheless, they are especially useful for the biomedical industry and suitable to make scaffolds, implants, diagnostics, surgical instruments, and drug delivery systems because of their biocompatible and/or biodegradable properties whereby their degradation is mainly due to hydrolysis or mediated by metabolic processes (Song et al., 2018). They are also used in the cosmetics industry, with an emphasis on tissue engineering, drug and gene delivery, wound healing, and bio-imaging. They can also be used to replace some existing plastics, for example, low-density polyethylene (LDPE) and poly-vinylidene chloride (PVDC) (Wani, 2021; George et al., 2020).

3.3 TYPES OF BIONANOCOMPOSITES

In today's research field, the common bionanocomposites that have been studied extensively are the bionanocomposite films, hydrogels, and aerogels. This section lays out the definition and gives examples of bionanocomposite films in the research field and how are they fabricated at the lab or industrial scale. This section covers the fundamental and basic questions:

1. What are bionanocomposite films, hydrogels, and aerogels?
2. What are examples of bionanocomposite films, hydrogels, and aerogels and their functionality?

3.3.1 BIONANOCOMPOSITE FILMS

Bionanocomposites represent an evolving class of materials characterized by the presence of two or more distinct phases. They typically consist of a biopolymer, such as proteins, lipids, polysaccharides, or nucleic acids, serving as the continuous phase, and a filler, like silicates, carbon nanotubes (CNTs), or metal oxide acting as the discontinuous phase. These fillers possess at least one dimension within the nanometer range, typically ranging from 1 to 100 nanometers (Zubair and Ullah (2020).

Since these bionanocomposite films are made from natural materials, they are considered stable ecological materials with all of the advantages of both biopolymer and nanocomposite structures. Silicate and clay nanoplatelets, titanium dioxide, SiO_2, carbon nanotubes, chitin, graphene, chitosan nanoparticles, cellulose-based nanofibers, starch nanocrystals, and other inorganics can be used as nanofillers for biopolymer matrices such as those as mentioned in Chapter 2 to enhance the performance of the entire composite, resulting in enhanced thermal, barrier, and mechanical characteristics (Sarfraz et al., 2021). Bionanocomposite films should outperform their individual component materials in terms of optical, chemical, and mechanical properties. Depending on the preferred application, bionanocomposites can be customized to achieve certain physical properties such as transparency, oxygen permeability, and water permeability; mechanical properties such as tensile and elongation properties; thermal properties with regard to thermal stability; and biodegradable properties.

The most common bionanocomposite films are edible bionanocomposite films and coatings and mulch bionanocomposite films. The functions for each of these films are further elaborated as follows (Unalan et al., 2014; Shankar & Rhim, 2018; Tuan Zainazor et al., 2020; Zubair & Ullah, 2020; Sarfraz et al., 2021):

- **Edible bionanocomposite films and coatings**
  - They are thin layers of an edible substance used in food application to preserve food products or a membrane of edible substance used in drugs.
  - They are divided into two types: monolayer and multilayer.
    - **Monolayer films:** By altering the surface of other materials, a monolayer film may be used in coating technology to enhance specific properties.
    - **Multilayer films:** They are made using a layer-by-layer formation technique. Hydrogen bonds and electronic interactions serve as driving forces between two layers deposited by solution dipping or spin coating in this technology. Non-electrostatic interactions, on the other hand, play a role as driving forces as two biobased materials are assembled layer by layer.
  - Since edible films and coatings are promising non-toxic and non-polluting commodities that are safe for human use, they are more suitable for use as food packaging and drug delivery systems.
  - For food packaging, edible films exhibit characteristics such as being fast to decompose and environmentally safe, as edible film is made from renewable and biodegradable agriculture sources. Excellent mechanical and barrier properties are needed for an ideal edible film for packaging.
  - Their primary functions as food packaging film are to prevent mechanical, physical, chemical, and microbiological harm to the food. They preserve food consistency; prevent moisture loss; prevent bacteria development; and serve as a barrier to oxygen, water vapor, carbon dioxide, and volatile compounds.

- For drug delivery systems, edible films act as a membrane or medium to deliver the drug to the targeted site or on a biological substrate.
- Edible films and coatings for drug delivery systems should possess properties such as sufficient shelf life, good spreadability, sufficient tensile strength, and non-toxicity and be non-irritant.

In recent years, nanotechnology has been a breakthrough for edible food packaging and drug delivery systems (X. He et al., 2019). The incorporation of nanotechnology with antimicrobial nanoencapsulation has great advantages owing to its excellent protective system that is able to stand against diverse biological and environmental changes (Figure 3.4). This particular trend involves two steps of the process. First, antimicrobial compounds are packed into carriers, and second, their size is reduced to the nanoscale (10^{-9} m) dimension. The types of nanofillers/nanoparticles that are commonly incorporated into the biopolymers include nanofibers (e.g. starch, cellulose, chitin, and chitosan), nanoclay (e.g. MMT, kaolinite, hectorite, bentonite, saponite, organically modified clay), and metallic nanoparticles (e.g. silver, zinc oxide, titanium oxide, copper, copper oxide, gold, and silica) (Becerril et al., 2020).

To truly comprehend the benefits of natural bionanocomposite systems, complete exfoliation and homogeneous dispersion of the nanoparticles in the biopolymer matrix are needed (Becerril et al., 2020). Various methods, such as solvent casting, in-situ polymerization, and melt manufacturing, have been widely used to create nanocomposite materials for a variety of uses, including antimicrobial packaging films (Bratovcic & Suljagic, 2019). These methods are further discussed in the next chapter.

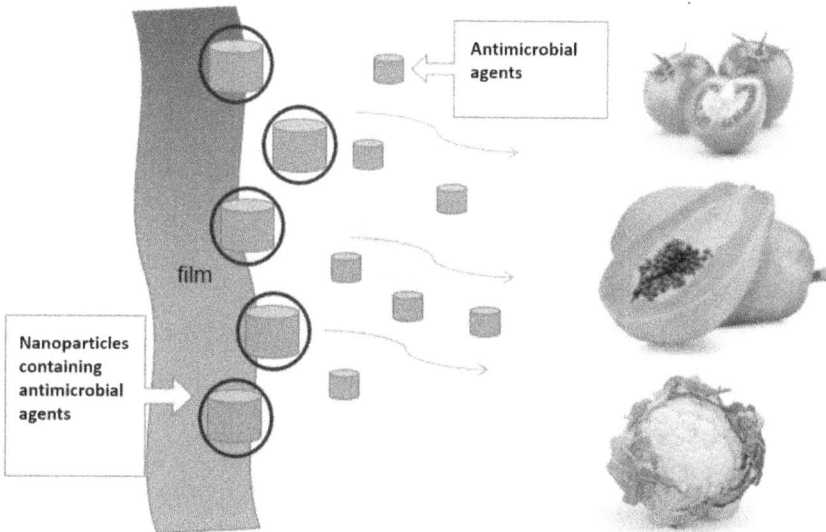

FIGURE 3.4 Nanoencapsulation of antimicrobial agents, which protects food from diverse changes.

The technology of nanoencapsulation in the form of nanotubes, nanorods, nano-capsules, nanoshells, and nanospheres applied on bionanocomposite films and coat-ings has brought many benefits to food products, as shown in Figure 3.5, owing to the high surface area of the carriers relative to their bulk equivalents (Nile et al., 2020). Nanoemulsions, biopolymeric nanocarriers, solid lipid nanoparticles (SLNs), elec-trospun nanofibers, and nanoliposomes are some of the typical nano-encapsulating structures used in the development of novel nano-based active antimicrobial packag-ing systems with controlled release functionality (Nile et al., 2020). The nanocarrier systems are either starch-, protein-, or lipid-based. Although there are many advan-tages of carbohydrate and protein-based nanocapsules, they lack the ability to truly scale up due to the need for various complicated chemical or heat treatments that are difficult to monitor. Lipid-based nanocarriers, however, can be mass-produced industrially and have higher encapsulation performance and lower toxicity (X. He et al., 2019; Nile et al., 2020).

In general, there are three models for bioactive integration into SLNs: the homo-geneous matrix model, bioactive-enriched shell model, and bioactive-enriched heart model (Tan & McClements, 2021). The formulation components (e.g. lipid, lipophilic, or hydrophilic bioactive compounds and surfactant) and the fabrication approach determine the type of model obtained (e.g. hot or cold homogenization) (Nile et al., 2020). When using the cold homogenization approach and integrat-ing very lipophilic actives into SLNs using the hot homogenization procedure, a homogeneous matrix is primarily obtained. In this process, the bioactive compound is released by a dissolution process (Nile et al., 2020). On the other hand, in the

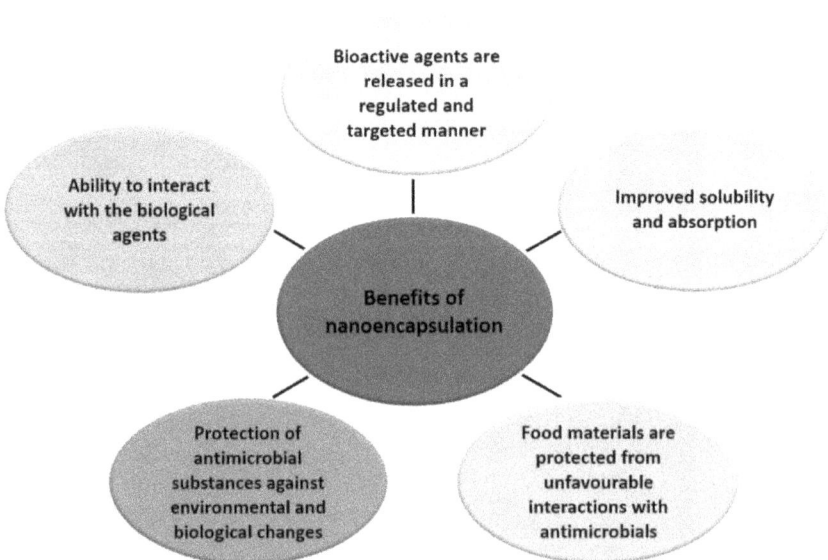

FIGURE 3.5 The benefits of nanoencapsulation for bionanocomposite films that contribute high surface area of the carriers relative to their bulk equivalents.

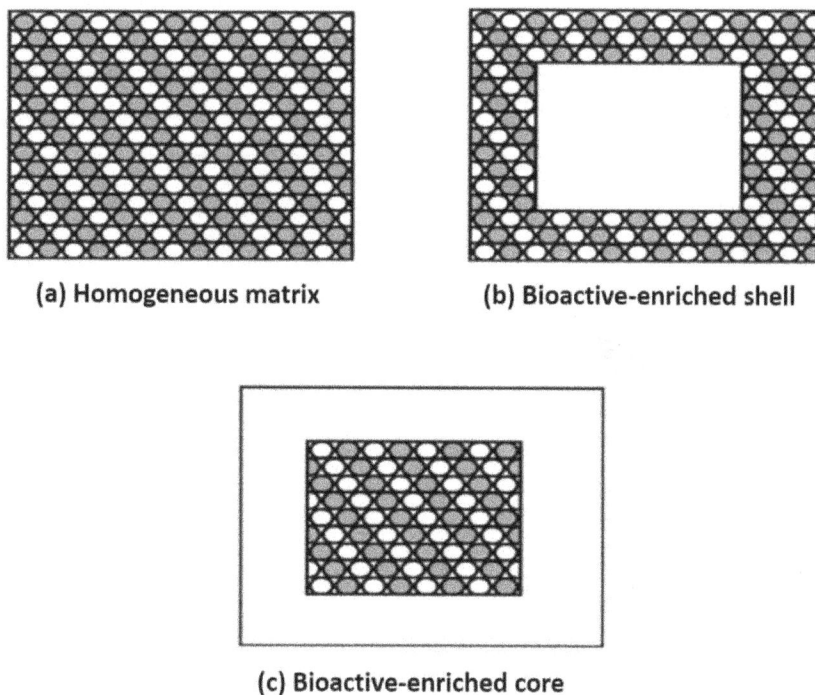

(a) Homogeneous matrix   (b) Bioactive-enriched shell

(c) Bioactive-enriched core

FIGURE 3.6   Models for bioactive integration into SLNs.

cooling process from the liquid oil droplet, phase separation happens during which a bioactive-enriched shell can be attained, while a bioactive-enriched core is formed when a bioactive compound starts with precipitation, which result in lesser components being encapsulated (Figure 3.6) (X. He et al., 2019; Tan & McClements, 2021).

Recently, Rizal et al. (2021) discovered that seaweed incorporated with lignin nanoparticles significantly enhanced moisture resistance, tensile strength, Young's modulus, elongation at breaks, and contact angle properties owing to high compatibility between the lignin and seaweed and strong interfacial interaction between the nanofiller and the matrix, as shown in Figure 3.7, where there is formation of a hydrogen bond between the hydroxyl groups of the nanoparticles and the matrix. The authors concluded that bioplastic films demonstrated significant functional properties, such as mechanical, thermal, and water barriers, that could be a successful choice to replace traditional petroleum-derived plastics in packaging material for a variety of applications.

Biopolymers of various types are used to modulate the properties of films. Biopolymers in high demand include gluten, polymerized pullulan, sodium alginate, pectin, gelatin, and dextrin (Fahmy et al., 2020; Shankar & Rhim, 2018; Sarfraz et al., 2021). They have the potential to enhance solubility and increase stability. Furthermore, certain biopolymers, such as pullulans, have high tensile strength and temperature tolerance. Bionanocomposite films and coatings are commonly used in

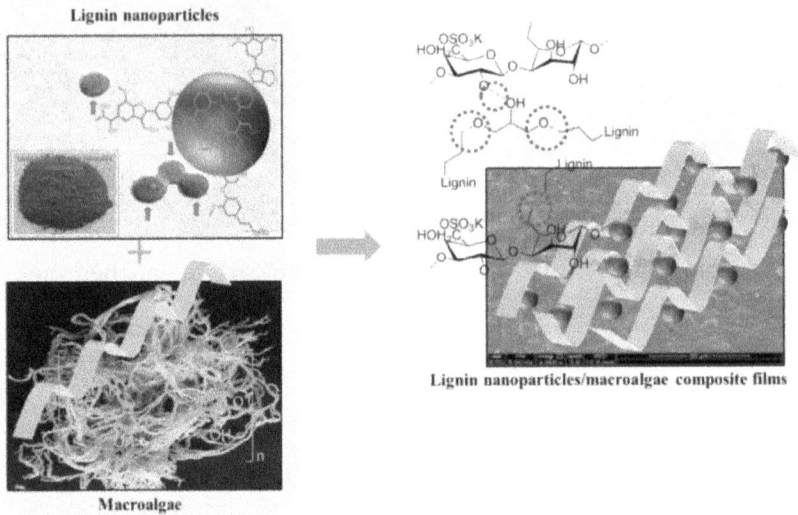

FIGURE 3.7 Incorporation of lignin nanoparticles in seaweed macroalgae. Reproduced from Rizal et al. (2021).

tissue engineering, pharmaceuticals, glass coating, food processing, wood coating, steel coating, medicine coating, and fruit coating, among other uses (Wani, 2021). However, depending on the demands of the final outcome, each film and coating application serves a variety of purposes. Another type of bionanocomposite film is mainly applied to agriculture field: mulch film. Mulch film is defined by some researchers as follows (Ayu et al., 2020; Surya et al., 2021):

- **Mulch film**
  - It is the covering layer on soil used by farmers in crop production systems to preserve and insulate vulnerable plant root systems from extreme weather conditions.
  - It acts as a buffer to increase the nutrient profile of the soil.
  - Mulch films also helps to prevent erosion, improve the soil's ability to retain moisture, reduce weed growth, and increase crop yield and precocity.

Plastic mulch comes in 1,000–4,000-foot-long rolls that are 4–6 feet high and 1.0–1.5 mil thick (Y. Zhang et al., 2020). It comes in a wide range of shades, from transparent to opaque. Colored mulches have recently been studied for their effect on pest control and plant yields. Reflective or silver mulches, for example, have been found to suppress onion thrip infestations (Cárcamo et al., 2021). Mulch films are usually applied once the fields have been leveled and smoothed, fertilizer has been added, and there is adequate soil moisture (Y. Zhang et al., 2020). Since the soil is warmed by heat conduction, good uniform soil contact is important when using

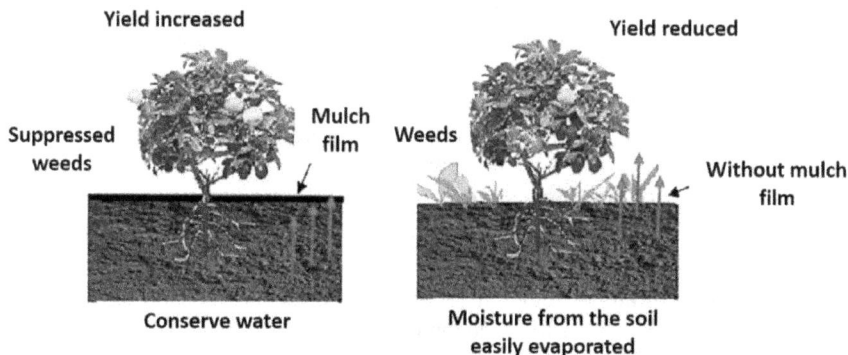

FIGURE 3.8 A contrast of mulch film and no mulch film on the soil surface. Reproduced from Surya et al. (2021).

black mulch. A mechanical mulch covering is the easiest way to spread mulch film evenly. Applying by hand is an alternative, but covering more than a half-acre can be tedious and time consuming (Ayu et al., 2020). As previously stated, numerous experimental trials have shown the effectiveness of plastic mulching, and measurements have been made to illustrate the difference with or without mulch (Surya et al., 2021), as can be seen in Figure 3.8.

Mulch films, especially biodegradable mulch films, have played an important role in the sector of agriculture. The primary functions of these mulch films are as follows (Surya et al., 2021; Y. Zhang et al., 2020; Ayu et al., 2020):

- **Enhanced soil structure:** Mulch films prevent soil from clumping together. Moisture and heat are easily trapped in soil. This would help to retain water in the soil. It also prevents the depletion of plant nutrients, which is very much suitable for drip irrigation. Furthermore, the plastic film deters people and pets from entering the field, enhancing the soil structure even more.
- **Soil protection:** Farmers use mulch for a variety of purposes, one of which is to help the soil maintain heat as the winter months approach. This is because the majority of plants are temperature dependent, and vegetables, in particular, cannot tolerate the cold in the winter. Plastic mulch films, for example, warm the soil by up to 5°F. During the colder months, plastic mulching evenly maintains soil temperature, insulating temperature-sensitive plants.
- **Weed control:** Plastic mulch films are able to suppress weed growth and do so over a wide area. Plastic mulch, as used in the greenhouse, prevents weeds from receiving the sunshine they require for photosynthesis. Weeds fail, as they are starved of sunshine, which saves the time and effort of manually removing weeds.
- **Fruit yield and growth**: Plastic mulching helps to start growing crops earlier in the season. Mulching is important for higher-quality fruits and vegetables, as it serves as a barrier, preventing the fruits from coming into

contact with the soil. This will prevent the transmission of diseases and rot. Furthermore, since the fruits are not in close contact with the soil, they mature even more cleanly.

* **Reduce root damage**: Since the soil is not disturbed, plant roots may expand and disperse deeply and effectively into the ground.

The most common form of mulch film is polyethylene film (Y. Zhang et al., 2020). Due to their non-biodegradability, these films pose a danger to the setting. As a result, researchers all over the world have been looking for ways to replace polyethylene film with biodegradable mulch films derived from biological sources. However, not all biodegradable mulch films decompose and become fertilizer (Surya et al., 2021). According to R. Li et al. (2020), some biodegradable mulch films have raised concerns due to their unintended consequences, where the remnants of these films break down into microplastics. These microplastic particles are believed to pose environmental and organismal hazards.

Due to increasing environmental concerns, researchers have developed biodegradable mulch films that can decompose quickly in the soil. This type of mulch film is typically made of biopolymer materials, specifically polysaccharide-based, protein-based, and polylactic acid (PLA)-based materials (Hayes et al., 2012; Cárcamo et al., 2021). To achieve the efficiency of commercial mulch films, nanotechnology has been used, in which nanofillers or nanoparticles are inserted into the polymer matrix with the main function to improve the physical and mechanical properties of the polymer matrix (Surya et al., 2021). Several studies have been conducted, which include thermoplastic starch (Yin et al., 2020), seaweed incorporated with calcium carbonate (Abdul Khalil et al., 2018), and PLA bionanocomposites of lignocellulosic nanoparticles derived from yerba mate residues (Arrieta et al., 2018).

3.3.2 BIONANOCOMPOSITE HYDROGELS

A hydrogel is a water-swollen, crosslinked polymeric network made by combining one or more monomers in a simple reaction to form a three-dimensional (3D) network of hydrophilic crosslinked polymer chains that can also be found as a colloidal gel in which water serves as the dispersion medium (Ullah et al., 2015; Parhi, 2017). Another meaning is a polymeric substance that can swell and hold a large amount of water within its structure but may not dissolve in water. Because of their high water content, they have a degree of versatility that is somewhat similar to natural tissue (Guo et al., 2020). The inclusion of hydrophilic groups such as $-NH_2$, $-COOH$, $-OH$, $-CONH_2$, $-CONH-$, and $-SO_3H$ attached to the polymeric backbone gives hydrogels their ability to absorb water, while crosslinks between network chains give them resistance to dissolution (Catoira et al., 2019). Hydrogels can be cationic, anionic, or acidic in accordance with their ionic charges on the bound groups. Covalent crosslinks, ionic forces, hydrogen bonds, affinity interactions, hydrophobic interactions, polymer crystallites, and physical entanglements of individual polymer chains or a mixture of these interactions hold the hydrophilic polymer chains together as water-swollen gels (Tang et al., 2019).

There are numerous types of hydrogels, which can be classified based on various factors, including their source, polymeric structure, physical and chemical composition, method of crosslinking, network electrical charges, and properties, as summarized in Figure 3.9 (Ahmed, 2015). In hydrogels, the presence of water is vital for the overall permeation of active ingredients in and out of the gel. Water in hydrogels can be divided into four major categories: bound water (primary and secondary), semi-bound water, interstitial water, and free water or bulk water, as illustrated in Figure 3.10 (Gun'ko et al., 2017).

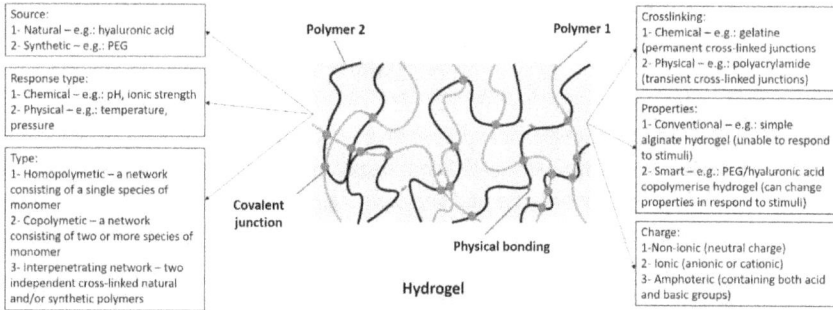

FIGURE 3.9 Various types of hydrogels categorized under different bases. Source, polymeric structure, physical and chemical composition, technique of crosslinking, network electrical charges, and characteristics are the various factors to classify hydrogels.

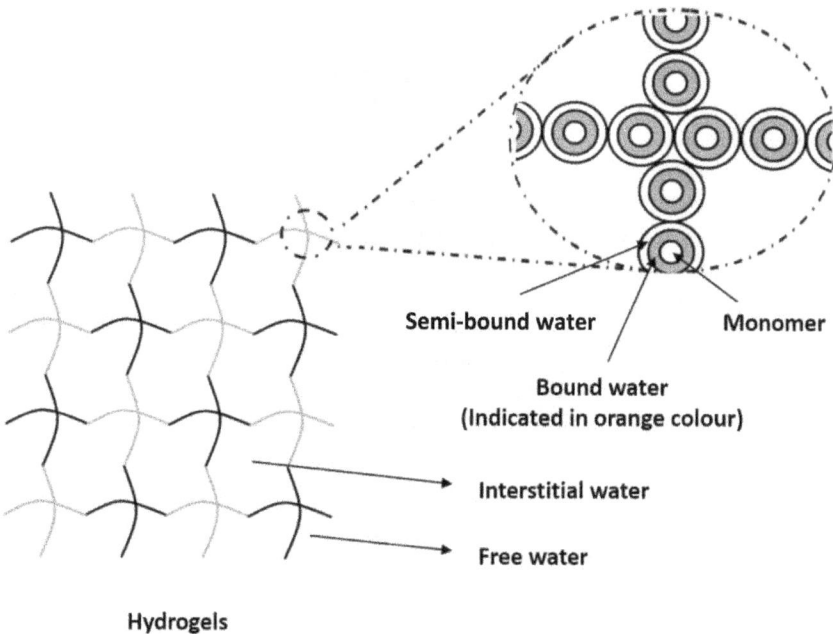

FIGURE 3.10 Types of water associated with hydrogels.

As polar hydrophilic groups come into contact with water, they are the first to be hydrated, resulting in the formation of "primary bound water". As the network expands, the hydrophobic groups will be exposed. These groups will then interact with water molecules to form hydrophobically binding vapor, also known as "secondary bound water" (Gun'ko et al., 2017). When primary and secondary bound water are combined, it is referred to as "total bound water". This water is an essential component of the hydrogel system and can only be isolated from it in exceptional circumstances (Gun'ko et al., 2017). Due to the osmotic pushing force of the network chains towards endless dilution, the network can consume more water. The covalent or physical crosslinks oppose this additional swelling, resulting in an elastic network retraction force. As a result, the hydrogel will hit a point of equilibrium swelling (Ahsan et al., 2021). The excess consumed water is referred to as "free water" or "bulk water", which is thought to fill the spaces between the network chains, as well as the centers of larger pores, macropores, or voids (Gun'ko et al., 2017). There is a water layer called "semi bound water" that exists between the bound water on the surface of the polymeric monomer and the free water. Interstitial water exists in the interstices of a hydrated polymeric network that is physically bound but not connected to a hydrogel network (Ahsan et al., 2021).

Nonetheless, if the network chain or crosslinks are degradable, the next focus is disintegration and/or breakdown, depending on the structure and composition of the hydrogel (Hamedi et al., 2018). Biodegradable hydrogels with labile bonds are thus advantageous in tissue engineering, wound healing, and drug delivery applications (Ahmed, 2015). These bonds can be found in the backbone of the polymer or in the crosslinks used to make the hydrogel. Under physiological conditions, labile bonds can be dissolved either enzymatically or chemically, with hydrolysis being the most common method. Apart from that, controlling the polarity, surface properties, mechanical properties, and swelling behavior of the hydrogel will affect its chemistry (Hamedi et al., 2018).

Several studies have been done on incorporating nanofillers/nanoparticles into the biopolymer matrix with the goal to enhance hydrogel properties. These include bionanocomposite hydrogels based on sodium alginate (SA) as a polymer matrix and graphene oxide (GO) nanosheets with zinc as a crosslinking agent (Sabater i Serra et al., 2020). Through interactions with low concentrations of zinc, robust and strongly entangled networks were obtained from GO nanosheets scattered in SA matrices. The combined properties of these nanocomposite hydrogels make them appealing biomaterials in the field of regenerative medicine and wound treatment (Sabater i Serra et al., 2020). Hydrogels are also well known for the treatment of wastewater. Chemical crosslinking of cellulose, carboxymethyl cellulose (CMC), and intercalated clay in a NaOH/urea aqueous solution exhibited superabsorbent hydrogels. These hydrogels had superabsorbent properties in purified water, better mechanical strength, and a high ability to remove microbes from wastewater than hydrogels made with only a polymer matrix (Qian, 2018).

Hydrogels are much more well known in the pharmaceutical. According to Parhi (2017), Mohite and Adhav (2017), and Hamedi et al. (2018), hydrogels have unique properties that make them ideal for pharmaceutical, medicinal, and biochemical uses. Among these features are:

- Since hydrogels are structurally and mechanically identical to the native extracellular matrix, they can serve as a supporting material for cells during tissue regeneration.
- Their hydrophilic and crosslinked properties contribute to their excellent biocompatibility.
- They are soft in feature, which encourages which encourages water uptake and hence the formation of hydrated but solid materials, similar to cells in the body.
- Low interfacial stress between the hydrogel surface and body fluid decreases protein adsorption and cell adhesion, lowering the likelihood of a negative immune response.
- The mucoadhesive and bioadhesive properties of certain polymers used in hydrogel formulations such as polyacrylic acid (PAA), polyethylene glycol (PEG), and polyvinyl alcohol (PVA) improve drug residence time on the skin/plasma membrane, resulting in increased tissue permeability.

While hydrogels have numerous advantages, such as those mentioned before, they also have a number of disadvantages, one of which is the difficulty in forming them into predetermined geometries. Another disadvantage of hydrogels is that their reaction time is relative to their size, which is caused by sluggish water molecule diffusion (Mohite & Adhav, 2017).

3.3.3 Bionanocomposite Aerogels

The term "aerogel" refers to a gel in which the liquid portion has been replaced by a gas (typically by the use of a supercritical drying technique) without affecting the overall structure, which also known as "open porous forms". The end result is a solid with an exceptionally low density and a number of unique properties, the most notable of which is its usefulness as a thermal insulator and its extremely low density (Ganesan et al., 2018). Due to its transparency and the way light scatters in it, it is also known as frozen smoke, solid smoke, or blue smoke. High porosity (usually greater than 80%), high thermal resistance ($0.005-0.1$ W.mK^{-1}), low density ($0.003-0.5$g/cm^3), low dielectric constant (κ value ranging from 1.0 to 2.0), low refractive index (≈ 1.05), and high specific surface area ($500-1200$ m^2/g) are among the critical properties that can be identified in an aerogel (Guastaferro et al., 2021).

Aerogels can be categorized into various types, with classification based on their visual characteristics, preparation techniques, unique microstructures, and chemical compositions. In past research, inorganic aerogels have been widely fabricated with the addition of nanofillers such as SiO_2, Al_2O_3, and TiO_2 in combination with synthetic polymers such as polyamide, polyurethane, and polyimide serving as the fundamental matrix for aerogel synthesis (Shi et al., 2021). However, concerns arise due to their negative environmental impact. Hence, more and more natural biopolymers, especially chitosan, cellulose, agar, agarose, pectin, and other polysaccharides and proteins, have been used to fabricate aerogels (Shi et al., 2021; Ganesan et al., 2018). Similar to hydrogels, the types of aerogels can be classified on various bases such as appearance, preparation, chemical structures, and microstructures (Thapliyal & Singh, 2014).

Aerogels can be manufactured in a variety of sizes and shapes in the forms of monolith, powder, and film since the initial liquid mixture can be adjusted during the gel forming process (Thapliyal & Singh, 2014; Ganesan et al., 2018). Figure 3.11 shows the common pathway of fabricating biopolymer-based aerogels.

The method starts with dissolving biopolymers in water or chemical solvents to make organic aerogels (i.e. polysaccharides are generally soluble in aqueous solutions). The polymer chains then rearrange themselves into an open porous network as a result of solution gelation. Chemical, enzymatic, or physical crosslinking may cause gelation (Guastaferro et al., 2021). Due to the existence of functional groups localized on the backbones of several biopolymers, Van der Waals forces or hydrogen bonding may form (Shi et al., 2021). The porous structure of polysaccharide-based aerogels is determined by the drying technique. After drying, three types of solid materials can be produced: xerogel, cryogel, and aerogel. If a rigid material is dried under atmospheric pressure and at room temperature for several days, it is referred to as a xerogel (Abdul Khalil et al., 2020). The resulting materials are termed cryogels, as water (ice) within the hydrogel is sublimated by freeze-drying (Ganesan et al., 2018; Guastaferro et al., 2021). These samples, on the other hand,

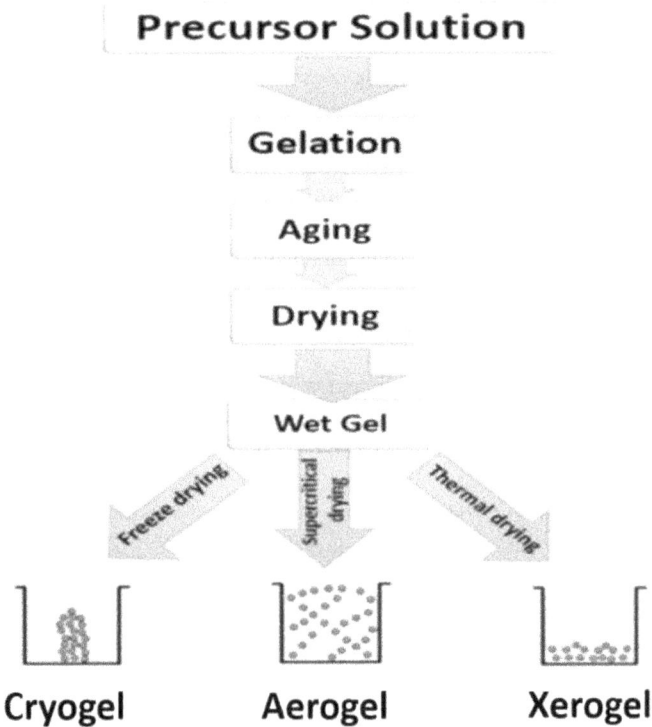

FIGURE 3.11   The pathway of fabricating biopolymer-based aerogels. Reproduced from Abdul Khalil et al. (2020).

have a macroporous structure with large, irregular pores. Supercritical CO_2 drying is commonly used to synthesize aerogels (Abdul Khalil et al., 2020; Guastaferro et al., 2021). This technique, when correctly executed by the operative pressure and temperature, is able to prevent the nanostructure from collapsing. Furthermore, it preserves the gel's excellent textural properties (Zhang & Zhao, 2020; Guastaferro et al., 2021).

The recent polysaccharide-based aerogels synthesized from natural ingredients show that their properties were on a par with or even better than those synthesized by synthetic polymer and silica. Among the significant properties possessed by natural biopolymer-based aerogels are high specific surface area, low thermal conductivity, biodegradability, biocompatibility, and sustainability (Shi et al., 2021).

Due to their unique characteristics, biopolymer-based aerogels have shown great potential in a wide range of applications. They can serve as effective thermal insulation, versatile drug delivery systems, valuable components for tissue engineering and regenerative medicine, efficient catalysts, sensitive sensors, reliable adsorbents, and essential raw materials for the production of carbon aerogels (Shi et al., 2021; Guastaferro et al., 2021). Leveraging their exceptional thermal, chemical, and functional properties, biopolymer-based aerogels find utility in separation processes and catalysis. They also serve as efficient drug delivery carriers due to their high specific surface area, which leads to rapid drug release upon contact with liquid media. Furthermore, both unmodified and drug-loaded biopolymer-based aerogels have been suggested for use as superabsorbents and for applications in wound healing due to their substantial pore volume and impressive swelling capabilities (Ganesan et al., 2018).

3.4 FABRICATION PROCESS OF BIOPOLYMER-BASED NANOCOMPOSITES

The functionality of various biopolymers can be consolidated by conjugating them with nanoparticles, and the resulting bionanocomposites show characteristics of both biopolymers and nanoparticles. The wide range of bionanocomposite architectures and properties has broadened their applications. This section covers an overview of techniques for synthesizing bionanocomposites in both in academic research and industrial settings.

The techniques to fabricate bionanocomposites are crucial, as they affect the properties of bionanocomposites and allow bionanocomposites to reach their full potential if the nanoparticles/nanofillers are distributed homogeneously (Bah et al., 2020; Unalan et al., 2014). One of the main issues in the manufacturing of polymer nanocomposites is the uniform and homogeneous dispersion of nanoparticles in the polymer matrix (Wei et al., 2010). Nanofillers have a propensity to assemble and shape micron-sized filler clusters, limiting nanoparticle dispersion into the polymer matrix and degrading the properties of nanocomposites (Dantas de Oliveira and Augusto Gonçalves Beatrice, 2019; Varghese et al., 2021). Therefore, many attempts have been made by researchers to spread nanofillers evenly and homogeneously in the polymer matrix using chemical reactions, polymerization reactions, and filler content surface modifications (Tanahashi, 2010). In the context of fabrication, the synthesis of bionanocomposites can be carried out following bottom-up strate-gies, which can be classified into a direct mixing approach, melt mixing/blending approach, in situ polymerization approach, and sol-gel approach.

The basic structure of the bionanocomposites obtained are intercalated (and/or flocculated) or exfoliated (or delaminated) microstructures, which depend on the approach and materials used (Fawaz & Mittal, 2014; Bhattacharya, 2016; Jafarbeglou et al., 2016), as shown in Figure 3.12. The intercalation process will alter the struc-ture of the host materials by expanding the interlayer spacing, weakening the van der Waals impact, and expanding the interlayer spacing, which inserts particles/fillers between the layers, while the exfoliation process breaks the van der Waals force between the adjacent layers by exerting external force, causing the separation

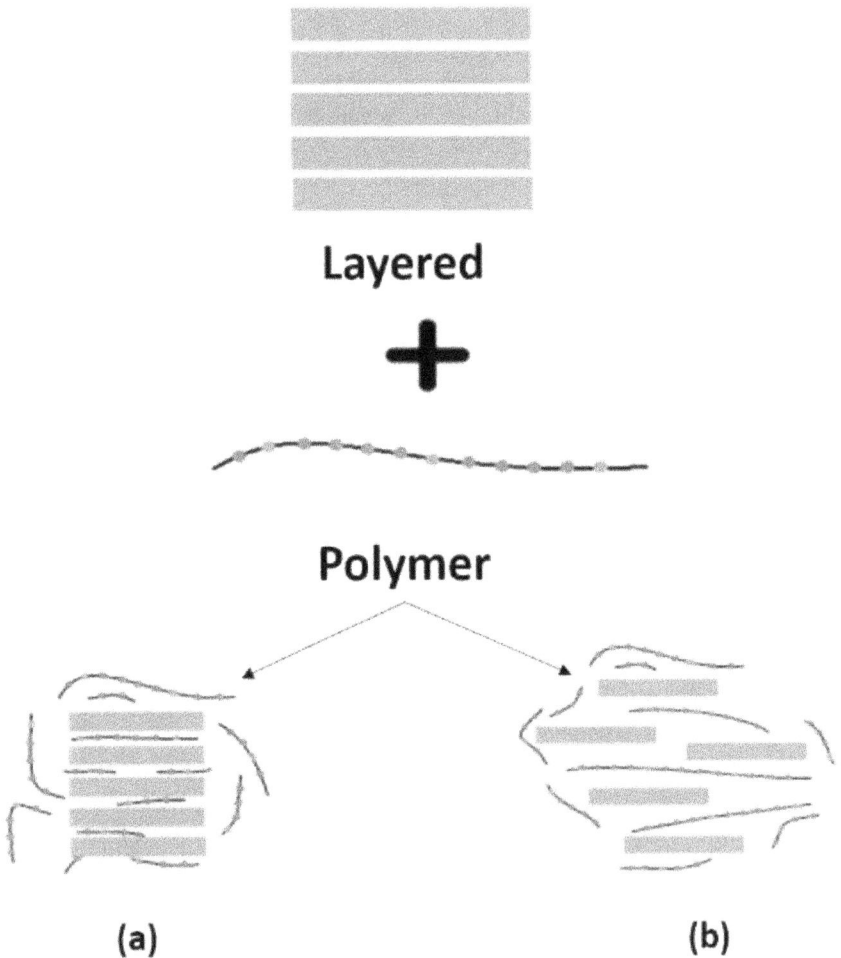

FIGURE 3.12 The basic structure of bionanocomposites obtained are (a) intercalated (and/or flocculated) or (b) exfoliated (or delaminated) microstructures.

of layers with individual layers dispersed within the polymer matrix (Backes et al., 2020; Bhattacharya, 2016).

Researchers have unveiled a multitude of manufacturing methods designed to overcome the limitations associated with biopolymers. Recently, they have strived to develop innovative manufacturing technologies suitable for both commercial and academic applications with the aim to streamline the synthesis of bionanocomposite materials, offering numerous benefits such as environmental friendliness, time efficiency, ease of operation, long-term sustainability, cost-effectiveness, exceptional physical and mechanical properties, and reliance on sustainable and eco-friendly resources (Bertolino et al., 2020; Ghorbani et al., 2021; Tuan Zainazor et al., 2020).

3.4.1 DIRECT MIXING APPROACH

For preparation of bionanocomposites based on insoluble biopolymers, the direct mixing process mixes a soluble biopolymer and the dispersed nanofillers or nanoparticles in a solvent (Tanahashi, 2010), as shown in Figure 3.13. Note that the biopolymer and nanoparticles are dissolved or dispersed separately in solvent (Müller et al., 2017). These two solutions can be mixed together to allow intercalation to take place whereby the biopolymer chains replace the solvent molecules within the interlayer spaces of the nanoparticles (Li & Zhong, 2011; Zhan et al., 2017). Mixing by agitating the solvent can be done through magnetic stirring, shear mixing, reflux, or sonication. It usually results in thickening and gel-like precipitation (Salehiyan & Ray, 2018; Rafiee & Shahzadi, 2019). Upon solvent evaporation, the intercalated or exfoliated structure remains, which gives bionanocomposites. The properties of the bionanocomposites are determined by the interaction between the constituents of the matrix and the filler (Awan et al., 2021).

Direct mixing is often utilized in biopolymers with nanoclay or nanosilicates for layer–layer assembly to form bionanocomposite films, coatings, or membranes (Salehiyan & Ray, 2018) (Awan et al., 2021). It starts with nanoparticles being dispersed in the soluble biopolymer mixture (Rafiee & Shahzadi, 2019). The interaction between the biopolymer solvent and the nanoparticles can cause a formation of stack

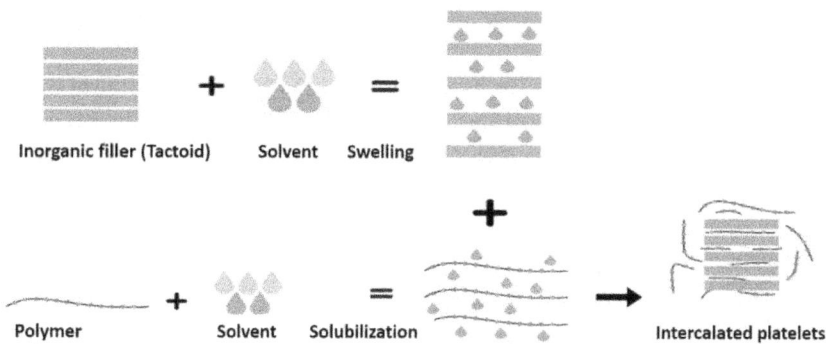

FIGURE 3.13 Direct mixing process for bionanocomposite fabrication.

or layer-by-layer structure due to the van der Waals forces between the biopolymer solvent and the nanoparticles (Tanahashi, 2010; Müller et al., 2017).

This approach is easy to execute and handle. Thus, many studies have used this approach to fabricate bionanocomposites, particularly in films (Leng et al., 2019; Li & Zhong, 2011; Awan et al., 2021). However, the disadvantage of this approach is the poor compatibility of the matrix and the nanofiller, which leads to heterogeneous bionanocomposites when the nanofillers are not distributed evenly (Rafiee & Shahzadi, 2019). Apart from that, nanoparticles have the tendency to agglomerate upon solvent evaporation (Salehiyan & Ray, 2018). Another limitation of this approach is when it comes to fabricating bulk material such as hydrogels, as it needs a large amount of solvent to do so, which is cost ineffective (B. Li & Zhong, 2011). This approach is more appropriate to use to prepare bionanocomposites, especially films based on water-soluble biopolymers such as PVOH and polysaccharide-based biopolymers, as these types of biopolymers consist of polar sites to interact with silicate surfaces and exfoliate in the water, forming colloidal particles (He et al., 2017; Na et al., 2020).

3.4.2 MELT BLENDING APPROACH

The melt blending approach applies heat to the matrix at a suitable temperature and transfers it in a molten state to mix with the fillers (Müller et al., 2017). Nanoparticles can be combined directly with molten polymer rather than using solvent as medium, as shown in Figure 3.14.

Many experiments have shown that polar interactions between polymers and the clay surface are important in achieving dispersion of particles (Tanahashi, 2010; Müller et al., 2017; R. Zhang & Zhao, 2020). The degree of nanoparticle dispersion or the compatibility between the matrix and the nanoparticles is influenced by two main factors, the processing conditions and the enthalpic interaction, where optimization of processing temperature and pressure is required. This method is very much affected by the sheer force applied, which needs to be monitored from time to time. This is because too much shear force can easily split the nanofibers into shorter fragments, hence decreasing the reinforcement effect and properties when high–aspect ratio nanoparticles are preferred (Jamróz et al., 2020). Melt blending provides a cost-effective path and is environmentally friendlier due to the absence of solvent in the process. It also has the ability to process via extrusion and injection molding, which can be applied in industry. This approach has the advantage for mass production (Müller et al., 2017). The only limitation for this approach is the degradation of biopolymers that leads to polymer chain cleavage

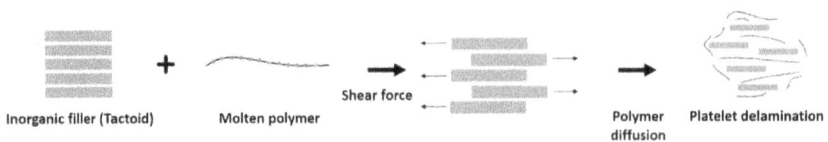

Inorganic filler (Tactoid)   Molten polymer   Shear force   Polymer diffusion   Platelet delamination

FIGURE 3.14 Melt blending approach for bionanocomposite fabrication.

due to mechanical shearing force or the temperature added during the process (Ke et al., 2012).

Biopolymers such as soy protein, wheat gluten, zein, gelatin, PLA, and PHA are among the materials that are suitable to be processed in the extruder (Mangaraj et al., 2019). It is also an adaptable approach for thermoplastic biopolymers (Müller et al., 2017). As for this approach, layered clay minerals are often used as the filler because they give good intercalation of stacked clay mineral sheets (Dlamini et al., 2019). This approach is convenient and easy to carry out in a lab or even at industrial scale. This approach has also been used to fabricate biopolymer/cellulose nanofiber (CNF) bionanocomposites (Safdari et al., 2017; Dufresne, 2018).

3.4.3 *In Situ* Polymerization

In situ polymerization begins with the processes to distribute inorganic nanoparticles to the monomer or monomer solution, followed by monomer polymerization, which usually traps the exfoliated particles in the resultant polymer matrix. In this approach, the nanoparticle is distributed in a liquid monomer or a monomer solution. This process takes place when the filler/particulates swell in the liquid monomer solution (Geng et al., 2018). The swelling is caused by the process where the low molecular weight monomer percolates through the interlayers (Habibi et al., 2014; Geng et al., 2018). Polymerization can be accomplished by the use of heat or radiation, the absorption of a suitable initiator, or the use of an artificial initiator or catalyst, allowing polymer formation to occur between the intercalated or exfoliated sheets (Rane et al., 2018), as shown in Figure 3.15.

Utilizing economical materials, the simplicity of automation, and the flexibility to integrate with diverse heating and curing techniques, this method possesses the capacity to attain the optimal polymer chain length and molecular weight with the enhancement of both particulate and matrix distribution systems. Notably, it outperforms direct blending and melt blending in achieving superior exfoliation (Arzac et al., 2014). However, some disadvantages of this preparation approach include minimal supply of available materials, a short time to complete the polymerization process, and the need for costly machinery (Mishra et al., 2014; Arzac et al., 2014). This approach is suitable for biopolymers and thermoplastic-based nanocomposites (Qin et al., 2020). Direct incorporation of well-dispersed nanoparticles in bulk polymer composites is possible with in situ polymerization. This approach is widely used in tissue engineering and food packaging applications (Doberenz et al., 2020; Fahmy et al., 2020).

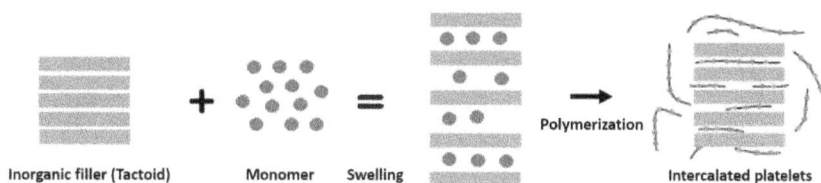

Inorganic filler (Tactoid)   Monomer   Swelling   Polymerization   Intercalated platelets

FIGURE 3.15   In situ polymerization process of the fabrication of bionanocomposites.

3.4.4 SOL-GEL APPROACH

The sol-gel technique, also known as the template synthesis approach, is a bottom-up technique that operates on the inverse theory of all previous approaches (Vaseghi & Nematollahzadeh, 2020). It is also known as the wet chemical procedure, which follows a sequential order, as displayed in Figure 3.16.

Sol is a colloidal suspension of solid nanoparticles in a monomer solution, and gel is the three-dimensional interconnecting network generated between the phases during the gelation process. Solid nanoparticles are scattered in the monomer solution in this process, resulting in a colloidal suspension of solid nanoparticles (sol). By polymerization followed by hydrolysis, they form an interconnected network between phases (gel) (Baig et al., 2021). As seen in Figure 3.17, the polymer nanoparticle 3D network expands across the liquid. The polymer acts as a nucleating agent, promoting the formation of layered crystals. As the crystals expand, the polymer seeps through the layers. The solvent can be removed by evaporation or by critical drying, depending on the desired type of nanocomposite. Solvent removed through evaporation usually results in xerogel formation, while solvent removed by critical drying usually results in the formation of an aerogel.

Research on the sol-gel approach is still lacking, and better comprehension of fundamental inorganic polymerization chemistry is required to expand the functionalities and potential applications of this approach (Abdul Khalil et al., 2020). As of now, the sol-gel approach can be applied with a small investment, though the disadvantages need to be considered (Donato et al., 2017). With all the approaches mentioned, the major complication of processing bionanocomposites is the agglomeration of nanoparticles, which is accompanied by a lack of dispersion in the target formulation. Agglomeration and aggregation are caused by differences in surface area and volume effect (Zare et al., 2017). Hence, it is crucial to determine the dispersion efficiency and quality by characterizing them via different established techniques (Unalan et al., 2014), including x-ray diffraction (XRD), scanning electron microscopy (SEM), transmission electron microscopy (TEM), infrared spectroscopy (IR), and atomic force microscopy (AFM).

Direct melt-compounding of the polymer with the fillers is the conventional practical approach for dispersing inorganic nanofillers in polymer matrices. The surface activity of the nanofillers, on the other hand, is exceedingly high. As a result,

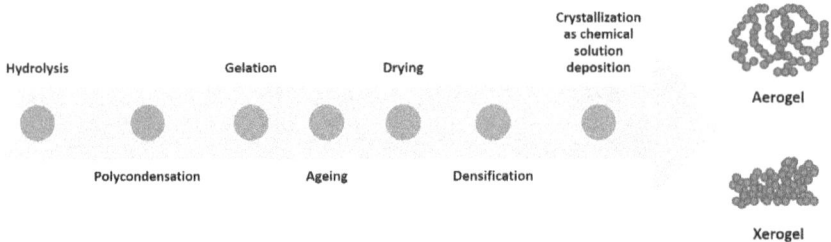

FIGURE 3.16 The generic flow of the sol-gel approach, also known as wet chemical procedure.

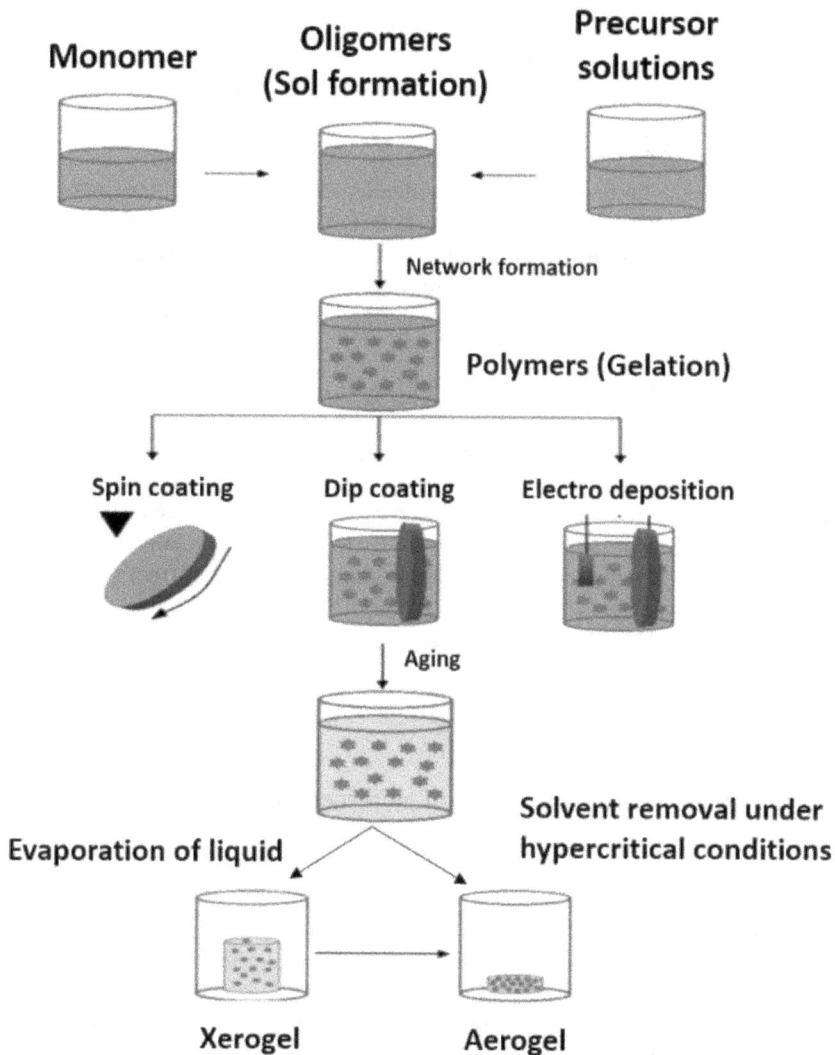

FIGURE 3.17   Sol-gel approach where the 3D polymer nanoparticle acts as a nucleating agent promoting the formation of layered crystals for the formation of xerogel and aerogel.

the particles appear to cluster closely. This is one of the most difficult issues in the production of filler/polymer nanocomposites. In light of this problem, a number of attempts have been made to distribute nanofillers uniformly in polymer matrices using methods involving organic modification of the surface or interlayer of nanofillers as well as a combination of sol-gel and/or polymerization reactions (Varghese et al., 2021). These developments, however, necessitate complicated chemical reactions, rendering them undesirable for industrial-scale manufacture of nanocomposites with a broad volume fraction spectrum of nanofillers and different combinations of filler and polymer content.

In terms of industrial-scale fabrication of high-performance particle/polymer nanocomposites, a simple method for dispersing inorganic nanoparticles into various polymers through direct melt-blending, which does not require complicated reactions, is much preferred. Apart from that, the melt blending approach has shown greater advantages in terms of cost, practicality, and eco-friendliness compared to the other approaches. Another factor to note is the optimization of conditions to achieve better dispersion and exfoliation in bionanocomposites (Zare et al., 2017), for instance, the condition of temperature, where degradation can be avoided by running at lower temperatures or using more thermal stable modifications.

3.5 THE PROPERTIES OF BIONANOCOMPOSITES

This section aims to cover the most generic properties that are used to evaluate properties such as physical, mechanical, and thermal properties and biodegradability. According to Dantas de Oliveira and Augusto Gonçalves Beatrice (2019), bionanocomposites have some advantages in terms of their properties, among them being lighter than conventional composites due to high degrees of stiffness and strength achieved from much less high-density material, better barrier properties as compared to neat biopolymers, potentially superior mechanical and thermal properties, and biodegradability characteristics.

3.5.1 PHYSICAL PROPERTIES

Although there are many other physical properties that are evaluated for bionanocomposites, optical, water and gas barrier, and antimicrobial properties are common physical measurements used to determine the physical properties of bionanocomposite films. Materials' optical properties are especially important in some industries, where they can influence either the efficiency of the material or the consumer's preference (Parola et al., 2016). Both aspects should be taken into account when developing a new bionanocomposite. For instance, in the food industry, UV radiation (wavelengths less than 340 nm) should be avoided because it can induce photooxidation in photosensitive foods such as meat, alcohol, and milk, resulting in color, flavor, and taste changes. On the other hand, high visible light transmittance (wavelengths between 340 and 800 nm) can be determined at the same time, as it helps users see through the packaging (visual inspection of processed food). The effect of bionanocomposite on the optical properties can be regulated in two ways (Unalan et al., 2014):

- Using appropriate methods and procedures in the overall manufacturing phase
- Selecting the most appropriate filler type

In the exfoliation process of the filler, for example, physical, chemical, or mechanical approaches may be more or less effective depending on the filler. As for selecting the most appropriate filler type, it merely depends on the final products. If the final substance must be transparent, synthetic laponite (LAP) would offer better performance compared to one with sheet-like properties such as MMT, which is mainly due to the chemical composition and the well-defined dimensions of LAP compared to MMT (Unalan et al., 2014). In terms of barrier properties, bionano-composites have excellent gas (e.g., O_2 and CO_2) and water vapor barrier properties (Dantas de Oliveira and Augusto Gonçalves Beatrice, 2019). According to previous studies, the form of fillers, aspect ratio of fillers, and nanocomposite composition all play a role in reducing gas permeability (Cader Mhd Haniffa et al., 2016; Mousa et al., 2016; Shankar & Rhim, 2018; Wani, 2021). Generally, bionanocomposites featuring completely exfoliated clay minerals characterized by high aspect ratios tend to exhibit the most robust gas barrier properties. The underlying principle of this barrier effect, as elucidated by the authors, revolves around the ability of nanoparticles or nanofillers to establish a convoluted pathway within the matrix's gallery structure (Dantas de Oliveira and Augusto Gonçalves Beatrice, 2019; Abdul Khalil et al., 2019). This phenomenon lengthens the transportation distance of gas or water permeate through the matrix, which results in higher barrier properties (Figure 3.18).

Apart from the optical and barrier properties, bionanocomposites incorporated with antimicrobial agents such as silver nanoparticles, chitosan, chitin, and nanoclay are effective in microbial growth inhibition, as antimicrobial carriers and antimicro-bial packaging films, and for enhancing shelf life due to the high surface-to-volume ratio and improved surface reactivity of nano-sized antimicrobial agents (Sharma et al., 2020; Shankar & Rhim, 2018). They are able to prevent more microorganisms than their larger-scale equivalents. Examples of bionanocomposites that have sig-nificantly exhibited antimicrobial properties were reviewed by Sharma et al. (2020). One example recorded that as silver nanoparticles and organoclay were integrated into gelatin, they demonstrated strong antimicrobial activity against food-borne pathogens such as *E. coli* and *L. monocytogenes*, with inhibition zones of 12 and 13 mm, respectively. Moreover, previous research has shown that oxidized starch/CuO

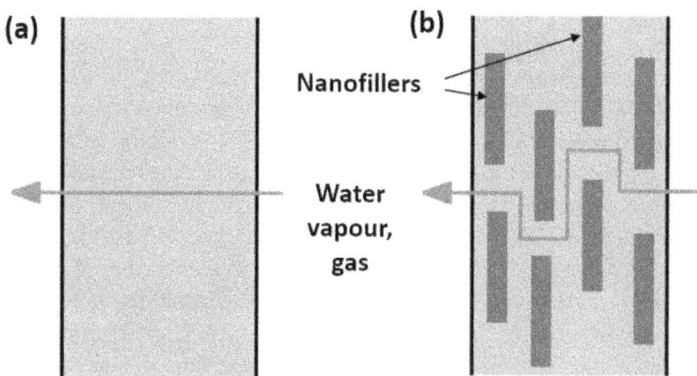

FIGURE 3.18 (a) Water vapor and gas passing through a clear path; b) water vapor and gas passing through a complex pathway.

bio-nanocomposite hydrogels can serve as an antibacterial and stimuli-responsive agent with possible colon-specific naproxen distribution in bionanocomposite hydrogels (Namazi et al., 2020).

3.5.2 MECHANICAL PROPERTIES

Mechanical properties are determined by evaluating the composite's stiffness and resistance to load exertion. To ensure product integrity and durability, composites must be able to withstand and last under such loads for a particular application (Rafiee & Shahzadi, 2019). There are three main factors that affect the mechanical properties of bionanocomposites often discussed in research works (Abdul Khalil et al., 2019; Jamróz et al., 2020; Yin et al., 2020; Awan et al., 2021; Surya et al., 2021). These three factors are filler content, size and aspect ratio of the filler, and well dispersion of fillers.

The mechanical properties of bionanocomposites are frequently found to be highly dependent on the filler material. Appropriate content of fillers is important to ensure the biopolymers are incorporated with nanofillers to demonstrate a significant improvement in the mechanical properties (Abdul Khalil et al., 2019). As for the incorporation of nanofillers, filler content lower than 5 wt.% is usually applied to prevent overloading and lead to filler agglomeration. The main reason for agglomeration could be due to the Van der Waals forces between the fillers (Costa et al., 2017). The size and aspect ratio of the fillers play an important role in affecting the mechanical properties of bionanocomposites. According to previous research, the tensile strength of hydroxypropyl methyl cellulose (HPMC)-based matrices integrated with smaller size AgNPs (40 nm) was 44.5% higher than the control (i.e. only contain biopolymers), whereas that incorporated with larger-size AgNPs (100 nm) was 26.5% higher than the control (De Moura et al., 2012).

Mechanical properties can be affected by the well dispersion of nanofillers/nanoparticles. This is to ensure no agglomeration occur in the bionanocomposite, which can lead to rigidity and weaker mechanical properties. In a bionanocomposite integrated with nanoclay, well dispersion of intercalated or exfoliated nanoclay resulted in significant changes in the mechanical performances. Another instance was found in cellulose/iron oxide bionanocomposites, where the well dispersion of 1 wt.% of iron nanoparticles showed a smooth morphology, resulting in $9.42 \pm 0.6\%$ and $45.16 \pm 0.4\%$ higher tensile strength and Young's modulus compared to the control (i.e. only cellulose) (Yadav, 2018). Well dispersion of fillers can be affected by the interfacial adhesion between the filler and the matrix. As articulated by Taib and Julkapli (2018), when two separate materials are blended, merged, or mixed, interfacial adhesion occurs. This combination can result in better material dispersion in the matrices. Typically, to improve interfacial adhesion, a mixture of materials with similar properties, such as hydrophilic fillers and hydrophilic matrices or hydrophobic and hydrophobic materials, should always be used, resulting in a close bond between the two materials. The high rigidity and aspect ratio of nanoclay, along with the strong affinity through interfacial interaction between the polymer matrix and dispersed nanoclay, can be credited to the improvement in mechanical properties of polymer nanocomposites.

3.5.3 THERMAL PROPERTIES

Thermal properties are the properties of a substance that are related to its heat conductivity. In other words, these are the characteristics that a substance exhibits as heat is applied to it. The thermal properties of bionanocomposites must be determined in order to guarantee that bionanocomposites can withstand the heat they exposed to. Bionanocomposites usually exhibit better thermal stability compared to pure biopolymers. Previous studies have shown that thermal and dimensional stabilities of bionanocomposites were improved with the incorporation of layered silicate clays; these are critical properties for large-scale production of thermoforming films at elevated temperatures without shrinkage after processing of food packaging materials. The dimensional stability of clay/bionanocomposites is improved due to the higher modulus of nanoclay and lower thermal expansion coefficient than the polymer matrix.

Through differential thermal analysis (DTA) analysis, alginate/copper oxide (CuO) bionanocomposites showed significant enhancement of thermal stability compared to pure alginate (Saravanakumar et al., 2020). Thermal stability was also found to be enhanced in ginger nanofiber (GNF)/starch bionanocomposites, which was mainly attributed to the well-ordered orientation of the GNF cellulose in the composite moiety. This alignment was caused by the chemical reaction with the starch molecules, which resulted in higher thermal stability than GNF alone (Jacob et al., 2018). PVA/2D halloysite nanotube (HNT) bionanocomposites exhibit enhanced thermal stability when compared to pristine PVA, as evident from the shift in the maximum degradation temperature to higher levels. This improvement is attributed to the effective role of HNTs as heat and mass transfer barriers. Additionally, the natural hollow tubular structures of HNTs function as entrapment zones for volatile particles, thus augmenting thermal stability by impeding mass transfer during the decomposition process (M. Mousa and Dong, 2020). In general, the incorporation of nanofillers in the biopolymer matrix was found to improve thermal stability because the dispersed nanofillers act as an insulator for heat transfer and a barrier for mass transfer to the volatile products produced during thermal decomposition, and nanofillers also protect the polymer from the action of oxygen, dramatically increasing thermal stability under oxidation.

3.5.4 BIODEGRADABLE PROPERTIES

One of the most essential properties of bionanocomposites is biodegradability. This is because this property helps to reduce critical waste problems. Polymer biodegradation can occur by any of the following mechanisms, which can occur alone or in combination (Vaezi et al., 2020):

- hydrolysis
- enzyme-catalyzed hydrolysis
- solubilization
- ionization/microbial degradation

In general, biodegradation of polymers happens in two stages: depolymerization and mineralization (Rao et al., 2019). Bionanocomposite packaging materials are expected to degrade quickly in the atmosphere after being disposed of (Vaezi et al., 2020). In general, it is understood that the biodegradability of biopolymer films is greatly enhanced after incorporation with nanoclays. This applies to polylactic acid, where the breakdown begins with hydrolysis of the ester bond, then breakup into oligomer fragments, solubilization of oligomer fragments, diffusion of soluble oligomers, and finally mineralization into CO_2 and H_2O (Rao et al., 2019).

Another study done by Vaezi et al. (2020) showed that starch-based nanocellulose (NCC) particles and MMT bionanocomposites showed better degradation compared to the control (i.e. pure starch). It was explained that this behavior was caused by the dual effect from NCC and MMT. The higher degradation rate was mainly attributed to the addition of a hydrophilic filler, the NCC. On the other hand, MMT can limit water diffusion by improving barrier properties, as a result delaying hydrolysis and improving degradation. The authors also suggested that all of the investigated bionanocomposites began the disintegration phase earlier and at a faster pace than pure CS films, implying their potential advantages in industrial applications where short biodegradation times are needed. Nevertheless, in some cases, nanofillers/nanoparticles are able to delay degradation. This was applied to starch/calcium carbonate bionanocomposites by Swain et al. (2018) and nanoclay/nanofibers/wood plastic by Saieh et al. (2019). These findings indicate that the structure of the nanoparticles and the existence of certain surface-modifying chemicals, such as quaternary ammonium cations, may influence the degree of biodegradation of bionanocomposites (Salehiyan & Ray, 2018; Saieh et al., 2019; Mangaraj et al., 2019). By fine-tuning the biodegradation rate, this property can be used to produce bionanocomposite materials with the desired biodegradability properties.

3.6 CONCLUSION

Bionanocomposites, hybrid nanostructured materials based on natural polymers, have garnered significant research interest across various fields and have found diverse applications ranging from regenerative medicine to food packaging. This surge in interest has led to a growing number of scientific publications, although studies have primarily approached the investigation of these biohybrid materials independently. However, there is potential to integrate bionanocomposites into a new interdisciplinary field at the intersection of materials science, life sciences, and nanotechnology. Two primary reasons have driven the utilization of biopolymers in synthesizing nanocomposites. First, incorporating these natural polymers enables the production of biodegradable materials, which are crucial for developing environmentally friendly alternatives that help combat plastic waste pollution. Second, biocompatibility is vital for applying these biohybrids in areas such as food packaging and tissue engineering for regenerative medicine. The future advancement of bionanocomposites lies in developing novel materials with enhanced properties and multi-functionality. This field of research holds immense potential due to the abundant and diverse availability of natural biopolymers and their advantageous combination with inorganic nanosized solids. The chapter underscores the importance of understanding the biodegradation behavior of biopolymers, particularly those sourced from microbial origins. Additionally, it highlights the significance of bionanocomposites derived from hydrogels and aerogels, explores the fabrication processes involved, and discusses their physical and mechanical properties.

REFERENCES

Abdul Khalil, H. P. S., Adnan, A. S., Yahya, E. B., Olaiya, N. G., Safrida, S., Hossain, Md S., Balakrishnan, V., Gopakumar, D. A., Abdullah, C. K., Oyekanmi, A. A., & Pasquini, D. (2020). A review on plant cellulose nanofibre-based aerogels for biomedical applications. *Polymers.* https://doi.org/10.3390/polym12081759

Abdul Khalil, H. P. S., Chong, E. W. N., Owolabi, F. A. T., Asniza, M., Tye, Y. Y., Rizal, S., Nurul Fazita, M. R., Mohamad Haafiz, M. K., Nurmiati, Z., & Paridah, M. T. (2019). Enhancement of basic properties of polysaccharide-based composites with organic and inorganic fillers: A review. *Journal of Applied Polymer Science.* https://doi.org/10.1002/app.47251

Abdul Khalil, H. P. S., Chong, E. W. N., Owolabi, F. A. T., Asniza, M., Tye, Y. Y., Tajarudin, H. A., Paridah, M. T., & Rizal, S. (2018). Microbial-induced CaCO3 filled seaweed-based film for green plasticulture application. *Journal of Cleaner Production, 199,* 150–163. https://doi.org/10.1016/j.jclepro.2018.07.111

Ahmed, E. M. (2015). Hydrogel: Preparation, characterization, and applications: A review. *Journal of Advanced Research, 6*(2), 105–121. https://doi.org/10.1016/j.jare.2013.07.006

Ahsan, A., Tian, W. X., Farooq, M. A., & Khan, D. H. (2021). An overview of hydrogels and their role in transdermal drug delivery. *International Journal of Polymeric Materials and Polymeric Biomaterials.* https://doi.org/10.1080/00914037.2020.1740989

Arrieta, M. P., Peponi, L., López, D., & Fernández-García, M. (2018). Recovery of yerba mate (Ilex paraguariensis) residue for the development of PLA-based bionanocomposite films. *Industrial Crops and Products, 111.* https://doi.org/10.1016/j.indcrop.2017.10.042

Arzac, A., Leal, G. P., Fajgar, R., & Tomovska, R. (2014). Comparison of the emulsion mixing and in situ polymerization techniques for synthesis of water-borne reduced graphene oxide/polymer composites: Advantages and drawbacks. *Particle and Particle Systems Characterization, 31*(1). https://doi.org/10.1002/ppsc.201300286

Awan, M. O., Shakoor, A., Saad Rehan, M., & Gill, Y. Q. (2021). Development of HDPE composites with improved mechanical properties using calcium carbonate and NanoClay. *Physica B: Condensed Matter, 606.* https://doi.org/10.1016/j.physb.2020.412568

Ayu, R. S., Khalina, A., Harmaen, A. S., Zaman, K., Nurrazi, N. M., Isma, T., & Lee, C. H. (2020). Effect of empty fruit brunch reinforcement in polybutylene-succinate/modified tapioca starch blend for agricultural mulch films. *Scientific Reports, 10*(1). https://doi.org/10.1038/s41598-020-58278-y

Backes, C., Abdelkader, A. M., Alonso, C., Andrieux-Ledier, A., Arenal, R., Azpeitia, J., Balakrishnan, N., Banszerus, L., Barjon, J., Bartali, R., Bellani, S., Berger, C., Berger, R., Ortega, M. M. B., Bernard, C., Beton, P. H., Beyer, A., Bianco, A., Bøggild, P., . . . & Bonaccorso, F. (2020). Production and processing of graphene and related materials. *2D Materials, 7*(2). https://doi.org/10.1088/2053-1583/ab1e0a

Bah, M. G., Hafiz Muhammad Bilal, and Jingtao Wang. 2020. Fabrication and application of complex microcapsules: A review. *Soft Matter, 3.* https://doi.org/10.1039/c9sm01634a

Baig, N., Kammakakam, I., Falath, W., & Kammakakam, I. (2021). Nanomaterials: A review of synthesis methods, properties, recent progress, and challenges. *Materials Advances.* https://doi.org/10.1039/d0ma00807a

Becerril, R., Nerín, C., & Silva, F. (2020). Encapsulation systems for antimicrobial food packaging components: An update. *Molecules.* https://doi.org/10.3390/molecules25051134

Bertolino, V., Cavallaro, G., Milioto, S., & Lazzara, G. (2020). Polysaccharides/halloysite nanotubes for smart bionanocomposite materials. *Carbohydrate Polymers*. https://doi.org/10.1016/j.carbpol.2020.116502

Bhattacharya, M. (2016). Polymer nanocomposites—A comparison between carbon nanotubes, graphene, and clay as nanofillers. *Materials*, *9*(4), 262. https://doi.org/10.3390/ma9040262

Borgonovo, C., & Apelian, D. (2011). Manufacture of aluminum nanocomposites: A critical review. *Materials Science Forum*. https://doi.org/10.4028/www.scientific.net/MSF.678.1

Bratovcic, A., & Suljagic, J. (2019). Micro- and nano-encapsulation in food industry. *Croatian Journal of Food Science and Technology*, *11*(1). https://doi.org/10.17508/cjfst.2019.11.1.17

Cader Mhd Haniffa, M. A., Ching, Y. C., Abdullah, L. C., Poh, S. C., & Chuah, C. H. (2016). Review of bionanocomposite coating films and their applications. *Polymers*. https://doi.org/10.3390/polym8070246

Cárcamo, L., Sierra, S., Osorio, M., Velásquez-Cock, J., Vélez-Acosta, L., Gómez-Hoyos, C., Castro, C., Zuluaga, R., & Gañán, P. (2021). Bacterial nanocellulose mulch as a potential greener alternative for urban gardening in the small-scale food production of onion plants. *Agricultural Research*, *10*(1). https://doi.org/10.1007/s40003-020-00479-y

Catoira, M. C., Fusaro, L., Di Francesco, D., Ramella, M., & Boccafoschi, F. (2019). Overview of natural hydrogels for regenerative medicine applications. *Journal of Materials Science: Materials in Medicine*, *30*(10). https://doi.org/10.1007/s10856-019-6318-7

Costa, P., Nunes-Pereira, J., Oliveira, J., Silva, J., Agostinho Moreira, J., Carabineiro, S. A. C., Buijnsters, J. G., & Lanceros-Mendez, S. (2017). High-performance graphene-based carbon nanofiller/polymer composites for piezoresistive sensor applications. *Composites Science and Technology*, *153*. https://doi.org/10.1016/j.compscitech.2017.11.001

Dantas de Oliveira, A., & Gonçalves Beatrice, C. A. (2019). Polymer nanocomposites with different types of nanofiller. In *Nanocomposites—recent evolutions*. https://doi.org/10.5772/intechopen.81329

Dlamini, D. S., Li, J., & Mamba, B. B. (2019). Critical review of montmorillonite/polymer mixed-matrix filtration membranes: Possibilities and challenges. *Applied Clay Science*. https://doi.org/10.1016/j.clay.2018.10.016

Doberenz, F., Zeng, K., Willems, C., Zhang, K., & Groth, T. (2020). Thermoresponsive polymers and their biomedical application in tissue engineering: A review. *Journal of Materials Chemistry B*. https://doi.org/10.1039/c9tb02052g

Donato, K., Matějka, L., Mauler, R., & Donato, R. (2017). Recent applications of ionic liquids in the sol-gel process for polymer—silica nanocomposites with ionic interfaces. *Colloids and Interfaces*, *1*(1). https://doi.org/10.3390/colloids1010005

Dong, M., Li, Q., Liu, H., Liu, C., Wujcik, E. K., Shao, Q., Ding, T., Mai, X., Shen, C., & Guo, Z. (2018). Thermoplastic polyurethane-carbon black nanocomposite coating: Fabrication and solid particle erosion resistance. *Polymer*, *158*. https://doi.org/10.1016/j.polymer.2018.11.003

Dufresne, A. (2018). Cellulose nanomaterials as green nanoreinforcements for polymer nanocomposites. *Philosophical Transactions of the Royal Society A: Mathematical, Physical and Engineering Sciences*. https://doi.org/10.1098/rsta.2017.0040

Fahim, I. S., Aboulkhair, N., & Everitt, N. M. (2018). Nanoindentation investigation on chitosan thin films with different types of nano fillers. *Journal of Materials Science Research*, *7*(2). https://doi.org/10.5539/jmsr.v7n2p11

Fahmy, H. M., Salah Eldin, R. E., Abu Serea, E. S., Gomaa, N. M., AboElmagd, G. M., Salem, S. A., Elsayed, Z. A., Edrees, A., Shams-Eldin, E., & Shalan, A. E. (2020).

Advances in nanotechnology and antibacterial properties of biodegradable food packaging materials. *RSC Advances*. https://doi.org/10.1039/d0ra02922j

Fawaz, J., & Mittal, V. (2014). Synthesis of polymer nanocomposites: Review of various techniques. In *Synthesis techniques for polymer nanocomposites*. https://doi.org/10.1002/9783527670307.ch1

Fu, S., Sun, Z., Huang, P., Li, Y., & Hu, N. (2019). Some basic aspects of polymer nanocomposites: A critical review. *Nano Materials Science*, *1*(1). https://doi.org/10.1016/j.nanoms.2019.02.006

Ganesan, K., Budtova, T., Ratke, L., Gurikov, P., Baudron, V., Preibisch, I., Niemeyer, P., Smirnova, I., & Milow, B. (2018). Review on the production of polysaccharide aerogel particles. *Materials*. https://doi.org/10.3390/ma11112144

Geng, S., Wei, J., Aitomäki, Y., Noël, M., & Oksman, K. (2018). Well-dispersed cellulose nanocrystals in hydrophobic polymers by: In situ polymerization for synthesizing highly reinforced bio-nanocomposites. *Nanoscale*, *10*(25). https://doi.org/10.1039/c7nr09080c

George, A., Sanjay, M. R., Srisuk, R., Parameswaranpillai, J., & Siengchin, S. (2020). A comprehensive review on chemical properties and applications of biopolymers and their composites. *International Journal of Biological Macromolecules*. https://doi.org/10.1016/j.ijbiomac.2020.03.120

Ghorbani, F., Kamari, S., Askari, F., Molavi, H., & Fathi, S. (2021). Production of NZVI—Cl nanocomposite as a novel eco—friendly adsorbent for efficient As(V) ions removal from aqueous media: Adsorption modeling by response surface methodology. *Sustainable Chemistry and Pharmacy*, *21*. https://doi.org/10.1016/j.scp.2021.100437

Gojayev, E. M., Aliyeva, S. V., Salimova, V. V., & Jabarov, S. H. (2020). The effect of UV irradiation on the dielectric properties of bionanocomposites with fillers of biological origin and metal nanoparticles. *Modern Physics Letters B*, *34*(17). https://doi.org/10.1142/S0217984920501869

Guastaferro, M., Reverchon, E., & Baldino, L. (2021). Polysaccharide-based aerogel production for biomedical applications: A comparative review. *Materials*, *14*(7). https://doi.org/10.3390/ma14071631

Gun'ko, V., Savina, I., & Mikhalovsky, S. (2017). Properties of water bound in hydrogels. *Gels*, *3*(4). https://doi.org/10.3390/gels3040037

Gunputh, U. F., & Le, H. (2020). A review of in-situ grown nanocomposite coatings for titanium alloy implants. *Journal of Composites Science*. https://doi.org/10.3390/jcs4020041

Guo, Y., Bae, J., Fang, Z., Li, P., Zhao, F., & Yu, G. (2020). Hydrogels and hydrogel-derived materials for energy and water sustainability. *Chemical Reviews*. https://doi.org/10.1021/acs.chemrev.0c00345

Gutiérrez, T. J., Toro-Márquez, L. A., Merino, D., & Mendieta, J. R. (2019). Hydrogen-bonding interactions and compostability of bionanocomposite films prepared from corn starch and nano-fillers with and without added Jamaica flower extract. *Food Hydrocolloids*, *89*. https://doi.org/10.1016/j.foodhyd.2018.10.058

Habibi, Y., Benali, S., & Dubois, P. (2014). *In situ polymerization of bionanocomposites*. https://doi.org/10.1142/9789814566469_0020

Hamedi, H., Moradi, S., Hudson, S. M., & Tonelli, A. E. (2018). Chitosan based hydrogels and their applications for drug delivery in wound dressings: A review. *Carbohydrate Polymers*, *199*(June), 445–460. https://doi.org/10.1016/j.carbpol.2018.06.114

Hayes, D. G., Dharmalingam, S., Wadsworth, L. C., Leonas, K. K., Miles, C., & Inglis, D. A. (2012). Biodegradable agricultural mulches derived from biopolymers. In K. Khemani, et al. (Eds.), *Degradable polymers and materials: Principles and practice* (2nd ed., pp. 201–223). American Chemical Society. https://doi.org/10.1021/bk-2012-1114.ch013

He, X., Deng, H., & Hwang, H. M. (2019). The current application of nanotechnology in food and agriculture. *Journal of Food and Drug Analysis*. https://doi.org/10.1016/j.jfda.2018.12.002

He, Z., Jiang, S., Li, Q., Wang, J., Zhao, Y., & Kang, M. (2017). Facile and cost-effective synthesis of isocyanate microcapsules via polyvinyl alcohol-mediated interfacial polymerization and their application in self-healing materials. *Composites Science and Technology*, *138*. https://doi.org/10.1016/j.compscitech.2016.11.004

Hussain, F., Hojjati, M., Okamoto, M., & Gorga, R. E. (2006). Review article: Polymer-matrix nanocomposites, processing, manufacturing, and application: An overview. *Journal of Composite Materials*, *40*(17). https://doi.org/10.1177/0021998306067321

Jacob, J., Haponiuk, J. T., Thomas, S., Peter, G., & Gopi, S. (2018). Use of ginger nanofibers for the preparation of cellulose nanocomposites and their antimicrobial activities. *Fibers*, *6*(4). https://doi.org/10.3390/fib6040079

Jafarbeglou, M., Abdouss, M., Shoushtari, A. M., & Jafarbeglou, M. (2016). Clay nanocomposites as engineered drug delivery systems. *RSC Advances*. https://doi.org/10.1039/c6ra03942a

Jamróz, E., Khachatryan, G., Kopel, P., Juszczak, L., Kawecka, A., Krzyściak, P., Kucharek, M., Bębenek, Z., & Zimowska, M. (2020). Furcellaran nanocomposite films: The effect of nanofillers on the structural, thermal, mechanical and antimicrobial properties of biopolymer films. *Carbohydrate Polymers*, *240*. https://doi.org/10.1016/j.carbpol.2020.116244

Jeevanandam, J., Barhoum, A., Chan, Y. S., Dufresne, A., & Danquah, M. K. (2018). Review on nanoparticles and nanostructured materials: History, sources, toxicity and regulations. *Beilstein Journal of Nanotechnology*. https://doi.org/10.3762/bjnano.9.98

Ke, K., Wang, Y., Liu, X. Q., Cao, J., Luo, Y., Yang, W., Xie, B. H., & Yang, M. B. (2012). A comparison of melt and solution mixing on the dispersion of carbon nanotubes in a poly(vinylidene fluoride) matrix. *Composites Part B: Engineering*, *43*(3), 1425–1432. https://doi.org/10.1016/j.compositesb.2011.09.007

Leng, Z., Tan, Z., Yu, H., & Guo, J. (2019). Improvement of storage stability of SBS-modified asphalt with nanoclay using a new mixing method. *Road Materials and Pavement Design*, *20*(7). https://doi.org/10.1080/14680629.2018.1465842

Li, B., & Zhong, W. H. (2011). Review on polymer/graphite nanoplatelet nanocomposites. *Journal of Materials Science*. https://doi.org/10.1007/s10853-011-5572-y

Li, R., Liu, Y., Sheng, Y., Xiang, Q., Zhou, Y., & Cizdziel, J. V. (2020). Effect of prothioconazole on the degradation of microplastics derived from mulching plastic film: Apparent change and interaction with heavy metals in soil. *Environmental Pollution*, *260*. https://doi.org/10.1016/j.envpol.2020.113988

Mangaraj, S., Yadav, A., Bal, L. M., Dash, S. K., & Mahanti, N. K. (2019). Application of biodegradable polymers in food packaging industry: A comprehensive review. *Journal of Packaging Technology and Research*, *3*(1). https://doi.org/10.1007/s41783-018-0049-y

Mishra, R., Militky, J., & Arumugam, V. (2019). Characterization of nanomaterials in textiles. *Nanotechnology in Textiles*. https://doi.org/10.1016/b978-0-08-102609-0.00005-5

Mishra, S. K., Tripathi, S. N., Choudhary, V., & Gupta, B. D. (2014). SPR based fibre optic ammonia gas sensor utilizing nanocomposite film of PMMA/reduced graphene oxide prepared by in situ polymerization. *Sensors and Actuators, B: Chemical*, *199*. https://doi.org/10.1016/j.snb.2014.03.109

Mohite, P. B., & Adhav, S. S. (2017). A hydrogels: Methods of preparation and applications. *International Journal of Advances in Pharmaceutics*, *6*(3).

Moura, M. R. de, Mattoso, L. H. C., & Zucolotto, V. (2012). Development of cellulose-based bactericidal nanocomposites containing silver nanoparticles and their use as active food packaging. *Journal of Food Engineering, 109*(3). https://doi.org/10.1016/j. jfoodeng.2011.10.030

Mousa, M. H., & Dong, Y. (2020). The role of nanoparticle shapes and structures in material characterisation of polyvinyl alcohol (PVA) bionanocomposite films. *Polymers, 12*(2). https://doi.org/10.3390/polym12020264

Mousa, M. H., Dong, Y., & Davies, I. J. (2016). Recent advances in bionanocomposites: Preparation, properties, and applications. *International Journal of Polymeric Materials and Polymeric Biomaterials.* https://doi.org/10.1080/00914037.2015.1103240

Müller, K., Bugnicourt, E., Latorre, M., Jorda, M., Sanz, Y. E., Lagaron, J. M., Miesbauer, O., Bianchin, A., Hankin, S., Bölz, U., Pérez, G., Jesdinszki, M., Lindner, M., Scheuerer, Z., Castelló, S., & Schmid, M. (2017). Review on the processing and properties of polymer nanocomposites and nanocoatings and their applications in the packaging, automotive and solar energy fields. *Nanomaterials.* https://doi.org/10.3390/nano7040074

Na, Y., Lee, J., Lee, S. H., Kumar, P., Kim, J. H., & Patel, R. (2020). Removal of heavy metals by polysaccharide: A review. *Polymer-Plastics Technology and Materials.* https://doi. org/10.1080/25740881.2020.1768545

Namazi, H., Pooresmaeil, M., & Hasani, M. (2020). Oxidized starch/CuO bio-nanocomposite hydrogels as an antibacterial and stimuli-responsive agent with potential colon-specific naproxen delivery. *International Journal of Polymeric Materials and Polymeric Biomaterials.* https://doi.org/10.1080/00914037.2020.1798431

Nile, S. H., Baskar, V., Selvaraj, D., Nile, A., Xiao, J., & Kai, G. (2020). Nanotechnologies in food science: Applications, recent trends, and future perspectives. *Nano-Micro Letters.* https://doi.org/10.1007/s40820-020-0383-9

Othman, S. H. 2014. "Bio-nanocomposite materials for food packaging applications: Types of biopolymer and nano-sized filler. *Agriculture and Agricultural Science Procedia, 2.* https://doi.org/10.1016/j.aaspro.2014.11.042

Palem, R. R., Ganesh, S. D., Kronekova, Z., Sláviková, M., Saha, N., & Saha, P. (2018). Green synthesis of silver nanoparticles and biopolymer nanocomposites: A comparative study on physico-chemical, antimicrobial and anticancer activity. *Bulletin of Materials Science, 41*(2). https://doi.org/10.1007/s12034-018-1567-5

Parhi, R. (2017). Cross-linked hydrogel for pharmaceutical applications: A review. *Advanced Pharmaceutical Bulletin.* https://doi.org/10.15171/apb.2017.064

Parola, S., Julián-López, B., Carlos, L. D., & Sanchez, C. (2016). Optical properties of hybrid organic-inorganic materials and their applications. *Advanced Functional Materials.* https://doi.org/10.1002/adfm.201602730

Poh, T. Y., Ali, N. A. B. M., Aogáin, M. M., Kathawala, M. H., Setyawati, M. I., Ng, K. W., & Chotirmall, S. H. (2018). Inhaled nanomaterials and the respiratory microbiome: Clinical, immunological and toxicological perspectives. *Particle and Fibre Toxicology, 15*(1). https://doi.org/10.1186/s12989-018-0282-0

Qian, L. (2018). *Cellulose-based composite hydrogels: Preparation, structures, and applications.* https://doi.org/10.1007/978-3-319-76573-0_23-1

Qin, Y., Summerscales, J., Graham-Jones, J., Meng, M., & Pemberton, R. (2020). Monomer selection for in situ polymerization infusion manufacture of natural-fiber reinforced thermoplastic-matrix marine composites. *Polymers, 12*(12). https://doi.org/10.3390/polym12122928

Rafiee, R., & Shahzadi, R. (2019). Mechanical properties of nanoclay and nanoclay reinforced polymers: A review. *Polymer Composites, 40*(2), 431–445. https://doi.org/10.1002/pc.24725

Rajak, D. K., Pagar, D. D., Kumar, R., & Pruncu, C. I. (2019). Recent progress of reinforcement materials: A comprehensive overview of composite materials. *Journal of Materials Research and Technology, 8*(6). https://doi.org/10.1016/j.jmrt.2019.09.068

Rane, A. V., Kanny, K., Abitha, V. K., & Thomas, S. (2018). Methods for synthesis of nanoparticles and fabrication of nanocomposites. *Synthesis of Inorganic Nanomaterials*, 121–139. https://doi.org/10.1016/b978-0-08-101975-7.00005-1

Rao, R. U., Venkatanarayana, B., & Suman, K. N. S. (2019). Enhancement of mechanical properties of PLA/PCL (80/20) blend by reinforcing with MMT nanoclay. In *Materials today: Proceedings* (Vol. 18). https://doi.org/10.1016/j.matpr.2019.06.280

Rizal, S., Alfatah, T., Abdul Khalil, H. P. S., Mistar, E. M., Abdullah, C. K., Olaiya, F. G., Sabaruddin, F. A., Ikramullah, & Muksin, U. (2021). Properties and characterization of lignin nanoparticles functionalized in macroalgae biopolymer films. *Nanomaterials*, *11*(3). https://doi.org/10.3390/nano11030637

Sabater I Serra, R., Molina-Mateo, J., Torregrosa-Cabanilles, C., Andrio-Balado, A., Meseguer Dueñas, J. M., & Serrano-Aroca, Á. (2020). Bio-nanocomposite hydrogel based on zinc alginate/graphene oxide: Morphology, structural conformation, thermal behavior/degradation, and dielectric properties. *Polymers*, *12*(3). https://doi.org/10.3390/polym12030702

Safdari, F., Carreau, P. J., Heuzey, M. C., & Kamal, M. R. (2017). Effects of poly(ethylene glycol) on the morphology and properties of biocomposites based on polylactide and cellulose nanofibers. *Cellulose*, *24*(7). https://doi.org/10.1007/s10570-017-1327-5

Saieh, S. E., Eslam, H. K., Ghasemi, E., Bazyar, B., & Rajabi, M. (2019). Physical and morphological effects of cellulose nano-fibers and nano-clay on biodegradable WPC made of recycled starch and industrial sawdust. *BioResources*, *14*(3). https://doi.org/10.15376/biores.14.3.5278-5287

Salehiyan, R., & Ray, S. S. (2018). Influence of nanoclay localization on structure—property relationships of polylactide-based biodegradable blend nanocomposites. *Macromolecular Materials and Engineering*, *303*(7). https://doi.org/10.1002/mame.201800134

Sanusi, O. M., Benelfellah, A., Bikiaris, D. N., & Hocine, N. A. (2021). Effect of rigid nanoparticles and preparation techniques on the performances of poly(lactic acid) nanocomposites: A review. *Polymers for Advanced Technologies*, *32*(2), 444–460. https://doi.org/10.1002/pat.5104

Saravanakumar, K., Sathiyaseelan, A., Mariadoss, A. V. A., Xiaowen, H., & Wang, M. H. (2020). Physical and bioactivities of biopolymeric films incorporated with cellulose, sodium alginate and copper oxide nanoparticles for food packaging application. *International Journal of Biological Macromolecules*, *153*. https://doi.org/10.1016/j.ijbiomac.2020.02.250

Sarfraz, J., Gulin-Sarfraz, T., Nilsen-Nygaard, J., & Pettersen, M. K. (2021). Nanocomposites for food packaging applications: An overview. *Nanomaterials*, *11*(1), 1–27. https://doi.org/10.3390/nano11010010

Sen, M. (2020). Nanocomposite materials. In *Nanotechnology and the environment*. IntechOpen. https://doi.org/10.5772/intechopen.93047

Shankar, S., & Rhim, J.-W. (2018). Bionanocomposite films for food packaging applications. *Reference Module in Food Science*. https://doi.org/10.1016/b978-0-08-100596-5.21875-1

Sharma, R., Jafari, S. M., & Sharma, S. (2020). Antimicrobial bio-nanocomposites and their potential applications in food packaging. *Food Control*, *112*(September 2019). https://doi.org/10.1016/j.foodcont.2020.107086

Shi, W., Ching, Y. C., & Chuah, C. H. (2021). Preparation of aerogel beads and microspheres based on chitosan and cellulose for drug delivery: A review. *International Journal of Biological Macromolecules*. https://doi.org/10.1016/j.ijbiomac.2020.12.214

Siqueira, G., Bras, J., & Dufresne, A. (2010). Cellulosic bionanocomposites: A review of preparation, properties and applications. *Polymers*. https://doi.org/10.3390/polym2040728

Song, R., Murphy, M., Li, C., Ting, K., Soo, C., & Zheng, Z. (2018). Current development of biodegradable polymeric materials for biomedical applications. *Drug Design, Development and Therapy.* https://doi.org/10.2147/DDDT.S165440

Surya, I., Chong, E. W. N., Abdul Khalil, H. P. S., Funmilayo, O. G., Abdullah, C. K., Sri Aprilia, N. A., Olaiya, N. G., Lai, T. K., & Oyekanmi, A. A. (2021). Augmentation of physico-mechanical, thermal and biodegradability performances of bio-precipitated material reinforced in eucheuma cottonii biopolymer films. *Journal of Materials Research and Technology, 12.* https://doi.org/10.1016/j.jmrt.2021.03.055

Swain, S. K., Pradhan, G. C., Dash, S., Mohanty, F., & Behera, L. (2018). Preparation and characterization of bionanocomposites based on soluble starch/nano CaCO3. *Polymer Composites, 39.* https://doi.org/10.1002/pc.24326

Taib, M. N. A. M., & Julkapli, N. M. (2018). Dimensional stability of natural fiber-based and hybrid composites. In *Mechanical and physical testing of biocomposites, fibre-reinforced composites and hybrid composites* (pp. 61–79). Elsevier. https://doi.org/10.1016/B978-0-08-102292-4.00004-7

Tan, C., & McClements, D. J. (2021). Application of advanced emulsion technology in the food industry: A review and critical evaluation. *Foods.* https://doi.org/10.3390/foods10040812

Tanahashi, M. (2010). Development of fabrication methods of filler/polymer nanocomposites: With focus on simple melt-compounding-based approach without surface modification of nanofillers. *Materials, 3*(3). https://doi.org/10.3390/ma3031593

Tang, S., Zhao, L., Yuan, J., Chen, Y., & Leng, Y. (2019). Physical hydrogels based on natural polymers. In *Hydrogels based on natural polymers.* https://doi.org/10.1016/B978-0-12-816421-1.00003-3

Thakur, V. K., Thakur, M. K., & Kessler, M. R. (Eds.). (2017). *Handbook of composites from renewable materials, nanocomposites: Science and fundamentals* (Vol. 7). John Wiley & Sons.

Thapliyal, P. C., & Singh, K. (2014). Aerogels as promising thermal insulating materials: An overview. *Journal of Materials, 2014.* https://doi.org/10.1155/2014/127049

Tuan Zainazor, T. C., Fisal, A., Goh, E. G., Che Sulaiman, N. F., & Sarbon, N. M. (2020). Emerging of bio-nano composite gelatine-based film as bio-degradable food packaging: A review. *Food Research.* https://doi.org/10.26656/FR.2017.4(4).365

Ullah, F., Hafi Othman, M. B., Javed, F., Ahmad, Z., & Akil, H. M. (2015). Classification, processing and application of hydrogels: A review. *Materials Science and Engineering C.* https://doi.org/10.1016/j.msec.2015.07.053

Unalan, I. U., Cerri, G., Marcuzzo, E., Cozzolino, C. A., & Farris, S. (2014). Nanocomposite films and coatings using inorganic nanobuilding blocks (NBB): Current applications and future opportunities in the food packaging sector. *RSC Advances.* https://doi.org/10.1039/c4ra01778a

Vaezi, K., Asadpour, G., & Sharifi, S. H. (2020). Bio nanocomposites based on cationic starch reinforced with montmorillonite and cellulose nanocrystals: Fundamental properties and biodegradability study. *International Journal of Biological Macromolecules, 146.* https://doi.org/10.1016/j.ijbiomac.2020.01.007

Varghese, N., Francis, T., Shelly, M., & Nair, A. B. (2021). Nanocomposites of polymer matrices: Nanoscale processing. In *Nanoscale processing.* https://doi.org/10.1016/b978-0-12-820569-3.00014-1

Vaseghi, Z., & Nematollahzadeh, A. (2020). Nanomaterials: Types, synthesis, and characterization. In *Green synthesis of nanomaterials for bioenergy applications* (pp. 23–82). Wiley. https://doi.org/10.1002/9781119576785.ch2

Wani, S. D. (2021). A review: Emerging trends in bionanocomposites. *International Journal of Pharmacy Research & Technology, 11*(1), 1–8.

Wei, L., Hu, N., & Zhang, Y. (2010). Synthesis of polymer-mesoporous silica nanocomposites. *Materials*. https://doi.org/10.3390/ma3074066

Yadav, M. (2018). Study on thermal and mechanical properties of cellulose/iron oxide bionanocomposites film. *Composites Communications, 10*. https://doi.org/10.1016/j.coco.2018.04.010

Yin, P., Chen, C., Ma, H., Gan, H., Guo, B., & Li, P. (2020). Surface cross-linked thermoplastic starch with different UV wavelengths: Mechanical, wettability, hygroscopic and degradation properties. *RSC Advances, 10*(73). https://doi.org/10.1039/d0ra07549c

Zabihi, O., Ahmadi, M., Nikafshar, S., Preyeswary, K. C., & Naebe, M. (2018). A technical review on epoxy-clay nanocomposites: Structure, properties, and their applications in fiber reinforced composites. *Composites Part B: Engineering*. https://doi.org/10.1016/j.compositesb.2017.09.066

Zare, Y., Rhee, K. Y., & Hui, D. (2017). Influences of nanoparticles aggregation/agglomeration on the interfacial/interphase and tensile properties of nanocomposites. *Composites Part B: Engineering, 122*. https://doi.org/10.1016/j.compositesb.2017.04.008

Zhan, C., Yu, G., Lu, Y., Wang, L., Wujcik, E., & Wei, S. (2017). Conductive polymer nanocomposites: A critical review of modern advanced devices. *Journal of Materials Chemistry C*. https://doi.org/10.1039/c6tc04269d

Zhang, R., & Zhao, Y. (2020). Preparation and electrocatalysis application of pure metallic aerogel: A review. *Catalysts*. https://doi.org/10.3390/catal10121376

Zhang, Y., Feng, R., Nie, W., Wang, F., & Feng, S. (2020). Plastic film mulch performed better in improving heat conditions and drip irrigated potato growth in Northwest China than in Eastern China. *Water (Switzerland), 12*(10). https://doi.org/10.3390/w12102906

Zubair, M., & Ullah, A. (2020). Recent advances in protein derived bionanocomposites for food packaging applications. *Critical Reviews in Food Science and Nutrition*. https://doi.org/10.1080/10408398.2018.1534800

4 Biopolymers in 3D Printing Technology

4.1 3D PRINTING TECHNOLOGIES

Three-dimensional (3D) printing, also known as additive manufacturing (AM), is a growing technology that has had a revolutionary impact on product fabrication for applications in several areas like healthcare and medicine, aeronautics, space, automotive, food, art, textile and fashion, architecture, and construction and has drawn increasing attention worldwide. Recently, the global market is moving towards the fourth industrial revolution (IR 4.0), which includes eight categories of digital transformation for the manufacturing industry: autonomous robots, simulated and augmented reality, the Internet of Things, cloud computing, cybersecurity, horizontal and vertical system integration, additive manufacturing, and big data analytics. In 2015, the Global Agenda Council on the Future of Software and Society at the World Economic Forum (WEF) anticipated that by 2025, we would witness the debut of the first 3D-printed car, 5% of consumer products being manufactured using 3D printers, and the pioneering transplantation of a 3D-printed liver.

Three-dimensional printing was first described by Charles Hull in 1986 (Wang et al., 2017). The generic 3D printing process must start with 3D computer-aided design (CAD) information, as shown in Figure 4.1. It is originally generated by a CAD program such as AutoDesk, AutoCAD, SolidWorks, or Creo Parametric. Three-dimensional printing is derived from a Standard Tessellation Language, or STereoLithography (STL), file by converting the file into a G-file via slicer software present in the 3D printer. Then, the G-file divides the 3D STL file into a sequence of two-dimensional (2D) horizontal cross-sections, which allows the 3D object to be printed, starting at the base, in consecutive layers of the desired material, essentially constructing the model from a series of 2D layers derived from the original CAD file (Azlin et al. 2022). Simply put, 3D printing is a process of joining materials from 3D computer model data layer by layer using filaments of various types and sizes to make objects. Three-dimensional printing is good at reducing product development times and costs, and most importantly, it can fabricate designs and features unmatched by other methods of manufacturing. Table 4.1 lists advantages over traditional types of manufacturing techniques (Attaran, 2017).

There are many types of 3D printing techniques, but no matter the technology involved, all are additive and build the object layer by layer (Jeffri et al. 2022). According to ASTM International Technical Committee F42, 3D printing techniques can be generally classified into four processes, material extrusion, vat

DOI: 10.1201/9781003416043-4

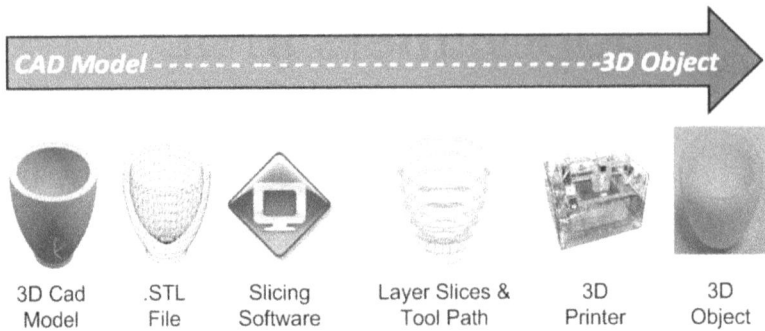

FIGURE 4.1 Stages in the 3D printing process start from the file generated by CAD until the desired product is obtained. Reproduced from Ilyas et al. (2022).

TABLE 4.1

Advantages over Traditional Types of Manufacturing Techniques

Area of Application	Advantages
Rapid prototyping	Reduce time to market by accelerating prototyping.
	Reduce the cost involved in product development.
	Make companies more efficient and competitive at innovation.
Production of spare parts	Reduce repair times.
	Reduce labor cost.
	Avoid costly warehousing.
Small volume manufacturing	Small batches can be produced cost efficiently.
	Eliminate the investment in tooling.
Customized unique items	Enable mass customization at low cost.
	Quick production of exact and customized replacement parts on site.
	Eliminate penalty for redesign.
Very complex work pieces	Produce very complex work pieces at low cost.
Machine tool manufacturing	Reduce labor cost.
	Avoid costly warehousing.
	Enable mass customization at low cost.
Rapid manufacturing	Direct manufacturing of finished components.
	Relatively inexpensive production of small numbers of parts.
Component manufacturing	Enable mass customization at low cost.
	Improve quality.
	Shorten supply chain.
	Reduce the cost involved in development.
	Help eliminate excess parts.
On site and on-demand manufacturing of customized replacement parts	Eliminate storage and transportation costs.
	Save money by preventing downtimes.
	Reduce repair costs considerably.
	Shorten supply chain.
	Reduce need for large inventory.
	Allow product lifecycle leverage.
Rapid repair	Significant reduction in repair time.
	Opportunity to modify repaired components to the latest design.

TABLE 4.2

Descriptions of Various 3D Printing Technologies

Processes	Technology
Material extrusion (fused filament fabrication, liquid deposition modelling)	Fused deposition modelling (FDM) • A material is melted and extruded in layers, one upon the other (this technique normally used in 3D printers at home).
Direct energy deposition (DED)	• An electron beam melts a metal wire to form an object layer by layer.
Powder bed fusion (selective laser sintering)	Selective laser sintering (SLS) • A bed of powder material is "sintered" (hardened) by a laser layer upon layer until a model is pulled out of it.
Binder jetting	• Powder is bonded by a binding material distributed by a movable inkjet unit layer by layer.
Vat photopolymerization (stereolithography)	Stereolithography (SLA) • A beam of concentrated ultraviolet light is focused on the surface of a vat filled with liquid photo-curable resin. The UV laser beam hardens slice by slice as the light hits the resin. When a projector beams the UV light through a mask onto the resin, it is called digital light processing (DLP).
Photopolymerzation (material jetting)	Polyjet process • A photopolymer liquid is precisely jetted out and then hardened with a UV light. The layers are stacked successively.
Sheet lamination (laminated object manufacturing, composite additive manufacturing)	Laminated object manufacturing (LOM) • Layers of adhesive-coated paper, plastic, or metal laminates are glued together and cut to shape with a knife or laser cutter.

photopolymerization, sheet lamination, and powder bed fusion. Table 4.2 shows a summary of various 3D printing techniques (Goh et al., 2019).

4.1.1 Fused Deposition Modelling

Fused deposition modelling (FDM) was first developed in the 1980s and was commercialized by Scott Crump of Stratasys Inc., USA, in the early 1990s (Mohan et al., 2017). An FDM machine (Figure 4.2) fabricates a 3D model by extruding thermoplastic filaments and depositing the semi-molten filaments onto the bed platform layer by layer and subsequently changing the print material, which enables more user control over device fabrication for experimental use. The materials used to build 3D models are moved down by two rollers to the nozzle tip of the extruder of a print head, where they are heated by temperature control units to a semi-molten state. The semi-molten materials are extruded out of the nozzle and solidified in the desired areas when the print head traces the design of each defined cross-sectional layer horizontally. The stage is then lowered, and another layer is deposited in the same manner, and these steps are repeated to fabricate a 3D structure in a layer-by-layer manner. Usually the outline of the part is printed

Material Extrusion

(a) Fused Filament Fabrication   Other Approaches

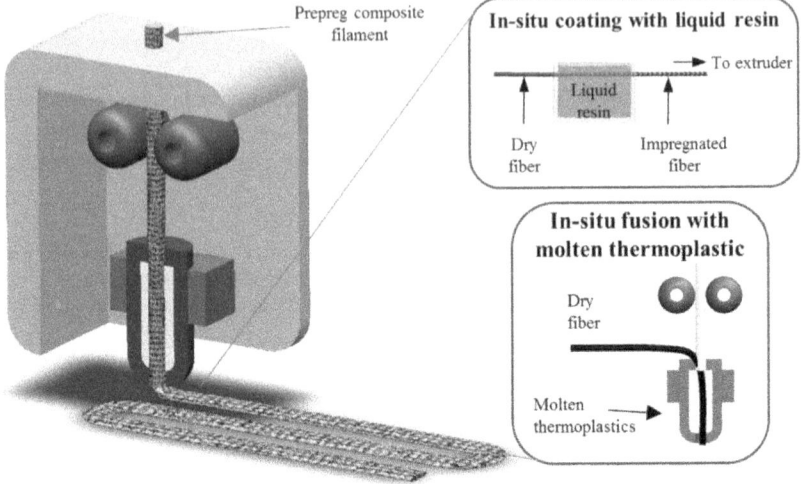

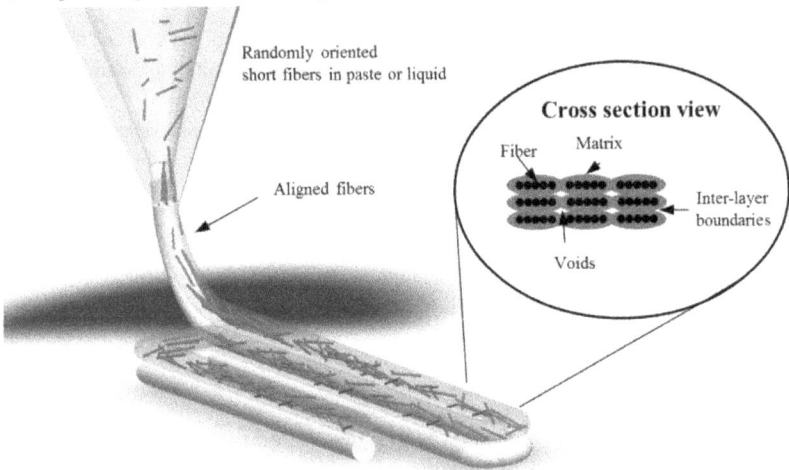

FIGURE 4.2   Material extrusion methods (a) fused filament fabrication and (b) liquid deposition modeling. Reproduced from Goh et al. (2019).

first, followed by the internal structures (2D planes) layer by layer. In FDM, complex geometric components of the 3D model are produced by converting a file containing the 3D model into 3D STL format using CAD software. Subsequently, the STL file is brought into computer-aided manufacturing (CAM) software, where it is transformed into a tangible representation of the 3D model. The CAM software then divides it into fine layers, each comprising tool paths that guide the

3D printing machine in depositing a continuous feedstock filament onto a surface. This gradual layer-by-layer approach is used to construct the 3D component (Owolabi et al., 2016).

The key benefit of FDM is that it is inexpensive, and technological advancements have made it easier for users to communicate with machines and optimize tool paths. Additionally, the most widely used materials on FDM devices, polylactic acid (PLA) and acrylonitrile butadiene styrene (ABS), are both commercially available at low prices. While the benefits of FDM are well known, due to component anisotropy and sometimes greater porosity, the mechanical properties of FDM printed parts are usually inferior to injection-molded parts. The degree of interfacial adhesion between discrete layers, tool (extrusion or print head) routes, and the inherent properties of the polymer in-plane cause anisotropy in AM part mechanical properties. (Wu et al., 2020).

4.1.2 LIQUID DEPOSITION MODELING

Liquid deposition modeling (LDM) is a 3D printing technique that uses fluid or paste as a feedstock. Materials are injected selectively using a syringe connected to a computerized numerical control unit in LDM. Thermoset resins are simpler to handle than thermoplastics using the LDM technique, since the material used in LDM is in liquid form at room temperature. However, in order to extrude smoothly from the nozzle, epoxy resin ink requires unique rheological and viscoelastic properties. It's worth noting that most ink dispersions made with these materials have a shear-thinning behavior, with a rising shear rate as viscosity decreases (Goh et al., 2019).

4.1.3 STEREOLITHOGRAPHY

Stereolithography (SLA), also known as the vat photopolymerization technique, was the first advanced manufacturing process that was well known for producing low-porosity printed parts (0–5%). The polymerization of liquid resin or monomer exposed to electromagnetic radiation such as UV laser or electron beam is the basis for this method. At room temperature, polymerization takes place point by point, line by line, and eventually layer by layer. The build platform is lowered to a depth equal to the layer/cure thickness below the liquid resin/monomer, and a concentrated laser beam is guided on the liquid surface to cure it. SLA has been used to build fiber-reinforced polymer composites (FRPCs) with reinforcements in the form of discontinuous fibers, continuous fibers, and fiber mats to date. A single layer/cross-section, per the CAD model, is completed by rastering the beam, and then the build platform is lowered by the layer thickness, and the process is repeated. Cure depth (25–500 m) and width must be managed with sufficient beam size and scan speed for efficient bonding (interlayer and interscan). Since 80% of polymerization occurs during the actual SLA process, parts are processed using heat or photo-curing to complete the curing and improve mechanical properties after the build process is completed (Balla et al., 2019; Goh et al., 2019). Figure 4.3 shows the vat photopolymerization technique.

Vat Photopolymerization

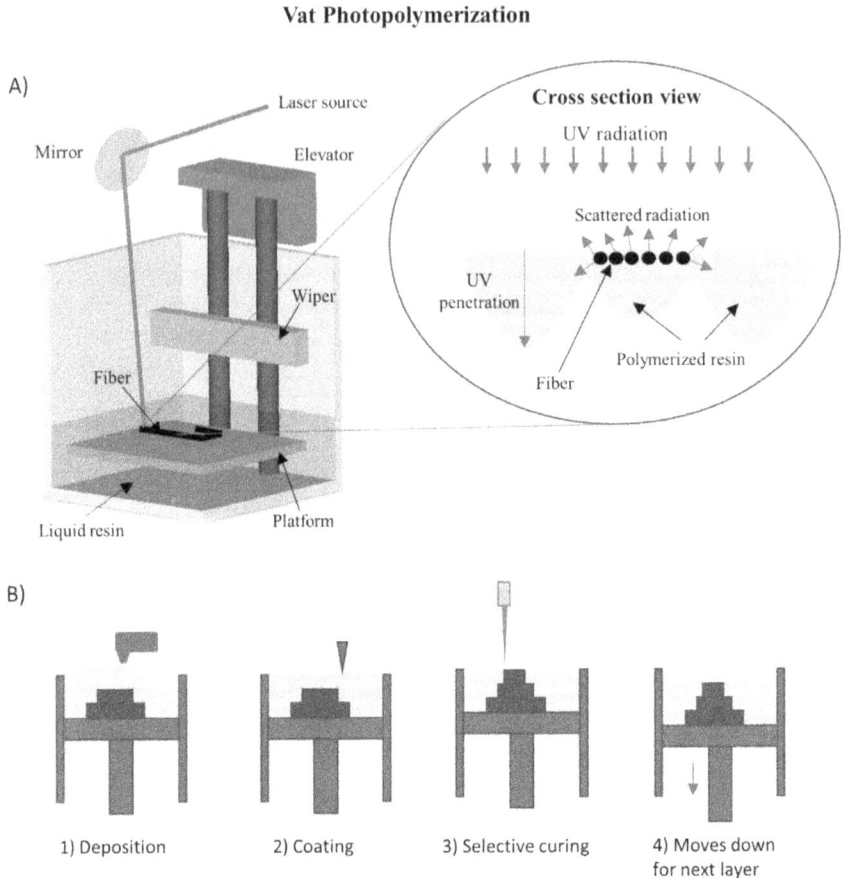

FIGURE 4.3 Vat photopolymerization process that creates 3D objects by selectively curing liquid resin through targeted light-activated polymerization. Reproduced from Goh et al. (2019).

4.1.4 Laminated Object Manufacturing

Laminated object manufacturing (LOM) (Figure 4.4a) is a method for creating a component from a stack of fiber sheets that incorporates additive and subtractive techniques. Each layer is cut with a laser, and then all of the layers are fused together with adhesive, pressure, and heat to minimize void material. Sheet material, which can be industrial prepreg sheets or any fiber preform, is used as the feedstock for LOM (Goh et al., 2019). There are two methodologies to this technique, LOM and composite-based additive manufacturing method (CBAM) (Figure 4.4b).

4.1.5 Composite-Based Additive Manufacturing

Composite-based additive manufacturing starts with an aqueous-based solution that is inkjet-deposited on each layer of fiber sheet. The fiber sheet is then coated in a

Sheet Lamination

(a) Laminated Object Manufacturing   (b) Composite-based Additive Manufacturing

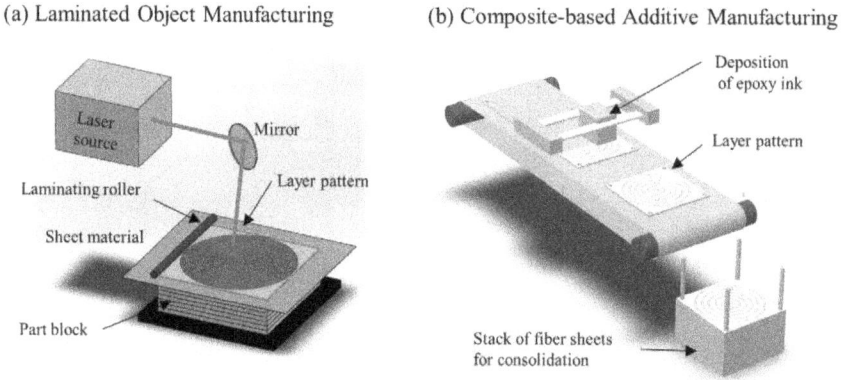

FIGURE 4.4 Sheet lamination process for creating a component from a stack of fiber sheets that incorporates additive and subtractive techniques (a) LOM and (b) CBAM. Reproduced from Goh et al. (2019).

thermoplastic powder film that only adheres to the aqueous solution. This procedure is repeated for all of the part's layers. The fiber sheets are stacked, compressed, and heated in the oven to fuse the matrix for consolidation after the excess powder is removed. Last, the component is sandblasted to remove any remaining fibers and expose the finished product. One of the benefits of the sheet lamination technique is its ability to manufacture high-strength pieces as opposed to traditional methods (Goh et al., 2019).

4.1.6 POWDER BED FUSION

Selective laser sintering (SLS) (Figure 4.5) is a well-known powder bed fusion method for polymers. In SLS, thermal energy is used to selectively fuse regions of a powder bed in powder bed fusion. A thin layer of loose powder is placed on a build platform with a spreader, typically in a controlled atmosphere build chamber. A high-power laser beam scanned (using an X–Y scanner) over the bed surface according to the CAD model cross-section is used to fuse this powder layer. The contact of the laser with the powder produces enough heat to melt the powder, resulting in a solid cross-section. Overhang frameworks can be supported by the unaffected loose powder. After spreading a fresh layer of powder on the build platform, the process is repeated for all cross-sections by raising and lowering the feed box and build platform by one layer/slice thickness (100 m), respectively. After all of the layers have been completed, the pieces are cooled in a controlled atmosphere chamber and loose powder is extracted (Balla et al., 2019).

4.1.7 BINDER JETTING

Binder jetting is a manufacturing process that employs a liquid bonding agent and a powder-based material. Three-dimensional artifacts are created with the help of a

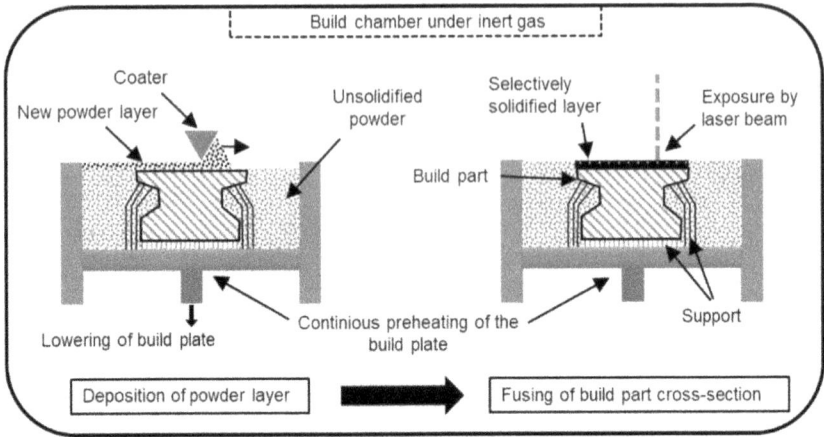

FIGURE 4.5 Powder bed fusion method, also known as selective laser sintering. Reproduced from Pannitz and Sehrt (2020).

print-head that selectively jets liquid agent according to the desired cross-section, gluing powder material together. Binder jetting can be used in tissue engineering to create advanced and complex scaffold structures because of this. Devices for binder jetting are low cost, and unused powder can be reused in subsequent processes (Goh et al., 2019). The binder jetting process is also quite similar to SLS. Layer thickness, powder size/shape/distribution, feed powder to layer thickness ratio, drop volume, binder saturation, binder viscosity, print head speed, number of printing passes/layer, spearing speed, drying temperature and time, and number of foundation layers are some of the significant process parameters in the binder jetting method (Balla et al., 2019).

4.2 3D PRINTING MATERIALS

The feedstock in additive manufacturing or 3D printing technologies must be formed into powder, sheet, filament, wire, or liquid, depending on the state that is compatible with the process. Polymers, metals, ceramics, and composites can all be used as 3D printing materials, depending on the 3D printing technology that is used.

4.2.1 THERMOPLASTIC POLYMERS

Thermoplastic polymers commonly used in 3D printing include polyamide (PA) or nylon, polycarbonate (PC), ABS, polymethyl methacrylate (PMMA), poly (lactic acid), polyethylene (PE), and polypropylene (PP). Material extrusion and powder bed fusion are two methods for processing thermoplastic polymers. Both methods depend on thermal layer adhesion, but the mechanisms are different. Material extrusion is usually done with amorphous thermoplastics, although powder bed fusion is best done with semicrystalline polymers (Balla et al., 2019; Alghamdi et al., 2021).

Amorphous thermoplastics are the preferred choice for material extrusion processes because of their unique melting characteristics. ABS and PLA polymers exhibit a wide range of softening temperatures, extending up to what is known as the glazing temperature. This glazing temperature is crucial, as it enables the transformation of these materials into a high-viscosity state, making them ideal for extrusion through nozzles with diameters ranging from 0.2 to 0.5 mm. Material extrusion processes produce supporting overhangs, which must be removed during post-processing. A more advanced solution involving a two-head device with support made of wax-based or PVA materials is used, and a lattice structure made of the same material but with a lighter nature and lower strength in relation to the part is used. The supports are melted or dissolved away during the post-processing stage. PLA model materials are supported by PVA, a water-soluble support material. There are normally voids between the deposited paths of extruded material. As a consequence, the mechanical properties may be weak, and anisotropy effects may be present (Bourell et al., 2017).

Powder bed fusion melts and fuses mainly semicrystalline powder feedstock using an infrared (IR) laser (typically a 10-mm CO_2 laser) or an IR or UV heat source (lamps). Polyamide 12 is the most widely used semicrystalline material for powder bed fusion (nylon). Polyamide 12 (nylon) has a melting point of about 35°C, which is higher than the crystallization temperature. The substance melted by the laser remains molten and in thermal equilibrium with the surrounding unmelted powder when the AM fabricator temperature is set between these peaks. After the build, recrystallization occurs uniformly, reducing residual stresses to a minimum. Because of the support offered by the surrounding powder cake, the powder bed fusion process does not involve overhang supports in the case of plastics. Multiple nested parts can be used in the construction. By tweaking the processing parameters, nearly fully dense, low-porosity artifacts can be developed. A post infiltration is needed for liquid pressure-tight applications (Bourell et al., 2017). Table 4.3 shows the various types of materials used in additive manufacturing, organized by process category and including commercial materials (Bourell et al., 2017).

4.2.2 THERMOSETTING POLYMERS

Monomers, oligomers, photoinitiators, and a number of other additives such as inhibitors, dyes, antifoaming agents, antioxidants, toughening agents, and other additives that help fine-tune the photopolymer's behavior and properties are common photopolymer materials used in AM (Gibson et al., 2015). Combinations of UV photoinitiators and acrylate monomers were the first photopolymers used in vat photopolymerization (Lovo et al., 2020). One of the limitations of acrylate resin is that ambient oxygen inhibits its polymerization reactions. Vinylethers were a form of monomer used in the early stages of resin growth. However, acrylate and vinylether resins shrank by 5 to 20%, causing residual stresses to accumulate as parts were built layer by layer.

Another thermoset resin that was introduced in the early 1990s was epoxy resin. Although complicating resin formulation, epoxy resin can be used to resolve several issues with other resins and provide significant benefits to the vat

TABLE 4.3

Current Commercial Materials Directly Processed by 3D Printing, by 3D Printing Process Category

Materials	Material Extrusion	Vat Polymerization	Material Jetting	Power Bed Fusion	Binder Jetting	Sheet Lamination	Directed Energy Deposition
Amorphous thermoplastics							
Acryonitrile butadiene styrene (ABS)	X						
Polycarbonate	X						
PC/ABS blend	X						
Polylactic acid (PLA)	X						
Polyetherimide (PEI)	X						
Polystyrene				X			
Semicrystalline thermoplastics							
Polyamide (nylon)				X			
Polypropylene				X			
Polyetheretherkeytone (PEEK)	X						
Thermoplastic polyurethane (elastomer)	X						
Thermoset							
Acrylics		X	X				
Acrylates		X	X				
Epoxies		X	X				
Aluminum alloys				X	X	X	X
Co-Cr alloys				X	X		X
Gold				X			
Nickel alloys				X	X		X
Silver				X			
Stainless steel				X	X	X	X
Titanium				X	X	X	X
Ti-6Al-4V				X	X	X	X
Tool steel				X	X		X

photopolymerization process. Epoxies are cationically polymerized photopolymers. As epoxy monomers are reacted, their rings open up, allowing other chemical bonds to form. Ring-opening is known to cause minimal volume shift since the number and types of chemical bonds are nearly identical before and after the reaction. As a result, epoxy-containing stereolithography (SLA) compound resins shrink less than acrylates and are less prone to warping and curling. Epoxies are present in almost all commercially available SLA resins in large quantities (Bourell et al., 2017) due to the favorable mechanical properties of the resulting resins. Suitable cationic photoinitiators, activated by UV radiation, include onium salts and metallocene salts.

Mixtures of acrylates, epoxies, and other oligomer materials make up commercial AM resins. Acrylates react quickly, while epoxies give the solid strength and durability. To form their respective polymer networks, acrylates polymerize dramatically, while epoxides polymerize cationically. The two types of monomers do not react with each other, but when combined, they form an interpenetrating polymer network (IPN). IPNs are a type of polymer blend in which both polymers are in network form and are created by two simultaneous reactions rather than a simple mechanical mixing phase. During the curing process, the acrylates and epoxies interact physically. The acrylate reaction increases photospeed, thus lowering the energy requirements for the epoxy reaction. In addition, the presence of acrylate monomers can reduce humidity's inhibitory effect on epoxy polymerization. The epoxy monomer, on the other hand, serves as a plasticizer during early acrylate monomer polymerization; the acrylate forms a network while the epoxy is still liquid. The chain propagation reaction is likely favored by this plasticizing effect, which increases molecular mobility. As a result, the acrylate polymerizes more widely in the presence of epoxy, resulting in higher molecular weights than in the neat acrylate monomer. Also, due to the viscosity increase induced by epoxy polymerization, the acrylate in the hybrid system has a lower sensitivity to oxygen than in the neat composition, which may result in reduced diffusion of atmospheric oxygen into the material (Bourell et al., 2017).

(Meth)acrylate-based resins are compatible with various commercial 3D printers and custom 3D printers. These resins have been used in a variety of applications, including shape memory polymer 3D printing, a siloxane-based hybrid polymer network, highly stretchable photopolymers, and functional biomaterials. PEGDA, UDMA, triethylene glycol dimethacrylate (TEGDMA), bisphenol A-glycidyl methacrylate (Bis-GMA), trimethylolpropane triacrylate (TTA), and bisphenol A ethoxylate diacrylate (BisEDA) are the most common (meth)acrylate monomer/oligomers used in 3D printing (Bagheri & Jin, 2019).

4.2.3 Metals

Powder bed fusion and directed energy deposition are the main powder-based AM processes that are commercially used to manufacture quality metal parts. A metal wire feed can also be used instead of a powder feed in direct energy deposition (DED). Binder jetting is also used to produce metal parts. Polymer matrix parts are made that need furnace de-binding and sintering and/or infiltration with a lower melting point metal (e.g., brass) to obtain dense metal parts. The set of common commercially available alloys is limited to pure titanium, Ti6Al4V, 316L stainless steel, 17-4PH stainless steel and 18Ni300 maraging steel, AlSi10Mg, CoCrMo, and nickel-based superalloys Inconel 718 and Inconel 625. This range is continually expanding with new entrants to the materials supply market. Precious metals such as gold, silver, or platinum have been processed indirectly by 3D printing of lost wax models but are currently also being directly used in selective laser melting (SLM) (Carlotto et al., 2014; Khan & Dickens, 2012). Several factors contribute to this limited metal palette. When fusion is involved, the metals generally must be weldable and castable to be successfully processed in AM. The small, moving melt pool is significantly smaller than the dimensions of the final part (typically on the order of 102–104 times smaller). This local hot zone in direct contact with a large colder area leads to large thermal gradients, causing significant thermal residual stresses and non-equilibrium micro-structures. For powdered feedstock, particles should preferably be spherical with a certain size distribution, which is different for powder bed fusion (PBF) and DED. The latter tends to be less sensitive to the dimensional qualities of the feedstock. A wire is also a suitable precursor material for certain DED processes, creating a larger melt pool than powder-based DED, allowing a higher production rate (Bourell et al., 2017).

4.3 BIOPOLYMERS IN 3D PRINTING

4.3.1 Poly (Lactic Acid)

ABS, a petroleum-based plastic, and PLA, a biobased plastic, are examples of thermoplastic polymers often used in FDM 3D printing. Yet PLA is the preferred polymer in FDM 3D printing since PLA comes from renewable resources and is biodegradable (Prasong et al., 2020). Also, ABS will cause environmental issues like the emission of volatile compounds such as styrene, and ABS releases unpleasant odors (Andrzejewski et al., 2020). Owing to global environmental issues, PLA is considered a possible alternative to replace petroleum-based polymers, making it an ideal choice for industrial applications. In term of appearance, PLA is preferable for its wide range of available colors, translucencies, and glossy feel, and it has a semi-sweet smell compared to ABS (Wijk & Wijk, 2015). Though the mechanical strength of PLA is lower than that of ABS, PLA is a simple linear molecular chain structure, making it possible to enhance the mechanical properties of modified PLA (Liu et al., 2019). Also, FDM machines are restricted to amorphous polymers or those with low levels of crystallinity, as they exhibit a low degree of polymer shrinkage, which is crucial to the accuracy of components produced (Stoof & Pickering, 2018).

Low thermal expansion is advantageous for mitigating the internal stresses that occur during cooling. This makes PLA a more favorable choice when compared to ABS, as PLA exhibits lower glass transition, melting, and printing temperatures than ABS. Consequently, PLA enables precise dimensional control during the printing process, resulting in printed products that closely match the original 3D model (Wijk & Wijk, 2015; Liu et al., 2019). This characteristic of PLA significantly reduces the likelihood of warping effects (Cardoso et al., 2020; Andrzejewski et al., 2020). During cooling, the materials stretch and slightly shrink until the printed

TABLE 4.4
Comparison of ABS and PLA Properties for 3D Printing

Polymer Type	ABS	PLA
Extrusion temperature (°C)	225–250	190–240
Bed temperature (°C)	80–110	20–55
Moisture	ABS with moisture will bubble and sputter when printed but easily dry	PLA with moisture will bubble and sputter when printed. It will not easily dry; can react with water and depolymerize at high temperatures
Heat	Less deformation due to heating	Product can deform because of heat
Smell	Plastic styrene smell	Corn-like sweet smell
Color	Less color brightness	Bright, shiny colors and smooth appearance
Hardness	Very sturdy and hard	Less sturdy than ABS
Fumes	Hazardous fumes	Non-hazardous fumes
Details	Higher layer height, less sharp printer corners, needs a heated printer bed for less warping	Higher max printer speed, lower layer height, sharper printed corners, less part warping
Lifetime	Longer-lifetime products	–
Environment	Non-biodegradable, made from oil	Biodegradable, made from sugar, corn, soybeans, or maize

product starts to stiffen at 110°C and 56°C for ABS and PLA, respectively, based on their glass transition temperature, and the remaining shrinking is resisted by the stiffness of the material. The stresses are internally stored instead of being relieved by the warmer material's ability to flow; thus, bending and warping occur. In addition, more warping is likely to happen for ABS compared to PLA because of its high glass transition temperature. Thus, printed PLA is relatively warm, but ABS is cool. In other words, without a heated bed or closed chamber, the bottom layers of the printed product will get cool fast, with shrinking and stiffening occurring at the same time for the ABS printed product (Andrzejewski et al., 2020; Bates-Green & Howei, 2017). PLA is the most popular polymer used among home printers, hobbyists, and universities other than in the industrial section. Table 4.4 shows a comparison between PLA and ABS in 3D printing (Cale Rauch, 2018).

4.3.2 POLY (LACTIC ACID)/POLY (BUTYLENEADIPATE-CO-TEREPHTHALATE) POLYMER BLENDS IN 3D PRINTING

Recently, growing interest in research has focused on the improvement of PLA mechanical properties by polymer blending for conventional techniques, but only a few studies have been done on 3D printed polymer blending of poly (lactic acid)/poly (butyleneadipate-co-terephthalate) (PLA/PBAT) (see Table 4.5). The performance of the FDM printable product is highly based on the printing parameters, such as bed temperature, nozzle temperature, print speed, and layer height, all of which can affect the properties and printability of the product printed by FDM (Diederichs et al., 2019). Prasong et al. (2020) and Andrzejewski et al. (2020) observed that the performance trends in mechanical properties for FDM 3D printing and conventional injection molding techniques using PLA/PBAT blends are similar, primarily hinging on the content of the PBAT phase. In this context, the addition of PBAT provides a practical means to enhance the brittleness characteristics of PLA. Prasong et al. (2020) recommended that the optimal PLA/PBAT blend for 3D printing falls within the range of 10 to 30 wt.% PBAT. However, in a study by Andrzejewski et al. (2020), they noted that a 70/30 wt.% PLA/PBAT blend demonstrated superior mechanical properties. This improvement was attributed to favorable printing orientation and printing speed, which exceeded those achieved through conventional molding techniques.

Both Prasong et al. (2020) and Andrzejewski et al. (2020) reported that printing quality is also affected by the viscosity of the polymer during printing, which is based on the printing temperature. According to Rahim et al. (2019), temperature is important in controlling the viscosity of polymer melting to ensure good flowability of the filament material during printing and thus result in good surface quality and optimal structural strength. High temperature may improve the mechanical properties of the printed sample, but it also affected the dimensional accuracy. Andrzejewski et al. (2020) mentioned the nozzle temperatures used to print PLA/PBAT blends were at 230°C for 10 to 20 wt.% PBAT content, and the temperature increased to 270°C for 30 wt.% PBAT content in a PLA/PBAT blend. However, it is said that the bed temperature has an important effect on the consistency of the printing layers and final mechanical properties of the printed sample compared to the nozzle temperature. Meanwhile, Liu et al. (2019) focused on cooling speed during printing, which is a crucial factor for interface bonding. Too slow a cooling speed causes bad surface quality, forming ability, filaments, and deformation shape of the printed product, whereas if the cooling speed is too fast, it causes insufficient time for the filament to solidify, poor diffusivity of PLA molecular chains on the interface, and bad interface bonding, resulting in poor mechanical properties. Thus, the printing parameters play an important role in performance of the printed product. Table 4.5 presents the mechanical properties of 3D-printed PLA/PBAT blends.

4.3.3   3D Printing of Native Cellulose-Based Materials

Cellulose, a homogeneous polysaccharide consisting of linear β-(1→4)-glucan with intra- and intermolecular hydrogel networks (Figure 4.3), is the most abundant renewable and biodegradable biopolymer worldwide. Development of cellulose-based materials as a new type of feedstocks for 3D printing technologies would open a new window to explore cellulose materials. Currently, the most intensively studied native cellulose-based 3D printing material is cellulose nanofibril (CNF) hydrogel. CNF hydrogel-based inks are biocompatible and can mimic the microenvironment of the extracellular matrix (ECM), which has been recently considered in biomedical

TABLE 4.5
Mechanical Properties of 3D-Printed PLA/PBAT Blends

PLA/PBAT (%)	Manufacturing Method	Filament Diameter (mm)	Nozzle Diameter (mm)	Bed and Nozzle Temperature (°C)	Printing Speed (mm/s)	Tensile Strength (MPa)	Tensile Modulus (GPa)	Elongation (%)	Impact Strength (J/m)	References
70/30	Twin-screw extrusion, capillary rheometer	1.75	0.4	45, 210	25	46.9	–	225.8	–	(Prasong et al., 2020)
70/30	Twin-screw extrusion	2	0.5	60, 270	10	48.8	2.289	37.3	327	(Andrzejewski et al., 2020)
70/30	Mixer	–	0.2	60, 200	70	22.9	–	–	–	(Lyu et al., 2020)
80/20	Twin-screw, single-screw extrusion	1.75	0.35	50, 210	40	44.8	2.513	34	5.5 kJ/m²	(Wang et al., 2020)

applications (Liu et al., 2016). However, the high hydrophilicity and inherently entangled state of CNF limit the increase of ink concentration at a given viscosity and storage modulus. Efforts have been made to directly dissolve or utilize native cellulose in other forms.

Native cellulose cannot be readily dissolved in common solvent systems due to the strong H-bond network and crystalline structure. Ionic liquids (ILs) are a new class of "green" cellulose solvents with low melting temperature. Dissolution of cellulose is achieved by disruption of the native cellulose H-bond network, with new H-bonds formed between cellulose and anions and hydrophobic interactions with cations. Anti-solvents, such as water, ethanol, and acetone, can be added to regenerate cellulose and recycle the ILs (Gupta & Jiang, 2015). Liu et al. (2019) dissolved different cellulose materials, including bacterial cellulose (BC), microcrystalline cellulose (MCC), and dissolving pulp, in an IL (EmimAc) with solid concentration up to 4% and applied it for direct ink writing (DIW) printing. Complex patterns of 2D structures and multilayered cylinder structures were printed and coagulated with water. However, the spatial resolution of the printed structure and recycling of ILs still need to be addressed.

Cellulose nanocrystals (CNCs) prepared from acid hydrolysis may offer advantages over cellulose IL dissolution and CNF in terms of the increase of concentration at a given rheology requirement for 3D printing. Siqueira et al. (2017) dispersed CNCs in water to prepare different concentrations of suspensions/gel inks (0.5–40%) for DIW and yielded cellulose-based structures with a high degree of CNC particle alignment along the printing direction, offering the opportunity to print cellulosic architectures with tailored mechanical properties.

Native cellulose can also be utilized in 3D food printing by serving as dietary fiber, bulk filling agents, rheological modifying ingredients, or reinforcing ingredients owing to its indigestibility, excellent mechanical performance, and high viscosity of native cellulose. Lille et al. (2018a) incorporated CNF gel in a 3D printing ink formula as a reinforcing ingredient for the development of healthy customized 3D printed foods. CNF was found to improve the shape stability of the printed structures and to decrease the hardness of the dried objects, revealing its potential for 3D printing healthy fiber-rich and structured foods.

4.3.4 3D Printing of Cellulose Derivative-Based Materials

The super-molecular structure of cellulose, that is, high crystallinity degree and rigid intra/intermolecular hydrogen bond networks, restricts its application. However, the abundant hydroxyl groups on the cellulose surface open opportunities for chemical modification or functionalization, such as esterification, etherification, selective oxidation, graft copolymerization, and intermolecular crosslinking reactions, bringing novel opportunities to exploit for various applications such as in food, cosmetics, medicine, and pharmacy (Liu et al., 2015).

The interfacial compatibility problem between polar native cellulose and the nonpolar polymer matrix is the key limitation for the application of native cellulose, which has excellent mechanical performance as a reinforcing agent in composite manufacture. To improve compatibility with PLA, surface modification of cellulose was conducted using a titanate coupling agent via coordination exchange (neoalkoxy) or solvolysis (monoalkoxy) (Murphy et al., 2016). The modified cellulose was found to improve dispersion and interfacial adhesion between the cellulose and the PLA, leading to higher mechanical strength of the composite filament than neat PLA alone. Porous scaffold prototypes were successfully printed with FDM using cellulose-reinforced PLA filaments, offering the opportunity for rapid production of fully degradable biocomposite 3D prototypes for applications in the biomedical, automotive, and construction sectors.

Similarly, to improve the dispersibility of CNC and interfacial bonding after DIW and UV polymerization, Siqueira et al. (2017) applied an acetylation reaction for CNC surface modification by using methacrylic anhydride. The modified CNC was dispersed in a mixture of photopolymerizable monomer and photoinitiator. After DIW printing and curing, the yielded structures were found to have a high degree of CNC particle alignment along the printing direction, providing the opportunity to print cellulosic architectures with programmable reinforcement along prescribed directions. (2,2,6,6-Tetramethylpiperidin-1-yl)oxyl (TEMPO)-mediated oxidation and carboxymethylation are two common approaches to prepare CNF by introducing negative charges on the cellulosic fiber surface followed by mechanical processing (Liu et al., 2014). The yielded CNF hydrogels with shear thinning behavior and high zero shear viscosity have been tested as inks for 3D bioprinting. Rees et al. (2015) prepared CNF hydrogels with TEMPO-mediated oxidation and a combination of carboxymethylation and periodate oxidation and used them to print 3D porous structures for wound dressing application.

Both CNF-based inks were found to inhibit bacterial growth, suggesting potential application as wound dressing materials. The CNF prepared from TEMPO-mediated oxidation failed to print constructs with acceptable resolution and desired tracks due to the limited low consistency (0.95%) but relatively high viscosity. The one prepared by carboxymethylation and periodate oxidation was suitable for use as a bioink in terms of high consistency (3.9%) and appropriate rheological properties. The printed 3D structures have fine tracks with open porosity and the potential to carry and release antimicrobial components for wound dressing. Carboxymethylated CNF with high surface charge and higher consistency (2%), which holds the 3D shape after deposition, has been investigated as ink to print 3D structures (Håkansson et al., 2016). By controlling the solidification process, such as air drying, air drying with surfactants, solvent exchange before drying, and freeze drying, 3D printed structures with the desired mechanical, surface texture, and porous structure can be obtained (Figure 4.4A). Incorporation of functional ingredients, such as conductive carbon nanotubes, offers additional functionality to the printed products. As shown in Figure 4.4B, C, conductive ink composed of carbon nanotubes and CNF can be printed in a CNF hydrogel matrix in different layers with fine resolution. After drying, two over-crossing conductive lines with decent electrical conductivity were separated by the insulation CNF layers, suggesting a potential for fabrication of sustainable commodities such as packaging, textiles, biomedical devices, and furniture with conductive parts.

Pattinson and Hart (2017) demonstrated the manufacture of fully dense cellulose-based materials using cellulose acetate by 3D printing. Cellulose acetate has 80% of the surface hydroxyl groups replaced by acetate groups, which effectively disrupted the hydrogen bonding network of native cellulose and made it possible to dissolve in acetone with high consistency (25–35%). Solid cellulose acetate structures with isotropic strength and high toughness were built upon the evaporation of acetone. By incorporating antimicrobial agents, direct 3D printing of cellulose acetate-based objects with tailored biochemical functionality (e.g. antimicrobial properties) can be achieved, enabling customization of medical instruments in a short time (e.g. forceps with customized tips; Figure 4.4D). Methylcellulose was utilized by Schütz et al. (2015) to enhance the temporary printability of a 3% alginate with limited viscosity. The addition of methylcellulose significantly enhanced the bioink viscosity, enabling accurate and easy 3D bioplotting of constructs with tailored architecture and high shape fidelity, after which the methylcellulose was released from the scaffolds during the following cultivation. Embedded mesenchymal stem cells in 3D printed methylcellulose-alginate scaffolds were found to maintain their differentiation potential with high viability. The temporary integration of methylcellulose into alginate-based bioink with low concentration allowed the generation of scaffolds with high shape fidelity and stability while keeping the advantages of low-concentrated alginate bioink for cell embedding.

A research team from Aalto University working within the project "Design Driven Value Chains in the World of Cellulose" studied 3D FDM printing of thermoplastic cellulose derivatives (Ali, 2015). The printability of the pure cellulose derivative was improved through plasticization, which reduced melt viscosity, improved the layer adhesion, and lowered glass transition temperature, allowing manufacture of a variety of cellulose-based 3D structures. Similarly, within the same project, oxidized cellulose hydrogel, cellulose-based plastics, and pulp fiber composites were studied for textile printing (Figure 4.4E, F) (Tenhunen et al., 2018). Structures with the desired look or feel, such as hard or soft, strong or brittle, stiff of flexible, or porous or dense texture, can be printed directly on substrates or fabrics or form self-standing structures and can find applications from medical science to personalized sportswear (Chaunier et al., 2018).

4.3.5 3D PRINTING OF CELLULOSE COMPOSITE-BASED MATERIALS

Incorporation of other ingredients into cellulose-based ink formulations may significantly improve ink properties such as processability, printability, mechanics, and bioactivity (Guvendiren et al., 2016). CNF hydrogels have ideal structural similarity to ECM and excellent rheological properties for 3D printing. However, pure CNF printed structures lack high shape fidelity (fine line resolution), sufficient mechanical properties for handling during transplantation, and long-term structure stability, limiting their application in the 3D printing of complex scaffolds in different biomedical applications (Gatenholm et al., 2016). Alginate also possesses shear thinning behavior at high shear forces, but the low viscosity at a zero shear rate makes it difficult to print 3D constructs. Inks composed of CNF and alginate keep the shear thinning behavior and high zero shear viscosity of CNF, while the alginate allows crosslinking by divalent ions to maintain long-term shape fidelity and structural integrity after 3D printing. Markstedt et al. (2015) optimized and evaluated the printability and biocompatibility of this composite ink. Anatomically shaped cartilage structures, such as a human ear and sheep meniscus, were 3D printed with high shape fidelity and stability (Figure 4.5A–C). Human nasoseptal chondrocytes were incorporated into the CNF-alginate composite ink for 3D bioprinting, and the results suggested that the shear forces during mixing and the crosslinking process could result in a decrease of cell viability. Although cell viability can be recovered after seven days of culture, cells embedded in the structures (e.g. thickness exceeds 150–200 µm) may gradually lose viability due to the limited diffusion of oxygen/gas, nutrients, growth factors, and waste product (Lee Ventola, 2014).

To overcome this issue, Martínez Ávila et al. (2016) printed a chondrocyte-laden patient-specific auricular construct with open porosity for oxygen, nutrients, and waste diffusion (Figure 4.5D). The bioprinted constructs showed excellent shape and size stability after bioprinting, and long-term 3D culture and the CNF-alginate bioink was found to support redifferentiation of human chondrocytes, re-establishing and maintaining their chondrogenic phenotype, suggesting the usefulness of the bioink for auricular cartilage tissue engineering and many other biomedical applications. Similarly, Nguyen et al. (2017) designed human-derived induced pluripotent stem cells (iPSCs) and irradiated human chondrocyte-laden CNF-alginate/hyaluronic acid composite inks to mimic cartilaginous tissue for potential cartilage lesion treatment. In the composite inks, the CNF was supposed to mimic the bulk collagen matrix, alginate stimulates proteoglycans, and hyaluronic acid serves to substitute for hyaluronic acid in native cartilage. Bioink of CNF-alginate was found to maintain the pluripotency of stem cells and support the new generation of cartilaginous tissue with collagen expression and high cell density, suggesting the bioprinting of iPSCs with CNF-alginate bioinks as a promising treatment to repair damaged cartilage. Alginate sulfate with the potential to support the chondrocyte phenotype has also been incorporated into CNF hydrogels to formulate bioinks for cartilage bioprinting (Müller et al., 2016). The composite bioink was found to promote bovine chondrocyte spreading, proliferation, and collagen II synthesis. However, the bioprinting process significantly compromised cell proliferation, especially in the case that used nozzles and valves with a small diameter and high extrusion pressure and stress.

Currently, CNF-alginate composite ink formulation has been successfully commercialized with the brand name of CELLINK and has been evaluated in vitro and in vivo with human chondrocytes, human dermal fibroblasts and keratinocytes, neural cells, mesenchymal stem cells derived from bone marrow, and adipose tissue and iPSC cells derived from chondrocytes (Gatenholm et al., 2016). For example, Henriksson et al. (2017) bioprinted the living adipocyte-laden CELLINK bioink and a bioink composed of nanocellulose and hyaluronic acid (CELLINK-H) for 3D bioprinting in adipose tissue engineering. The adipocytes laden in 3D-printed structures, especially the CELLINK-H, produced better adipogenic differentiation and a more mature cell phenotype with larger lipid droplets than conventional 2D culture systems, suggesting a promising method for adipose tissue engineering. A clinical study was recently conducted by transplanting a bioprinted cell-laden CELLINK bioink scaffold in a subcutaneous pocket of mice (Möller et al., 2017) (Figure 4.5E). The histological, immunohistochemical, and mechanical analysis suggested that the 3D-bioprinted scaffolds have excellent structural integrity, shape fidelity, and good mechanical properties after 60 days of implantation, and the scaffolds can result in cartilage synthesis, suggesting a potential application of CELLINK bioinks in 3D bioprinting cartilage tissue for application in reconstructive surgery (Liu et al., 2019).

4.3.6   3D PRINTING OF HEMICELLULOSE-BASED MATERIALS

Hemicelluloses, which mainly include xylans, glucomannans, arabinans, galactans, and glucans, are a type of heterogeneous polysaccharide and are widely distributed in biomass. Hemicelluloses have long been studied and widely used in different areas, such as food and feed, medicine and pharmaceutics, and papermaking due to their biocompatibility, biodegradability, nontoxicity, and specific therapeutic activities (Liu et al., 2015). However, utilization of hemicelluloses either natively or their functionalized products as feedstocks for 3D printing has rarely been studied.

Recently, a spruce wood hemicellulose (O-acetyl galactoglucomannan, GGM) was utilized to partially replace a synthetic PLA as feedstock in 3D FDM printing (Xu et al., 2018). The blends of hemicelluloses and PLA with a varied ratio up to 25% of GGM were evenly mixed with a solvent casting approach and were extruded into filaments by hot melt extrusion. 3D scaffold prototypes were successfully printed from the composite filaments by FDM 3D printing (Figure 4.3). As a pioneer exploration, this study demonstrated the feasibility of applying biorenewable hemicelluloses as a novel material candidate for FDM 3D printing, which has explored a new route to utilize hemicellulose-based biopolymers in 3D printing for versatile applications in, but not limited to, biomedical devices.

By mimicking the hemicelluloses' natural affinity to cellulose, Markstedt et al. (2017) mixed a tyramine-substituted xylan with the CNF to introduce a crosslinking property of the all-wood-based inks for 3D printing. The tyramine substitution degree of xylan-tyramine and the mixture ratio were found to influence the printability and crosslinking density of the all-wood-based inks and the swelling properties of the printed structure, providing the opportunity to tune the mechanical and structural properties of the printed object, which might even transfer to 4D printing (Liu et al., 2019).

4.3.7   3D PRINTING OF STARCH-BASED MATERIALS

Starch is another type of abundant polysaccharide produced by higher plants as energy storage and is composed of linear amylose and highly branched amylopectin with structures of α-$(1\rightarrow4)$ linked glucan and the $-(1\rightarrow4)$ linked glucan with α-$(1\rightarrow6)$ branch linkages, respectively (Fraser-Reid et al., 2008). Utilization of starch or starch-based polymers, such as thermoplastic starch and PLA, as feedstocks in 3D printing technologies for a variety of applications has drawn increasing interest from both academic research and in industry applications owing to its renewability, biodegradability, biocompatibility, and high abundance in nature (Davachi & Kaffashi, 2015)

A three-powder blend ink consisting of 50 wt.% corn starch, 30 wt.% dextran, and 20 wt.% gelatin and water-based binders was adapted for 3D printing of porous scaffolds for tissue engineering applications (Lam et al., 2002). After 3D printing, post-processing by heat treatment (dried at 100°C for 1 h) was applied to sinter the three blend powdered particles together via necking connections, enhancing the strength of the scaffolds and also the water resistance. Although the fabricated porous scaffolds, which consisted of natural polymers and a water-based binder, showed potential in tissue engineering, further biocompatibility study is needed.

In 3D food printing, starch is commonly utilized as a thickening/gelling agent or rheological modifier and also serves as an important carbohydrate source. Incorporation of starch into the ink formulation improves ink printability by offering the ink shear thinning behavior and also shape stability of the printed structure. For example, starch has been applied to facilitate the 3D printing of sodium caseinate (Schutyser et al., 2018), mashed potatoes (Liu et al., 2018), lemon juice gel (Yang et al., 2018), and protein- and fiber-rich food materials (Lille et al., 2018b) by working as thickening/gelling agent.

Similar to cellulose, the structure of starch also has abundant hydroxyl groups, offering numerical chemical modification/functionalization possibilities to produce starch derivatives, starch-based polymers, or plastics for 3D printing (BeMiller & Whistler, 2009). For instance, a thermoplastic starch (TPS)/acrylonitrile-butadienestyrene copolymer was prepared by using a compatibilizer (styrene maleic anhydride) to generate hydrogen bonds as well as strong van der Waals force between TPS and ABS and was subjected to filament extrusion and FDM 3D printing. The results revealed that the TPS/ABS filaments and printed 3D structures had superior mechanical properties and thermal resistance compared to that of commercial ABS, suggesting that the potential of using biomass polymeric materials and their derivatives with excellent 3D printing processibilities and physical performance would be highly promising in the future (Kuo et al., 2016).

Through fermentation with micro-organisms and polymerization processes, PLA can be produced from starch in a sustainable and green approach. Owing to a variety of benefits, including biocompability, nontoxicity, biodegradability, ease of processing, low cost, excellent mechanics, and the green features of its synthesis routes from renewable resources, PLA makes up the majority of the FDM 3D printing feedstock and has drawn increasing interest for biomedical applications, such as scaffolds, implants, prostheses, drug-based delivery systems, and anatomical model fabrication (Saini et al., 2016; Tyler et al., 2016). Depending on the ratio of PLLA and PDLA, the mechanical performance and crystalline structure of the polymers can be tailored for specific requirements of different biomedical applications. For example, sutures and orthopedic devices that need high mechanical strength and toughness can be fabricated with a high ratio of PLLA, while drug delivery systems that require a porous and monophasic matrix can be produced with amorphous PDLA (Shah Mohammadi et al., 2014).

While the advantages of PLA make it a promising material for various applications in the biomedical and pharmaceutical fields, it's essential to acknowledge its limitations. PLA's hydrophobic nature, limited ultimate elongation strain, absence of cell motif sites, production of small particles, and release of acidic byproducts during degradation, which can trigger foreign-body inflammatory reactions, are factors that must be considered. These drawbacks may potentially lead to clinical complications and impose restrictions on its utilization in the field of biomedicine (Suganuma et al., 1991). Therefore, increasing efforts have been made to enhance the hydrophilic properties, to increase the cell motif sites, to introduce bioactivity, to improve the ultimate elongation strain, and to address the formation of acidic biodegradation products (Manavitehrani et al., 2016; Ulery et al., 2011). For example, Wang et al. (2016) printed PLA-based scaffolds treated with a cold atmospheric plasma (CAP) modification for enhancing hydrophilicity and cell motif sites. The ends of the PLA chains ($-CH_3$) on the surface were reacted with reactive oxygen species (peroxides, superoxides, hydroxyl radicals, and atomic oxygen) and converted to hydrophilic groups, such as $-CH_2OH$, $-CHO$, and $-COOH$.

Meanwhile, the CAP treatment raised the surface temperature of the PLA to ca. 40°C, which partially melted and reformed the polymer, allowing conversion from microscale to nanoscale surface features. Results showed that the enhanced surface hydrophilicity and nanoscale roughness by CAP surface modification significantly promoted both osteoblast (bone-forming cells) and mesenchymal stem cell attachment and proliferation, suggesting promising applications in bone tissue engineering (Wang et al., 2016).

Along with physical treatment, surface functionalization with bioactive molecules is another approach for improving biomaterial bioactivity. For instance, Kao et al. (2015) functionalized 3D-printed PLA scaffolds via a mussel-inspired surface coating with polydopamine to accelerate protein adsorption and the cell cycle of human adipose–derived stem cells. Similarly, surface coating of PLA scaffolds by physical sorption and covalent bonding of collagen and polymers (polyethyleneoxide-polypropyleneoxide copolymers) or incorporation of bioactive glass has also been adapted to improve scaffold bioactivity and to tune cell response (Serra et al., 2013). The surface-coated or modified PLA scaffolds are demonstrated to regulate cell organization, adhesion, and proliferation and to induce osteogenesis and angiogenesis differentiation, providing a very promising tool to regulate stem cell behavior, and they may serve as an effective stem cell delivery carrier for bone tissue engineering (Lin & Fu, 2016).

Modification or functionalization of PLA to tune the mechanical performance of 3D printed PLA-based scaffolds also play a keys role in its biomedical application. Senatov et al. (2016) designed and fabricated a PLA-based porous scaffold with a shape memory effect by FDM 3D printing for potential application as a self-fitting implant in bone defect replacement. PLA, which serves as a bioresorbable matrix, and the bioactive filler hydroxyapatite (HA) powder (15 wt.%) were mixed in a screw extruder, and the PLA/HA composite filaments were extruded for FDM 3D printing. The 3D-printed PLA-based porous scaffolds were shown to have the shape memory effect after heat activation, which changed the polymer chain mobility. The scaffolds were found to be able to withstand up to three compression–heating–compression cycles without delamination. However, the higher activation temperature (70°C) for the shape memory effect compared to the body temperature is an obstacle for the application, and therefore further study to lower the activation temperature is needed (Senatov et al., 2016). Similarly, Zhang et al. (2016) prepared a PLA/HA (15 wt.%) composite scaffold with a pore size of 500 μm and 60% porosity by a mini-deposition system (FDM) (Jiang et al., 2012). Incorporation of HA can not only enhance the mechanical performance of the scaffolds (Jiang et al., 2012) but also offer scaffolds better biocompatibility, biodegradability, and osteoinductive activity than pure PLA scaffolds to enhance bone formation, showing potential application in bone tissue engineering as a promising candidate for bone defect repair (Liu et al., 2019; Zhang et al., 2016).

4.3.8 3D PRINTING OF ALGAE-BASED MATERIALS

Algae-based materials, such as alginate, agarose, and carrageen, extracted from brown and red algae, represent another group of naturally derived biopolymers from marine biomass for 3D printing ink formulation, especially for 3D bioprinting (Axpe & Oyen, 2016). Alginate is one of the most widely explored seaweed anionic polysaccharides and is composed of β-(1→4)-linked D-mannuronic acid (M block) and α-L-guluronic acid (G block) units arranged with varying proportions of GG, MG, and GM blocks (Liu et al., 2015). It has been recognized as the most commonly employed material as bioink formulation in 3D bioprinting for a variety of biomedical and pharmaceutical applications, such as wound healing, cartilage repair, bone regeneration, and drug delivery, owning to the benefits of nontoxicity, biocompatibility, biodegradability, non-antigenicity, and ease and mildness of gelation (Chimene et al., 2016; Liberski, 2016). By mixing it with multivalent cations, such as Ca^{2+}, alginate is ionically crosslinked to form a hydrogel, offering an ECM-mimicking environment for cell incorporation and survival, good printability, structural integrity, and mechanical stiffness. By adjusting or using alginate with different G/M ratios, molar mass, solid content, and cell density, the rheological, mechanical, and macro- and micro-structural properties of the alginate-based bioink can be tuned to fit the requirements of different bioprinting methodologies and biomedical applications (Grigore et al., 2014; Luo et al., 2015).

However, some undesirable features, including poor long-term structural integrity and mechanical properties, slow and uncontrolled degradation kinetics, and the bioinert property (non–cell-adhesion by itself), should be carefully considered in the design and development of alginate-based bioink for bioprinting (Chung et al., 2013). For instance, the poor mechanical performance could be addressed by combining ionic and covalent crosslinking (Duan et al., 2014; Kesti et al., 2015; Rutz et al., 2015); by the incorporation of other components such as hydroxyapatite (Bendtsen et al., 2017), polycaprolactone (Daly et al., 2016), β-tricalcium phosphate (Diogo et al., 2014), and nanocellulose (Möller et al., 2017); and by assisting deposition of a sacrificial layer (e.g. PEG) that can be removed after printing (Armstrong et al., 2016). The slow degradation kinetics of native alginate can be addressed by oxidation with sodium peroxide or periodate (Jia et al., 2014), by controlling the alginate molar mass and distribution (Boontheekul et al., 2005), by enzyme-assisted degradation (e.g. alginate lyase) (Wong et al., 2000), or by incubation with sodium citrate that accelerates the dissolving of alginate hydrogel by chelating calcium ions from the $CaCl_2$ crosslinked alginate hydrogel (Wu et al., 2016). The bioinert property of native alginate can be addressed by conjugation of cell attachment ligands (e.g. peptide sequence Arg-Gly-Asp, i.e. RGD) or with incorporation of other biomacromolecules, such as fibrin, collagens, gelatin, chitosan, hyaluronic acid, avidin protein, and various growth factors to offer the desired bioactivities (Leppiniemi et al., 2017). Agar and its purified agarose are galactose-based polysaccharides derived from red algae and seaweed such as *Gracilaria*, and *Gelidium* agarose is a linear polysaccharide composed of alternating (1→4)-linked (3,6)-anhydro-α-L-galactose and (1→3)-linked β-D-galactose, while agar is a mixture of the predominant agarose and agaro-pectin composed of partially sulfated (1→3)-linked D-galactose (Serwer, 1983). These galactose-based polysaccharides, especially agarose, have proved suitable for bioprinting owning to their cytocompatibility, stable mechanical properties, and mild temperature-dependent gelling conditions with a starting gelling temperature of 37°C and full crosslinking at 32.7°C (Malda et al., 2013). Different cell lines have been encapsulated into the agar, agarose, and their composite hydrogels with other polymers such as alginate, collagen, matrigel, and chitosan as bioinks in bioprinting for potential applications in tissue engineering, regenerative medicine, and drug discovery (Navarro & Garcia, 2016; Wei et al., 2015).

4.3.9 3D Printing of Chitosan-Based Materials

Chitosan, a cationic polysaccharide, is composed of β-(1→4)-linked nacetyl-D-glucosamine and deacetylated D-glucosamine units (Figure 4.6C). It is the deacetylated form of chitin that naturally exists in the exoskeletons of crustaceans (crab, shrimp shells, and insects), as well as the cell wall of fungi and yeast, and it is the second richest natural polymer in nature after cellulose (Dutta, 2016; Rinaudo, 2006). Chitosan-based biomaterials with unique advantages have drawn increasing attention for application in medical and pharmaceutical areas, such as tissue engineering, drug delivery, and wound healing, owing to their biocompatibility, bioresorbability, biodegradability, antimicrobial activity, mucoadhesivity, and low toxicity, as well as natural abundance (Elviri et al., 2017). Generally, chitosan is soluble in acidic conditions, such as 2% acetic acid, and the amine groups in chitosan are protonated to confer poly-cationic behavior. Chitosan chains expand into a semi-rigid rod conformation due to ionic repulsion between the charged groups (NH_3^+), and thus chitosan ink exhibits shear thinning behavior, which is beneficial for the extrusion-based 3D printing (Wu et al., 2017). Chitosan or chitosan-based composites in the form of hydrogels or pastes, filaments, or strips have been applied in a variety of 3D printing technologies for potential bone tissue engineering, cartilage tissue regeneration, and drug screening.

These printing approaches include FDM (Wu, 2016); stereolithography SLA (Morris et al., 2017); and different nozzle extrusion-based 3D printing technologies such as double-nozzle assembling (Shengjie et al., 2009), rapid prototyping robotic dispensing (Ang et al., 2002), direct printing (Almeida et al., 2014), and low-temperature manufacturing (Elviri et al., 2017). However, poor mechanical resistance and degradability in physiological conditions, as well as acidic dissolution, make it difficult to incorporate living cells into chitosan-based inks in bioprinting for biomedical applications. Neutralization and gelation in alkaline conditions (Almeida et al., 2014; Elviri et al., 2017), derivatization, crosslinking, or combination with other polymers are required prior to cell seeding or implantation (Pandey et al., 2017). For instance, hydroxyapatite, pectin, and genipin have been incorporated into chitosan for mechanical reinforcement of chiton-based inks or 3D-printed scaffolds in bone tissue engineering (Demirtaş et al., 2017). Combination with other natural or synthetic polymers, such as alginate, gelatin, fibrinogen, laminin, and thermoplastics, would potentially offer many opportunities to overcome these challenges and may allow the introduction of desired properties (Liu et al., 2019).

4.4 CONCLUSIONS

The emergence of 3D printing as a groundbreaking technology holds the promise of revolutionizing traditional product fabrication. Nevertheless, a significant challenge that needs to be addressed before its widespread adoption in various sectors is the scarcity of high-quality printing materials that are also environmentally friendly. Among the abundant and renewable sources of biobased materials, such as lignocellulosic materials; seaweed materials; and exoskeleton materials from crustaceans like crabs, shrimp shells, and insects, naturally derived biopolymers offer great potential for diverse 3D printing technologies. By modifying or combining these biopolymers with other ingredients to enhance their processability and the functionality of printed products, we can expand their application areas. Future trends in 3D printing material development will likely involve utilizing nature-derived biopolymers, either in their original form or as functionalized products, as feedstocks for various 3D printing technologies. Recent advancements and proof-of-concept demonstrations have confirmed the technical feasibility of developing naturally derived biopolymers as feedstock formulations. However, there are still several challenges to overcome, including material processability; degradation; and chemical, biological, and mechanical properties, before fully realizing the potential of biopolymers. Once these hurdles are addressed, 3D printing technology is poised to greatly facilitate the widespread adoption of naturally derived biopolymers in various fields. Consequently, it is foreseeable that future material development in 3D printing will progressively shift towards the utilization of these environmentally friendly feedstocks derived from nature.

REFERENCES

Alghamdi, S. S., John, S., Choudhury, N. R., & Dutta, N. K. (2021). Additive manufacturing of polymer materials: Progress, promise and challenges. *Polymers*, *13*(5), 1–39. https://doi.org/10.3390/polym13050753

Ali, H. (2015). *Cellulose based packaging with Thermocell plastic film*. www.vttresearch.com/media/news/cellulose-turning-into-a-supermaterial-of-the-future

Almeida, C. R., Serra, T., Oliveira, M. I., Planell, J. A., Barbosa, M. A., & Navarro, M. (2014). Impact of 3-D printed PLA- and chitosan-based scaffolds on human monocyte/macrophage responses: Unraveling the effect of 3-D structures on inflammation. *Acta Biomaterialia*, *10*(2), 613–622. https://doi.org/10.1016/j.actbio.2013.10.035

Andrzejewski, J., Cheng, J., Anstey, A., Mohanty, A. K., & Misra, M. (2020). Development of toughened blends of poly(lactic acid) and poly(butylene adipate-co-terephthalate) for 3D printing applications: Compatibilization methods and material performance evaluation. *ACS Sustainable Chemistry and Engineering*, *8*(17), 6576–6589. https://doi.org/10.1021/acssuschemeng.9b04925

Ang, T. H., Sultana, F. S. A., Hutmacher, D. W., Wong, Y. S., Fuh, J. Y. H., Mo, X. M., Loh, H. T., Burdet, E., & Teoh, S. H. (2002). Fabrication of 3D chitosan-hydroxyapatite scaffolds using a robotic dispensing system. *Materials Science and Engineering C*, *20*(1–2), 35–42. https://doi.org/10.1016/S0928-4931(02)00010-3

Armstrong, J. P. K., Burke, M., Carter, B. M., Davis, S. A., & Perriman, A. W. (2016). 3D bioprinting using a templated porous bioink. *Advanced Healthcare Materials*, *5*(14), 1724–1730. https://doi.org/10.1002/adhm.201600022

Attaran, M. (2017). The rise of 3-D printing: The advantages of additive manufacturing over traditional manufacturing. *Business Horizons*, *60*(5), 677–688. https://doi.org/10.1016/j.bushor.2017.05.011

Axpe, E., & Oyen, M. L. (2016). Applications of alginate-based bioinks in 3D bioprinting. *International Journal of Molecular Sciences*, *17*(12). https://doi.org/10.3390/ijms17121976

Azlin, M. N. M., Ilyas, R. A., Zuhri, M. Y. M., Sapuan, S. M., Harussani, M. M., Sharma, S., Nordin, A. H., & Afiqah, A. N. (2022). 3D printing and shaping polymers, composites, and nanocomposites: a review. *Polymers*, *14*(1), 180.

Bagheri, A., & Jin, J. (2019). Photopolymerization in 3D printing. *ACS Applied Polymer Materials*, *1*(4), 593–611. https://doi.org/10.1021/acsapm.8b00165

Balla, V. K., Kate, K. H., Satyavolu, J., Singh, P., & Tadimeti, J. G. D. (2019). Additive manufacturing of natural fiber reinforced polymer composites: Processing and prospects. *Composites Part B: Engineering*, *174*(March), 106956. https://doi.org/10.1016/j.compositesb.2019.106956

Bates-Green, K., & Howie, T. (2017). Materials for 3D printing by fused deposition. *Technical Education in Additive Manufacturing and Materials*, 1–21.BeMiller, J. N., & Whistler, R. L. (Eds.). (2009). *Starch: Chemistry and technology*. Academic Press.

Bendtsen, S. T., Quinnell, S. P., & Wei, M. (2017). Development of a novel alginate-polyvinyl alcohol-hydroxyapatite hydrogel for 3D bioprinting bone tissue engineered scaffolds. *Journal of Biomedical Materials Research—Part A*, *105*(5), 1457–1468. https://doi.org/10.1002/jbm.a.36036

Boontheekul, T., Kong, H. J., & Mooney, D. J. (2005). Controlling alginate gel degradation utilizing partial oxidation and bimodal molecular weight distribution. *Biomaterials*, *26*(15), 2455–2465. https://doi.org/10.1016/j.biomaterials.2004.06.044

Bourell, D., Kruth, J. P., Leu, M., Levy, G., Rosen, D., Beese, A. M., & Clare, A. (2017). Materials for additive manufacturing. *CIRP Annals—Manufacturing Technology*, *66*(2), 659–681. https://doi.org/10.1016/j.cirp.2017.05.009

Cardoso, P. H. M., Coutinho, R. R. T. P., Drummond, F. R., da Conceição, M. do N., & Thiré, R. M. da S. M. (2020). Evaluation of printing parameters on porosity and mechanical properties of 3D printed PLA/PBAT blend parts. *Macromolecular Symposia*, *394*(1), 1–12. https://doi.org/10.1002/masy.202000157

Carlotto, A., Loggi, A., Sbornicchia, P., Zito, D., Maggian, D., Molinari, A., & Cristofolini, I. (2014). AM: Technologies: Optimization of SLM technology: Main parameters in the production of gold and platinum jewelry. *European Congress and Exhibition on Powder Metallurgy. European PM Conference Proceedings* (p. 1). The European Powder Metallurgy Association.

Chaunier, L., Guessasma, S., Belhabib, S., Della Valle, G., Lourdin, D., & Leroy, E. (2018). Material extrusion of plant biopolymers: Opportunities & challenges for 3D printing. *Additive Manufacturing*, *21*(January 2017), 220–233. https://doi.org/10.1016/j.addma.2018.03.016

Chimene, D., Lennox, K. K., Kaunas, R. R., & Gaharwar, A. K. (2016). Advanced bioinks for 3D printing: A materials science perspective. *Annals of Biomedical Engineering*, *44*(6), 2090–2102. https://doi.org/10.1007/s10439-016-1638-y

Chung, J. H. Y., Naficy, S., Yue, Z., Kapsa, R., Quigley, A., Moulton, S. E., & Wallace, G. G. (2013). Bio-ink properties and printability for extrusion printing living cells. *Biomaterials Science, 1*(7), 763–773. https://doi.org/10.1039/c3bm00012e

Daly, A. C., Cunniffe, G. M., Sathy, B. N., Jeon, O., Alsberg, E., & Kelly, D. J. (2016). 3D bioprinting of developmentally inspired templates for whole bone organ engineering. *Advanced Healthcare Materials, 5*(18), 2353–2362. https://doi.org/10.1002/adhm.201600182

Davachi, S. M., & Kaffashi, B. (2015). Polylactic acid in medicine. *Polymer—Plastics Technology and Engineering, 54*(9), 944–967. https://doi.org/10.1080/03602559.2014.979507

Demirtaş, T. T., Irmak, G., & Gümüşderelioğlu, M. (2017). A bioprintable form of chitosan hydrogel for bone tissue engineering. *Biofabrication, 9*(3). https://doi.org/10.1088/1758-5090/aa7b1d

Diederichs, E. V., Picard, M. C., Chang, B. P., Misra, M., Mielewski, D. F., & Mohanty, A. K. (2019). Strategy to improve printability of renewable resource-based engineering plastic tailored for FDM applications. *ACS Omega, 4*(23), 20297–20307. https://doi.org/10.1021/acsomega.9b02795

Diogo, G. S., Gaspar, V. M., Serra, I. R., Fradique, R., & Correia, I. J. (2014). Manufacture of β-TCP/alginate scaffolds through a Fab@home model for application in bone tissue engineering. *Biofabrication, 6*(2). https://doi.org/10.1088/1758-5082/6/2/025001

Duan, B., Kapetanovic, E., Hockaday, L. A., & Butcher, J. T. (2014). Three-dimensional printed trileaflet valve conduits using biological hydrogels and human valve interstitial cells. *Acta Biomaterialia, 10*(5), 1836–1846. https://doi.org/10.1016/j.actbio.2013.12.005

Dutta, P. K. (2016). *Chitin and chitosan for regenerative medicine.* https://doi.org/10.1007/978-81-322-2511-9

Elviri, L., Bianchera, A., Bergonzi, C., & Bettini, R. (2017). Controlled local drug delivery strategies from chitosan hydrogels for wound healing. *Expert Opinion on Drug Delivery, 14*(7), 897–908. https://doi.org/10.1080/17425247.2017.1247803

Elviri, L., Foresti, R., Bergonzi, C., Zimetti, F., Marchi, C., Bianchera, A., Bernini, F., Silvestri, M., & Bettini, R. (2017). Highly defined 3D printed chitosan scaffolds featuring improved cell growth. *Biomedical Materials (Bristol), 12*(4). https://doi.org/10.1088/1748-605X/aa7692

Fraser-Reid, B., Tatsuta, K., & Thiem, J. (2008). *Glycoscience: Chemistry and chemical biology.* Springer. https://doi.org/10.1201/9781315371399-3

Gatenholm, P., Martinez, H., Karabulut, E., Amoroso, M., Kölby, L., Markstedt, K., Gatenholm, E., & Henriksson, I. (2016). *3D printing and biofabrication* (pp. 1–23). https://doi.org/10.1007/978-3-319-40498-1

Gibson, I., Rosen, D., & Stucker, B. (2015). Vat photopolymerization processes. In *Additive manufacturing technologies* (pp. 63–106). Springer.

Goh, G. D., Yap, Y. L., Agarwala, S., & Yeong, W. Y. (2019a). Recent progress in additive manufacturing of fiber reinforced polymer composite. *Advanced Materials Technologies, 4*(1), 1–22. https://doi.org/10.1002/admt.201800271

Grigore, A., Sarker, B., Fabry, B., Boccaccini, A. R., & Detsch, R. (2014). Behavior of encapsulated MG-63 cells in RGD and gelatine-modified alginate hydrogels. *Tissue Engineering—Part A, 20*(15–16), 2140–2150. https://doi.org/10.1089/ten.tea.2013.0416

Gupta, K. M., & Jiang, J. (2015). Cellulose dissolution and regeneration in ionic liquids: A computational perspective. *Chemical Engineering Science, 121*, 180–189. https://doi.org/10.1016/j.ces.2014.07.025

Guvendiren, M., Molde, J., Soares, R. M. D., & Kohn, J. (2016). Designing biomaterials for 3D printing murat. *ACS Biomaterials Science & Engineering, 2*(10), 704–711. https://doi.org/10.1053/j.gastro.2016.08.014.CagY

Håkansson, K. M. O., Henriksson, I. C., de la Peña Vázquez, C., Kuzmenko, V., Markstedt, K., Enoksson, P., & Gatenholm, P. (2016). Solidification of 3D printed nanofibril hydrogels into functional 3D cellulose structures. *Advanced Materials Technologies*, *1*(7). https://doi.org/10.1002/admt.201600096

Henriksson, I., Gatenholm, P., & Hägg, D. A. (2017). Increased lipid accumulation and adipogenic gene expression of adipocytes in 3D bioprinted nanocellulose scaffolds. *Biofabrication*, *9*(1). https://doi.org/10.1088/1758-5090/aa5c1c

Jeffri, N. I., Nurul Fazita, M. R., Leh, C. P., Hashim, R., Mohamad Haafiz, M. K., Abdullah, C. K., . . . & Kosugi, A. (2022, October). Processing and characterization of poly (hydroxybutyrate) (PHB) and poly (butylene-co-adipate-terephthalate) (PBAT) blends for fused deposition modeling (FDM) 3D printing. In *Asian Workshop on Polymer Processing* (pp. 17–31). Springer Nature Singapore.

Jia, J., Richards, D. J., Pollard, S., Tan, Y., Rodriguez, J., Visconti, R. P., Trusk, T. C., Yost, M. J., Yao, H., Markwald, R. R., & Mei, Y. (2014). Engineering alginate as bioink for bioprinting. *Acta Biomaterialia*, *10*(10), 4323–4331. https://doi.org/10.1016/j.actbio.2014.06.034

Jiang, W., Shi, J., Li, W., & Sun, K. (2012). Morphology, wettability, and mechanical properties of polycaprolactone/hydroxyapatite composite scaffolds with interconnected pore structures fabricated by a mini-deposition system. *Polymer Engineering and Science*, *52*(11), 2396–2340. https://doi.org/10.1002/pen

Kao, C. T., Lin, C. C., Chen, Y. W., Yeh, C. H., Fang, H. Y., & Shie, M. Y. (2015). Poly(dopamine) coating of 3D printed poly(lactic acid) scaffolds for bone tissue engineering. *Materials Science and Engineering C*, *56*, 165–173. https://doi.org/10.1016/j.msec.2015.06.028

Kesti, M., Müller, M., Becher, J., Schnabelrauch, M., D'Este, M., Eglin, D., & Zenobi-Wong, M. (2015). A versatile bioink for three-dimensional printing of cellular scaffolds based on thermally and photo-triggered tandem gelation. *Acta Biomaterialia*, *11*(1), 162–172. https://doi.org/10.1016/j.actbio.2014.09.033

Khan, M., & Dickens, P. (2012). Selective laser melting (SLM) of gold (Au). *Rapid Prototyping Journal*, *18*(1), 81–94. https://doi.org/10.1108/13552541211193520

Kuo, C. C., Liu, L. C., Teng, W. F., Chang, H. Y., Chien, F. M., Liao, S. J., Kuo, W. F., & Chen, C. M. (2016). Preparation of starch/acrylonitrile-butadiene-styrene copolymers (ABS) biomass alloys and their feasible evaluation for 3D printing applications. *Composites Part B: Engineering*, *86*, 36–39. https://doi.org/10.1016/j.compositesb.2015.10.005

Lam, C. X. F., Mo, X. M., Teoh, S. H., & Hutmacher, D. W. (2002). Scaffold development using 3D printing with a starch-based polymer. *Materials Science and Engineering C*, *20*(1–2), 49–56. https://doi.org/10.1016/S0928-4931(02)00012-7

Lee Ventola, C. (2014). Medical applications for 3D printing: Current and projected uses. *Pharmacy and Therapeutics*, *39*(10), 704–711.

Leppiniemi, J., Lahtinen, P., Paajanen, A., Mahlberg, R., Metsä-Kortelainen, S., Pinomaa, T., Pajari, H., Vikholm-Lundin, I., Pursula, P., & Hytönen, V. P. (2017). 3D-printable bioactivated nanocellulose-alginate hydrogels. *ACS Applied Materials and Interfaces*, *9*(26), 21959–21970. https://doi.org/10.1021/acsami.7b02756

Liberski, A. R. (2016). Three-dimensional printing of alginate: From seaweeds to heart valve scaffolds. *QScience Connect*, *2016*(2). https://doi.org/10.5339/connect.2016.3

Lille, M., Nurmela, A., Nordlund, E., Metsä-Kortelainen, S., & Sozer, N. (2018a). Applicability of protein and fiber-rich food materials in extrusion-based 3D printing. *Journal of Food Engineering*, *220*, 20–27. https://doi.org/10.1016/j.jfoodeng.2017.04.034

Lille, M., Nurmela, A., Nordlund, E., Metsä-Kortelainen, S., & Sozer, N. (2018b). Applicability of protein and fiber-rich food materials in extrusion-based 3D printing. *Journal of Food Engineering*, *220*, 20–27. https://doi.org/10.1016/j.jfoodeng.2017.04.034

Lin, C. C., & Fu, S. J. (2016). Osteogenesis of human adipose-derived stem cells on poly(dopamine)-coated electrospun poly(lactic acid) fiber mats. *Materials Science and Engineering C, 58*, 254–263. https://doi.org/10.1016/j.msec.2015.08.009

Liu, J., Cheng, F., Grénman, H., Spoljaric, S., Seppälä, J., Eriksson, J. E., Willför, S., & Xu, C. (2016). Development of nanocellulose scaffolds with tunable structures to support 3D cell culture. *Carbohydrate Polymers, 148*, 259–271. https://doi.org/10.1016/j.carbpol.2016.04.064

Liu, J., Korpinen, R., Mikkonen, K. S., Willför, S., & Xu, C. (2014). Nanofibrillated cellulose originated from birch sawdust after sequential extractions: A promising polymeric material from waste to films. *Cellulose, 21*(4), 2587–2598. https://doi.org/10.1007/s10570-014-0321-4

Liu, J., Sun, L., Xu, W., Wang, Q., Yu, S., & Sun, J. (2019). Current advances and future perspectives of 3D printing natural-derived biopolymers. *Carbohydrate Polymers, 207*(November 2018), 297–316. https://doi.org/10.1016/j.carbpol.2018.11.077

Liu, J., Willför, S., & Xu, C. (2015). A review of bioactive plant polysaccharides: Biological activities, functionalization, and biomedical applications. *Bioactive Carbohydrates and Dietary Fibre, 5*(1), 31–61. https://doi.org/10.1016/j.bcdf.2014.12.001

Liu, Z., Wang, Y., Wu, B., Cui, C., Guo, Y., & Yan, C. (2019). A critical review of fused deposition modeling 3D printing technology in manufacturing polylactic acid parts. *International Journal of Advanced Manufacturing Technology, 102*(9–12), 2877–2889. https://doi.org/10.1007/s00170-019-03332-x

Liu, Z., Zhang, M., Bhandari, B., & Yang, C. (2018). Impact of rheological properties of mashed potatoes on 3D printing. *Journal of Food Engineering, 220*, 76–82. https://doi.org/10.1016/j.jfoodeng.2017.04.017

Lovo, J. F. P., de Camargo, I. L., Erbereli, R., Morais, M. M., & Fortulan, C. A. (2020). Vat photopolymerization additive manufacturing resins: Analysis and case study. *Materials Research, 23*(4). https://doi.org/10.1590/1980-5373-MR-2020-0010

Luo, Y., Zhai, D., Huan, Z., Zhu, H., Xia, L., Chang, J., & Wu, C. (2015). Three-dimensional printing of hollow-struts-packed bioceramic scaffolds for bone regeneration. *ACS Applied Materials and Interfaces, 7*(43), 24377–24383. https://doi.org/10.1021/acsami.5b08911

Lyu, Y., Chen, Y., Lin, Z., Zhang, J., & Shi, X. (2020). Manipulating phase structure of biodegradable PLA/PBAT system: Effects on dynamic rheological responses and 3D printing. *Composites Science and Technology, 200*(August), 108399. https://doi.org/10.1016/j.compscitech.2020.108399

Malda, J., Visser, J., Melchels, F. P., Jüngst, T., Hennink, W. E., Dhert, W. J. A., Groll, J., & Hutmacher, D. W. (2013). 25th anniversary article: Engineering hydrogels for biofabrication. *Advanced Materials, 25*(36), 5011–5028. https://doi.org/10.1002/adma.201302042

Manavitehrani, I., Fathi, A., Badr, H., Daly, S., Shirazi, A. N., & Dehghani, F. (2016). Biomedical applications of biodegradable polyesters. *Polymers, 8*(1). https://doi.org/10.3390/polym8010020

Markstedt, K., Escalante, A., Toriz, G., & Gatenholm, P. (2017). Biomimetic inks based on cellulose nanofibrils and cross-linkable xylans for 3D printing. *ACS Applied Materials & Interfaces, 9*(46), 40878–40886.

Markstedt, K., Mantas, A., Tournier, I., Martínez Ávila, H., Hägg, D., & Gatenholm, P. (2015). 3D bioprinting human chondrocytes with nanocellulose-alginate bioink for cartilage tissue engineering applications. *Biomacromolecules, 16*(5), 1489–1496. https://doi.org/10.1021/acs.biomac.5b00188

Martínez Ávila, H., Schwarz, S., Rotter, N., & Gatenholm, P. (2016). 3D bioprinting of human chondrocyte-laden nanocellulose hydrogels for patient-specific auricular cartilage regeneration. *Bioprinting, 1–2*, 22–35. https://doi.org/10.1016/j.bprint.2016.08.003

Mohan, N., Senthil, P., Vinodh, S., & Jayanth, N. (2017). A review on composite materials and process parameters optimisation for the fused deposition modelling process. *Virtual and Physical Prototyping*, *12*(1), 47–59. https://doi.org/10.1080/17452759.2016.1274490

Möller, T., Amoroso, M., Hägg, D., Brantsing, C., Rotter, N., Apelgren, P., Lindahl, A., Kölby, L., & Gatenholm, P. (2017). In vivo chondrogenesis in 3D bioprinted human cell-laden hydrogel constructs. *Plastic and Reconstructive Surgery—Global Open*, *5*(2), 1–7. https://doi.org/10.1097/GOX.0000000000001227

Morris, V. B., Nimbalkar, S., Younesi, M., McClellan, P., & Akkus, O. (2017). Mechanical properties, cytocompatibility and manufacturability of chitosan: PEGDA hybrid-gel scaffolds by stereolithography. *Annals of Biomedical Engineering*, *45*(1), 286–296. https://doi.org/10.1007/s10439-016-1643-1

Müller, T. M., Sandini, I. E., Rodrigues, J. D., Novakowiski, J. H., Basi, S., & Kaminski, T. H. (2016). Combinação de métodos de inoculação de Azospirillum brasiliense com adubação nitrogenada de cobertura aumenta produtividade de milho. *Ciencia Rural*, *46*(2), 210–215. https://doi.org/10.1590/0103-8478cr20131283

Murphy, A., Casey, A., Byrne, G., Chambers, G., & Howe, O. (2016). Silver nanoparticles induce pro-inflammatory gene expression and inflammasome activation in human monocytes. *Journal of Applied Toxicology*, *36*(10), 1311–1320. https://doi.org/10.1002/jat.3315

Navarro, G., & Garcia, I. (2016). Study of tissue printing parameters for generating complex tissue constructs. *Journal of Tissue Science & Engineering*, *7*(2), 2–4. https://doi.org/10.4172/2157-7552.1000175

Nguyen, D., Hgg, D. A., Forsman, A., Ekholm, J., Nimkingratana, P., Brantsing, C., Kalogeropoulos, T., Zaunz, S., Concaro, S., Brittberg, M., Lindahl, A., Gatenholm, P., Enejder, A., & Simonsson, S. (2017). Cartilage tissue engineering by the 3D bioprinting of iPS cells in a nanocellulose/alginate bioink. *Scientific Reports*, *7*(1), 1–10. https://doi.org/10.1038/s41598-017-00690-y

Owolabi, G., Peterson, A., Habtour, E., Riddick, J., Coatney, M., Olasumboye, A., & Bolling, D. (2016). Dynamic response of acrylonitrile butadiene styrene under impact loading. *International Journal of Mechanical and Materials Engineering*, *11*(3), 1–8. https://doi.org/10.1186/s40712-016-0056-0

Pandey, A. R., Singh, U. S., Momin, M., & Bhavsar, C. (2017). Chitosan: Application in tissue engineering and skin grafting. *Journal of Polymer Research*, *24*(8). https://doi.org/10.1007/s10965-017-1286-4

Pannitz, O., & Sehrt, J. (2020). Transferability of process parameters in laser powder bed fusion processes for an energy and cost efficient manufacturing. *Sustainability*, *12*, 1565. https://doi.org/10.3390/su12041565

Pattinson, S. W., & Hart, A. J. (2017). Additive manufacturing of cellulosic materials with robust mechanics and antimicrobial functionality. *Advanced Materials & Technologies*, *1803336*, 1600084. https://doi.org/10.1002/adma

Prasong, W., Muanchan, P., Ishigami, A., Thumsorn, S., Kurose, T., & Ito, H. (2020). Properties of 3D printable poly(lactic acid)/poly(butylene adipate-co-terephthalate) blends and nano talc composites. *Journal of Nanomaterials*, *16*. https://doi.org/10.1155/2020/8040517

R. A., I., Sapuan, S., M. M., H., Norizan, M. N., Mohamed Yusoff, M. Z., Sharma, S., & Azlin, M. N. M. (2022). 3D printing and shaping polymers, composites, and nanocomposites: A review. *Polymers*, *14*. https://doi.org/10.3390/polym14010180

Rahim, T. N. A. T., Abdullah, A. M., & Md Akil, H. (2019). Recent developments in fused deposition modeling-based 3D printing of polymers and their composites. *Polymer Reviews*, *59*(4), 589–624. https://doi.org/10.1080/15583724.2019.1597883

Rauch, C. (2018). *Effect of infill angle on tensile strength of solid 3D printed carbon fibre reinforced Acrylonitrile Butadiene Styrene (ABS)*. Ball State University.

Rees, A., Powell, L. C., Chinga-Carrasco, G., Gethin, D. T., Syverud, K., Hill, K. E., & Thomas, D. W. (2015). 3D bioprinting of carboxymethylated-periodate oxidized nanocellulose constructs for wound dressing applications. *BioMed Research International, 2015*. https://doi.org/10.1155/2015/925757

Rinaudo, M. (2006). Chitin and chitosan: Properties and applications. *Progress in Polymer Science (Oxford), 31*(7), 603–632. https://doi.org/10.1016/j.progpolymsci.2006.06.001

Rutz, A. L., Hyland, K. E., Jakus, A. E., Burghardt, W. R., & Shah, R. N. (2015). A multi-material bioink method for 3D printing tunable, cell-compatible hydrogels. *Advanced Materials, 27*(9), 1607–1614. https://doi.org/10.1002/adma.201405076

Saini, P., Arora, M., & Kumar, M. N. V. R. (2016). Poly(lactic acid) blends in biomedical applications. *Advanced Drug Delivery Reviews, 107*, 47–59. https://doi.org/10.1016/j.addr.2016.06.014

Schutyser, M. A. I., Houlder, S., de Wit, M., Buijsse, C. A. P., & Alting, A. C. (2018). Fused deposition modelling of sodium caseinate dispersions. *Journal of Food Engineering, 220*, 49–55. https://doi.org/10.1016/j.jfoodeng.2017.02.004

Schütz, K., Placht, A.-M., Paul, B., Brüggemeier, S., Gelinsky, M., & Lode, A. (2015). Three-dimensional plotting of a cell-laden alginate/methylcellulose blend: Towards biofabrication of tissue engineering constructs with clinically relevant dimensions. *Journal of Tissue Engineering and Regenerative Medicine, 11*(5), 1574–1587. https://doi.org/10.1002/term

Senatov, F. S., Niaza, K. V., Zadorozhnyy, M. Y., Maksimkin, A. V., Kaloshkin, S. D., & Estrin, Y. Z. (2016). Mechanical properties and shape memory effect of 3D-printed PLA-based porous scaffolds. *Journal of the Mechanical Behavior of Biomedical Materials, 57*, 139–148. https://doi.org/10.1016/j.jmbbm.2015.11.036

Serra, T., Planell, J. A., & Navarro, M. (2013). High-resolution PLA-based composite scaffolds via 3-D printing technology. *Acta Biomaterialia, 9*(3), 5521–5530. https://doi.org/10.1016/j.actbio.2012.10.041

Serwer, P. (1983). Agarose gels: Properties and use for electrophoresis. *Electrophoresis, 4*(6), 375–382. https://doi.org/10.1002/elps.1150040602

Shah Mohammadi, M., Bureau, M. N., & Nazhat, S. N. (2014). Polylactic acid (PLA) biomedical foams for tissue engineering. In *Biomedical foams for tissue engineering applications*. Woodhead Publishing Limited. https://doi.org/10.1533/9780857097033.2.313

Shengjie, L., Xiong, Z., Wang, X., Yan, Y., Liu, H., & Zhang, R. (2009). Direct fabrication of a hybrid cell/hydrogel construct by a double-nozzle assembling technology. *Journal of Bioactive and Compatible Polymers, 24*(3), 249–265. https://doi.org/10.1177/0883911509104094

Siqueira, G., Kokkinis, D., Libanori, R., Hausmann, M. K., Gladman, A. S., Neels, A., Tingaut, P., Zimmermann, T., Lewis, J. A., & Studart, A. R. (2017). Cellulose nanocrystal inks for 3D printing of textured cellular architectures. *Advanced Functional Materials, 27*(12). https://doi.org/10.1002/adfm.201604619

Stoof, D., & Pickering, K. (2018). Sustainable composite fused deposition modelling filament using recycled pre-consumer polypropylene. *Composites Part B: Engineering, 135*, 110–118. https://doi.org/10.1016/j.compositesb.2017.10.005

Suganuma, J., Alexander, H., Traub, J., & Ricci, J. L. (1991). Biological response of intramedullary bone to poly-L-lactic acid. *MRS Proceedings, 252*, 339–343. https://doi.org/10.1557/proc-252-339

Tenhunen, T. M., Moslemian, O., Kammiovirta, K., Harlin, A., Kääriäinen, P., Österberg, M., Tammelin, T., & Orelma, H. (2018). Surface tailoring and design-driven prototyping of fabrics with 3D-printing: An all-cellulose approach. *Materials and Design*, *140*, 409–419. https://doi.org/10.1016/j.matdes.2017.12.012

Tyler, B., Gullotti, D., Mangraviti, A., Utsuki, T., & Brem, H. (2016). Polylactic acid (PLA) controlled delivery carriers for biomedical applications. *Advanced Drug Delivery Reviews*, *107*, 163–175. https://doi.org/10.1016/j.addr.2016.06.018

Ulery, B. D., Nair, L. S., & Laurencin, C. T. (2011). Biomedical applications of biodegradable polymers. *Journal of Polymer Science, Part B: Polymer Physics*, *49*(12), 832–864. https://doi.org/10.1002/polb.22259

Wang, M., Favi, P., Cheng, X., Golshan, N. H., Ziemer, K. S., Keidar, M., & Webster, T. J. (2016). Cold atmospheric plasma (CAP) surface nanomodified 3D printed polylactic acid (PLA) scaffolds for bone regeneration. *Acta Biomaterialia*, *46*, 256–265. https://doi.org/10.1016/j.actbio.2016.09.030

Wang, S., D'hooge, D. R., Daelemans, L., Xia, H., De Clerck, K., & Cardon, L. (2020). The transferability and design of commercial printer settings in PLA/PBAT fused filament fabrication. *Polymers*, *12*(11), 1–20. https://doi.org/10.3390/polym12112573

Wang, X., Jiang, M., Zhou, Z., Gou, J., & Hui, D. (2017). 3D printing of polymer matrix composites: A review and prospective. *Composites Part B: Engineering*, *110*, 442–458. https://doi.org/10.1016/j.compositesb.2016.11.034

Wei, J., Wang, J., Su, S., Wang, S., Qiu, J., Zhang, Z., Christopher, G., Ning, F., & Cong, W. (2015). 3D printing of an extremely tough hydrogel. *RSC Advances*, *5*(99), 81324–81329. https://doi.org/10.1039/c5ra16362e

Wijk, A. V., & Wijk, I. V. (2015). *3D printing with biomaterials: Towards a sustainable and circular economy*. IOS Press.

Wong, T. Y., Preston, L. A., & Schiller, N. L. (2000). Alginate lyase: Review of major sources and enzyme characteristics, structure-function analysis, biological roles, and applications. *Annual Review of Microbiology*, *54*, 289–340. https://doi.org/10.1146/annurev.micro.54.1.289

Wu, C. S. (2016). Modulation, functionality, and cytocompatibility of three-dimensional printing materials made from chitosan-based polysaccharide composites. *Materials Science and Engineering C*, *69*, 27–36. https://doi.org/10.1016/j.msec.2016.06.062

Wu, H., Fahy, W. P., Kim, S., Kim, H., Zhao, N., Pilato, L., Kafi, A., Bateman, S., & Koo, J. H. (2020). Recent developments in polymers/polymer nanocomposites for additive manufacturing. *Progress in Materials Science*, *111*(April 2019). https://doi.org/10.1016/j.pmatsci.2020.100638

Wu, Q., Maire, M., Lerouge, S., Therriault, D., & Heuzey, M. C. (2017). 3D printing of microstructured and stretchable chitosan hydrogel for guided cell growth. *Advanced Biosystems*, *1*(6), 1–6. https://doi.org/10.1002/adbi.201700058

Wu, Z., Su, X., Xu, Y., Kong, B., Sun, W., & Mi, S. (2016). Bioprinting three-dimensional cell-laden tissue constructs with controllable degradation. *Scientific Reports*, *6*(September 2015), 1–10. https://doi.org/10.1038/srep24474

Xu, W., Pranovich, A., Uppstu, P., Wang, X., Kronlund, D., Hemming, J., Öblom, H., Moritz, N., Preis, M., Sandler, N., Willför, S., & Xu, C. (2018). Novel biorenewable composite of wood polysaccharide and polylactic acid for three dimensional printing. *Carbohydrate Polymers*, *187*(January), 51–58. https://doi.org/10.1016/j.carbpol.2018.01.069

Yang, F., Zhang, M., Bhandari, B., & Liu, Y. (2018). Investigation on lemon juice gel as food material for 3D printing and optimization of printing parameters. *LWT*, *87*, 67–76. https://doi.org/10.1016/j.lwt.2017.08.054

Zhang, H., Mao, X., Du, Z., Jiang, W., Han, X., Zhao, D., Han, D., & Li, Q. (2016). Three dimensional printed macroporous polylactic acid/hydroxyapatite composite scaffolds for promoting bone formation in a critical-size rat calvarial defect model. *Science and Technology of Advanced Materials*, *17*(1), 136–148. https://doi.org/10.1080/14686996.2016.1145532

5 Applications of Biopolymer Blends and Biopolymer-Based Nanocomposites

5.1 INTRODUCTION

Throughout human history, biopolymers have served various purposes, such as food, clothing, and materials for furniture. However, starting in the 1950s, petroleum-based polymers emerged as the primary source for the development and production of commercial products, such as plastics, which are now ubiquitous. In light of growing environmental concerns and limited fossil fuel resources, researchers have shifted their focus towards finding alternatives to petroleum-based products. The turning point came with the signing of the environmental treaty on December 11, 1997, which became law on February 16, 2005. This treaty addressed pressing environmental issues, including greenhouse gas emissions, climate change, and the adverse impacts on ecosystems stemming from toxic waste, pollution, resource depletion, and environmental degradation (Attaran et al., 2017; Abdul Khalil et al., 2023).

The substitution of biobased polymers for synthetic plastic products has emerged as a significant trend across numerous industrial sectors. This shift is regarded as a viable alternative to synthetic polymers, given that biobased polymers are derived from renewable resources. They offer a positive environmental impact and can naturally replenish through the Earth's cycles. The renewable nature of biopolymers leads to the development of new sustainable products driven from renewable feedstock with many advantages, including disintergrability and degradability, improved possibilities for recycling, a lack of toxic components, and high biocompatibility (Luzi et al., 2019). Biopolymers can be from a living organism or materials that need to be polymerized but come from renewable resources. Biopolymers that come from living organisms include polysaccharides (starch, cellulose), protein (wheat gluten, soy protein, gelatin), and polyester-like poly-alkoxy alkanoates (PHAs) which are produced from bacteria. The latter includes polylactic acid (PLA); polycaprolactone; polyhydroxyalkanoates; polyethene glycol; and aliphatic polyesters such as PBS, PVA, and polyurethanes (PUs) (Attaran et al., 2017; Chassenieux et al., 2013; Tang et al., 2012).

Despite the positives claimed regarding biopolymers, they also suffer from drawbacks that may limit their application, such as high production cost and limited mechanical and thermomechanical properties concerning synthetic polymers (Kumar et al., 2017; Luzi et al., 2019). Consequently, developing

DOI: 10.1201/9781003416043-5

eco-friendly polymeric blends can overcome these limitations (Luzi et al., 2019). During the last three decades, a new method for polymeric material preparation by blending two or more materials for various uses has been observed. Blending biopolymers with other biopolymers is a promising technique for preparing polymers with "tailor-made" properties (functional properties and biodegradability). The combination of biopolymers with biodegradable synthetic polymers is also considered a trend, as it provides a way to reduce the overall cost of the materials and offer a method of modifying both properties and degradation rates (Asyraf et al., 2023).

Biopolymer-based nanocomposites are advanced materials formed from a fascinating interdisciplinary area that consists of biology, materials science, and nanotechnology. The superiority of these new materials comes from the various choices of biopolymers and reinforcements available, including clays, nanocellulose, metal nanoparticles, and graphene. The interaction of nano-scale fillers acts as a bridge in the polymer matrix and leads to the enhancement of the mechanical properties of the nanocomposites. This combination of these materials also adds a new dimension based on biocompatibility and biodegradability. Therefore, bio-nanocomposites are of great interest for biomedical technologies, including tissue engineering, medical implants, dental applications, and controlled drug delivery (Chassenieux et al., 2013). Moreover, the utilization of biopolymer-based nanocomposites extends across a range of sectors, including construction, automotive, aerospace, biomedical, cosmetics, and packaging. These industries have embraced and extensively explored the application of nanotechnology as a viable strategy to enhance the specific properties necessary for their respective domains. The application of nano-scale fillers as reinforcing materials helps turn biopolymers into biocomposites with superior mechanical strength and other characteristics essential for specific applications (Luzi et al., 2019; Pacheco-Torgal, 2016).

5.2   BUILDING AND CONSTRUCTION

Biopolymers find extensive application in the realm of building and construction. In certain scenarios, biopolymers present unique benefits when contrasted with their synthetic counterparts. Notably, biopolymers are gaining wider environmental acceptance, particularly within the context of interior home construction.

5.2.1   SUPERPLASTICIZERS

Chemical admixtures are essential and important for manufacturing modern concrete and have been a focus of development in high-tech fields. Superplasticizers are widely used to produce flowable, strong, and durable Portland cement in the presence of plasticizers. The addition of a superplasticizer helps enhance properties, especially after the material hardens (Xun et al., 2020). The utilization of biopolymers in concrete has a long history, with instances of natural occurrences. Historically, the exploration of biopolymer applications dates back to the construction techniques pioneered by the Romans. During their flourishing era, the Romans made noteworthy advancements, including the development of a cementitious substance known as *opus caementitium*, which served as the fundamental building material for iconic structures like the Roman Colosseum. Marcus Vitruvius Pollio (84–10 BC) described the invention of construction and materials in his famous encyclopedia, *De acrhitectura decem*, which explained the role of admixtures in improving building materials, for example, the use of air lime mortars with the addition of vegetable fat (Giavarini et al., 2006). The Romans were celebrated for their innovative techniques aimed at enhancing building materials. They ingeniously incorporated dried blood as an air-entraining agent, while biopolymers, such as proteins, played a role as set retarders for gypsum. Similarly, the Chinese employed substances like egg white, fish oil, and blood in the construction of the Great Wall due to their reinforcing properties (F. Pacheco-Torgal, 2016; Plank, 2005). In 1507, mortars based on lime mixed with small amounts of vegetable oil added during the slaking process were applied in the construction of the Portuguese fortress *Nossa Senhora da Conceicao*, located on Gerum island, Ormuz, Persian Gulf (Pacheco-Torgal, 2014; Pacheco Torgal & Jalali, 2011).

Later in the 20th century, lignosulfonates, a biopolymer for concrete plastification, were introduced. This marked a pivotal moment in the use of biopolymers, specifically lignosulfonates, for plasticizing ordinary portland cement (OPC) concrete. This was the first functional polymer in construction to be used on a large scale (Plank, 2005). OPC concrete, a typical civil engineering construction material, is the most-used material on earth. To date, around 15% of total OPC concrete production contains chemical admixtures to modify the properties. Example of biopolymers used in concrete include lignosulfonate, starch, chitosan, pine root, extract, protein hydrolysate, and vegetable oils. The addition of polymers in Portland cement offers improved mechanical properties. The function of polymer addition is actually to reduce permeability, diminish the number of large pores, refine them, and hinder the propagation of cracks. Also, the addition of polymer helps to better organize the microstructure of concrete. The final products will become more elastic, tough, and resilient. Similarly, this mechanism also represents the function of biopolymers. Concrete with higher toughness and resilience is required for structures designed for relevant thermal variations or dynamic loads followed by fatigue (Bezerra, 2016).

Also, biotech admixtures received increasing attention due to their high biosynthesis rate compared to plant-based biopolymers. These admixtures include sodium gluconate, xanthan gum, curdlan, and gellan gum (F. Pacheco-Torgal, 2016). Table 5.1 shows a list of biopolymers and synthetic polymers used for chemical admixtures. Although OPC and dry-mix mortars use mostly biopolymers, a great diversity of biomixtures for over 500 different products is currently being used by building material industries and is expected to grow with the expansion of urban land.

The greatest application of biopolymers in the construction industry relies mostly on inorganic binders and coatings of building materials that mostly focus on the field of rheology control (dispersing/thinning or thickening) and water retention. Biopolymer admixtures also serve many other purposes, includes dispersing/thinning effects, viscosity enhancement, water retention, set acceleration and retardation, air-entrainment, de-foaming, hydrophobing and adhesion, and film-forming.

TABLE 5.1
Major Milestones in Chemical Admixture Technology for Construction

Year of Introduction	Admixture Chemistry	Function	Type of Admixture
1920s	Lignosulfonate	Concrete plasticizer	Biopolymers
1940s	Lignite	Bentonite thinner	Biopolymer
1960s	Xanthan gum	Viscosifier	Biopolymer
1962	Melamine, naphthalene condensaters	Concrete superplasticizer	Synthetic polymer
1970s	Cellulose ethers	Water-retention agent	Biopolymer
1980s	Vinyl-sulfonate copolymers	Water-retention agent	Biopolymer
1980s	Polycarboxylate copolymers	Concrete superplasticizer	Synthetic polymer
1990s	Polyaspartic acid	Biodegradable dispersant, retarder	Biopolymer

According to Bezerra et al. (2016), these biopolymers are usually utilized in these forms:

- Powder form: biopolymers that can be either added to cement or diluted water for concrete preparation. Example: chitin, chitosan, and so on.
- Liquid form: biopolymers that are usually diluted in water for concrete preparation. Example: latex materials (rubber, avelos, *Araucaria*, dilutan, welan, xanthan, gelan, gutta-percha, guar), and so on.
- Fiber form: biopolymers that have undergone the biopolymerization process that will increase the tensile strength of the concrete.

The function of biopolymers as plasticizers continues to be explored by researchers to enhance the properties of concrete. For example, a biopolymer-based admixture to enhance the viscosity properties of concrete was studied by Leon-Martinez et al. (2014) using nopal mucilage and marine brown algae extract. In this study, the authors reported the success of nopal mucilage and marine brown algae to increase the viscosity of concrete. The study discovered concrete prepared with 0.25% of nopal mucilage produced excessive bleeding and segregations, whereas concrete with 0.25% of marine brown algae extract seemed to be the best admixture, presenting good spread, low segregation, and bleeding with high air content (<6%). This finding was proved by the proposed structure (Figure 5.1) and presented a complex steric effect provided by the presence of the alginate chains that deviate cement particles.

In a separate discussion, a biobased superplasticizer derived from starch was used in an attempt to replace the function of polycarboxylate ether (PCE) in the slump of a metakaolin-based geopolymer mortar (Tutal et al., 2020). The study aimed to assess the potential of sodium silicate pentahydrate (SSP) as an effective additive in geopolymer applications, focusing on aspects such as slump, hardening, compressive and flexural strength, shrinkage, and porosity. The findings indicated that SSP

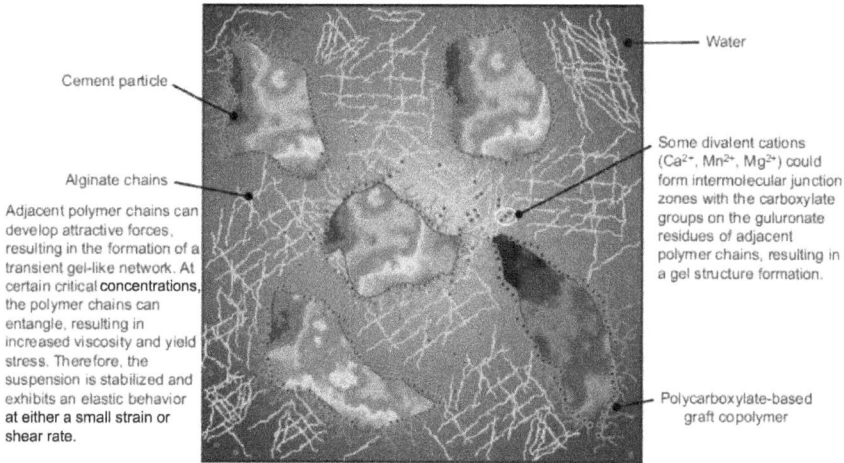

FIGURE 5.1 Schematic representations of network formation between marine brown algae extract polymer chains in cement paste. Reproduced from León-Martínez et al. (2014).

positively influenced the slump of metakaolin-based geopolymer mortar, although no significant improvement was observed with PCE. Additionally, SSP exhibited a minor impact on porosity, promoting the formation of more gel pores, which positions SSP as a promising plasticizing agent for metakaolin-based geopolymer mortar and concrete.

5.2.2 INSULATED MATERIALS

The carbon footprint of buildings is influenced by the choice of materials, manufacturing processes, and insulation performance. Notably, the building envelope plays a crucial role in energy consumption, especially in regulating indoor temperature for occupant comfort. One significant advancement in this realm is the utilization of structural insulated panels (SIPs) in home construction. An SIP consists of two outer skins sandwiching an inner core of insulating materials, as shown in Figure 5.2. Commonly, petroleum-based polymer foam is usually applied as an insulating material that includes polyurethane (PU) and expanded polystyrene (PS) due to their intrinsic properties. Nowadays, the transition to renewable materials for material resources has increased interest in bio-resource options. Biobased polymers such as poly (caprolactone) (PCL), polyhydroxyalkanoates, polybutylene succinate (PBS), and PLA have been considered for insulation purposes to reduce the carbon footprint of buildings (Frank & Billington, 2008; Oluwabunmi et al., 2020).

The application of biodegradable materials had been applied commercially by Biopolymer Network Limited, with their products known as ZealaFoam. Bioplastic products derived from biomass sources are commercially available in the form of PLA beads processed via a green blowing agent using carbon dioxide, which can be safely disposed of by industrial composting or burning. Their products are available in a wide range of applications, one of which is material for structural insulations

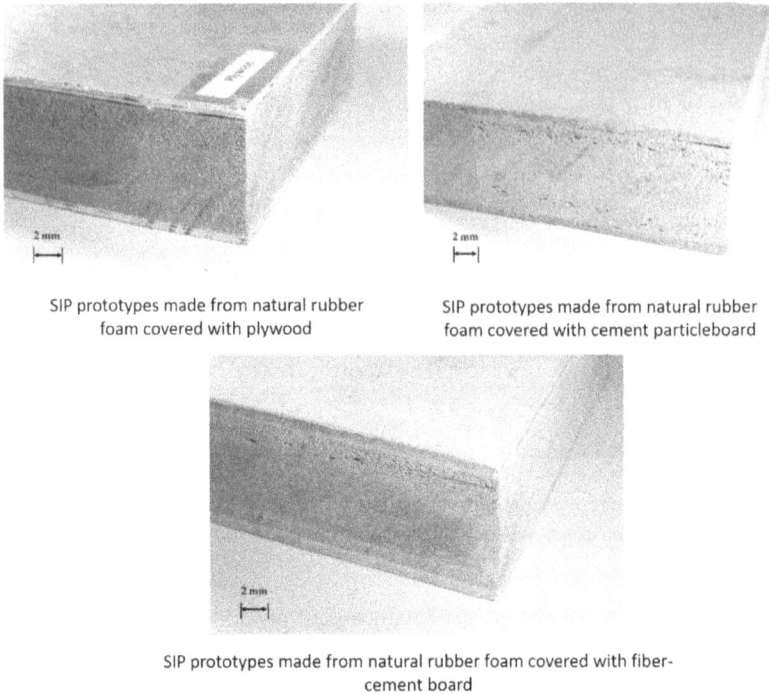

SIP prototypes made from natural rubber foam covered with plywood

SIP prototypes made from natural rubber foam covered with cement particleboard

SIP prototypes made from natural rubber foam covered with fiber-cement board

FIGURE 5.2 Structural insulated panels with different types of insulating materials. Reproduced from Thongcharoen et al. (2021).

("ZealaFoam," n.d.). In a study conducted by Oluwabunmi et al. (2020), research was undertaken to explore compostable biobased foam made from PLA using the eco-friendly blowing agent CO_2. The study aimed to develop materials suitable for zero-energy buildings. The introduction of micro-cellulose fibers (MCFs) in this research had several noteworthy effects. It resulted in reduced open porosity, increased bulk density, decreased expansion ratios, and smaller cell sizes. Furthermore, the incorporation of MCF led to a lowered glass transition temperature in PLA, while the thickness of the cell walls also decreased.

One significant outcome of this study was the improved biodegradability of the foam due to the presence of cellulose. Moreover, when the foam was tested in an energy simulation on a model house, it was found to enhance heating and cooling efficiency by 12%. As a result, this biobased foam shows promise as a viable building material for applications in the construction of energy-efficient structures.

Another advance in biopolymer-based material used in building and construction applications is aerogel. Aerogel derived from biopolymers has been utilized as a thermal insulation material due to its significantly low thermal conductivity, which can be as low as 0.01–0.03 W/m.K (similar to that of air at ambient conditions, with a value of 0.026 W/m.K) (Wicklein et al., 2015; Zhu et al., 2019). The ultralow value of thermal conductivity is attributed to (1) the high porosity and tortuosity of the solid nanostructure of aerogel; (2) the effective suppression of thermal radiation; and (3) the size of the pores that means a free path of the gas phase, which efficiently reduces the thermal convection contribution (Zhao et al., 2018). However, biopolymer aerogel has strong hydrophilicity, which is not conducive to thermal insulation applications. Therefore, appropriate biopolymer blending or the addition of inorganic substances can enhance the properties of biopolymer aerogels.

The applications of aerogel to produce lightweight anisotropic foams with thermal insulating and fire-retardant properties using nanocellulose-graphene oxide-based aerogels were studied by Wicklein et al. (2015). Highly porous foams were produced via freeze casting from a colloidal suspension of cellulose nanofibrils (CNFs) and graphene oxide (GO). The CNF-GO aerogel was mechanically stiff in the freezing direction and able to sustain a considerable load. Aerogel foams also showed fire-retardant properties with excellent combustion resistance and exhibited a thermal conductivity of 15 mWm/mK, which is about half that of expanded polystyrene (EPS). Even at 30°C and 85% humidity, the foam retained more than half of its initial strength. This finding can help to motivate development of high-performance thermal insulating materials for enhanced energy efficiency and reduction of the environmental impact of buildings.

The utilization of biopolymer aerogel for ultra-low thermal conductivity has been the focus of researchers. Zhu et al. (2019), in their study, reported the production of a simple, green, and low-cost aerogel from Konjac glucomannan (KGM)-reinforced silicon dioxide (SiO_2) via a facile freeze-drying process like that shown in Figure 5.3. The obtained novel KGM-SiO_2 aerogel showed remarkable compressive strength (δ_{max} = 1.65 MPa), a high specific surface area of 416.1 m²/g, high hydrophobicity

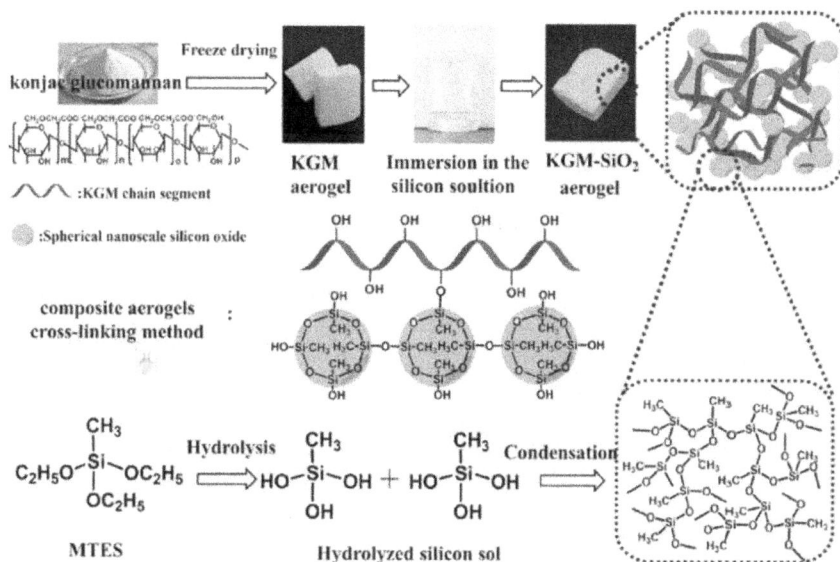

FIGURE 5.3 Illustration of all synthetic steps of KGM-SiO_2 composite aerogel. Reproduced from Zhu et al. (2019).

($\theta = 146°$), and low thermal conductivity of 0.032–0.039 W/m K. Thus, the high performance of KGM-SiO$_2$ aerogel could further expand thermal insulation application in sustainable development and energy-saving buildings.

5.3   MARINE AND COASTAL CONSTRUCTION

Biopolymers have been applied in geotechnical engineering for the construction and maintenance of coastal protection infrastructures. Marine soils usually have low undrained shear strength and are susceptible to consolidation caused by infrastructure-induced loads (e.g. harbors, airports, and energy plants) due to their fine-grained soil content (silt and clay) and high in situ water content (Phetchuay et al., 2016). Therefore, ground improvement techniques are needed to enhance the consolidation behavior, shear and compressive strength, and flocculation sedimentation to enable construction on marine areas. Previously, cement-based materials have been applied to stabilize marine soil. Portland cement and lime have been widely used to improve shear strength and reduce the compressibility of soil. However, the application of cement leads to various environmental problems such as carbon dioxide emissions, groundwater contamination, and ocean and ecosystem pollution (Chang et al., 2016; Kwon et al., 2019).

Studies have proved that biopolymers are an environmentally friendly soil reinforcement. Biopolymers interact with soil particles, mainly clay, via direct ionic bonding (hydrogen bonding); thereby the soil strength and stiffness can be improved (Chang & Cho, 2014, 2019). On the other hand, cationic biopolymers have the potential for coagulating soil suspension by accelerating the sedimentation process by forming bridges between clay with high water content. A study using xanthan gum biopolymer on soil strengthening to improve soil properties, including aggregate stability, strength, and erosion resistance, was done by Chang et al. (2015). The findings showed that xanthan gum reacted directly with the charged surfaces of clay particles, and the strengthening effect was dependent on four factors: type of soil, hydration level, xanthan gum content, and mixing method. The xanthan accumulated and aligned as threads and textiles that reinforced soil clods. The mechanism is presented in Figure 5.4.

Another study was done by Kwon et al. (2019) using xanthan gum; an anionic biopolymer; and ε-polylysine, a cationic biopolymer, to treat the soft marine soil. Xanthan gum showed potential as a soil-strengthening material by affecting the geotechnical properties of soft marine soil by enhancing its viscosity and bonding with soil particles and was most efficiently applied at a shallow depth. Meanwhile, ε-polylysine biopolymer showed potential as a soil-coagulating material because of its cationic properties. ε-polylysine coagulates soil by forming a bridge, thus increasing floc size and accelerating settling velocity, aligning soil in a dense way.

5.4   GEOTEXTILES

Geosynthetics are thin, flexible material sheets applied in civil and environmental engineering and include geotextiles, geonets, geogrids, geomembranes, geosynthetic clay liners, geofoams, geocells, and geocomposites. Among all products, geotextiles

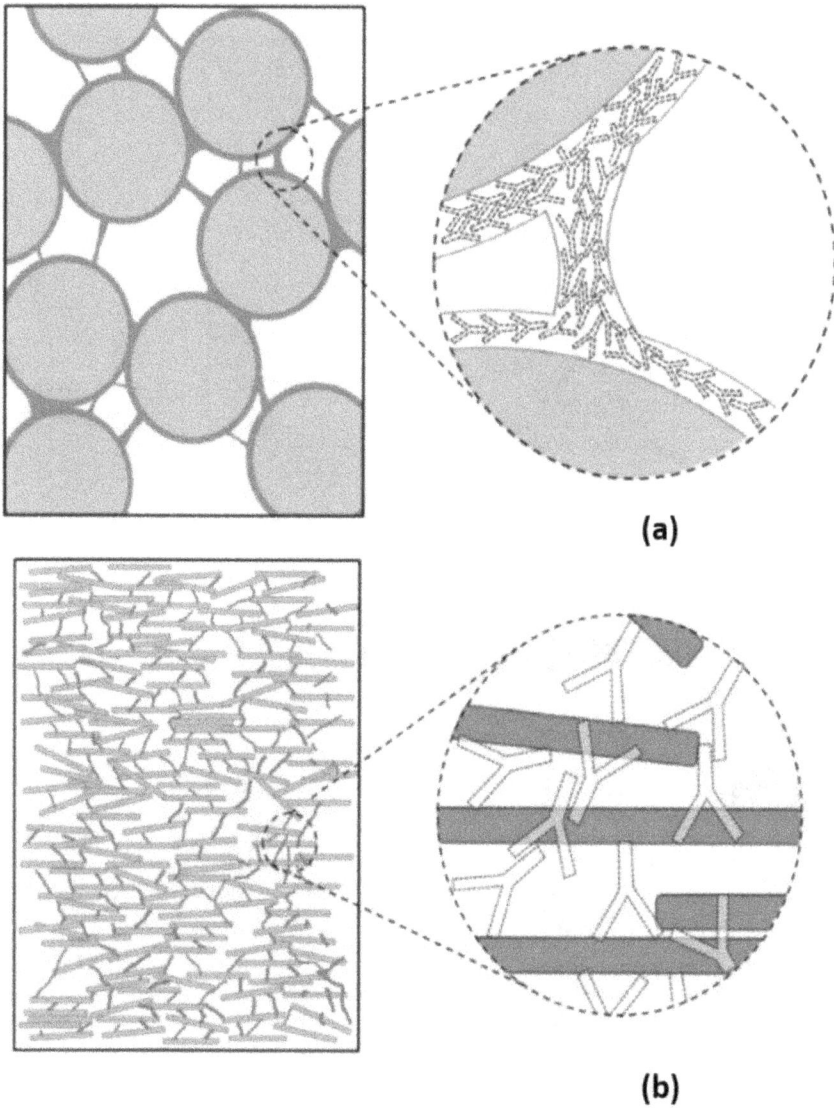

FIGURE 5.4 Xanthan gum-soil interaction model. (a) Coarse and fine soil. (b) Xanthan gum forms threads, providing reinforcement with soil clods. Reproduced from Chang et al. (2015).

are the largest group consisting of polymers and additives to improve stability. Geotextiles (Figure 5.5) are usually used with foundation, soil, rock, earth, or any geotechnical engineering-related material as an integral part of a structure or system. In general, the functions of geotextiles are: (1) drainage, where redundant water is gathered or removed from the soils or walls; (2) separation, to avoid mixture of two different foundations of different properties, usually required in road and railway

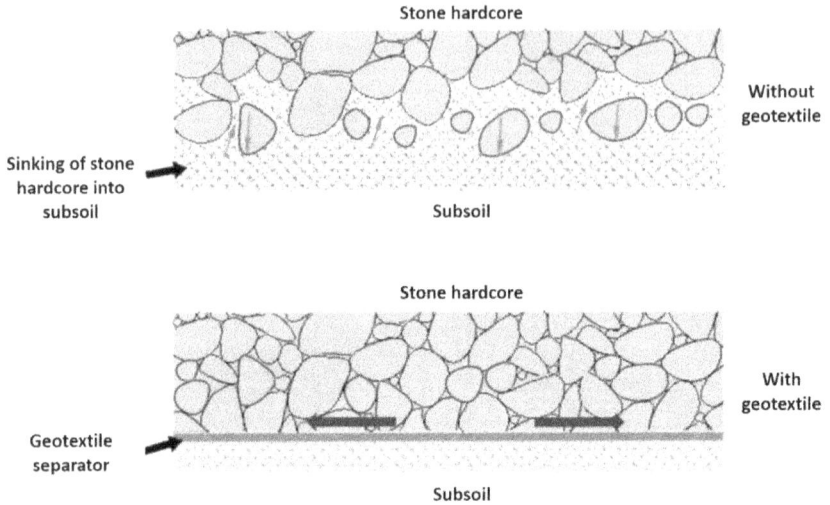

FIGURE 5.5   The function of geotextiles. Reproduced from Wu et al. (2020).

Woven geotextile   **Non-woven geotextile**

FIGURE 5.6   Woven and non-woven geotextiles.

construction; (3) filtration, where particles are retained in the geotextile as the water passes through; (4) reinforcement of unstable or highly exposed environments; and (5) protection against mechanical damage (Prambauer, Wendeler, Weitzenböck et al., 2019).

There are two main types of geotextiles, non-woven and woven, which can be installed on or in the ground depending on the application (Figure 5.6). Specifically, non-woven fiber sheets, which are the biggest category of geotextiles, are utilized in filtration and drainage applications due to their 3D structure, which results in high porosity and permeability with varying pore size. Meanwhile, woven geotextiles possess superior mechanical properties and are usually applied in the field of soil reinforcement and stabilization (Li et al., 2016; Prambauer, Wendeler, Weitzenböck et al., 2019; Wiewel & Lamoree, 2016).

Most applied materials in the field are non-degradable polymers, including polypropylene, polyolefin, or polyester, which contribute to environmental problems concerning soil pollution and accumulation of micro-plastics. To overcome the disadvantages of synthetic polymers, biopolymers have become the main alternative. Biopolymers are biobased, meaning they are constructed from natural fibers and are biodegradable, capable of being broken down, especially into innocuous products by the action of living things, such as microorganisms. Similar to all polymers, biopolymers are easier to process and obviate the weakness in physical properties. Additives help in the improvement of biopolymers, but some additives impart side effect, for instance, the processing modifier for PLA, often reducing the biodegradability of the products. Sometimes a compound of natural origin also leads to increased resistance against rotting. However, the biodegradability of biopolymers is not always seen as undesirable. Rather than using additives to prevent it from degrading, natural geotextiles are used in applications where the ecosystem will take over its function in time (Prambauer, Wendeler, Weitzenböck et al., 2019; Wiewel & Lamoree, 2016).

The existing biodegradable geotextiles that have been applied are mainly made of jute and coir fibers that have already been on the market. Geo-Synthetics, LLC, is one of the largest distributors, fabricators, and installers of geosynthetic materials. Their products, GEOCOIR and DeKowe, are biodegradable geotextiles mainly used for slope stabilization and protection of stream banks, shorelines, and wetlands. These products are mainly made from the natural fibers coir and jute (Geo-Synthetics, n.d.). However, there is an issue with these kinds of products, as natural fibers are quite likely to take up a high amount of water, which leads to highly significant weight gain and premature degradation (Prambauer, Wendeler, Weitzenböck et al., 2019).

Several biopolymers can be suitable candidates for use as environmentally friendly geotextiles, especially when natural fibers fail to fulfil their function effectively. There is continuous study regarding the application of biopolymers in geotextile applications. For example, Jeon (2016), in his study, compared the function of biodegradable PLA biopolymers and PLA/poly(butylene adipate-co-terephthalate) (PBAT) biopolymer blends on environmental adaptability as green geosynthetics for civil engineering applications. PLA/PBAT blends showed improvement in mechanical performance and proposed suitable biodegradable evaluation using experimental data analysis. Through the prediction using an Arrhenius plot accelerated the experimental data, the author predicts long-term biodegradable behavior and suggested more evaluation items are needed considering actual field installation conditions. In a separate study, Prambauer et al. (2019) found the advantages of short-lived geotextiles in soil and erosion protection applications, which enable improvement of seeding conditions on the exposed surface. In this study, four different bio-polyesters were produced via the melt spinning process and were characterized to investigate the influence of moisture and UV light radiation on mechanical stability. The findings showed promising results of bio-polyesters based on their fracture strength, even after 750 hours of exposure.

5.5 AGRICULTURE

Natural polymers' roles as biodegradable and eco-friendly materials include their applications in the agriculture industry, as widely reported in agricultural sectors. Biopolymers are preferred, as they can degrade naturally in soil by microorganisms in ambient conditions, releasing enzymes capable of cleaving the linkage of macromolecules. Biopolymers also can enhance the water retention ability of soil. In agriculture, water is one of the important factors, as it can affect crops and fruit growth and the productivity of both. The utilization of biopolymers helps in maintaining the balance within the ecosystem, including soil, plants, air, and water (Mohamady Ghobashy, 2020).

5.5.1 BIODEGRADABLE MULCHES

Since the 1930s and 1940s, polymers have grown rapidly in agriculture applications in the form of films for greenhouse coverings, fumigation, and mulching. Later in the 1950s, plastic mulching was introduced to agriculture to increase vegetable and fruit crop productivity by reducing the production of weeds and preventing soil erosion. However, applications of degradable plastics are needed, especially for mulches and agricultural planting containers that allow degradable plastics to be combined with other biodegradable materials and converted into useful soil-improving materials (Hayes et al., 2012; Raj et al., 2011). Conventional mulches are films made from non-renewable plastics, usually polyethylene (PE), with thicknesses in the range of 0.5 and 1.25 mm. However, concerns regarding product cost, disposal, and labor for removal from the field have been raised regarding the applications of non-renewable plastics. The common disposal method for conventional mulches is by burning them, which is a global concern, as it can release airborne pollutants such as dioxane, while disposal in landfills can lead to an increased level of potential pesticides due to absorbed agents that accompany mulches and at the same time decrease the available landfill space (Hayes et al., 2012).

Biopolymer-based mulches have become the most popular alternatives to conventional mulches these days. These mulches help plant growth and then photodegrade in the fields, thus reducing the cost needed for removal. The application of mulches also helps to improve the plant growth rate due to their ability to maintain moisture, reduce weeds, and increase soil temperatures (Raj et al., 2011). The biodegradability properties of the additives and monomers in bio-plastic mulches into agricultural soil are very important, as they have potential environmental impacts. During their use, three differential stages can be established (refer to Figure 5.7): (1) mulch storage and installation in the field where biodegradable mulches remain stable in convenient conditions, including in dry conditions, low temperature, and indoor storage; (2) mulch during the crop cycle, where the mulches are exposed to conditions that may affect their structures and properties, which enables the release of fragments and chemicals into the soil; and (3) mulch after crop harvesting, where the mulches were tilled into the soil and continue to biodegrade until their complete mineralization to CO_2 and H_2O (Serrano-Ruiz et al., 2021).

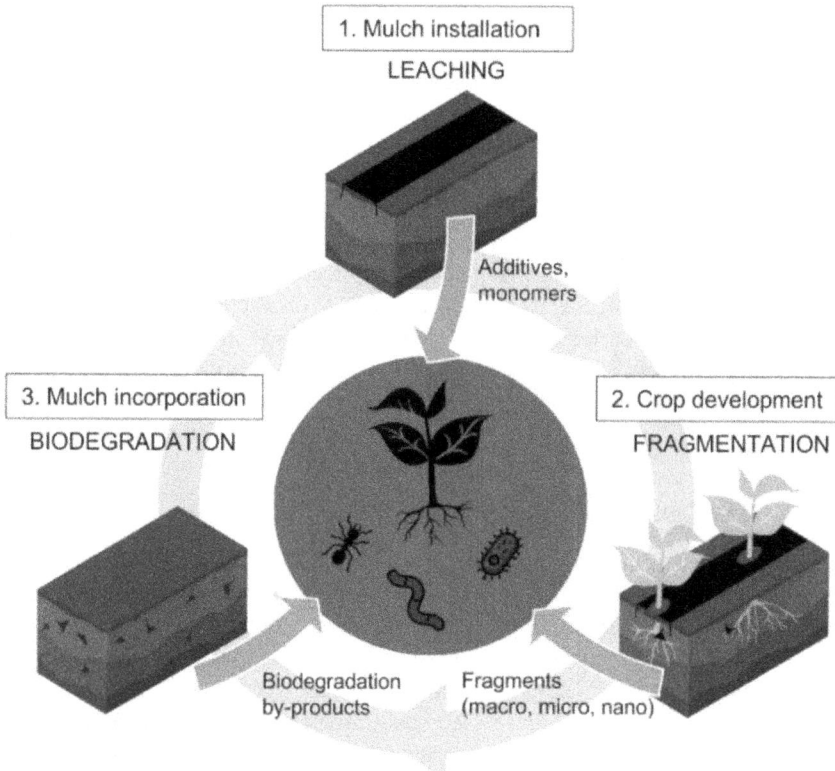

FIGURE 5.7 Agricultural cycle of biodegradable plastic mulch films during application. Reproduced from Serrano-Ruiz et al. (2021).

There are a number of commercial biodegradable mulches available on the market using biopolymers and polymer blends, such as Biocycle, Bio-Flex, Biomax TPS, Biosafe, Eco-Flex, GreenBio, Ingeo, and Mirel (Hayes et al., 2012). The biodegradable mulch product manufactured by DuPont (USA), referred to as Biomax TPS, uses a combination of high amylose starch and thermoplastic starch. DuPont also developed a PBAT-related copolymeric product as a biodegradable toughener for PLA film, denoted Biomax. The biodegradable mulch called Ingeo is manufactured by NatureWorks (USA) and used the application of PLA incorporated with starch and PBS to enhance plant growth, drain the landscape, and control the weeds ("NatureWorks," n.d.). Meanwhile, Ecoflex, manufactured by BASF, consists of partially synthetic copolymer PBAT blended with starch biopolymers (Hayes et al., 2012).

A biodegradable film product from FKUR, Willich (Germany), denoted Bio-Flex, consists of blended of PLA and another co-polyester blend. A study has been done comparing the effectiveness of Bio-Flex with conventional PE mulch for strawberry cultivation (Giordano et al., 2020). This study proved the efficiency of biodegradable mulch films, where strawberries showed higher calcium content, which may encourage growers toward eco-friendly agricultural practices.

However, these agricultural mulches were available in the form of thin films; therefore, biodegradable polymer feedstock may need additives to further enhance deficiencies by the addition of nucleating agents, plasticizers, performance additives, and lubricants. Many advanced studies have been done to develop truly biodegradable mulches with many properties superior to those of the mulches available in the market. For example, Hasan et al. (2019) conducted a study where they created a biodegradable plasticulture application using a red seaweed biofilm that integrated microbial-induced calcium carbonate (M-CaCO$_3$) alongside commercial calcium carbonate (C-CaCO$_3$).The red seaweed–reinforced M-CaCO$_3$ bio-polymer film was found to have better water barrier, hydrophobicity, and biodegradability compared to conventional mulch film and bio-polymer film–reinforced C-CaCO$_3$.

5.5.2 Seed Coatings

Enhancement of seeds can be achieved via many techniques, including seed priming, coating, and conditioning to improve seed delivery during planting; to increase seed germination, stand uniformity, and seedling growth; and to suppress disease. Seed enhancement through seed coating can provide micro- and macronutrients or biostimulant materials to increase germination, seedling vigor, and stand establishment. Recently, seed coating technology has been applied via various approaches depending on the shape and size of the seed and the type of film coating, and seed encrusting has also been widely applied for coating or treatment procedures in the seed industry to enhance plant and seedling performance (Qiu et al., 2020).

Nowadays, biopolymers are very popular as seed coating materials that serve as soil stabilizers, seed protectors, yield enhancers, and plant growth regulators. Biopolymers are preferred, as they can dry quickly, dissolve rapidly in water, form an effective water-soluble film, readily adhere to seeds, minimize the required dose of fungicides, and provide excellent control of plant diseases, thereby contributing to enhancing plant productivity (Raj et al., 2011). Among all polysaccharide biopolymers, chitin and chitosan are common polysaccharide-based natural polymers used for agricultural purposes. Apart from being biocompatible and biodegradable, chitosan also possesses antimicrobial, antiviral, antibacterial, and antifungal properties and is available from regenerating resources like waste seafood shells. The composition in chitin and chitosan is known to elicit a variety of defense responses in host plants in response to microbial infections and thus has a good ability to reduce the negative impact of disease on the yield and quality of crops (Raj et al., 2011). This was proved by the findings of Rakesh (2017), where seeds covered by chitosan biopolymer showed better germination and growth parameters compared to synthetic polymers in both castor and groundnuts.

Additives like biostimulants can also be added to seed coating formulations to increase their germination properties. A biostimulant seed treatment using a blended

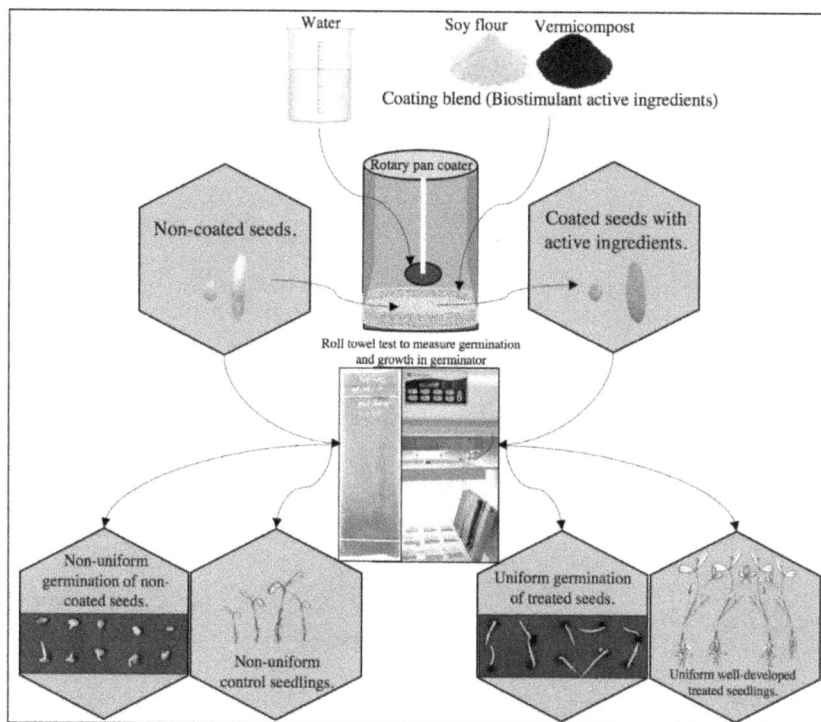

FIGURE 5.8 Seed coating technology using biostimulant compounds of soy flour and vermicompost for sustainable agriculture. Reproduced from Qiu et al. (2020).

biopolymer of soy flour with diatomaceous earth, micronized vermicompost, and concentrated vermicompost extract on red clover (*Trifolium pratense* L.) and perennial ryegrass (*Lolium pereme* L.) was studied by Qiu et al. (2020). The germination and growth of coated and non-coated red clover and perennial ryegrass were observed after 7 and 10 days, respectively. The comparison study between non-coated and coated seeds shows the difference between germination rate and uniformity, as shown in Figure 5.8.

Meanwhile, agrichemical delivery and seedling development can be further enhanced by using biopolymer-based nanoplatforms. Xu et al. (2020) utilized nanofibers synthesized via electrospinning of biopolymer blends of gelatin and cellulose acetate and enabled tunable agrichemical release by modulating the polymer composition and hydrophilicity of nanofibers (Figure 5.9(a)). Through investigation in greenhouse study (Figure 5.9(b)), the nanofiber seed coating enabled precise delivery of agrichemicals at a very low amount, while the Cu-release nanofiber coatings appeared to promote seed germination, especially in diseased media conditions, and increased seedling biomass for both tomato and lettuce in healthy media conditions.

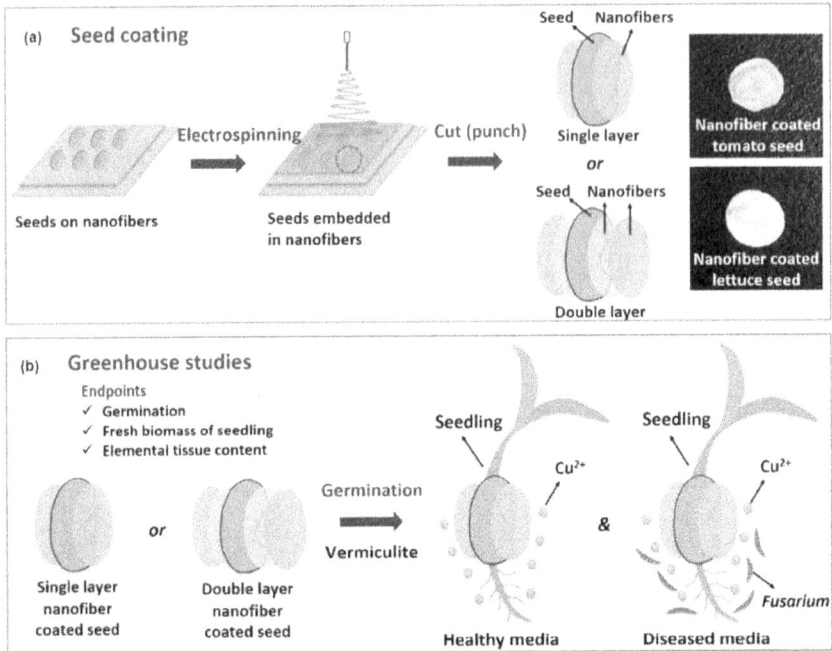

FIGURE 5.9   (a) Seed coating using electrospun Cu^{2+}-loaded nanofibers on tomato and lettuce seeds and (b) germination of nanofiber-coated seeds in the presence or absence of a fungal pathogen. Reproduced from Xu et al. (2020).

5.5.3   SOIL CONDITIONERS

Agriculture is an important sector of the economy, but this industry consumes a significant amount of water. It is very important to find suitable techniques and efforts to mitigate the use of water that affects the growth, survival, and yield of crops. Various applications of biopolymers in soil conditioners give different improvements in water availability and irrigation potency for plants by affecting the water-holding properties of the soil. At the same time, biopolymers are also used as a medium to release nutrients needed for soil into crops (Guilherme et al., 2015; Michalik & Wandzik, 2020). For example, Zainescu et al. (2011) used biopolymers derived from leather waste in an attempt to remediate degraded/eroded soils and enhance greenhouse and field plant growth.

To date, hydrogel materials have been utilized as soil conditioners and yield enhancers due to their ability to retain both water and nutrients and then release them over an extended period (Michalik & Wandzik, 2020). Biopolymers and biopolymer blends are commonly used to develop materials with high durability and resistance towards all forms of degradation, usually in the form of a hydrogel. Hydrogels are 3D-crosslinked polyelectrolyte polymers that can absorb a large amount of water, up to 100 times their weight. These characteristics particularly give advantages to agricultural applications such as permeability, rigidity, and transparency and make them suitable to be used as soil conditioners. Hydrogels can be considered good soil conditioners and yield enhancers, as they are capable of retaining both water and nutrients and releasing them over an extended period (Michalik & Wandzik, 2020; Mohamady Ghobashy, 2020).

For example, a biopolymer is more oxygen permeable in forming an agricultural polysaccharide hydrogel (APH) that imparts benefits to aerobic bacteria and help increase nutrient availability by providing fixed nitrogen suitable for soil conditioner applications (Mohamady Ghobashy, 2020). Polysaccharides are commonly used for the preparation of hydrogels considering the presence of hydrophilic functional groups that can absorb water and are easy to modify by grafting copolymerization reactions or crosslinking using chemical and physical methods (Hidangmayum et al., 2019; Sharif et al., 2018). Michalik and Wandzik (2020) used chitosan-based hydrogels due to their antimicrobial properties and availability as soil conditioner materials to deliver and control the release of active ingredients to soils without contaminating them.

Biopolymers have also been utilized to treat soil contamination. Arabyarmohammadi et al. (2018) developed chitosan-based nanocomposites incorporated with biochar and nanoclay to simultaneously immobilize Cu, Pb, and Zn metal ions within contaminated soil and water. Soil contamination can adversely affect the agricultural industry by reducing soil quality for crops. Biochar was chosen, as it can efficiently improve soil quality with its high stability and resistance to degradation in the soil system, while nanoclay was used to improve the physical properties of the chitosan biopolymer. The study found that chitosan/clay/biochar bionanocomposites were able to immobilize heavy metals, making the bionanocomposites an efficient metal sorbent in mine-impacted acidic water and soils.

5.5.4 ORGANIC FERTILIZERS

The increasing productivity of field crops, along with the use of high-quality seed materials, reflects the special role of fertilizer. Optimum fertilizer application can help increase the utilization of nutrients by cultivated plants, thus increasing the profitability of the entire plant industry. The applications of biopolymers in agricultural production in the fertilizer systems of cultivated plants have been widely reported. For example, the inclusion of a starch biopolymer compound in a system of mineral nutrition of plants with an appropriate fertilizer mixture can have a positive impact in the modern plant industry. According to Bamatov et al. (2019), only a 5% inclusion of starch biopolymer in the mineral system can significantly increase the coefficient of realization of the genetic potential of winter wheat varieties to 11–18%. The concentration of mineral nitrogen and mobile compounds including phosphorus and potassium were observed to increase by 29.2%, 25%, and 17.5%, respectively.

To date, fertilizer manufacturers have concentrated to produce eco-friendly slow-release (SRF) and controlled-release fertilizers (CRFs). SRFs are molecules with lower solubility than traditional fertilizers, whereas CRFs depend only on diffusion through coatings and not directly on biodegradation and thus are more efficient at controlling the release of nutrients. The main advantages of slow-release and controlled-released fertilizers include (1) extending the durability of fertilizers over time, (2) lowering the amount of fertilizer applications, (3) reducing the cost of typical covert application of traditional fertilizers, and (4) reducing environmental pollution by limiting the applications of fertilizers released in soil (Gil-Ortiz et al., 2020).

Studies on controlled release fertilizer have been widely reported. A study was done on controlled-release urea (CRU) coated with modified starch and double-coated with geopolymer with the aim to enhance the release characteristics and prolong nitrogen release compared to conventional products. According to Azeem et al. (2020), fluidizing, air temperature, and spray rate play important roles in determining the release characteristic of the CRU. The longevity and kinetics of CRU depend on the film thickness and the uniformity of the coating. The optimum process conditions also affect the scale production of CRU when using a viscous material like starch.

Another example of biopolymer utilization in a controlled-released fertilizer was investigated by Gil-Ortiz et al. (2020) using two-lignin coated fertilizers enriched with humic substances and seaweed extract, denoted CFRa and CRFb, respectively. These advanced controlled-release fertilizers were identified as significantly beneficial for wheat crop management, enabling simplified single applications as a basal dressing, streamlining crop handling. Application of these CRFs limits the amount of nitrogen (N) in formulations, leading to a reduction of cost; minimizes N losses; and reduces common contamination problems that usually occur when applying traditional fertilizers.

Biopolymers in the form of hydrogels are also applied as a coating for environmentally friendly fertilizer to improve soil water retention capacity. Slow-release fertilizer hydrogels are another type of agricultural material that is able to regulate water and nutrients in one system. These hydrogels possess the combined abilities of superabsorbent hydrogels and fertilizers that are able to improve the soil quality and increase fertilizer efficiency (Chen et al., 2018; Guilherme et al., 2015; Ramli, 2019).

5.5.5 PESTICIDES

Chemical pesticides are applied to control plant pathogens; however, the continuous application of these compounds has two major drawbacks: concern of the public regarding contamination of perishable agricultural products with pesticide residue and proliferation of resistance in the pest population (Badawy & Rabea, 2011). Therefore, there is a growing emphasis on environmentally friendly products to overcome the harms of synthetic pesticides by using nonchemical treatments for pest control. Among these strategies, the application of biopolymers such as chitosan as a safe alternative to hazardous pesticides with negligible risk to human health and the environment has been reported.

Owing to the biodegradability, non-toxicity, and antimicrobial properties of chitosan, this biopolymer is widely used as an antimicrobial agent either alone or blended with other biopolymers. Chitosan and its derivatives also contain plant

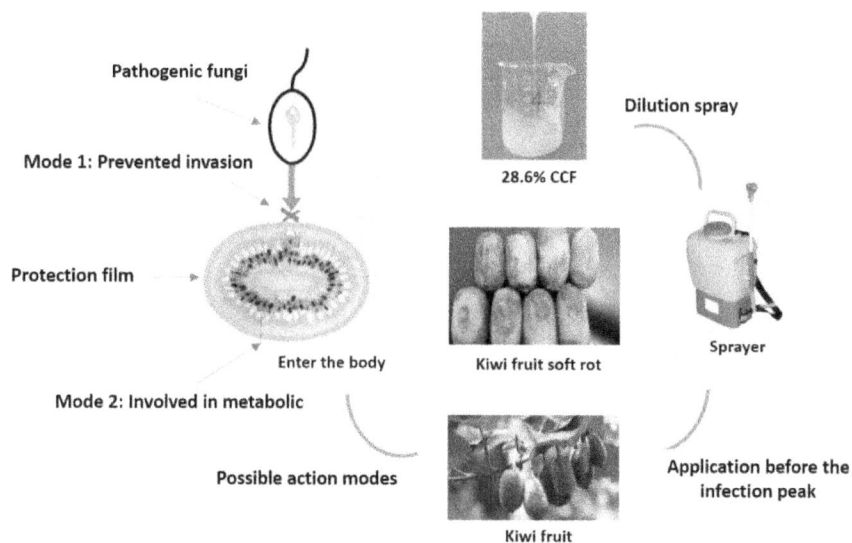

FIGURE 5.10 The mechanism of CCF application before pathogen infection for control of soft rot in kiwi fruit. Reproduced from Zhang et al. (2020).

defense-eliciting functions, which enables them to be applied as a useful pesticide in the control of plant disease(Xia et al., 2011). Numerous studies on the antimicrobial activity of chitosan against pathogens and pests in crops have been reported (Badawy & Rabea, 2011; Rabea et al., 2009; Sharma et al., 2019; Xing et al., 2015; Zhang et al., 2020). For example, Zhang et al. (2020) used chitosan composites incorporating dextrin, ferulic acid, calcium, and auxiliaries, denoted CCF, and sprayed them on kiwi fruit to prevent pathogen infections and effectively control postharvest soft rot in kiwi fruit (Figure 5.10). It was found that CCF had excellent control against soft rot, which might be associated with its film-forming property and antifungal activity, which enable it to prevent infection and induce plant defense mechanisms. Also, concentrations of 0.71–1.42 g/L CCF were optimal for field applications to overcome the onset of disease symptoms in plants caused by pathogen infections of *B. dothidea* and *Phomopsis* sp.

Additionally, nano-enabled agrochemicals, referred to as agronanochemicals, highlight the application of biopolymers, especially in the development of integrated management of pests and disease in crops. Agronanochemicals can help overcome environmental contamination issues regarding excessive usage of conventional agrochemicals by lowering the toxicity of agrochemicals; enhancing agrochemical uptake; improving solubility and stability; and reducing volatilization, leaching, and run-off of agrochemicals that can affect health and the environment (Duhan et al., 2017). As presented in Figure 5.11, chitosan nanoparticles can be produced via many techniques and used for various applications. The active ingredients loaded or encapsulated in chitosan nanoparticles enable potent biocidic, non-toxic, and biocompatible chitosan to be applied as a protection wall, protecting plants from the

Strategies for production of chitosan nanoparticles

Application of chitosan nanoparticles in agriculture

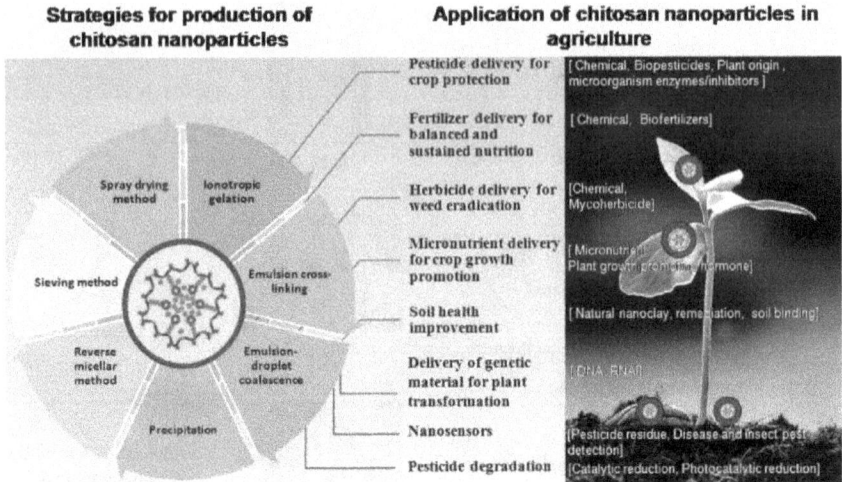

Pesticide delivery for crop protection [Chemical, Biopesticides, Plant origin, microorganism enzymes/inhibitors]

Fertilizer delivery for balanced and sustained nutrition [Chemical, Biofertilizers]

Herbicide delivery for weed eradication [Chemical, Mycoherbicide]

Micronutrient delivery for crop growth promotion [Micronutrient, Plant growth promoting hormone]

Soil health improvement [Natural nanoclay, remediation, soil binding]

Delivery of genetic material for plant transformation [DNA, RNA]

Nanosensors [Pesticide residue, Disease and insect pest detection]

Pesticide degradation [Catalytic reduction, Photocatalytic reduction]

Spray drying method / Ionotropic gelation / Sieving method / Emulsion cross-linking / Reverse micellar method / Emulsion-droplet coalescence / Precipitation

FIGURE 5.11   Strategies of chitosan nanoparticle production and their application in agriculture. Reproduced from Kashyap et al. (2015).

toxic effect of agrochemicals they contain (Maluin & Hussein, 2020; Michalik & Wandzik, 2020; Sharif et al., 2018).

5.6 PACKAGING

Polymer plastics or synthetic plastics are preferred for use in the packaging sector due to their combination of superior properties such as transparency, strength ability, flexibility, thermal performance, permeability, and simple sterilization methods (Luzi et al., 2019; Tang et al., 2012). However, despite plastic contributions in the packaging sector, environmental effects related to plastic waste are becoming a global concern. It is estimated that between 2030 and 2050, plastic waste based on the packaging sector alone will increase two- or three-fold.

Research efforts are increasingly focusing on developing biodegradable packaging materials sourced from renewable resources, with a primary goal of reducing environmental impacts by seeking alternatives to petroleum-based plastics (Tang et al., 2012). Their sustainability and greener image with improved technical properties are additional factors for them to be applied in the packaging industry. Biopolymers are commonly used for food and non-food packaging in the form of coatings, packaging materials, and encapsulation matrices for food and non-food products. The application of biopolymers provides a unique solution to enhance product safety and shelf life while reducing the overall carbon footprint related to packaging (Mohan et al., 2016). Big brands like Danone (Actimel, Activia, Volvic), Coca Cola (plant bottler), PepsiCo, Heinz, Tetra Pak, and L'Occitane have already introduced bioplastic packaging to help in reducing non-biodegradable plastic consumption (Reichert et al., 2020).

The commercially viable materials in biobased packaging include biodegradable polyesters and thermoplastics like starch, PLA, PHA, and other biopolymers that are able to be processed by conventional equipment. Starch and PLA biopolymers are the most attractive types of biodegradable polymers because of their properties and availability. PLA, for example, has attracted particular interest in biodegradable packaging due to its excellent transparency and relatively good water resistance (Mohan et al., 2016). Market research analysis forecast an increase in biopolymers applied in packaging consumed in Europe from a total of 1743.9 million m^2 in 2016 to 2427.1 million m^2 in 2021 (Reichert et al., 2020). However, the inherent rigidity and difficulty of processing with conventional equipment are the main drawbacks of biobased materials. Also, the hydrophilic nature of most biopolymers affects their application as high-end products. The poor mechanical and water barrier properties limit their industrial use. Therefore, consistent research has been done to develop techniques on improving mechanical properties and reducing hydrophilicity and thus increase the water vapor barrier properties of these biobased packaging materials.

Developing green and eco-friendly polymeric blends with improved properties and advancement can overcome these limitations. Blending polymers can enhance their biodegradability, changing their degradation rates and modulating the cost of the obtained materials. Due to the susceptibility of biobased polymer to moisture and low properties in water barrier and gas diffusion, the effort of blending biopolymers with synthetic polymers shows great potential in solving some of the problems. Hybrid blends in commercial packaging play a crucial role in imparting the necessary properties and functionality to a wide range of materials. While these combinations may not be entirely environmentally friendly, they have the potential to strike a balance between environmental, economic, and social considerations. It is essential to assess the life cycle impact of these materials and make informed choices by selecting compatible and more sustainable alternatives (Attaran et al., 2017; Luzi et al., 2019). Luzi et al. (2019) listed commercial examples of hybrid biopolymer blends with conventional plastics (e.g. low-density polyethylene (LLDPE), polypropylene (PP), high-impact polystyrene (HIPS)) that are already available on the market, particularly using starch biopolymers. Big companies have commercially applied biopolymer blend products with conventional polymers, including; Careplast, Teknor Apex, BIOP, Biograde, Biome Bioplastic, and Cardia Bioplastic.

Recently, a new class of materials, bio-nanocomposites, have drawn interest and opened up new possibilities for the food industry. Packaging materials based on biobased polymer nanocomposites derived from polysaccharides and lipid- or protein-based matrices possess good mechanical, thermal, biodegradable, chemical resistance, antimicrobial, and gas barrier properties (for oxygen, moisture, flavors, and lipids) with the aid of nanofillers (Kumar et al., 2017; Sharma et al., 2020).

5.6.1 Food Packaging

The food packaging industry represents one of the most rapidly expanding segments within the realm of plastic packaging. Approximately 40% of all plastic products are dedicated to food packaging. This stems from the necessity to employ robust materials that ensure food safety by preventing contamination while also maintaining freshness and durability for safe delivery to customers. According to Attaran et al. (2017), the main aim of food packaging is to protect food from the penetration

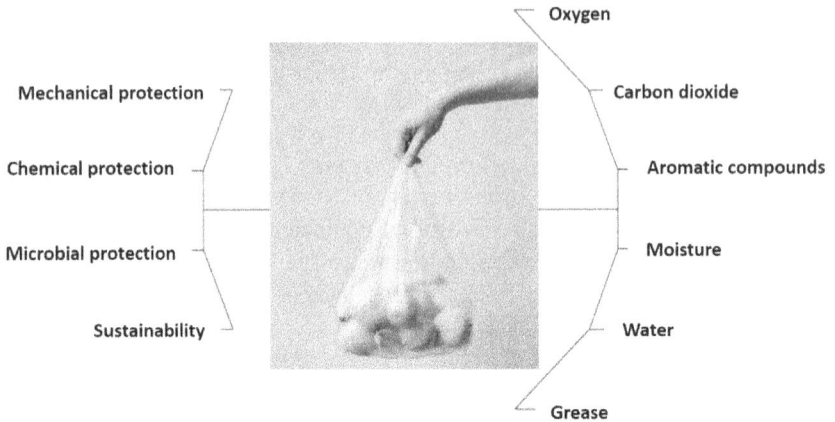

FIGURE 5.12 Desirable properties for food packaging.

of harmful and dangerous substances such as oxygen, moisture, microbes, and light and extend shelf life during storage and distribution. Other functions related to food packaging are product attraction, recyclability, sustainability, and disposability. Figure 5.12 summarizes the main desirable properties for food packaging.

The application of biopolymers in food packaging takes a variety of forms, such as films, coatings, cups, bottles, sheets, trays, tubs, and others. Several companies have already implemented the application of biobased materials in their plastic products. BIOME Bioplastic (United Kingdom) is one example of plastic production companies that practice the application of 100% biodegradable and compostable natural materials (plant starches from potato and corn) in their food packaging products that can be processed at high temperature, including via injection molding, sheet extrusion, and thermoforming ("BIOME Bioplastic," n.d.).

Food packaging in the form of a film is highly recommended with biodegradable coatings. Biodegradable films are thin layers of materials that are suitable for use with food and can be placed on/between food parts (Jafarzadeh et al., 2020). Biopolymers derived from plant by-products like polysaccharides (starch, chitosan, cellulose, carrageenan), proteins (wheat gluten whey protein, soy protein, fish protein, and pea protein), or a combination of both biopolymers are the most-used biopolymers in film preparation (Hasan et al., 2019; Jafarzadeh et al., 2020). Giro et al. (2020) used bacterial exopolysaccharides (xanthan) as a food film coating for fish and meat products. Their study revealed the application of biobased film showed a reduction in oxygen access, completely inhibited the growth of aerobic microorganisms, and also extended the shelf life of meat products.

To protect the freshness, safety, and quality of food from contamination by foodborne pathogenic microorganisms, antimicrobial and antioxidant packaging has been introduced. With the incorporation of microbial substances in food contact materials, the growth of microorganisms can be inhibited, which leads to increased shelf life of food products (Sharma et al., 2020). Generally, antimicrobial agents can be divided into two main groups: organic and inorganic. Biopolymers, such as

chitosan, chitin, PLA, cellulose, and starch, are organic materials commonly reinforced with antimicrobial agents in food packaging. These materials exhibit high
sensitivity to harsh processing conditions, including elevated temperatures and pres-
sures (Hosseinnejad & Jafari, 2016; Huang et al., 2019). Chitosan, in particular,
stands out among these biopolymers due to its inherent antimicrobial properties,
making it a popular choice for antimicrobial food packaging (Sharma et al., 2020).

To enhance the mechanical and moisture sensitivity properties of the packaging
film, the blending of different biopolymers has been studied. Al-Hassan and Norziah
(2012) used blended film with a combination of different ratios of sago starch and fish
gelatin, plasticized with glycerol or sorbitol. The starch/gelatin ratios of 3:1, 4:1, and
5:1 were observed to have good flexibility. Tulamandi et al. (2016) and Li et al. (2017)
used soy protein and blended it with gelatin and chitosan, respectively, to enhance
the mechanical properties and water resistance of soy-based films. Meanwhile, in
separate studies, soy protein-based film was improved by the addition of zein (N.
Wang et al., 2016) and cellulose nanocrystals (NCCs) (Yu et al., 2018), giving rise to
more stretchable and mechanically resistant film.

Meanwhile, microbial biopolymer-based nanocomposites rely on the type of
nanofiller, with microbial properties incorporated in the nanocomposite system.
Meira et al. (2016) applied halloysite and nisin in a starch biopolymer to improve the
mechanical properties and observed microbial activity against *L. monocytogenesm*,
Clostridium perfringens, and *S. aureus* in packaging meant for soft cheese. Gorrasi
and Bugatti (2016) used nanohybrid layered double hydroxide salicylate reinforced
in pectin biopolymer film for fresh apricots and observed improved elongation at
breaks for the pectin film, with improved water vapor barrier and extended shelf life.

5.6.2 Non-Food Packaging

Generally, the packaging industry focuses on creating lighter products to reduce the
raw materials used, transportation cost, and waste volume, whereas consumers and
producers focus on recyclable, environmentally friendly, and non-fossil based packaging solutions (Helanto et al., 2019). The application of biopolymer-based packaging materials other than food packaging includes plastic bags, wrapping plastics, and
coating plastics for paperboard packaging. These applications commonly call for
necessary mechanical strength and required barrier performance with biodegradability value for each product.

The common problem of paperboard packaging is the barrier resistance and wettability of papers. Petroleum-based plastics such as polyethylene, waxes, and flour
derivatives are commonly used for coating. However, due to unfavorable properties
towards the environment, alternatives are used. Biopolymers like polysaccharides,
proteins, lipids, and polyester can be used as a new pathway for fully biobased paper
coatings. Blending these biopolymers is encouraged to get the favorable barrier resistance needed for packaging materials (Rastogi & Samyn, 2015).

According to Rastogi and Samyn (2015), a variety of biopolymers have been
applied as coatings for paperboard, encompassing polysaccharides such as starch
and cellulose derivatives, chitosan, and alginates, as well as proteins like casein,
whey, collagen, soy, and gluten, in addition to lipids, including beeswax, free fatty

acids, and polyester (PHA, PLA, etc.). Nevertheless, many biobased polymers rely on ambient humidity, which can impact their barrier properties against substances like water, water vapor, air, oil, and grease, as well as their mechanical characteristics. Consequently, it is advisable to blend biobased polymers with other biopolymer matrices and introduce micro- to nanoscale fillers to enhance their performance, particularly with regard to hydrophobicity and processing.

For example, Kansal et al. (2020) successfully invented a double-layer coating system on Kraft paper using starch and zein biopolymer to replace the application of commonly used polyfluorinated (PFAS) silicone oil and plastic-coated paper. The mechanism of the coatings is shown in Figure 5.13. The characterization of the double-layer coating was done with SEM analysis. Meanwhile, the mechanical and thermal stability of the coating was maintained, which has the potential to be commercially applied in packaging materials. Meanwhile Song et al. (2014) used modified nanocellulose (CNF) added to a PLA-based composite film. The bionanocomposite was subsequently employed in the cast-coating process to coat the paper surface, resulting in a noteworthy reduction in the paper's water vapor transmission rate (WVTR). This development suggests the potential use of these materials in environmentally friendly packaging solutions.

Despite its barrier properties, the mechanical, thermal, and biodegradability properties of plastic packaging are also important in applications of biobased plastic products. Despite being derived from naturally occurring long-chain molecules, biobased materials also differ from synthetic polymers, as they can be decomposed by bacteria, fungi, or other organisms into natural metabolic products (Wojnowska-Baryła et al., 2020). There are three major commercialized biobased polymers: biobased polyethylene terephthalate (beverage bottles), polylactic acid (single-use cups, single-use cutlery, packaging films), and starch plastics (clips, mulch film, and carrier bags). Several companies have applied packaging made from biopolymers. For example, Novamont S.p.A. from Italy used material composed of corn starch and oil known as MATER BI for the production of films, bags, trays, cups, additives, foamed packaging material, extruded material, and injection-molded materials ("MATER BI, Novamont," n.d.). Meanwhile, PLA has been utilized by NatureWorks LLC (USA), which converted Ingeo PLA (tailored PLA for specific performance) into market-available products such as cards, cartons, and several types of non-food packaging (food serviceware, blister packaging, etc.; "NatureWorks," n.d.).

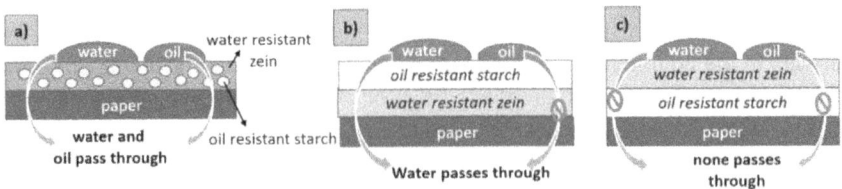

FIGURE 5.13 Biopolymer coatings on the paper substrate. (a) A blend of zein and starch allows oil and water to pass through to the paper, (b) a dual-layer coating of top-layer starch and bottom-layer zein permits the passage of water, and (c) a dual-layer coating of top-layer zein and bottom-layer starch offers a barrier against both oil and water (Kansal et al., 2020).

Also, biopolymer-reinforced nanofiller composite products have been widely studied for packaging applications. Studies have focused on improving current biopolymers with the addition of organic and inorganic nanoscale filler. For instance, PLA, a material extensively employed in the production of biopolymer products such as agricultural film, plastic bags, food containers, garbage bags, and food packaging, typically exhibits high brittleness, thereby restricting its utility. To enhance its usability while preserving its biodegradable attributes, it is crucial to incorporate fillers and make necessary modifications to PLA in order to ensure the product's quality remains intact. Qian et al. (2018) used bamboo cellulose nanowhiskers treated with silane to toughen the mechanical and thermal properties of the PLA matrix. Meanwhile, Elhussieny et al. (2020) prepared a composite film from chitosan extracted from raw shrimp shells and added nano rice straw fibers with biodegradable properties. The authors found that the mechanical, thermal, and biological properties of chitosan biocomposites increased with the addition of rice straw fibers, and it was suitable to be used as packaging bags with biodegradable properties.

5.7   COSMETICS

Cosmetics are substances or preparations designed for application to the external surfaces of the human body, including mucous membranes in the oral cavity and teeth. These products are meant to be rubbed, poured, sprinkled, sprayed, introduced, or applied onto the human body or its parts with the purpose of cleansing, enhancing beauty, promoting attractiveness, or altering one's appearance. In essence, cosmetics serve the following functions: modifying body odors, transforming physical appearance, purifying the skin and hair while maintaining their health, and providing fragrance and skin protection. Nowadays, there are various types of cosmetics available on the market, including skincare, haircare, lip care, and eye care (Manikrao Donglikar & Laxman Deore, 2016; Mitura et al., 2020).

5.7.1   Skin Care

The skin is the largest organ of the body, making up 16% of human body mass, and it is the main part that has cosmetics applied to it. The application of cosmetics provides physical and chemical protection from various environmental influences. The knowledge of the anatomy of the skin and the reaction of each product applied is very important. According to Mitura et al. (2020), the skin plays a pivotal role in shaping external experiences, giving rise to a distinct and recognizable image that can be easily identified by others. The skin is composed of three main layers: (1) the epidermis, which is the top of the skin; (2) the dermis, the second layer of the skin, which is much thicker; and (3) subcutaneous fat as the bottom layer. The epidermis is the outermost layer of the skin with the role of protecting the organs against invasive elements such as microorganisms, harmful chemicals, radiation, and electric flow (Dias-Ferreira et al., 2020). Sebum has an important function on the skin, including as moisturizer and enhancing barrier properties, and it might act as an antifungal and antibacterial agent. If the skin barrier is damaged, reconditioning is required, which is when moisturizers need to function temporarily until the skin and barrier are re-established. Moisturizers assist skin repair by creating an environment suitable for regeneration by reducing the loss of water and creating a barrier on the skin (Mitura et al., 2020).

In cosmetic preparation, several materials have been utilized and are commercially available. Those employed in cosmetic formulations are selected according to regulations, which may vary by region or/and country. Among these materials, the application of synthetic polymers and biopolymers has been widely reported (Mitura et al., 2020). Polymers have played a significant role in cosmetic formulations. They are used in the form of film-formers in hair cosmetics, mascara, nail enamels, and transfer-resistant color cosmetics. Polymers are also utilized as thickeners and rheology modifiers in emulsions, gels, and hair colorants and emulsifiers in lotions, sunscreen, hair color and conditioner, moisturizers, emollients, dispersants, and waterproofers. However, today, polymers based on natural materials are preferred, as they can offer biocompatibility and biodegradability.

A variety of biopolymers such as cellulose exist widely in nature. Some biopolymers contain the same ingredients as the main components of skin and hair, which can naturally be used as regenerative ingredients. Biopolymers such as collagen, chitosan, elastin, keratin, and silk fibroin can be obtained in nature and are available as waste from food products such as feathers, animal hair, animal skin, crustacean shells, fish scales, and bones (Figure 5.14). These biopolymers have been widely

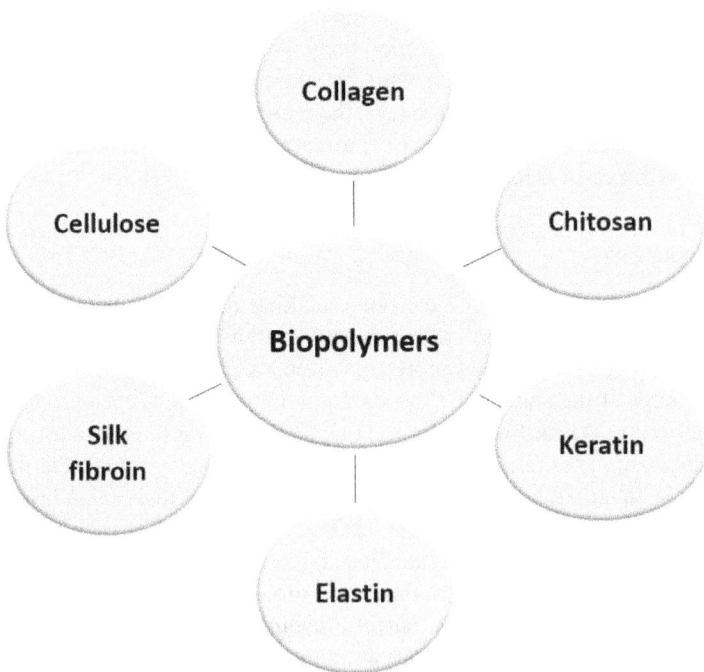

FIGURE 5.14 Biopolymers extracted from waste of food products like feathers, animal hair, animal skin, crustacean shells, fish scales, and bones used for cosmetic applications.

applied in cosmetic formulations as moisturizers and thickening agents (Mitura et al., 2020).

Cosmetic products also can be derived from combinations of two or more biopolymers in the form of hydrogels. In cosmetic applications, hydrogels are usually applied in skincare, hair care, and oral care products. Products from hydrogels can be prepared from a variety of biopolymers, including collagen, gelatin, hyaluronic acid, alginate, chitosan, xanthan gum, pectin, starch, cellulose, and its derivatives (Mitura et al., 2020). Though synthetic polymers are often used due to their reproducibility, biopolymers can offer better biocompatibility properties through hydrogel products.

5.7.2 Face Masks/Sheets

Among facial cosmetics, a facial mask is the most commonly used product for skin rejuvenation. Facial masks are formulated with a set of active ingredients, including various types of vitamins (e.g. A, B2, B3, B7, and C), proteins (e.g. collagen and elastin), minerals (e.g. copper, iodine, and zinc), antioxidants (e.g. polyphenols and tocopherols), and other nutraceuticals (e.g. carotenoids, coenzyme Q10, and green tea), which are claimed to improve skin health and appearance. The method of application of facial masks requires the end product to be semisolid, for example, an emulsion, cream, hydrogel, or sheet without an oily texture. Face masks or sheets combine the advantages of being affordable and easy to apply; however, they contain substances like parabens, phthalate esters, and fragrances that might induce allergic reactions (Nilforoushzadeh et al., 2018).

In general, there are many types of facial masks based on their distinct texture, polymeric composition, shape, and structure, including hydrogel masks, rinse-off masks, peel-off masks, bio-cellulose masks, foil sheet masks, knit cotton masks, pulp masks, ampoule sheet masks, and bubble sheet masks, as shown in Figure 5.15. The sheet mask is a common mask available on the market, and the rinse-off masks include several types such as moisturizing, cleansing, toning, exfoliating, and waxy and mud masks. Other than that, peel-off masks are also available in the form of a film that can easily be peeled off. Various materials such as herbal soap, moisturizers, plasticizers, fragrances, and preservatives can be embedded in the mask. Another typical mask is a hydrogel mask that is usually used for sensitive skin, with a cooling and soothing effect (Nilforoushzadeh et al., 2018).

A peel-off mask is one of the cosmetic types of gel facial skincare. It will form a layer of elastic transparent film which can be removed after drying without rinsing. The physical quality of the peel-off mask is influenced by the composition of the ingredients used, especially PVA. PVA is a common material used in peel-off masks, as it has adhesive properties that allow it to have the peel-off effect and be easily peeled after drying (Birck et al., 2014). The application of biopolymers in gel peel-off masks was studied by Ridwanto et al. (2019) using a combination of chitosan, xanthan gum, and carboxymethyl cellulose (CMC). The combination of chitosan/xanthan gum showed the best viscosity effect for a gel peel-off mask compared to other formulas.

Facial masks produced from hydrogels can be prepared by blending two or more polymers or/and biopolymers. Hydrogel masks are usually used for sensitive skin,

Hydrogel Mask	Rinse-off mask	Peel-off mask
Bio-cellulose mask	Foil -Sheet mask	Ampoule sheet maks

Bubble sheet mask	Cotton pulp maks

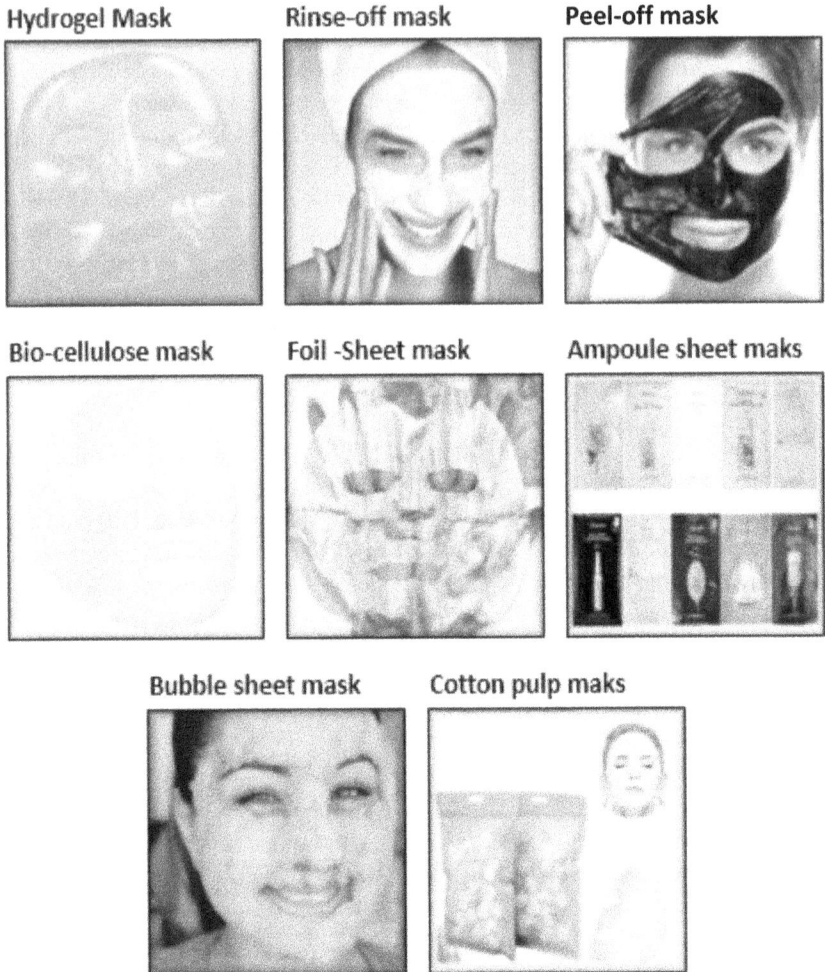

FIGURE 5.15   Different types of beauty masks.

with cooling and soothing effects. Hydrogel masks also promote a hydration effect for the skin, restore elasticity, and promote anti-ageing performance using formulations including biopolymers like glycerine, collagen, hyaluronic acid, ceramides, and essential oils (Dias-Ferreira et al., 2020; Mitura et al., 2020; Nilforoushzadeh et al., 2018). Many products and studies have shown the variations of the functions of biopolymers in the form of hydrogel masks (Dias-Ferreira et al., 2020; Mitura et al., 2020; A. Sionkowska, Lewandowska et al., 2014; Alina Sionkowska, 2015). The invention of a cosmetic facial mask using *Thanaka* heartwood (native to Myanmar) cellulose hydrogel embedded in the cotton facial mask was done by Cho and Kobayashi (2021) for facial treatment, dehydration of the skin, and maintenance of the freshness of the skin.

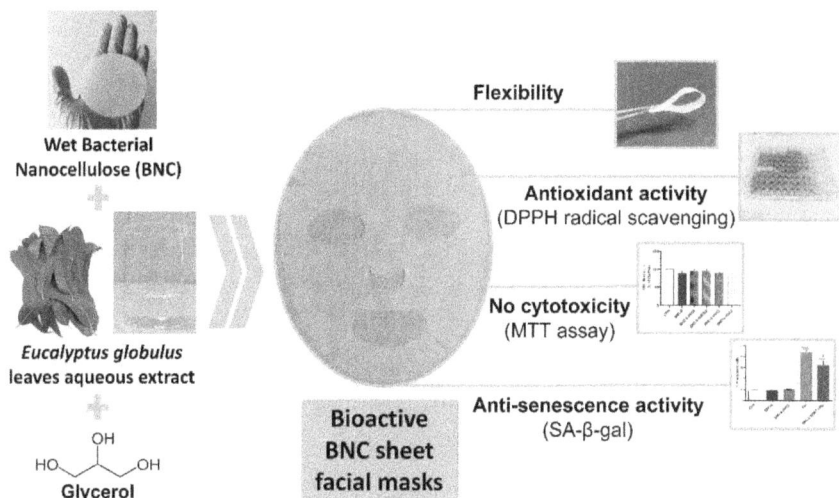

FIGURE 5.16 Prototype of a bioactive mask from bacterial nanocellulose membranes enriched with *Eucalyptus globulus Labill* leaf aqueous extract for anti-aging. Reproduced from Almeida et al. (2022).

Applications of nanocellulose as an ingredient in facial masks have also been reported widely. Amorim et al. (2020) studied a facial mask made from bacterial cellulose incorporated with propolis extract as a hydrating and anti-inflammatory sheet mask for skin prone to acne and inflammation. The polymer blend, called BioMask, was presented as an ideal film with beneficial characteristics to be used in the cosmeceutical industry. Meanwhile, in a separate study, a silk sericin-releasing bacterial nanocellulose gel was used as a bioactive material in a facial mask and exhibited proper biological features for facial treatment (Aramwit & Bang, 2014). Figure 5.16 shows the prototype of a bioactive mask from bacterial nanocellulose membranes enriched with *Eucalyptus globulus Labill* leaf aqueous extract for anti-aging applications.

5.7.3 UV Sunscreens

Sunscreen is a cosmetic product that protects skin from damage by sunlight radiation. The function of sunscreen is to either absorb or reflect radiation to protect skin from the harmful effects of radiation. Oral sunscreen products and their components are now available on the market as a means of internal protection against skin damage. Today, a single sunscreen product can offer a range of health benefits beyond mere cosmetic enhancement (Manikrao Donglikar & Laxman Deore, 2016).

Sun exposure is a natural phenomenon in human life. Protection against chronic sunlight exposure needs to considered, including by using shaded spaces, textile hats, and sunscreen. Ultraviolet rays are known as prominent radiation hitting Earth. UV radiation is fragmented into three types, UVA (320–400 nm), UVB (280–320 nm), and UVC (100–280 nm). Exposure to UV radiation may induce acute or chronic injuries. Upon exposure, the cell experiences multiple complex reactions, which lead to apparent changes in the physiology of the body such as photokeratitis (snow blindness), tanning, erythema (sunburn), increased melanogenesis, and immunosuppression (Cerqueira-Coutinho et al., 2015; Dias-Ferreira et al., 2020).

Sunscreen plays a crucial role in reducing and controlling skin exposure to sunlight and UV rays. Sunscreen, as a cosmetic product, can be either organic or inorganic in nature. The sun protection factor (SPF) is determined by comparing the time of UVB exposure required to induce erythema with sunscreen to the time of UVB exposure needed to induce erythema on unprotected skin. Also, to be considered water resistant, sunscreen must last 40–80 minutes. The Food and Drug Administration (FDA) categorized three distinct groups of sunscreen for legal authorization: UVA filters, UVB filters, and physical blockers. The efficiency of sunscreen in protecting skin from UV radiation is also associated with the number of products applied to the skin and the capacity of the formulation to make a uniform film. Sunscreen formulation must exhibit pseudo-plastic behavior, which is a shear-thinning profile characterized by a decrease of viscosity with increasing shear strain, which is typical of non-Newtonian fluids (Dias-Ferreira et al., 2020).

In general, biopolymers can be used in sunscreen in three separate ways (Figure 5.18): (1) as a thickening agent in a sunscreen gel, (2) in the aqueous phase of emulsified protectors, and (3) in the UV filter delivery system. The biopolymer can be found as a hydrophilic thickener. Natural materials like polyphenols (flavonoids, tannins), carotenoids, anthocyanidins, a few vitamins, fixed oils, volatile oils from vegetables, fruits, medicinal plants parts (leaves, flowers, fruits, berries), algae, and lichens are more effective than synthetic chemicals due to their long-term beneficial effects, especially against free radical–generated skin damage, along with UV ray blocking (Manikrao Donglikar & Laxman Deore, 2016).

Biopolymers have also been studied as skin protectors in the form of gels. Morsy et al. (2017) studied an antibacterial sunscreen gel using a combination of chitosan and hydroxyapatite. Hydroxyapatite is used as an inorganic sunscreen agent due to its excellent opacity and reflectivity. On the other hand, the utilization of chitosan–hydroxyapatite gel has emerged as a novel approach for the development of antibacterial sunscreens. Also, chitosan has been studied and could improve the stability of sunscreen formulation for at least six months, remaining photostable when irradiated in a solar simulator and effective as a UV blocker (Cerqueira-Coutinho et al., 2015; Ma et al., 2010; Wang et al., 2009). Chitosan was also found to promote retention of the formulation in the epidermis, which helps to increase the safety of the sunscreen formulation. In a separate study, Peres et al. (2017) produced a UV filter incorporated with hydrolyzed collagen and found that the protein promoted a better distribution of the sunscreen on the skin by forming a homogenous film with enhanced formulation sensory aspects. Meanwhile, xanthan gum also was used as a thickening agent in an experiment done by Wang et al. (2014) to prepare emulsion-based lotions. The addition of xanthan gum helps to show percutaneous permeation on insect repellent using *N, N*-diethyl-*m*-toluamide and a typical UV filter using oxybenzone.

The development of sunscreen has garnered growing interest in incorporating micro/nanoparticles into sunscreen ingredients to enhance UV-blocking performance. For example, Suh et al. (2019) used polylactic branched hyper polyglycerol

(PLA-HPG) to demonstrate a delivery platform for bioadhesive sunscreen by encapsulating small molecules of UVB filter padimate O. The composed materials showed enhanced UV protection comparable to FDA-approved sunscreen. In a separate discussion, biopolymer nanocapsules were developed to encapsulate a sunscreen agent (Alvarez-Román et al., 2001). This study found that the lipophilic sunscreen, Parsol MCX (OMC), was successfully encapsulated in poly-ε-caprolactone nanocapsules that enabled it to provide partial protection against UV-induced erythema in a manner significantly better than a conventional gel.

5.8 HAIRCARE PRODUCTS

Hair, like skin, is a complex biological system that performs a specific function. Hair can be strategically divided into two distinct parts: the hair follicle, deeply buried in the skin, and the visible hair fiber. Hair fiber consists of a protein called keratin and a small number of lipids (Figure 5.17). Many factors affect hair structure, including ageing, lengthening, and environmental factors such as pollution and sunlight. This modification also affects the properties of natural hair. This damage can negatively affect the hair, scalp, and even the consumer's health. Thus, appropriate cosmetic chemistry can support hair condition using appropriate chemical compounds to improve the function and repair damaged hair, as hair fibers cannot be restored to their original structure (Cruz et al., 2016; Mitura et al., 2020).

Hair product formulations usually consist of combinations of two or more types of biopolymers. Appropriate mixing of both types of biopolymers is very important in cosmetic formulations to impart improvement to the appearance and manageability of hair. A common application of biopolymer in hair products is conditioner. Conditioners reduce friction, detangle the hair, minimize frizz, improve combability, restore hydrophobicity, enhance shine, and increase smoothness and manageability. The mechanism by which conditioners work is to provide hair manageability by decreasing static electricity and reducing friction among hair fibers. Conditioners are also unable to seal the gaps that expose the cortex to environmental damage (Cruz et al., 2016).

Sionkowska et al. (Sionkowska, 2015; Sionkowska & Płanecka, 2013; Sionkowska, Płanecka et al., 2014) used chitosan/silk fibroin blends and collagen/keratin blends.

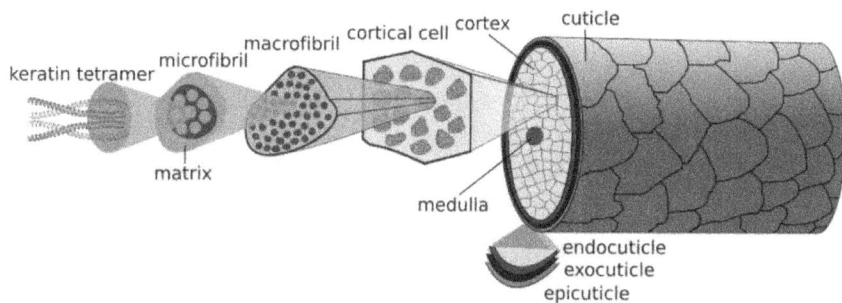

FIGURE 5.17 Cross-section of hair. Reproduced from Cruz et al. (2016).

Thin films were obtained via solvent evaporation of polymer blends of chitosan/silk fibroin, and polymeric blends of collagen/keratin biopolymers were prepared via the casting solution method. The surface properties of both films were found to have better hydrophobicity with the reduction of the polar component of surface free energy. The authors claimed film production from the biopolymer blends of chitosan/silk fibroin and collagen/keratin was suitable to apply in hair care products such as conditioners and sprays. By appropriate mixing of these biopolymers in one cosmetic formulation, one can get improvement in the appearance and manageability of hair.

5.9 BIOMEDICAL APPLICATIONS

Biopolymers have aroused great interest in biomedical applications, including tissue engineering, pharmaceutical carriers, and medical devices. The application of natural-based polymers is much preferred in most biomedical products based on their biocompatibility, biodegradability, and biochemical and biomechanical behavior. With the low cost, minimum effect on the environment, and low pollution, biopolymers with specific properties can be utilized to develop various kinds of biomedical products with interesting new properties (Chen et al., 2008; Indrani et al., 2017; Alina Sionkowska, 2015). Biopolymers have been used as implantables for a broad range of biomedical applications, including cardiology, cartilage, vasculature, bone, wound healing, drug delivery, and prosthetic dentistry (Lin et al., 2017). The application of biopolymers and their main applications in the medical field are summarized in Table 5.2 (Abdelhak, 2019).

Polymers play a significant role compared to metals and ceramic due to their flexible and adjustable chemical and physical design. Polymers are also able to have enhanced performance according to production techniques and the possibility to be tailored based on the requirements of the applications. As for biopolymers, the combination of knowledge of biomaterials and the medical field enabled the new development of methods and devices to improve human health such as bone and joint replacement, heart valves, and dental implants and also to repair damaged tissues

TABLE 5.2
Main Application of Biopolymers in the Medical Field

Biopolymer	Applications
Polyhydroxyalkanoates (PHA)	Suture, galenic, vascular implant, medical clothing accessories, osteosynthesis
Polyglycolides (PGA)	Suture, clip, staple, and adhesive
Polylactides (PLA)	Orthopedic fixation, artificial ligaments and tendons, tissue regeneration
Polyglactine (PLA-PGA)	Suture, orthopedic fixation, screw and pins, ligaments and tendons
Cellulose	Drug encapsulation, cell implantation
Poly-Lysine	Encapsulation of drugs, biosensor, and bactericide
Polyspartates	Medication encapsulation, suture, artificial skin

(Lin et al., 2017). Collagen and chitosan are biopolymers that have been widely used in biomedical applications. The combination of these biopolymers has been the focus of researchers, as these biopolymers are miscible in both solution and solid state, which enables them to produce new materials for potential biomedical applications (Alina Sionkowska, 2015).

5.9.1 IMPLANT AND SCAFFOLD TISSUE ENGINEERING

Tissue engineering is one part of the biomedical field that focuses on the improvement or regeneration of damaged tissue and organs that have lost their function. Lin et al. (2017) listed three general strategies that have been adopted by tissue engineering: (1) implantation of isolated cells or cell substitutes, (2) delivery of tissue-inducing substances, and (3) the use of tissue scaffolds, where cells are seeded onto a substrate to create an implantable tissue construct. The tissue engineering concept is shown in Figure 5.18. Tissue scaffolds are porous structures that provide mechanical support for cells and are responsible for the regeneration of neo-tissues. The scaffold must mimic the natural extracellular matrix (ECM) and create an environment similar to that of the ECM in the human body. In other words, it can be said that the biocompatibility of the biobased material is very important and it must be satisfied before it can be applied to the body. Other than that, the design of scaffolds also plays an important role in the end applications, including standard biocompatible properties with the host tissue structure and ability to mechanically support the neo-tissue regeneration process (Lin et al., 2017).

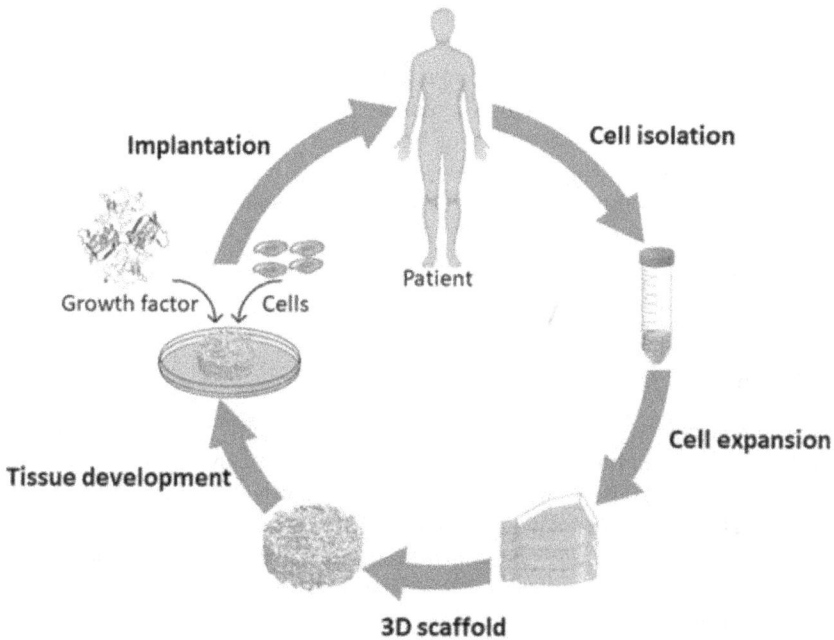

FIGURE 5.18 The concepts of tissue engineering. Reproduced from Asadian et al. (2020).

To date, natural-based polymers that have been studied to be used for 3D scaffolds are collagen, silk, keratin, elastin, chitosan, alginate, and so on (Lin et al., 2017; Sundar et al., 2020). Biopolymers based on natural materials are much recommended, as they provide important features like non-toxic behavior and the ability to mimic properties as cell tissues. For example, collagen is one of the common natural polymers that can be derived from fish, bovine, porcine, or chicken sources. Fish or marine collagen is widely used due to its superior properties and is usually used in burn; neoplastic; reconstructive, oral, peripheral nerve, and tendon surgeries; wound repair and healing; nerve regeneration; extrahepatic bile duct regeneration; and myocardial repair (Gao et al., 2011; Han et al., 2014; Li et al., 2012; Sundar et al., 2020; Yan et al., 2010).

Bone tissue engineering is one of the strategies using scaffold tissue that uses biodegradable polymeric matrices in combination with cells to provide mechanical support to the bone, at the same time promoting cell proliferation, differentiation, and tissue ingrowth. Aravamudhan et al. (2018) utilized cellulose acetate infused with Type I collagen to produce a natural polymer-based micro-nano structured matrix. They then combined this matrix with bone marrow stromal cells (BMSC) and assessed its performance in treating critical-sized bone defects. The authors concluded natural polymeric cellulose acetate-reinforced collagen performed better than commercial polylactic-co-glycolic acid (PLGA) of similar dimensions in inducing healing of critical-sized bone defects. The addition of BMSC enhanced bone and collagen formation two-fold on biopolymers as compared to PLGA.

Tissue engineering scaffolds also can be applied for organ and tissue transplantation. Naumenko et al. (2016) employed biopolymer blends of chitosan, gelatin, and agarose as the main materials and reinforced them with 3–6 wt.% clay nanotubes to form porous hydrogel nanocomposites. This hydrogel was designed as implantable 3D cell scaffolds to study the biocompatibility and biodegradability of the resulting scaffolds *in vitro* and *in vivo*. The results reported significant improvement in mechanical stability and wettability of biopolymer scaffolds without harming the cell in growth *in vitro*, while the *in vivo* study showed biocompatibility of the nanocomposite scaffolds in the organisms of rats with a slight inflammatory effect but no rejection of implants. Full restoration of blood supply was also observed after six weeks. These findings confirm the great potential of biopolymer-based nanocomposites to be applied as potential sustained drug delivery.

5.9.2 Wound Healing

Wound management is a significant healthcare challenge, particularly due to the intricacies of the healing process, which are often compounded by factors such as diabetes and infections. Wound healing is a multifaceted and constantly evolving process that relies on a cascade of biological reactions influenced by the nature of the injury. Wound dressings play a pivotal role in facilitating wound healing and minimizing scar formation. The major function of wound dressing is to protect the wound from microorganism deposition and dehydration, as shown in Figure 5.19 (Varaprasad et al., 2020). The applications of biopolymers as wound care materials are in their excellent properties of biocompatibility, ability to support cell growth,

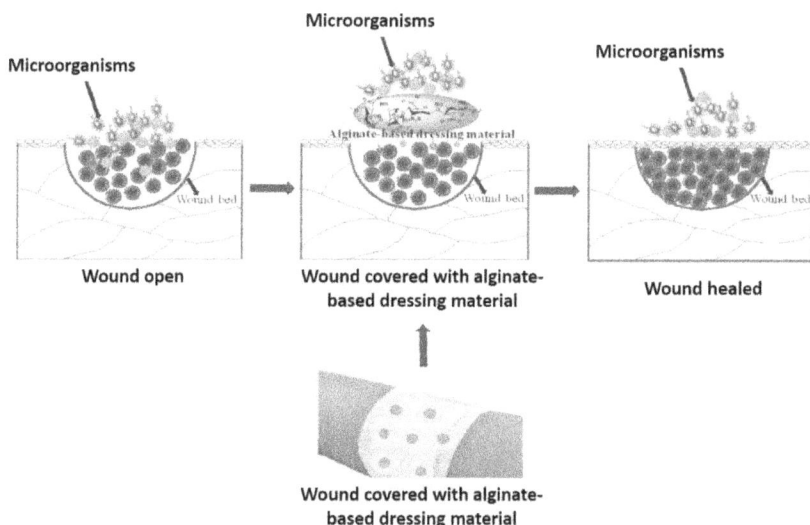

FIGURE 5.19 Schematic images of wound dressing function to protect human skin from wound infection bacteria (Varaprasad et al., 2020).

regenerative potential, biodegradability, and durability. Biopolymers have been widely used for healing process, including collagen, cellulose, chitosan, alginate, hyaluroma, fucoidan, and carrageenan, as they possess good antibacterial, anti-inflammation, proliferative, and other targeted actions for specific cell properties (Sahana & Rekha, 2018). Types of biopolymers, their biological roles, and types of wounds are listed in Table 5.3.

Wound dressing can be in any forms such as gauzes, gels, hydrogels, or hydrocolloids. It can be molded into hydrogels or scaffolds and blended with other polymers, which can impart good mechanical strength, biomimetic properties, and several other desired features needed for wound healing materials. Among them, hydrogels are the most promising, as they can provide a moist environment for the wound site, help in the removal of wound exudates, prevent infection, and give a suitable environment for tissue regeneration (Aswathy et al., 2020). There are various forms of hydrogel wound dressing available on the market, for example, the Neoheal hydrogel sheet that contains PEG, polyvinylpyrrolidone, and agar crosslinked via electron beams. These products contain 90% water and are able to treat ulcers, abrasions, burns, bedsores, and other chronic wounds. Hydrogel wound healing products are also available in the form of films, for example, Suprasorb G, manufactured by Lohmann & Rauscher Global. The hydrogel film is composed of a mixture of acrylic polymer, polyethylene, and phenoxyethanol, with 70% water content and is suitable to treat dry wounds, lower leg ulcers, pressure ulcers, and second-degree burns. Another product is hydrogel-impregnated gauze from DermaRite Industries. DermaGauze contains acrylate polymers and is suitable for acute and chronic partial- or full-thickness wounds. Finally, commercially available hydrogel wound dressings in gel form

TABLE 5.3

Characteristics of Biopolymers, Their Biological Role in the Healing Process and Types of Wounds

Biopolymer	Monomer Units and Linkage	Sources	Biological Role	Wound Types	Advantages	Disadvantages	References
Collagen	Amino acids linked by amide linkage	Bovine, porcine, fish (marine) collagen	• Induces fibroblast proliferation • Induces secretion of ECM components by fibroblast • Chemotic for macrophages	Bedsores, minor burns, foot ulcers, chronic wounds, low to heavy exudation wounds, surgical wounds	Cost-effective, high water holding capacity, high tensile strength	Cannot be used for patients allergic to birds, cattle, or swine Cannot be used for third-degree burns or dry wounds Does not have antibacterial activity	(Chattopadhyay & Raines, 2014; Gould, 2016; Marzec & Pietrucha, 2018; Naomi et al., 2020; Sun et al., 2018)
Cellulose	β-D-glucose linked by β-1–4, glycosidic linkage	Plant cell wall and bacteria	• Retention of moisture • Absorption of exudate	Burns, chronic wounds, plastic/reconstructive surgeries	Antibacterial	Cannot be used for second- and third-degree burns	(Ahn et al., 2018; El Fawal et al., 2018; Kucińska-Lipka et al., 2015; Mohamad et al., 2014; Pourali et al., 2018; Yuyu Qiu et al., 2016; Sulaeva et al., 2015; Wagenhäuser et al., 2016)
Alginic acid	β-D mannuronic acid and α-1,4-guluronic acid linkages	Brown algae	• Stimulated monocytes • Induces fibroblast proliferation and migration	Moderately to heavily draining wounds, surgical incisions, sinus tract, cavity wounds, infected wounds	Hemostatic, increases angiogenesis, enhances cell migration	Cannot be used for eschar and dry wounds or third-degree burns	(Chiu et al., 2008; Gao et al., 2019; Lozeau et al., 2020; Stubbe et al., 2019; M. Zhang & Zhao, 2020; Zhu et al., 2019)

Name	Composition	Source	Properties	Wound type	Effects		References
Hyaluronic acid	D-glucuronic acid and N-acetyl-D-glucosamine linked by β-1,4 and β-1,3 glycosidic linkages	Animal origin	• Stimulates fibroblast and keratinocyte proliferation and migration • Anti-inflammatory	Acute wounds, partial- and full-thickness wounds	Flexible, highly biocompatible, bacteriostatic	—	(Hsu et al., 2019; Y. C. Huang et al., 2019; Makvandi et al., 2016; Marin et al., 2020; Neuman et al., 2020; Taskan et al., 2015; Teh et al., 2012; S. Wu et al., 2017; Chenting Zhang et al., 2021)
Chitosan	N-acetyl glucosamine linked by β-1,4 glycosidic linkages	Exoskeleton of crabs, mollusks, insects, fungal cell walls	• Induces fibroblast and keratinocyte migration and proliferation	Acute wounds, pressure ulcers	Antimicrobial, promotes cell proliferation	—	(Anitha et al., 2014; Cai & Li, 2020; Croisier & Jérôme, 2013; Elviri et al., 2017; Memic et al., 2019; Mittal et al., 2018; Mndlovu et al., 2020; Mohandas et al., 2018; Muthukumar et al., 2019; Saatchi et al., 2020)
Fucoidan	α-L-Fucose linked by α-1,3 glycosidic linkages	Brown algae	• Angiogenic, mitogenic to fibroblasts and keratinocytes	Burn wounds	Slow blood clotting, prevents cancer cells, protects from inflammation and oxidative stress	—	(Barbosa et al., 2019; Benbow et al., 2020; Senthilkumar et al., 2017; Sezer & Cevher, 2011; Zeng & Huang, 2018)

include ActivHeal, DermaSyn, NU-GEL, Purilon, and WounDres (Aswathy et al., 2020).

However, most hydrogel wound dressing available on the market is based on synthetic polymers. Polysaccharide-based hydrogels have superior potential in wound healing applications, attributed to their high water absorption ability. Polysaccharide hydrogels possess porous swelling to impart great advantages at the interface of a negative pressure system such as wound dressing (Aduba & Yang, 2017; Zhu et al., 2019). Among all polysaccharides, chitosan has become an attractive wound dressing material due to its biocompatibility, non-toxic absorption, biodegradability, hemostaticity, and ability to be a substrate for cell attachment. In 2002, Ishihara et al. (2002) developed a chitosan hydrogel crosslinked through UV treatment. This hydrogel was found to effectively stanch bleeding at wound sites while promoting tissue granulation and epithelialization in rat models. Park et al. (2009) developed a bFgF-loaded chitosan hydrogel system able to accelerate wound repair in chronic ulcers.

The application of biopolymer nanofibers has also been reported in wound healing applications. Stojkovska et al. (2018) developed possibilities of alginate and silver nanoparticles for wound treatment that investigated second-degree thermal burns. Daily applications of alginate-incorporated silver nanoparticles in the form of a colloid solution, wet microfiber, and dry conditions were compared to commercial silver sulfadiazine, a commercial Ca-alginate wound dressing containing silver ions, and untreated controls. The study found the wound dressing containing alginate and silver nanoparticles showed faster healing in treating wounds, which completely healed after 19–21 days compared to the control samples, which took more than 25 days. The alginate-silver nanoparticle wound treatment also induced enhanced healing that had the potential for therapeutic applications in wound treatment.

Other promising biopolymer-based nanomaterials have been utilized for specific biomedical applications as wound dressings. Bacterial cellulose (BC) is a biopolymer produced by bacteria that possesses several advantages, including purity, high porosity, permeability, elevated water uptake capacity, and mechanical robustness. BC can also be modified according to the required antibacterial response and possible local drug delivery features (Portela et al., 2019). Due to its intrinsic features, especially in wound dressing applications and particularly for burn wounds, to date, BC has been exploited and applied commercially. Several manufacturers have commercially employed BC in their products, mainly for better healing rates and the antibacterial properties of the dressing. For instance, the company Biofill has introduced several BC-based products for wound dressing applications, which include (1) Biofill: This product serves as temporary skin or substitute for ulcers and burn wounds. It is designed to reduce pain, lower the risk of infection, and expedite the healing process. (2) BioProcess: Designed for ulcers and burn wounds, this product offers antibacterial properties and accelerates the healing rate. (3) Gengiflex: Tailored for use in dental implants and grafting procedures, Gengiflex focuses on periodontal tissue health. It aims to reduce inflammatory responses and streamline surgical steps. Additionally, Biofill offers other products like Bionext and Xcell, which target ulcers, burns, and lacerations. These products are designed to alleviate pain, prevent infections, and facilitate rapid healing (Picheth et al., 2017).

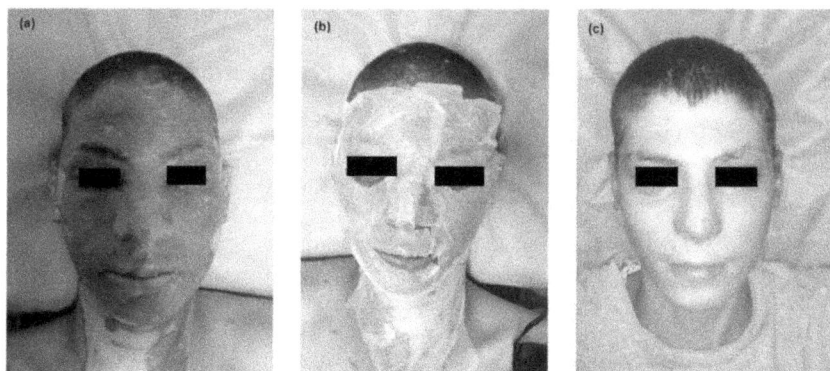

FIGURE 5.20 Healing evolution within 28 days. (a) Before cleaning, (b) BC sheets applied on the wound, and (c) the wound healed after 28 days (Aboelnaga et al., 2018).

To date, the production and commercialization of BC wound dressing has been continuously explored to increase its performance. For example, a study done by Łódź University of Technology (Poland) reported the production of a BC-based wound dressing product, Celmat. The product has been applied to several clinical trials that demonstrated BC wound dressing efficiency on wounds after being cleaned with normal saline and free from any bullae or debris. The application of microbial cellulose, specifically Epiprotect sheets, was conducted under strict aseptic conditions to support and expedite the wound healing process. The wound was observed to heal after 28 days, as shown in Figure 5.20 (Aboelnaga et al., 2018).

5.10 PHARMACEUTICAL APPLICATIONS

Nowadays, medical and pharmaceutical industries are turning their focus to natural polymers due to their biodegradability and resorbability. The role of biopolymers varies depending on the mechanism of drug release and its form. Applications of biopolymers have been established for many years as excipients, immediate-release forms through the oral route, controlling the manufacturing process and protecting the drug from degradation during storage. Biopolymers play the main role as excipients, as they can help in the formulation of the drugs and improve their effectiveness. Biopolymers offer the advantage of mitigating the toxicity and extending the release kinetics of active ingredients. Simultaneously, the use of biopolymers can help maintain a sustained level of active compounds in the bloodstream throughout the day, thereby enhancing overall efficiency. As a result, the need for frequent daily intake can be reduced (Abdelhak, 2019).

Various biopolymers impart different functions depending on their chemical properties, biocompatibility, and bioresorbability properties. Their mechanical strength is also very important to determine the expected applications in this field. The types of biopolymers and their functions in the pharmaceutical field are summarized in Table 5.4.

TABLE 5.4

Types of Biopolymers and Applications in the Pharmaceutical Field

Biopolymer	Applications	References
Cellulose	• Cellulose powder: dispersal and stabilizer in emulsion and suspension • Microcrystalline cellulose and nanocellulose: sustained drug delivery • Cellulose acetate: gastro-resistant and enteric-soluble coatings suitable to treat intestinal problems Applications: Bio-adhesive and mucoadhesive drugs, delivery systems, coating processes, extended-release (ER) solid dosage forms, extended-release polymeric matrices, osmotic drug delivery systems, thickening and stabilizing agents, fillers in solid dosage form, binders in granulation process, disintegration agents, and taste-masking agents Examples of cellulose drugs: Amarel 4 mg tablets, Metopimazine (Vogalene 15 mg) capsule	(Abdelhak, 2019; Shokri & Adibki, 2013; Sun et al., 2018)
Bacterial cellulose	Ability to keep water and permeability to water vapor Applications: contact lenses, electro-conductive composites hydrogel biosensors, wound dressing and biomembranes, artificial blood vessels	(Abdelhak, 2019; Lee & Park, 2017; Mohammedi, 2017)
Cellulose ethers	Used in the formulation of dosage forms and healthcare products, controlled-release drugs	(Abdelhak, 2019; Shokri & Adibki, 2013)
Starch	Carriers for controlled release of drugs and other bioactive agents Example: Vogalene (metopimazine) 15 mg capsule and Doliprane Lib (paracetamol) 500 mg	(Kaur et al., 2007; Martins & Rodrigues, 2012)
Amylose	Non-toxic filaments and fiber for medical sutures, treatment of stomach ulcers, increases antibiotic ability to enter lymphatic system, making it suitable for parenteral applications, decreases fluid loss and decreases diarrhea duration for adolescents and adults with cholera, delivery bioactive agents, protein and peptide to treat several diseases affecting the colon and for systematic absorption	(Chourasia & Jain, 2004; Franz, 1989; Ramakrishna et al., 2000; Ravin et al., 1962; Zhang et al., 2016)

TABLE 5.4 (*Continued*)
Types of Biopolymers and Applications in the Pharmaceutical Field

Biopolymer	Applications	References
Amylopectin	Undercoat material in enteric-coated pellets, enhances the acid resistance properties of simulated gastric juice	(Jivraj et al., 2000)
Alginate	Excipients, encapsulates drugs of fragile biological substances (enzymes, microorganisms, animal or human cells), dental additives (toothpaste) Example: Arthodont 1%, Gingaval paste, Homedent toothpaste	(Srivastava et al., 2012; Tønnesen & Karlsen, 2002)
Alginic acid	Disintegrating agent in tablets Examples: Haldol 1 mg tablets, Acticarbine tablets, Adrexan 40 mg tablets, Fasigyne 500 mg tablets, Avlocardyl 40 mg tablets, Logryx 100 mg tablets, Aparoxal 100 mg tablets, Propanolol Ratio 40 mg tablets, Dipiperon 40 mg tablets, Spasmaverine 40 mg tablets, Halivite 5% cream	(Abdelhak, 2019)
Sodium alginate	Extended-release tablets, base for cosmetic and medicated gels Example: Cream (Calmiphase, Halivite 5%), tablets (Isoptine 240 mg, Verapamil GNR 240 mg, Verapamil Merck LP 240 and 120 tablets), Gel (Sodium alginate 5.0 g, Glycerine 5.0 g, Sodium benzoate 1.0 g, Purified water 89.0 g)	(Srivastava et al., 2012; Tønnesen & Karlsen, 2002)
Xanthan gum	Binders in tablet formulations, suspending agents, emulsifiers and stabilizers in toothpaste, and ointment sustained release agents	(Dumitriu, 2005; Goswami & Naik, 2014; Wiederschain, 2007)
Guar gum	Binding and disintegrating agents in tablets, viscosifying agents and adjuvants for the release of controlled drugs	(Abdelhak, 2019)
Chitosan and chitin	Absorbable material with release control of active ingredients, formation of macrocapsule gels with anionic polymers, excipients, drug delivery, encapsulation of drugs, release of drugs	(Abdelhak, 2019; C, 2017; Cheung et al., 2015; Khor, 2010)
PLA	Bioabsorbable sutures, orthopedic implants, controlled release systems of drugs	(Abd Alsaheb et al., 2015)

In recent times, researchers have directed their attention towards the development of nano-biomedicine, which holds significant promise for advancing pharmaceutical applications. The utilization of nanomaterials serves to enhance targeting of specific tissues, enables access to deep molecular targets, and allows precise control over drug release. The application of nanomaterials in the pharmaceutical realm spans a wide spectrum, encompassing areas such as drug delivery, gene therapy, separation and purification, tissue engineering, DNA probes, nanoscale biochips, and microsurgical technology (Uddin et al., 2016).

5.10.1 DRUG DELIVERY

A drug delivery system (DDS) can be explained as a formulation or device that enables the transport of a therapeutic agent in the body by controlling the rate, time, place, and release of the drugs to the target site in the body. Excipient development plays a major role in pharmaceutical drug delivery because it influences formulation development and the drug delivery process in various ways. Polymeric drug delivery systems have attracted great interest for controlled delivery, as they impart optimized drug loading and releasing properties (Gopi et al., 2018; Saikia & Gogoi, 2015).

The utilization of synthetic polymers has shown drawbacks such as imposing dose reduction or treatment delay, or the given therapy is not continuous. The application of biopolymers has caught the attention of researchers and manufacturers for protective delivery systems for drugs due to their low toxicity, biodegradability, stability, and renewable nature (Gopi et al., 2018). The application of biopolymers in a drug delivery systems can be varied according to its end used. Developing better drug delivery methodologies, strategies, and policies is very important to reduce unwanted changes in stability and the organoleptic properties of the bioactive compound and improve its stability during product development.

The applications of biopolymers in drug delivery systems can be defined through several methods in the formulations that control the rate and period of the drug and target specific areas of the body for delivery of bioactive molecules. First is the phase separation technique to encapsulate bioactives for delivery in a controlled manner. The process known as complex coacervation, involving liquid–liquid phase separation, occurs when the concentration of one biopolymer surpasses a critical threshold. This biopolymer interacts with another biopolymer of either similar electrical charges or neutral, where no associative interaction takes place. This technique effectively segregates two liquids within a colloidal system, with the more concentrated phase becoming the coacervate, while the other phase maintains its status as an equilibrium solution (Saikia & Gogoi, 2015). A study on complex coacervation on casein/gum tragacanth mixture by Jain et al. (2016) found it was able to develop β-carotene–loaded microcapsule formulations and increase the stability of micronutrients after encapsulation. The microcapsules could be novel carriers for the safe and sustained release of micronutrients.

The application of a biopolymer-based drug delivery system also develops in the form of composites. Galkina et al. (2015) developed a drug delivery system based on cellulose nanofibers reinforced with titania nanocomposites grafted with three different types of model drugs, diclofenac sodium, penicillin-D, and phosphomycin,

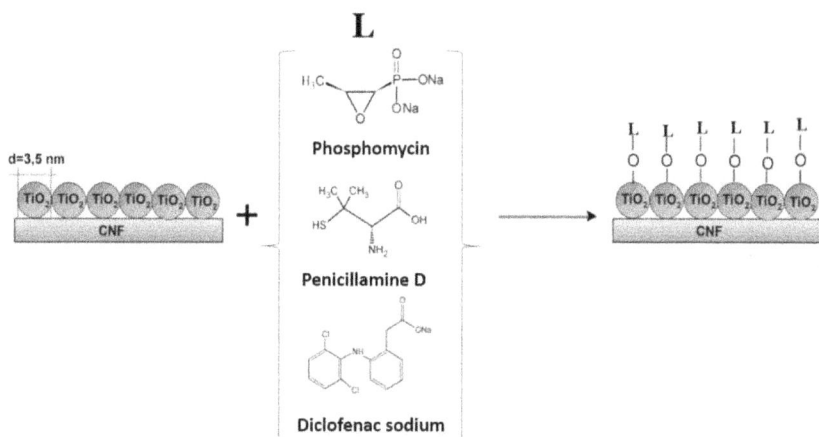

FIGURE 5.21   Schematic interaction of nanocomposites based on cellulose nanofiber and TiO$_2$ with different types of drugs (Galkina et al., 2015).

and displayed distinctly different controlled long-term release profiles. Through investigation, the experiment confirmed the uniform distribution of drugs within nanocellulose film with titania as a binding agent between cellulose nanofiber and drug molecules, enabling slow and controlled release, as shown in Figure 5.21. Drug release studies showed a long-term release profile with different kinetics of release depending on the medicine used. The fastest release was observed for the more soluble painkillers, followed by anti-inflammatories, and the slowest release was by strongly chemisorbed antibiotic agents. Therefore, nanocomposites can potentially be applied in transdermal drug delivery patches as anesthetics and wound-dressing materials.

Similar approaches also adapted to the process of encapsulation. In this process, capsules conceived at the nano- or microscale act as compartments in which the system can effectively protect and deliver an active agent with specific kinetics (Gopi et al., 2018). Biopolymeric encapsulation of drugs has received an encouraging response owing to its potential role as carrier of biomolecules for drug delivery. The method of encapsulation of drugs in biopolymeric materials was adopted by Dozie-Nwachukwu et al. (2017) for targeted chemotherapeutic agents. Prodigiosin loaded in chitosan microspheres was prepared via water in oil (w/o) emulsion techniques using glutaraldehyde as a crosslinker. Prodigiosin is a secondary metabolite extracted from the bacteria *Serratia marcescens*, which exhibits anti-cancer properties. The morphology properties of the microspheres were characterized along with their encapsulation efficiency, drug loading, swelling ratio, and drug release kinetics. The results reported the potential of encapsulation of microspheres, which is suitable for localized chemotherapy applications.

The development of technology has brought in advanced strategies, particularly in drug delivery systems. To date, biopolymers have been produced from genetically modified cells as cellular factories to be used in targeted drug delivery to specific cells and organs. For instance, an antibody is a biopolymer that comprises four peptide chains designed to aid immune systems. Among them, a monoclonal antibody (mAb) is a unique species of antibody with affinity to only one specific part of an antigen, which can be exploited as sensitive probes to guide target cells and organs for drug targeting. With this approach, cytotoxic drugs can be delivered to specific cells like cancer cells and minimize the damage to normal cells (Gopi et al., 2018).

5.11 AUTOMOTIVE

Biopolymer or plastics have been dominantly replacing metals in cars for the past 50 years as they impart better properties such as cheaper resources and lighter weight (better fuel efficiency), with high performance, for example, minimal corrosion, design freedom, flexibility in integrating components, safety, comfort, potential recyclability, and composability. The application of plastic can be distinguished in three major areas in automotive applications: exterior car parts, interior car parts, and interior textiles. Commonly, in many major automotive products, plastic represents almost 20% of the car weight, and the three top plastic materials that contribute to that application are PP, PU and polyvinyl chloride (PVC) (Barrett, 2019). Figure 5.22 shows Henry Ford with the first plastic-bodied car, called "Soybean Car", in 1941.

The application of biobased plastics in the automotive industry was first reported in the 1930s. The utilization of natural-based products was pioneered by the American Henry Ford and was later followed by manufacturers from Japan. The most widely applied bioplastics in the automotive industry are (1) biocomposites: the mixture of

FIGURE 5.22   Henry Ford with the first plastic-bodied "Soybean Car" on August 13, 1941.

synthetic plastic with natural fibers; (2) bio polyamides: engineering thermoplastics usually used in minor parts of car engines; (3) PLA and PLA-based composites: usually used in under-the-hood components and other interior parts; and (4) biobased PP: largely used in modern cars for interior and exterior parts. Axel Barret (2019) summarized the timeline of application of bioplastics by global car manufacturers as listed in Table 5.5.

Ongoing innovation within the automotive industry, both among manufacturers and suppliers, is driven by the dual objectives of meeting customer demands and fostering environmental sustainability. For Röchling Automotive, the application of biobased plastics represents the next step towards resource-conserving production. In collaboration with partners, Röchling Automotive has successfully invented

TABLE 5.5
Timeline of Bioplastic Applications by Car Manufacturers

Manufacturer	Applications
Ford	• 1930s: Soybean bioplastics were invented in an attempt to reduce the weight of cars and increase mileage efficiency • 1941: The first car body made from soybean bioplastics was unveiled • 2000: Ford starts testing and using renewable fiber-based polymers • 2003: Ford uses PLA fibers for canvas roof and carpet mats for Model U • 2010: Ford starts using headliners from soy foam blends, which are used to adhere to the inside roof of automobiles. Ford also starts using bio-PU foams in all North American vehicles, bio-PP from wheat straw for interior storage beans in the Ford Flex, and bio-PP from coconut for load floors in the Focus BEV • 2018: Testing new bioplastic made from agave plant fibers
General Motors (GM)	GM uses wood-based bio-PP for the seatbacks of the Cadillac Deville and flax-based bio-PP for trim and shelving in the Chevrolet Impala • 2001: Chrysler is the first automaker to use EcoCor, a biocomposite containing kenaf, hemp, and PP, for door panels of the Sebring
Mazda	• 2006: Starts developing bioplastics under brand name Mazda Biotechmaterial. Teijin and Mazda co-develop Biofront, the first mass-produced stereocomplexed PLA used in car seat fabric (Mazda Premacy), floor mats, pillar covers, door trim, front panels, and ceiling materials Mazda claims to be the first to reach 80% plant-derived content in interior fitting (Premacy Hydrogen RE Hybrid) and 100% plant-derived bio-fabric for seat covers • 2014: Mazda Motor Corporation collaborates with Mitsubishi Chemicals to produce bioplastics • 2015: Mazda MX-5 incorporates a biobased material in its interior components, comprising 88% corn-derived content and 12% petroleum-based materials. • 2016: Mazda introduces a roadster featuring a bioplastic coating, which effectively mitigates the emission of harmful paint substances through the utilization of bioplastics.

(Continued)

TABLE 5.5 (*Continued*)
Timeline of Bioplastic Applications by Car Manufacturers

Manufacturer	Applications
Toyota	Toyota has been using bio-polyester, bio-PET, and PLA. They claim to use sugarcane-based PET for in-vehicle liners and other interior surfaces. Toyota SAI and Prius were the first models to feature bioplastic applications such as headliners, sun visors, and floor mats
	• 2003: Toyota uses PLA fibers and fabric for floor mats and PLA/kenaf biocomposites for the cover of that spare wheel and translucent roofs of the Prius and Raum
	• 2003–2005: Toyota sets a goal to have 20% of all plastic components in their vehicles produced from bioplastics by 2015. In certain Toyota cars, the interior fabrics are now composed of as much as 60% bio-polyester content.
	Mixed PLA/PET for upholstery material on the door and luggage area trims, PLA/PP for injection-molded parts such as scuff plates and interior trims.
	Toyota also uses soy-based seat cushions for many of their cars (Prius, Corolla, Matrix, RAV4, and Lexus 350)
	Toyota also uses Dupont SoronaEP for vent louvres of Toyota Prius A Alpha that contain 20–37% starch-based polymer
	• 2017: Toyota uses Denso bioplastic for its navigation system
Mitsubishi	• 2006: Mitsubishi Chemicals and Faurecia (FR) co-develop a bioplastic known as BioMat (made from bio-succinic supplied from BioAmber and bio-PBS) used for automotive interior parts such as door panels, trim and strips, structural instrument panels, console inserts, air ducts, and door panel inserts
Fiat	• 2011: Fiat and Dupont win the Society of Plastics Engineers Automotive Innovation Award in the environmental category for the use of castor oil-based Zytel RS polyamide 1010 in some fuel lines. Zytel RS line is a renewable sourced long chain of nylon products (60–100% biomaterial)
	• 2012: Fiat uses castor oil-derived polyamides and soya-derived polyurethanes in more than 1 million vehicles. Fiat cars for Brazil contain polyurethane seat foams with 5% soy polyol
Lexus	The LEXUS CT200h contains bio-PE made from bamboo and corn in its luggage compartment, speakers, and floor mats
	The LEXUS HS250 uses biobased parts for the luggage trim upholstery, cowl side trim, seat cushions, door scuff plates, and toolbox area
Daimler and Mercedes Benz	Daimler mixes kenaf, flax, and sisal in plastics for its door linings
	Mercedes-Benz's Biome concept car uses organic bio fiber grown in a lab. It is stronger than steel, lighter than metal, and compostable at the end of its life
	A Mercedes-Benz model has bio-PE from flax in the engine, transmission cover, and underbody panels. The engine cover is made of EcopaXX, a 70% biobased polyamide produced by DSM
BMW	BMW uses wood-based biocomposites produced from Johnson Controls for its panels, making them 20% lighter. Some BMWs contain up to 24 kg of flax and sisal
Porsche	Porsche begins designing cars with bodies from hemp-composite materials
Sportscars	2019: An American 3D printed a homemade Lamborghini with PLA and other plastics
	Japanese create the first supercar made from cellulose and agricultural waste

a biopolymer that will cut the production of CO_2 emissions. The biopolymer derived from lactic acid and polylactides was chosen and was obtained by fermenting sugar. The raw materials are readily available, depending on the region, such as corn (USA), starch (Asia), and sugarcane (rest of the world). The production of biopolymers was able to reduce almost 90% of CO_2 emissions in comparison to conventional plastics and save around 480 kg of CO_2 for a mid-range car (Barillari & Chini, 2020).

To date, new studies are continuously being done on biopolymer-based products to enhance their current properties, for example, the application of nano-scale materials to improve the available properties as well as to impart additional functionalities such as conductivity and flame retardancy. PLA has been reported to show poor crystallization ability and low thermal resistance, which limits its industrial applications (Rasal et al., 2010). The study on improving PLA properties by the addition of nanomaterials was done by Xu (2020). Properties like heat resistance, heat distortion temperature, crystallization rate, and impact resistance were reported to improve with the addition of nanocellulose in PLA, but the author claims some modifications and suitable production techniques are needed to process PLA/nanocellulose composites and overcome the dispersion difficulties of nanoparticles in a PLA matrix.

5.12 CONCLUSIONS

There is a growing urgency to develop novel biobased products to reduce the widespread dependence on fossil fuel. This continuous innovation is essential in all industries. Today, the concepts of improving sustainability and lowering the carbon footprint have been very important to manufacturers and suppliers; therefore, alternatives and new inventions are required. As the coronavirus pandemic emerged in early 2020, the call for increased innovation became more pronounced. Governments and regulatory agencies worldwide responded with a slew of measures aimed at mitigating the economic impact on various industries. The adoption of biopolymers has emerged as a compelling solution to reduce reliance on synthetic plastics, owing to their renewable sources, widespread availability, and cost-effectiveness. Their biodegradability and non-toxic properties further bolster their credentials as environmentally sustainable materials, making them a pivotal choice for products that prioritize sustainability and a positive impact on the environment. Despite their remarkable properties, there are some noticeable limitations of biopolymers, including poor moisture resistance, dimensional stability, thermal decomposition temperature, fire resistance, UV resistance, and biological resistance, which limit the growth of its potential applications. The hybridization of biopolymers with other biobased or synthetic polymer blending systems also showed development in various areas that need to be focused on. The development of biopolymer-reinforced natural fibers and nanomaterials to form biocomposites and nanocomposites has shown significant opportunities for enhancing the properties of biopolymer products that make them able to be applied in various industries.

REFERENCES

Abd Alsaheb, R. A., Aladdin, A., Othman, N. Z., Abd Malek, R., Leng, O. M., Aziz, R., & El Enshasy, H. A. (2015). Recent applications of polylactic acid in pharmaceutical and medical industries. *Journal of Chemical and Pharmaceutical Research*, 7(12).

Abdelhak, M. (2019). A review: Application of biopolymers in the pharmaceutical formulation. *Journal of Advances in Bio, 1*(1), 15–25. http://doi.org/10.5281/zenodo.2577643

Abdul Khalil, H. P.S., Yahya, E. B., Jummaat, F., Adnan, A. S., Olaiya, N. G., Rizal, S., Abdullah, C. K., Pasquini, D., & Thomas, S. (2023). Biopolymers based aerogels: A review on revolutionary solutions for smart therapeutics delivery. *Progress in Materials Science, 131*, 101014.

Aboelnaga, A., Elmasry, M., Adly, O. A., Elbadawy, M. A., Abbas, A. H., Abdelrahman, I., Salah, O., & Steinvall, I. (2018). Microbial cellulose dressing compared with silver sulphadiazine for the treatment of partial thickness burns: A prospective, randomised, clinical trial. *Burns, 44*(8). https://doi.org/10.1016/j.burns.2018.06.007

Aduba, D. C., & Yang, H. (2017). Polysaccharide fabrication platforms and biocompatibility assessment as candidate wound dressing materials. *Bioengineering, 4*(1), 1–16. https://doi.org/10.3390/bioengineering4010001

Ahn, S., Chantre, C. O., Gannon, A. R., Lind, J. U., Campbell, P. H., Grevesse, T., O'Connor, B. B., & Parker, K. K. (2018). Soy protein/cellulose nanofiber scaffolds mimicking skin extracellular matrix for enhanced wound healing. *Advanced Healthcare Materials, 7*(9). https://doi.org/10.1002/adhm.201701175

Al-Hassan, A. A., & Norziah, M. H. (2012). Starch-gelatin edible films: Water vapor permeability and mechanical properties as affected by plasticizers. *Food Hydrocolloids, 26*(1). https://doi.org/10.1016/j.foodhyd.2011.04.015

Almeida, T., Moreira, P., Sousa, F. J., Pereira, C., Silvestre, A. J. D., Vilela, C., & Freire, C. S. R. (2022). Bioactive bacterial nanocellulose membranes enriched with Eucalyptus globulus Labill. Leaves aqueous extract for anti-aging skin care applications. *Materials, 15*(5). https://doi.org/10.3390/ma15051982

Alvarez-Román, R., Barré, G., Guya, R. H., & Fessi, H. (2001). Biodegradable polymer nanocapsules containing a sunscreen agent: Preparation and photoprotection. *European Journal of Pharmaceutics and Biopharmaceutics, 52*(2), 191–195. https://doi.org/10.1016/S0939-6411(01)00188-6

Amorim, J. D. P., Junior, C. J. G. S., Costa, A. F. S., Nascimento, H. A., Vinhas, G. M., & Sarrubo, L. A. (2020). BioMask, a polymer blend for treatment and healing of skin prone to acne. *Chemical Engineering Transactions, 79*. https://doi.org/10.3303/CET2079035

Anitha, A., Sowmya, S., Kumar, P. T. S., Deepthi, S., Chennazhi, K. P., Ehrlich, H., Tsurkan, M., & Jayakumar, R. (2014).+ Chitin and chitosan in selected biomedical applications. *Progress in Polymer Science.* https://doi.org/10.1016/j.progpolymsci.2014.02.008

Arabyarmohammadi, H., Darban, A. K., Abdollahy, M., Yong, R., Ayati, B., Zirakjou, A., & van der Zee, S. E. A. T. M. (2018). Utilization of a novel chitosan/clay/biochar nanobiocomposite for immobilization of heavy metals in acid soil environment. *Journal of Polymers and the Environment, 26*(5), 2107–2119. https://doi.org/10.1007/s10924-017-1102-6

Aramwit, P., & Bang, N. (2014). The characteristics of bacterial nanocellulose gel releasing silk sericin for facial treatment. *BMC Biotechnology, 14*(1). https://doi.org/10.1186/s12896-014-0104-x

Aravamudhan, A., Ramos, D. M., Nip, J., Kalajzic, I., & Kumbar, S. G. (2018). Micro-nanostructures of cellulose-collagen for critical sized bone defect healing. *Macromolecular Bioscience, 18*(2). https://doi.org/10.1002/mabi.201700263

Asadian, M., Chan, K. V., Norouzi, M., Grande, S., Cools, P., Morent, R., & De Geyter, N. (2020). Fabrication and plasma modification of nanofibrous tissue engineering scaffolds. *Nanomaterials.* https://doi.org/10.3390/nano10010119

Aswathy, S. H., Narendrakumar, U., & Manjubala, I. (2020). Commercial hydrogels for biomedical applications. *Heliyon, 6*(4), e03719. https://doi.org/10.1016/j.heliyon.2020.e03719

Asyraf, M. R. M., Syamsir, A., Ishak, M. R., Sapuan, S. M., Nurazzi, N. M., Norrrahim, M. N. F., Ilyas, R. A., Khan, T., & Rashid, M. Z. A. (2023). Mechanical properties of hybrid lignocellulosic fiber-reinforced biopolymer green composites: A review. *Fibers and Polymers, 24*(2), 337-353.

Attaran, S. A., Hassan, A., & Wahit, M. U. (2017). Materials for food packaging applications based on bio-based polymer nanocomposites. *Journal of Thermoplastic Composite Materials, 30*(2), 143–173. https://doi.org/10.1177/0892705715588801

Azeem, B., KuShaari, K., Naqvi, M., Kok Keong, L., Almesfer, M. K., Al-Qodah, Z., Naqvi, S. R., & Elboughdiri, N. (2020). Production and characterization of controlled release urea using biopolymer and geopolymer as coating materials. *Polymers, 12*(2), 400. https://doi.org/10.3390/polym12020400

Badawy, M. E. I., & Rabea, E. I. (2011). A biopolymer chitosan and its derivatives as promising antimicrobial agents against plant pathogens and their applications in crop protection. *International Journal of Carbohydrate Chemistry*, 1–29. https://doi.org/10.1155/2011/460381

Bamatov, I. M., Rumyantsev, E. V., & Kh Zanilov, A. (2019). The influence of biopolymer modification of mineral fertilizers on main agrochemical parameters of soil. In *IOP conference series: Earth and environmental science* (Vol. 315, p. 052059). IOP Publishing. https://doi.org/10.1088/1755-1315/315/5/052059

Barbosa, A. I., Coutinho, A. J., Costa Lima, S. A., & Reis, S. (2019). Marine polysaccharides in pharmaceutical applications: Fucoidan and chitosan as key players in the drug delivery match field. *Marine Drugs, 17*(12). https://doi.org/10.3390/md17120654

Barillari, F., & Chini, F. (2020). Biopolymers—sustainability for the automotive value-added chain. *Development Materials-ATZ Worldwide*, 36–39.

Barrett, A. (2019, November 26). History of bioplastics in the automotive industry symphony d2p is leader in antibacterial & antiviral plastics read free content, no subscription needed. *BioplasticNews.Com*, pp. 1–11. https://bioplasticsnews.com/2019/11/26/history-bioplastics-automotive-car-industry/

Benbow, N. L., Karpiniec, S., Krasowska, M., & Beattie, D. A. (2020). Incorporation of FGF-2 into pharmaceutical grade fucoidan/chitosan polyelectrolyte multilayers. *Marine Drugs, 18*(11). https://doi.org/10.3390/md18110531

Bezerra, U. T. (2016). Biopolymers with superplasticizer properties for concrete. In F. Pacheco-Torgal, V. Ivanov, N. Karak, & H. Jonkers (Eds.), *Biopolymers and biotech admixtures for eco-efficient construction materials* (pp. 195–220). Woodhead Publishing Limited. https://doi.org/10.1016/B978-0-08-100214-8.00010-5

BIOME Bioplastic. (n.d.). Retrieved February 26, 2021, from https://biomebioplastics.com/

Birck, C., Degoutin, S., Tabary, N., Miri, V., & Bacquet, M. (2014). New crosslinked cast films based on poly(vinyl alcohol): Preparation and physico-chemical properties. *Express Polymer Letters, 8*(12). https://doi.org/10.3144/expresspolymlett.2014.95

C, D. (2017). Opinion about advances of chitosan in pharmaceutical field: From past to now. *Modern Applications in Pharmacy & Pharmacology, 1*(1). https://doi.org/10.31031/mapp.2017.01.000503

Cai, H., & Li, G. (2020). Efficacy of alginate-and chitosan-based scaffolds on the healing of diabetic skin wounds in animal experimental models and cell studies: A systematic review. *Wound Repair and Regeneration.* https://doi.org/10.1111/wrr.12857

Cerqueira-Coutinho, C., Santos-Oliveira, R., dos Santos, E., & Mansur, C. R. (2015). Development of a photoprotective and antioxidant nanoemulsion containing chitosan as an agent for improving skin retention. *Engineering in Life Sciences, 15*(6), 593–604. https://doi.org/10.1002/elsc.201400154

Chang, I., & Cho, G.-C. (2014). Geotechnical behavior of a beta-1,3/1,6-glucan biopolymer-treated residual soil. *Geomechanics and Engineering, 7*(6), 633–647. https://doi.org/10.12989/gae.2014.7.6.633

Chang, I., & Cho, G.-C. (2019). Shear strength behavior and parameters of microbial gellan gum-treated soils: From sand to clay. *Acta Geotechnica*, *14*(2). https://doi.org/10.1007/s11440-018-0641-x

Chang, I., Im, J., & Cho, G. C. (2016). Introduction of microbial biopolymers in soil treatment for future environmentally-friendly and sustainable geotechnical engineering. *Sustainability*, *8*(251). https://doi.org/10.3390/su8030251

Chang, I., Im, J., Prasidhi, A. K., & Cho, G. (2015). Effects of Xanthan gum biopolymer on soil strengthening. *Construction and Building Materials*, *74*, 65–72. https://doi.org/10.1016/j.conbuildmat.2014.10.026

Chassenieux, C., Durand, D., Jyotishkumar, P., & Thomas, S. (2013). Biopolymer: State of the art, new challenges and opportunities. In S. Thomas, D. Durand, C. Chassenieux, & P. Jyotishkumar (Eds.), *Handbook of biopolymer-based materials: From blends and composites to gels and complex networks* (pp. 1–6). Wiley-VCH Verlag. https://doi.org/10.1002/9783527652457

Chattopadhyay, S., & Raines, R. T. (2014). Review collagen-based biomaterials for wound healing. *Biopolymers*. https://doi.org/10.1002/bip.22486

Chen, J., Lü, S., Zhang, Z., Zhao, X., Li, X., Ning, P., & Liu, M. (2018). Environmentally friendly fertilizers: A review of materials used and their effects on the environment. *Science of the Total Environment*, *613–614*, 829–839. https://doi.org/10.1016/j.scitotenv.2017.09.186

Chen, Z., Mo, X., He, C., & Wang, H. (2008). Intermolecular interactions in electrospun collagen-chitosan complex nanofibers. *Carbohydrate Polymers*, *72*(3). https://doi.org/10.1016/j.carbpol.2007.09.018

Cheung, R. C. F., Ng, T. B., Wong, J. H., & Chan, W. Y. (2015). Chitosan: An update on potential biomedical and pharmaceutical applications. *Marine Drugs*. https://doi.org/10.3390/md13085156

Chiu, C. T., Lee, J. S., Chu, C. S., Chang, Y. P., & Wang, Y. J. (2008). Development of two alginate-based wound dressings. *Journal of Materials Science: Materials in Medicine*, *19*(6). https://doi.org/10.1007/s10856-008-3389-2

Cho, C., & Kobayashi, T. (2021). Advanced cellulose cosmetic facial masks prepared from Myanmar thanaka heartwood. *Current Opinion in Green and Sustainable Chemistry*, *27*. https://doi.org/10.1016/j.cogsc.2020.100413

Chourasia, M. K., & Jain, S. K. (2004). Polysaccharides for colon targeted drug delivery. *Drug Delivery: Journal of Delivery and Targeting of Therapeutic Agents*. https://doi.org/10.1080/10717540490280778

Croisier, F., & Jérôme, C. (2013). Chitosan-based biomaterials for tissue engineering. *European Polymer Journal*. https://doi.org/10.1016/j.eurpolymj.2012.12.009

Cruz, C. F., Costa, C., Gomes, A. C., Matamá, T., & Cavaco-Paulo, A. (2016). Human hair and the impact of cosmetic procedures: A review on cleansing and shape-modulating cosmetics. *Cosmetics*, *3*(3), 1–22. https://doi.org/10.3390/cosmetics3030026

Dias-Ferreira, J., Fernandes, A. R., Soriano, J. L., Naveros, B. C., Severino, P., da Silva, C. F., & Souto, E. B. (2020). Skin rejuvenation: Biopolymers applied to UV sunscreens and sheet masks. In M. Agostini de Moraes, C. Ferrerira da Silva, & R. S. Vieira (Eds.), *Biopolymer membranes and films* (pp. 309–330). Elsevier. https://doi.org/10.1016/B978-0-12-818134-8.00013-4

Dozie-nwachukwu, S. O., Danyuo, Y., Obayemi, J. D., Odusanya, O. S., Malatesta, K., & Soboyejo, W. O. (2017). Extraction and encapsulation of prodigiosin in chitosan microspheres for targeted drug delivery Cross section of Micro. *Materials Science & Engineering C*, *71*, 268–278. https://doi.org/10.1016/j.msec.2016.09.078

Duhan, J. S., Kumar, R., Kumar, N., Kaur, P., Nehra, K., & Duhan, S. (2017). Nanotechnology: The new perspective in precision agriculture. *Biotechnology reports.* https://doi. org/10.1016/j.btre.2017.03.002

Dumitriu, S. (2005). *Polysaccharides: Structural diversity and functional versatility* (2nd ed.). CRC Press.

El Fawal, G. F., Abu-Serie, M. M., Hassan, M. A., & Elnouby, M. S. (2018). Hydroxyethyl cellulose hydrogel for wound dressing: Fabrication, characterization and in vitro evaluation. *International Journal of Biological Macromolecules, 111.* https://doi.org/10.1016/j. ijbiomac.2018.01.040

Elhussieny, A., Faisal, M., D'Angelo, G., Aboulkhair, N. T., Everitt, N. M., & Fahim, I. S. (2020). Valorisation of shrimp and rice straw waste into food packaging applications. *Ain Shams Engineering Journal, 11*(4). https://doi.org/10.1016/j.asej.2020.01.008

Elviri, L., Bianchera, A., Bergonzi, C., & Bettini, R. (2017). Controlled local drug delivery strategies from chitosan hydrogels for wound healing. *Expert Opinion on Drug Delivery.* https://doi.org/10.1080/17425247.2017.1247803

Frank, C., & Billington, S. (2008). *Energy-efficient biodegradable foams for structural insulated panels.* Polymer.

Franz, G. (1989). Polysaccharides in pharmacy: Current applications and future concepts. *Planta Medica.* https://doi.org/10.1055/s-2006-962078

Galkina, O. L., Ivanov, V. K., Agafonov, A. V., Seisenbaeva, G. A., & Kessler, V. G. (2015). Cellulose nanofiber-titania nanocomposites as potential drug delivery systems for dermal applications. *Journal of Materials Chemistry B, 3*(8), 1688–1698. https://doi. org/10.1039/c4tb01823k

Gao, J., Liu, J., Gao, Y., Wang, C., Zhao, Y., Chen, B., Xiao, Z., Miao, Q., & Dai, J. (2011). A myocardial patch made of collagen membranes loaded with collagen-binding human vascular endothelial growth factor accelerates healing of the injured rabbit heart. *Tissue Engineering—Part A, 17*(21–22), 2739–2747. https://doi.org/10.1089/ten.tea. 2011.0105

Gao, Y., Zhang, X., & Jin, X. (2019). Preparation and properties of minocycline-loaded carboxymethyl chitosan gel/alginate nonwovens composite wound dressings. *Marine Drugs, 17*(10). https://doi.org/10.3390/md17100575

Geo-Synthetics. (n.d.). *Coir and jute biodegradable geotextiles.* www.geo-synthetics.com/ construction-products/erosion-control/biodegradable-geotextiles/

Giavarini, C., Ferretti, A. S., & Santarelli, M. L. (2006). Mechanical behaviour and porperties. In S. K. Kourkoulis (Ed.), *Fracture and failure of natural building stones: Applications in the restoration of ancient monuments* (pp. 107–120). Springer. https:// doi.org/10.1007/978-1-4020-5077-0_7

Gil-Ortiz, R., Naranjo, M. Á., Ruiz-Navarro, A., Caballero-Molada, M., Atares, S., García, C., & Vicente, O. (2020). New eco-friendly polymeric-coated urea fertilizers enhanced crop yield in wheat. *Agronomy, 10*(3), 438. https://doi.org/10.3390/agronomy 10030438

Giordano, M., Amoroso, C. G., El-Nakhel, C., Rouphael, Y., De Pascale, S., & Cirillo, C. (2020). An appraisal of biodegradable mulch films with respect to strawberry crop performance and fruit quality. *Horticulturae, 6*(3). https://doi.org/10.3390/ horticulturae6030048

Giro, T. M., Beloglazova, K. E., Rysmukhambetova, G. E., Simakova, I. V., Karpunina, L. V., Rogojin, A. A., Kulikovsky, A. V., & Andreeva, S. V. (2020). Xanthan-based biodegradable packaging for fish and meat products. *Foods and Raw Materials, 8*(1). https:// doi.org/10.21603/2308-4057-2020-1-67-75

Gopi, S., Amalraj, A., Sukumaran, N. P., Haponiuk, J. T., & Thomas, S. (2018). Biopolymers and their composites for drug delivery: A brief review. *Macromolecular Symposia, 380*(1), 1–14. https://doi.org/10.1002/masy.201800114

Gorrasi, G., & Bugatti, V. (2016). Edible bio-nano-hybrid coatings for food protection based on pectins and LDH-salicylate: Preparation and analysis of physical properties. *LWT— Food Science and Technology, 69*. https://doi.org/10.1016/j.lwt.2016.01.038

Goswami, S., & Naik, S. (2014). Natural gums and its pharmaceutical application. *Journal of Scientific and Innovative Research JSIR, 3*(31).

Gould, L. J. (2016). Topical collagen-based biomaterials for chronic wounds: Rationale and clinical application. *Advances in Wound Care*. https://doi.org/10.1089/wound.2014.0595

Guilherme, M. R., Aouada, F. A., Fajardo, A. R., Martins, A. F., Paulino, A. T., Davi, M. F. T., Rubira, A. F., & Muniz, E. C. (2015). Superabsorbent hydrogels based on polysaccharides for application in agriculture as soil conditioner and nutrient carrier: A review. *European Polymer Journal, 72*, 365–385. https://doi.org/10.1016/j.eurpolymj.2015.04.017

Han, X., Zhang, W., Gu, J., Zhao, H., Ni, L., Han, J., Zhou, Y., Gu, Y., Zhu, X., Sun, J., Hou, X., Yang, H., Dai, J., & Shi, Q. (2014). Accelerated postero-lateral spinal fusion by collagen scaffolds modified with engineered collagen-binding human bone morphogenetic protein-2 in rats. *PLoS ONE, 9*(5). https://doi.org/10.1371/journal.pone.0098480

Hasan, M., Chong, E. W. N., Jafarzadeh, S., Paridah, M. T., Gopakumar, D. A., Tajarudin, H. A., Thomas, S., & Abdul Khalil, H. P. S. (2019). Enhancement in the physico-mechanical functions of seaweed biopolymer film via embedding fillers for plasticulture application-A comparison with conventional biodegradable mulch film. *Polymers, 11*(2). https://doi.org/10.3390/polym11020210

Hayes, D. G., Dharmalingam, S., Wadsworth, L. C., Leonas, K. K., Miles, C., & Inglis, D. A. (2012). Biodegradable agricultural mulches derived from biopolymers. In K. Khemani, et al. (Eds.), *Degradable polymers and materials: Principles and practice* (2nd ed., pp. 201–223). American Chemical Society. https://doi.org/10.1021/bk-2012-1114.ch013

Helanto, K., Matikainen, L., Talj, R., & Rojas, O. J. (2019). Bio-based polymers for sustainable packaging and biobarriers: A critical review. *BioResources, 14*(2), 4902–4951. https://doi.org/10.15376/biores.14.2.Helanto

Hidangmayum, A., Dwivedi, P., Katiyar, D., & Hemantaranjan, A. (2019). Application of chitosan on plant responses with special reference to abiotic stress. *Physiology and Molecular Biology of Plants*. https://doi.org/10.1007/s12298-018-0633-1

Hosseinnejad, M., & Jafari, S. M. (2016). Evaluation of different factors affecting antimicrobial properties of chitosan. *International Journal of Biological Macromolecules*. https://doi.org/10.1016/j.ijbiomac.2016.01.022

Hsu, Y. Y., Liu, K. L., Yeh, H. H., Lin, H. R., Wu, H. L., & Tsai, J. C. (2019). Sustained release of recombinant thrombomodulin from cross-linked gelatin/hyaluronic acid hydrogels potentiate wound healing in diabetic mice. *European Journal of Pharmaceutics and Biopharmaceutics, 135*. https://doi.org/10.1016/j.ejpb.2018.12.007

Huang, T., Qian, Y., Wei, J., & Zhou, C. (2019). Polymeric antimicrobial food packaging and its applications. *Polymers*. https://doi.org/10.3390/polym11030560

Huang, Y. C., Huang, K. Y., Lew, W. Z., Fan, K. H., Chang, W. J., & Huang, H. M. (2019). Gamma-irradiation-prepared low molecular weight hyaluronic acid promotes skin wound healing. *Polymers, 11*(7). https://doi.org/10.3390/polym11071214

Indrani, D. J., Lukitowati, F., & Yulizar, Y. (2017). Preparation of chitosan/collagen blend membranes for wound dressing: A study on FTIR spectroscopy and mechanical properties. In *IOP conference series: Materials science and engineering* (Vol. 202). https://doi.org/10.1088/1757-899X/202/1/012020

Ishihara, M., Nakanishi, K., Ono, K., Sato, M., Kikuchi, M., Saito, Y., Yura, H., Matsui, T., Hattori, H., Uenoyama, M., & Kurita, A. (2002). Photocrosslinkable chitosan as a dressing for wound occlusion and accelerator in healing process. *Biomaterials*, *23*(3). https://doi.org/10.1016/S0142-9612(01)00189-2

Jafarzadeh, S., Jafari, S. M., Salehabadi, A., Nafchi, A. M., Uthaya Kumar, U. S., & Abdul Khalil, H. P. S. (2020, June). Biodegradable green packaging with antimicrobial functions based on the bioactive compounds from tropical plants and their by-products. *Trends in Food Science and Technology*. Elsevier Ltd. https://doi.org/10.1016/j.tifs.2020.04.017

Jain, A., Thakur, D., Ghoshal, G., Katare, O. P., & Shivhare, U. S. (2016). Characterization of microcapsulated β-carotene formed by complex coacervation using casein & gum tragacanth. *International Journal of Biological Macromolecules*. Elsevier B.V. https://doi.org/10.1016/j.ijbiomac.2016.01.117

Jeon, H.-Y. (2016). Environmental adaptability of green geosynthetics as sustainable materials for civil engineering applications. In *Geosynthetics, forging a path to bona fide engineering materials* (Vol. 4, pp. 318–325). American Society of Civil Engineers. https://doi.org/10.1061/9780784480182.028

Jivraj, M., Martini, L. G., & Thomson, C. M. (2000). An overview of the different excipients useful for the direct compression of tablets. *Pharmaceutical Science and Technology Today*. https://doi.org/10.1016/S1461-5347(99)00237-0

Kansal, D., Hamdani, S. S., Ping, R., & Rabnawaz, M. (2020). Starch and zein biopolymers as a sustainable replacement for PFAS, silicone oil, and plastic-coated paper. *Industrial and Engineering Chemistry Research*, *59*(26), 12075–12084. https://doi.org/10.1021/acs.iecr.0c01291

Kashyap, P. L., Xiang, X., & Heiden, P. (2015). Chitosan nanoparticle based delivery systems for sustainable agriculture. *International Journal of Biological Macromolecules*. https://doi.org/10.1016/j.ijbiomac.2015.02.039

Kaur, L., Singh, J., & Liu, Q. (2007). Starch—A potential biomaterial for biomedical applications. In *Nanomaterials and nanosystems for biomedical applications*. https://doi.org/10.1007/978-1-4020-6289-6_5

Khor, E. (2010). Medical applications of chitin and chitosan. In *Chitin, chitosan, oligosaccharides and their derivatives*. https://doi.org/10.1201/ebk1439816035-c30

Kucińska-Lipka, J., Gubanska, I., & Janik, H. (2015). Bacterial cellulose in the field of wound healing and regenerative medicine of skin: Recent trends and future prospectives. *Polymer Bulletin*. https://doi.org/10.1007/s00289-015-1407-3

Kumar, N., Kaur, P., & Bhatia, S. (2017). Advances in bio-nanocomposite materials for food packaging: A review. *Nutrition and Food Science*, *47*(4), 591–606. https://doi.org/10.1108/NFS-11-2016-0176

Kwon, Y. M., Chang, I., Lee, M., & Cho, G. C. (2019). Geotechnical engineering behavior of biopolymer-treated soft marine soil. *Geomechanics and Engineering*, *17*(5), 453–464. https://doi.org/10.12989/gae.2019.17.5.453

Lee, S. E., & Park, Y. S. (2017). The role of bacterial cellulose in artificial blood vessels. *Molecular and Cellular Toxicology*. https://doi.org/10.1007/s13273-017-0028-3

León-Martínez, F. M., Cano-Barrita, P. F. D. J., Lagunez-Rivera, L., & Medina-Torres, L. (2014). Study of nopal mucilage and marine brown algae extract as viscosity-enhancing admixtures for cement based materials. *Construction and Building Materials*, *53*. https://doi.org/10.1016/j.conbuildmat.2013.11.068

Li, J. H., Hsieh, J. C., Lou, C. W., Hsieh, C. T., Pan, Y. J., Hsing, W. H., & Lin, J. H. (2016). Needle-punched thermally-bonded eco-friendly nonwoven geotextiles: Functional properties. *Materials Letters*, *183*. https://doi.org/10.1016/j.matlet.2016.07.074

Li, K., Jin, S., Liu, X., Chen, H., He, J., & Li, J. (2017). Preparation and characterization of chitosan/soy protein isolate nanocomposite film reinforced by Cu nanoclusters. *Polymers, 9*(7). https://doi.org/10.3390/polym9070247

Li, Q., Tao, L., Chen, B., Ren, H., Hou, X., Zhou, S., Sun, X., Dai, J., & Ding, Y. (2012). Extrahepatic bile duct regeneration in pigs using collagen scaffolds loaded with human collagen-binding bFGF. *Biomaterials, 33*(17). https://doi.org/10.1016/j.biomaterials.2012.03.003

Lin, S. T., Kimble, L., & Bhattacharyya, D. (2017). *Polymer Blends and Composites for Biomedical Applications.* https://doi.org/10.1007/978-3-662-53574-5_7

Litwiniuk, M., Krejner, A., & Grzela, T. (2016). Hyaluronic acid in inflammation and tissue regeneration. *Wounds, 28*(3), 78–88.

Lozeau, L. D., Grosha, J., Smith, I. M., Stewart, E. J., Camesano, T. A., & Rolle, M. W. (2020). Alginate affects bioactivity of chimeric collagen-binding LL37 antimicrobial peptides adsorbed to collagen-alginate wound dressings. *ACS Biomaterials Science and Engineering, 6*(6). https://doi.org/10.1021/acsbiomaterials.0c00227

Luzi, F., Torre, L., Kenny, J. M., & Puglia, D. (2019). Bio- and fossil-based polymeric blends and nanocomposites for packaging: Structure-property relationship. *Materials, 12*(3). https://doi.org/10.3390/ma12030471

Ma, G., Qian, B., Yang, J., Hu, C., & Nie, J. (2010). Synthesis and properties of photosensitive chitosan derivatives(1). *International Journal of Biological Macromolecules, 46*(5). https://doi.org/10.1016/j.ijbiomac.2010.02.009

Makvandi, P., Caccavale, C., Della Sala, F., Zeppetelli, S., Veneziano, R., & Borzacchiello, A. (2020). Natural formulations provide antioxidant complement to hyaluronic acid-based topical applications used in wound healing. *Polymers, 12*(8). https://doi.org/10.3390/POLYM12081847

Maluin, F. N., & Hussein, M. Z. (2020). Chitosan-based agronanochemicals as a sustainable alternative in crop protection. *Molecules, 25*(7), 1–22. https://doi.org/10.3390/molecules25071611

Manikrao Donglikar, M., & Laxman Deore, S. (2016). Sunscreens: A review. *Pharmacognosy Journal, 8*(3), 171–179. https://doi.org/10.5530/pj.2016.3.1

Marin, S., Popović-Pejičić, S., Radošević-Carić, B., Trtić, N., Tatić, Z., & Selaković, S. (2020). Hyaluronic acid treatment outcome on the post-extraction wound healing in patients with poorly controlled type 2 diabetes: A randomized controlled split-mouth study. *Medicina Oral Patologia Oral y Cirugia Bucal, 25*(2). https://doi.org/10.4317/medoral.23061

Martins, E., & Rodrigues, A. (2012). Starch: From food to medicine. In *Scientific, health and social aspects of the food industry.* https://doi.org/10.5772/38678

Marzec, E., & Pietrucha, K. (2018). Efficacy evaluation of electric field frequency and temperature on dielectric properties of collagen cross-linked by glutaraldehyde. *Colloids and Surfaces B: Biointerfaces, 162.* https://doi.org/10.1016/j.colsurfb.2017.12.005

MATER BI, Novamont. (n.d.). Retrieved February 26, 2021, from www.novamont.com/eng/mater-bi

Meira, S. M. M., Zehetmeyer, G., Scheibel, J. M., Werner, J. O., & Brandelli, A. (2016). Starch-halloysite nanocomposites containing nisin: Characterization and inhibition of Listeria monocytogenes in soft cheese. *LWT—Food Science and Technology, 68.* https://doi.org/10.1016/j.lwt.2015.12.006

Memic, A., Abudula, T., Mohammed, H. S., Joshi Navare, K., Colombani, T., & Bencherif, S. A. (2019). Latest progress in electrospun nanofibers for wound healing applications. *ACS Applied Bio Materials.* https://doi.org/10.1021/acsabm.8b00637

Michalik, R., & Wandzik, I. (2020). A mini-review on chitosan-based hydrogels with potential for sustainable agricultural applications. *Polymers*, *12*(10), 1–16. https://doi.org/10.3390/polym12102425

Mittal, H., Ray, S. S., Kaith, B. S., Bhatia, J. K., Sukriti, Sharma, J., & Alhassan, S. M. (2018). Recent progress in the structural modification of chitosan for applications in diversified biomedical fields. *European Polymer Journal*. https://doi.org/10.1016/j.eurpolymj.2018.10.013

Mitura, S., Sionkowska, A., & Jaiswal, A. (2020). Biopolymers for hydrogels in cosmetics: Review. *Journal of Materials Science: Materials in Medicine*, *31*(6). https://doi.org/10.1007/s10856-020-06390-w

Mndlovu, H., Du Toit, L. C., Kumar, P., Choonara, Y. E., Marimuthu, T., Kondiah, P. P. D., & Pillay, V. (2020). Bioplatform fabrication approaches affecting chitosan-based interpolymer complex properties and performance as wound dressings. *Molecules*, *25*(1). https://doi.org/10.3390/molecules25010222

Mohamad, N., Mohd Amin, M. C. I., Pandey, M., Ahmad, N., & Rajab, N. F. (2014). Bacterial cellulose/acrylic acid hydrogel synthesized via electron beam irradiation: Accelerated burn wound healing in an animal model. *Carbohydrate Polymers*, *114*. https://doi.org/10.1016/j.carbpol.2014.08.025

Mohamady Ghobashy, M. (2020). The application of natural polymer-based hydrogels for agriculture. In Y. Chen (Ed.), *Hydrogels based on natural polymers* (pp. 329–356). Elsevier. https://doi.org/10.1016/B978-0-12-816421-1.00013-6

Mohammedi, Z. (2017). Structure, properties and medical advances for biocellulose applications: A review. *American Journal of Polymer Science and Technology*, *3*(5). https://doi.org/10.11648/j.ajpst.20170305.12

Mohan, S., Oluwafemi, O. S., Kalarikkal, N., Thomas, S., & Songca, S. P. (2016). Biopolymers—application in nanoscience and nanotechnology. In S. B. B. K. Perveen (Ed.), *Recent advances in biopolymers* (pp. 47–72). IntechOpen. https://doi.org/10.5772/62225

Mohandas, A., Deepthi, S., Biswas, R., & Jayakumar, R. (2018). Chitosan based metallic nanocomposite scaffolds as antimicrobial wound dressings. *Bioactive Materials*. https://doi.org/10.1016/j.bioactmat.2017.11.003

Morsy, R., Ali, S. S., & El-Shetehy, M. (2017). Development of hydroxyapatite-chitosan gel sunscreen combating clinical multidrug-resistant bacteria. *Journal of Molecular Structure*, *1143*. https://doi.org/10.1016/j.molstruc.2017.04.090

Muthukumar, T., Song, J. E., & Khang, G. (2019). Biological role of gellan gum in improving scaffold drug delivery, cell adhesion properties for tissue engineering applications. *Molecules*, *24*(24). https://doi.org/10.3390/molecules24244514

Naomi, R., Ratanavaraporn, J., & Fauzi, M. B. (2020). Comprehensive review of hybrid collagen and silk fibroin for cutaneous wound healing. *Materials*. https://doi.org/10.3390/ma13143097

NatureWorks. (n.d.). Retrieved February 26, 2021, from www.natureworksllc.com/

Naumenko, E. A., Guryanov, I. D., Yendluri, R., Lvov, Y. M., & Fakhrullin, R. F. (2016). Clay nanotube-biopolymer composite scaffolds for tissue engineering. *Nanoscale*, *8*(13). https://doi.org/10.1039/c6nr00641h

Neuman, M. G., Nanau, R. M., Oruña-Sanchez, L., & Coto, G. (2015). Hyaluronic acid and wound healing. *Journal of Pharmacy and Pharmaceutical Sciences*, *18*(1). https://doi.org/10.18433/j3k89d

Nilforoushzadeh, M. A., Amirkhani, M. A., Zarrintaj, P., Salehi Moghaddam, A., Mehrabi, T., Alavi, S., & Mollapour Sisakht, M. (2018). Skin care and rejuvenation by cosmeceutical facial mask. *Journal of Cosmetic Dermatology*, *17*(5), 693–702. https://doi.org/10.1111/jocd.12730

Oluwabunmi, K., D'Souza, N. A., Zhao, W., Choi, T. Y., & Theyson, T. (2020). Compostable, fully biobased foams using PLA and micro cellulose for zero energy buildings. *Scientific Reports*, *10*(1), 1–20. https://doi.org/10.1038/s41598-020-74478-y

Pacheco-Torgal, F. (2014). Eco-efficient construction and building materials research under the EU Framework Programme Horizon 2020. *Construction and Building Materials*. https://doi.org/10.1016/j.conbuildmat.2013.10.058

Pacheco-Torgal, F. (2016). Introduction to biopolymers and biotech admixtures for eco-efficient construction materials. In F. Pacheco-Torgal, V. Ivanov, N. Karak, & H. Jonkers (Eds.), *Biopolymers and biotech admixtures for eco-efficient construction materials* (pp. 1–10). Woodhead Publishing Limited. https://doi.org/10.1016/B978-0-08-100214-8.00001-4

Pacheco-Torgal, F., & Jalali, S. (2011). *Eco-efficient construction and building materials*. https://doi.org/10.1007/978-0-85729-892-8

Park, C. J., Clark, S. G., Lichtensteiger, C. A., Jamison, R. D., & Johnson, A. J. W. (2009). Accelerated wound closure of pressure ulcers in aged mice by chitosan scaffolds with and without bFGF. *Acta Biomaterialia*, *5*(6). https://doi.org/10.1016/j.actbio.2009.03.002

Peres, D. D., Hubner, A., De Oliveira, C. A., De Almeida, T. S., Kaneko, T. M., Consiglieri, V. O., Pinto, C. A. S. de O., Velasco, M. V. R., & Baby, A. R. (2017). Hydrolyzed collagen interferes with in vitro photoprotective effectiveness of sunscreens. *Brazilian Journal of Pharmaceutical Sciences*, *53*(2). https://doi.org/10.1590/s2175-97902017000216119

Phetchuay, C., Horpibulsuk, S., Arulrajah, A., Suksiripattanapong, C., & Udomchai, A. (2016). Strength development in soft marine clay stabilized by fly ash and calcium carbide residue based geopolymer. *Applied Clay Science*, *127–128*, 134–142. https://doi.org/10.1016/j.clay.2016.04.005

Picheth, G. F., Pirich, C. L., Sierakowski, M. R., Woehl, M. A., Sakakibara, C. N., de Souza, C. F., Martin, A. A., da Silva, R., & de Freitas, R. A. (2017). Bacterial cellulose in biomedical applications: A review. *International Journal of Biological Macromolecules*. https://doi.org/10.1016/j.ijbiomac.2017.05.171

Plank, J. (2005). Applications of biopolymers in construction engineering. In *Biopolymers online* (pp. 29–93). Wiley Online Library. https://doi.org/10.1002/3527600035.bpola002

Portela, R., Leal, C. R., Almeida, P. L., & Sobral, R. G. (2019). Bacterial cellulose: A versatile biopolymer for wound dressing applications. *Microbial Biotechnology*. https://doi.org/10.1111/1751-7915.13392

Pourali, P., Razavianzadeh, N., Khojasteh, L., & Yahyaei, B. (2018). Assessment of the cutaneous wound healing efficiency of acidic, neutral and alkaline bacterial cellulose membrane in rat. *Journal of Materials Science: Materials in Medicine*, *29*(7). https://doi.org/10.1007/s10856-018-6099-4

Prambauer, M., Wendeler, C., Hofmann, H., & Burgstaller, C. (2019). *Biodegradable geotextiles—The use of biopolymers in short-lived soil protections*. https://doi.org/10.32075/17ECSMGE-2019-0976

Prambauer, M., Wendeler, C., Weitzenböck, J., & Burgstaller, C. (2019). Biodegradable geotextiles—An overview of existing and potential materials. *Geotextiles and Geomembranes*, *47*(1), 48–59. https://doi.org/10.1016/j.geotexmem.2018.09.006

Qian, S., Sheng, K., Yu, K., Xu, L., & Fontanillo Lopez, C. A. (2018). Improved properties of PLA biocomposites toughened with bamboo cellulose nanowhiskers through silane modification. *Journal of Materials Science*, *53*(15), 10920–10932. https://doi.org/10.1007/s10853-018-2377-2

Qiu, Y., Amirkhani, M., Mayton, H., Chen, Z., & Taylor, A. G. (2020). Biostimulant seed coating treatments to improve cover crop germination and seedling growth. *Agronomy*, *10*(2), 1–14. https://doi.org/10.3390/agronomy10020154

Qiu, Y., Qiu, L., Cui, J., & Wei, Q. (2016). Bacterial cellulose and bacterial cellulose-vaccarin membranes for wound healing. *Materials Science and Engineering C, 59.* https://doi.org/10.1016/j.msec.2015.10.016

Rabea, E. I., Badawy, M. E. I., Steurbaut, W., & Stevens, C. V. (2009). In vitro assessment of N-(benzyl)chitosan derivatives against some plant pathogenic bacteria and fungi. *European Polymer Journal, 45*(1). https://doi.org/10.1016/j.eurpolymj.2008.10.021

Raj, S. N., Lavanya, S. N., Sudisha, J., & Shetty, H. S. (2011). Applications of biopolymers in agriculture with special reference to role of plant derived biopolymers in crop protection. In S. Kalia & L. Averous (Eds.), *Biopolymers: Biomedical and environmental applications* (pp. 459–481). Scrivener Publishing LLC. https://doi.org/10.1002/9781118164792.ch16

Rakesh, P. (2017). Effect of biopolymers and synthetic seed coating polymers on castor and groundnut seed. *International Journal of Pure & Applied Bioscience, 5*(4), 2043–2048. https://doi.org/10.18782/2320-7051.5788

Ramakrishna, B. S., Venkataraman, S., Srinivasan, P., Dash, P., Young, G. P., & Binder, H. J. (2000). Amylase-resistant starch plus oral rehydration solution for cholera. *New England Journal of Medicine, 342*(5). https://doi.org/10.1056/nejm200002033420502

Ramli, R. A. (2019). Slow release fertilizer hydrogels: A review. *Polymer Chemistry, 10*(45), 6073–6090. https://doi.org/10.1039/c9py01036j

Rasal, R. M., Janorkar, A. V., & Hirt, D. E. (2010). Poly(lactic acid) modifications. *Progress in Polymer Science (Oxford).* https://doi.org/10.1016/j.progpolymsci.2009.12.003

Rastogi, V. K., & Samyn, P. (2015). Bio-based coatings for paper applications. *Coatings, 5*(4), 887–930. https://doi.org/10.3390/coatings5040887

Ravin, L. J., Baldinus, J. G., & Mazur, M. L. (1962). Effect of sulfate content of several anionic polymers on in vitro activity of pepsin. *Journal of Pharmaceutical Sciences, 51*(9). https://doi.org/10.1002/jps.2600510909

Reichert, C. L., Bugnicourt, E., Coltelli, M. B., Cinelli, P., Lazzeri, A., Canesi, I., Braca, F., Martínez, B. M., Alonso, R., Agostinis, L., Verstichel, S., Six, L., De Mets, S., Gómez, E. C., Ißbrücker, C., Geerinck, R., Nettleton, D. F., Campos, I., Sauter, E., . . . Schmid, M. (2020). Bio-based packaging: Materials, modifications, industrial applications and sustainability. *Polymers, 12.* https://doi.org/10.3390/polym12071558

Ridwanto, R., Lubis, M., Syahputra, R., & Inriyani, R. (2019). Utilisation of biopolymer combination as a material for making gel peel off mask. In *Roceedings of the 5th annual international seminar on trends in science and science education, AISTSSE 2018.* https://doi.org/10.4108/eai.18-10-2018.2287373

Saatchi, A., Arani, A. R., Moghanian, A., & Mozafari, M. (2020). Cerium-doped bioactive glass-loaded chitosan/polyethylene oxide nanofiber with elevated antibacterial properties as a potential wound dressing. *Ceramics International.* https://doi.org/10.1016/j.ceramint.2020.12.078

Sahana, T. G., & Rekha, P. D. (2018). Biopolymers: Applications in wound healing and skin tissue engineering. *Molecular biology reports.* https://doi.org/10.1007/s11033-018-4296-3

Saikia, C., & Gogoi, P. (2015). Chitosan: A promising biopolymer in drug delivery applications. *Journal of Molecular and Genetic Medicine, s4.* https://doi.org/10.4172/1747-0862.s4-006

Senthilkumar, K., Ramajayam, G., Venkatesan, J., Kim, S. K., & Ahn, B. C. (2017). Biomedical applications of fucoidan, seaweed polysaccharides. In *Seaweed polysaccharides: Isolation, biological and biomedical applications.* https://doi.org/10.1016/B978-0-12-809816-5.00014-1

Serrano-Ruiz, H., Martin-Closas, L., & Pelacho, A. M. (2021). Biodegradable plastic mulches: Impact on the agricultural biotic environment. *Science of the Total Environment, 750.* https://doi.org/10.1016/j.scitotenv.2020.141228

Sezer, A. D., & Cevher, E. (2011). Fucoidan: A versatile biopolymer for biomedical applications. In *Studies in mechanobiology, tissue engineering and biomaterials* (Vol. 8). https://doi.org/10.1007/8415_2011_67

Sharif, R., Mujtaba, M., Rahman, M. U., Shalmani, A., Ahmad, H., Anwar, T., Tianchan, D., & Wang, X. (2018). The multifunctional role of chitosan in horticultural crops: A review. *Molecules*. https://doi.org/10.3390/molecules23040872

Sharma, A., Sood, K., Kaur, J., & Khatri, M. (2019). Agrochemical loaded biocompatible chitosan nanoparticles for insect pest management. *Biocatalysis and Agricultural Biotechnology*, *18*. https://doi.org/10.1016/j.bcab.2019.101079

Sharma, R., Jafari, S. M., & Sharma, S. (2020). Antimicrobial bio-nanocomposites and their potential applications in food packaging. *Food Control*, *112*(September 2019). https://doi.org/10.1016/j.foodcont.2020.107086

Shokri, J., & Adibki, K. (2013). Application of cellulose and cellulose derivatives in pharmaceutical industries. *Cellulose—Medical, Pharmaceutical and Electronic Applications*. https://doi.org/10.5772/55178

Sionkowska, A. (2015). The potential of polymers from natural sources as components of the blends for biomedical and cosmetic applications. *Pure and Applied Chemistry*, *87*(11–12), 1075–1084. https://doi.org/10.1515/pac-2015-0105

Sionkowska, A., Lewandowska, K., Planecka, A., Szarszewska, P., Krasinska, K., Kaczmarek, B., & Kozlowska, J. (2014). Biopolymer blends as potential biomaterials and cosmetic materials. *Key Engineering Materials*, *583*(September), 95–100. https://doi.org/10.4028/www.scientific.net/KEM.583.95

Sionkowska, A., & Płanecka, A. (2013). Surface properties of thin films based on the mixtures of chitosan and silk fibroin. *Journal of Molecular Liquids*, *186*, 157–162. https://doi.org/10.1016/j.molliq.2013.07.008

Sionkowska, A., Płanecka, A., Lewandowska, K., & Michalska, M. (2014). The influence of UV-irradiation on thermal and mechanical properties of chitosan and silk fibroin mixtures. *Journal of Photochemistry and Photobiology B: Biology*, *140*(September), 301–305. https://doi.org/10.1016/j.jphotobiol.2014.08.017

Song, Z., Xiao, H., & Zhao, Y. (2014). Hydrophobic-modified nano-cellulose fiber/PLA biodegradable composites for lowering water vapor transmission rate (WVTR) of paper. *Carbohydrate Polymers*, *111*. https://doi.org/10.1016/j.carbpol.2014.04.049

Srivastava, A., Aaisa, J., Kumar TA, T., Ginjupalli, K., & Upadhya P, N. (2012). Alginates: A review of compositional aspects for dental applications. *Trends in Biomaterials & Artificial Organs*, *26*(1).

Stojkovska, J., Djurdjevic, Z., Jancic, I., Bufan, B., Milenkovic, M., Jankovic, R., Miskovic-Stankovic, V., & Obradovic, B. (2018). Comparative in vivo evaluation of novel formulations based on alginate and silver nanoparticles for wound treatments. *Journal of Biomaterials Applications*, *32*(9). https://doi.org/10.1177/0885328218759564

Stubbe, B., Mignon, A., Declercq, H., Van Vlierberghe, S., & Dubruel, P. (2019). Development of gelatin-alginate hydrogels for burn wound treatment. *Macromolecular Bioscience*, *19*(8). https://doi.org/10.1002/mabi.201900123

Suh, H.-W., Lewis, J., Fong, L., Ramseier, J. Y., Carlson, K., Peng, Z.-H., Yin, E. S., Saltzman, W. M., & Girardi, M. (2019). Biodegradable bioadhesive nanoparticle incorporation of broad-spectrum organic sunscreen agents. *Bioengineering & Translational Medicine*, *4*(1), 129–140. https://doi.org/10.1002/btm2.10092

Sulaeva, I., Henniges, U., Rosenau, T., & Potthast, A. (2015). Bacterial cellulose as a material for wound treatment: Properties and modifications: A review. *Biotechnology Advances*. https://doi.org/10.1016/j.biotechadv.2015.07.009

Sun, B., Zhang, M., Shen, J., He, Z., Fatehi, P., & Ni, Y. (2018). Applications of cellulose-based materials in sustained drug delivery systems. *Current Medicinal Chemistry*, 26(14). https://doi.org/10.2174/0929867324666170705143308

Sun, L., Gao, W., Fu, X., Shi, M., Xie, W., Zhang, W., Zhao, F., & Chen, X. (2018). Enhanced wound healing in diabetic rats by nanofibrous scaffolds mimicking the basketweave pattern of collagen fibrils in native skin. *Biomaterials Science*, 6(2). https://doi.org/10.1039/c7bm00545h

Sundar, G., Joseph, J., Prabhakumari, C., John, A., & Abraham, A. (2020). Natural collagen bioscaffolds for skin tissue engineering strategies in burns: A critical review. *International Journal of Polymeric Materials and Polymeric Biomaterials*, 1–12. https://doi.org/10.1080/00914037.2020.1740991

Tang, X. Z., Kumar, P., Alavi, S., & Sandeep, K. P. (2012). Recent advances in biopolymers and biopolymer-based nanocomposites for food packaging materials. *Critical Reviews in Food Science and Nutrition*, 52(5), 426–442. https://doi.org/10.1080/10408398.2010.500508

Taskan, M. M., Balci Yuce, H., Karatas, O., Gevrek, F., Isiker Kara, G., Celt, M., & Sirma Taskan, E. (2020). Hyaluronic acid with antioxidants improve wound healing in rats. *Biotechnic and Histochemistry*. https://doi.org/10.1080/10520295.2020.1832255

Teh, B. M., Shen, Y., Friedland, P. L., Atlas, M. D., & Marano, R. J. (2012). A review on the use of hyaluronic acid in tympanic membrane wound healing. *Expert Opinion on Biological Therapy*. https://doi.org/10.1517/14712598.2012.634792

Thongcharoen, N., Khongtong, S., Srivaro, S., Wisadsatorn, S., Chub-Uppakarn, T., & Chaowana, P. (2021). Development of structural insulated panels made from wood-composite boards and natural rubber foam. *Polymers*, 13(15). https://doi.org/10.3390/polym13152497

Tønnesen, H. H., & Karlsen, J. (2002). Alginate in drug delivery systems. *Drug Development and Industrial Pharmacy*. https://doi.org/10.1081/DDC-120003853

Tulamandi, S., Rangarajan, V., Rizvi, S. S. H., Singhal, R. S., Chattopadhyay, S. K., & Saha, N. C. (2016). A biodegradable and edible packaging film based on papaya puree, gelatin, and defatted soy protein. *Food Packaging and Shelf Life*, 10. https://doi.org/10.1016/j.fpsl.2016.10.007

Tutal, A., Partschefeld, S., Schneider, J., & Osburg, A. (2020). Effects of bio-based plasticizers, made from starch, on the properties of fresh and hardened metakaolin-geopolymer mortar: Basic investigations. *Clays and Clay Minerals*, 68(5), 413–427. https://doi.org/10.1007/s42860-020-00084-8

Uddin, I., Venkatachalam, S., Mukhopadhyay, A., & Amil Usmani, M. (2016). Nanomaterials in the pharmaceuticals: Occurrence, behaviour and applications. *Current Pharmaceutical Design*, 22(11), 1472–1484. https://doi.org/10.2174/1381612822666160118104727

Varaprasad, K., Jayaramudu, T., Kanikireddy, V., Toro, C., & Sadiku, E. R. (2020). Alginate-based composite materials for wound dressing application: A mini review. *Carbohydrate Polymers*. https://doi.org/10.1016/j.carbpol.2020.116025

Wagenhäuser, M. U., Mulorz, J., Ibing, W., Simon, F., Spin, J. M., Schelzig, H., & Oberhuber, A. (2016). Oxidized (non)-regenerated cellulose affects fundamental cellular processes of wound healing. *Scientific Reports*, 6. https://doi.org/10.1038/srep32238

Wang, J., Jin, X., & Chang, D. (2009). Chemical modification of chitosan under high-intensity ultrasound and properties of chitosan derivatives. *Carbohydrate Polymers*, 78(1). https://doi.org/10.1016/j.carbpol.2009.03.032

Wang, N., Gao, Y. Z., Wang, P., Yang, S., Xie, T. M., & Xiao, Z. G. (2016). Effect of microwave modification on mechanical properties and structural characteristics of soy protein isolate and zein blended film. *Czech Journal of Food Sciences*, *34*(2). https://doi.org/10.17221/442/2015-CJFS

Wang, T., Miller, D., Burczynski, F., & Gu, X. (2014). Evaluation of percutaneous permeation of repellent DEET and sunscreen oxybenzone from emulsion-based formulations in artificial membrane and human skin. *Acta Pharmaceutica Sinica B*, *4*(1). https://doi.org/10.1016/j.apsb.2013.11.002

Wicklein, B., Kocjan, A., Salazar-Alvarez, G., Carosio, F., Camino, G., Antonietti, M., & Bergström, L. (2015). Thermally insulating and fire-retardant lightweight anisotropic foams based on nanocellulose and graphene oxide. *Nature Nanotechnology*, *10*(3), 277–283. https://doi.org/10.1038/nnano.2014.248

Wiederschain, G. Y. (2007). Polysaccharides. Structural diversity and functional versatility. *Biochemistry (Moscow)*, *72*(6). https://doi.org/10.1134/s0006297907060120

Wiewel, B. V., & Lamoree, M. (2016). Geotextile composition, application and ecotoxicology—A review. *Journal of Hazardous Materials*. https://doi.org/10.1016/j.jhazmat.2016.04.060

Wojnowska-Baryła, I., Kulikowska, D., & Bernat, K. (2020). Effect of bio-based products on waste management. *Sustainability (Switzerland)*, *12*(5), 1–12. https://doi.org/10.3390/su12052088

Wu, H., Yao, C., Li, C., Miao, M., Zhong, Y., Lu, Y., & Liu, T. (2020). Review of application and innovation of geotextiles in geotechnical engineering. *Materials*. https://doi.org/10.3390/MA13071774

Wu, S., Deng, L., Hsia, H., Xu, K., He, Y., Huang, Q., Peng, Y., Zhou, Z., & Peng, C. (2017). Evaluation of gelatin-hyaluronic acid composite hydrogels for accelerating wound healing. *Journal of Biomaterials Applications*, *31*(10). https://doi.org/10.1177/0885328217702526

Xia, W., Liu, P., Zhang, J., & Chen, J. (2011). Biological activities of chitosan and chitooligosaccharides. *Food Hydrocolloids*, *25*(2). https://doi.org/10.1016/j.foodhyd.2010.03.003

Xing, K., Zhu, X., Peng, X., & Qin, S. (2015). Chitosan antimicrobial and eliciting properties for pest control in agriculture: A review. *Agronomy for Sustainable Development*. https://doi.org/10.1007/s13593-014-0252-3

Xu, T., Ma, C., Aytac, Z., Hu, X., Ng, K. W., White, J. C., & Demokritou, P. (2020). Enhancing agrichemical delivery and seedling development with biodegradable, tunable, biopolymer-based nanofiber seed coatings. *ACS Sustainable Chemistry and Engineering*, *8*(25), 9537–9548. https://doi.org/10.1021/acssuschemeng.0c02696

Xu, Z. (2020). Recently progress on polylactide/nanocellulose nanocomposites. *IOP Conference Series: Materials Science and Engineering*, *772*(1). https://doi.org/10.1088/1757-899X/772/1/012006

Xun, W., Wu, C., Leng, X., Li, J., Xin, D., & Li, Y. (2020). Effect of functional superplasticizers on concrete strength and pore structure. *Applied Sciences (Switzerland)*, *10*(10). https://doi.org/10.3390/app10103496

Yan, X., Chen, B., Lin, Y., Li, Y., Xiao, Z., Hou, X., Tan, Q., & Dai, J. (2010). Acceleration of diabetic wound healing by collagen-binding vascular endothelial growth factor in diabetic rat model. *Diabetes Research and Clinical Practice*, *90*(1). https://doi.org/10.1016/j.diabres.2010.07.001

Yu, Z., Sun, L., Wang, W., Zeng, W., Mustapha, A., & Lin, M. (2018). Soy protein-based films incorporated with cellulose nanocrystals and pine needle extract for active packaging. *Industrial Crops and Products*, *112*. https://doi.org/10.1016/j.indcrop.2017.12.031

Zainescu, G. A., Mihalache, M., Voicu, P., Constantinescu, R., Ilie, L., & Obrişcă, M. (2011). Biopolymers systems from leather wastes for degraded soils remediation. *Scientific Papers—Series A, Agronomy*, *54*, 60–68.

ZealaFoam. (n.d.). Retrieved March 7, 2021, from www.biopolymernetwork.com/content/Zealafoam/90.aspx

Zeng, H. Y., & Huang, Y. C. (2018). Basic fibroblast growth factor released from fucoidan-modified chitosan/alginate scaffolds for promoting fibroblasts migration. *Journal of Polymer Research*, 25(3). https://doi.org/10.1007/s10965-018-1476-8

Zhang, C., Chao, L., Zhang, Z., Zhang, L., Li, Q., Fan, H., Zhang, S., Liu, Q., Qiao, Y., Tian, Y., Wang, Y., & Hu, X. (2021). Pyrolysis of cellulose: Evolution of functionalities and structure of bio-char versus temperature. *Renewable and Sustainable Energy Reviews*, 135(110416). https://doi.org/10.1016/j.rser.2020.110416

Zhang, C., Long, Y., Li, J., Li, M., Xing, D., An, H., Wu, X., & Wu, Y. (2020). A chitosan composite film sprayed before pathogen infection effectively controls postharvest soft rot in kiwifruit. *Agronomy*, 10(2). https://doi.org/10.3390/agronomy10020265

Zhang, L., Sang, Y., Feng, J., Li, Z., & Zhao, A. (2016). Polysaccharide-based micro/nanocarriers for oral colon-targeted drug delivery. *Journal of Drug Targeting*. https://doi.org/1 0.3109/1061186X.2015.1128941

Zhang, M., & Zhao, X. (2020). Alginate hydrogel dressings for advanced wound management. *International Journal of Biological Macromolecules*. https://doi.org/10.1016/j.ijbiomac.2020.07.311

Zhao, S., Malfait, W. J., Guerrero-Alburquerque, N., Koebel, M. M., & Nyström, G. (2018). Biopolymer aerogels and foams: Chemistry, properties, and applications. *Angewandte Chemie—International Edition*. https://doi.org/10.1002/anie.201709014

Zhu, J., Hu, J., Jiang, C., Liu, S., & Li, Y. (2019). Ultralight, hydrophobic, monolithic konjac glucomannan-silica composite aerogel with thermal insulation and mechanical properties. *Carbohydrate Polymers*. https://doi.org/10.1016/j.carbpol.2018.11.073

Zhu, T., Mao, J., Cheng, Y., Liu, H., Lv, L., Ge, M., Li, S., Huang, J., Chen, Z., Li, H., Yang, L., & Lai, Y. (2019). Recent progress of polysaccharide-based hydrogel interfaces for wound healing and tissue engineering. *Advanced Materials Interfaces*, 6(17), 1–22. https://doi.org/10.1002/admi.201900761

6 Starch-Based Films with Essential Oils for Antimicrobial Food Packaging

6.1 INTRODUCTION

Food packaging plays a primary role in the protection of the food product from the influence of the external environment. The major goal of food packaging is to hold food in the best economical way, satisfying both industrial and consumer requirements, ensuring food safety, and minimizing environmental effects. Advances in food packaging research led to the development of active packaging and intelligent packaging. Active packaging is a novel method used to prolong the shelf life of perishable foods and maintain or improve the quality and safety of prepared foods due to its interaction with the product. Also, active packaging has the potential to replace the addition of active compounds into foods, reduce the movement of particles from packaging materials to food, and get rid of industrial processes that can cause the introduction of pathogenic microorganisms into the product (Schaefer & Cheung, 2018).

In the 20th century, the production of petroleum-based food packaging witnessed a significant increase, driven by its affordability, robust mechanical properties, and versatile applications. However, growing concern over the negative impacts of plastics on the environment has encouraged the development of alternative packaging that uses renewable sources as raw material with promising characteristics. For example, biopolymers stand out as renewable feedstock to supply the demand for sustainable, green, biodegradable packaging with antimicrobial characteristics (Ayu Rafiqah et al., 2021; Norrrahim et al., 2021). Polysaccharides, proteins, and lipids are well known in the production of films and coatings for food packaging (Kamarudin et al., 2022). Poly(lactic acid) (PLA) (Cheng et al., 2021), poly(hydroxybutyrate) (PHB) (Manikandan et al., 2020), and starch-based materials (Lauer & Smith, 2020) stand as prominent examples of commercially viable biopolymers utilized in the manufacturing of food packaging films. Nevertheless, the key challenge in utilizing these polymer materials lies in meeting the essential criteria for superior mechanical strength and barrier properties while maintaining cost-effectiveness in the market.

Starch is present in cereals, roots, tubers, fruits, and vegetables. Due to its biocompatibility, biodegradability, and availability, starch has been widely used in the production of edible films and coatings for food packaging. Corn starch and tapioca starch, for example, are considered the most consolidated bioplastic alternatives

 DOI: 10.1201/9781003416043-6

for this application (Rodrigues et al., 2020; Wang & Zhang, 2021). Significant advancements in the characteristics of starch films have been documented through the development of composites, polymer blends, the incorporation of active or intelligent features via additives (Fernandes et al., 2013), and various chemical and physical enhancements (Abbas et al., 2019). The antibacterial and anti-oxidation properties and UV, oxygen, and water vapor barriers can reduce the reproduction of microorganisms and prolong the shelf life of packaged foods. Intelligent starch-based films with pH-, temperature-, magnetic field-, glucose-, and enzyme-responsive characteristics can be used to monitor the freshness of foods and control the delivery of functional ingredients and drugs.

Edible films and coatings aim to prolong the shelf life of food products, providing microbial protection, better sensory perception, and reduction of antioxidant activity. To further enhance their performance, edible films can also contain active components, which not only improve food quality and safety but can also enhance the physicochemical properties of biopolymer-based films (Liu & Yang, 2019). For example, the addition of essential oils like *Eucalyptus globulus* in the production of films has resulted in products with better barrier and optical properties, in addition to incorporating antioxidant and antibacterial activities due to the migration of their active compounds (Azadbakht et al., 2018).

This chapter provides a thorough analysis of the impact of essential oils as additives on the main properties of starch-based films for the production of food packaging. Initially, theoretical aspects of starch and essential oils are briefly discussed in order to direct the reader toward the structure–properties relationships that are subsequently addressed. Throughout the chapter, an overview analysis of material properties and applications is also presented. The objective is to highlight the significant recent progress on sustainable starch-based films with essential oils added and also help the reader to identify opportunities and gaps related to this topic.

6.2 CHARACTERISTICS OF STARCH

Starch, a polysaccharide synthesized by plants and predominantly found in fruits, roots, tubers, legumes, and cereals, typically constitutes a range of 25 to 90% (McClements & Öztürk, 2021). It is a semi-crystalline polymer with about 1,000 to 2,000,000 (Copeland et al., 2009) glucose monomers linked by α-1,4 glycosidic bonds. There are three reactive hydroxyl groups on the glucose unit of the starch chain, with a primary and two secondary hydroxyl groups, which can act as anchor points for modification of chemistries. In the food industry, starch is used to improve the quality of final products, being able to confer functional properties and modify the texture and consistency of different food products (Ali et al., 2016). Starch granules are semi-crystalline and composed predominantly of two polysaccharides: amylose, with a molecular weight ranging from 105 to 106 Da, and amylopectin, of higher molecular weight (from 107 to 109 Da) and a large number of short branches (Maulana et al., 2022). With exceptions, amylopectin constitutes the principal mass component of starch, while amylose constitutes 15 to 30% (Bertoft, 2017). Figure 6.1 outlines the amylose and amylopectin branched chains from starch structures.

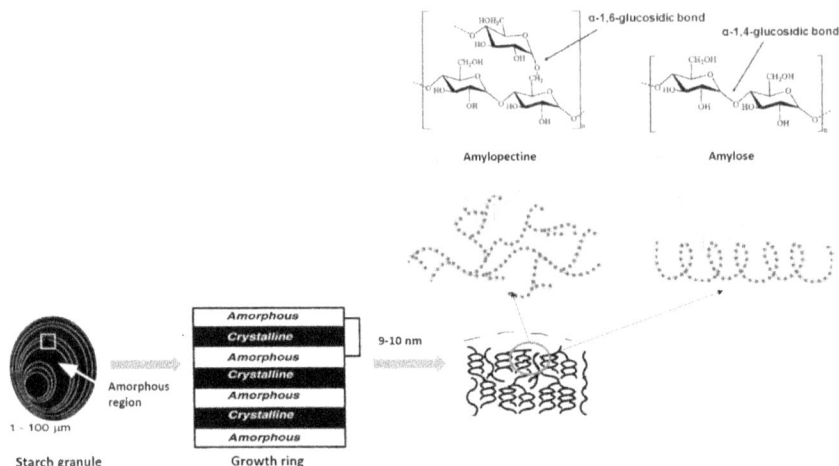

FIGURE 6.1 Schematic representation of the components of starch. Reproduced from Buleon et al. (1998) and Vianna et al. (2021).

Amylose is an essentially linear polymer with a helical structure, composed of, on average, 2 to 12×10^3 glucose units that are linked to each other by an α-1,4 glycosidic bond. The average composition of amylopectin is 4 to 35×10^6 glucose units in α-1,4 glycosidic bonds, connected by branching points (corresponding to 5% of the molecule) and linked to α-1,6, resulting in a complex structure (Khatami et al., 2021). Amylopectin retains a cluster-like organization, and some of its branches are long enough to form double helices with each other and crystallite clusters (Bertoft & Nilsson, 2017). The average degree of polymerization of amylose is less than 5000, while that of amylopectin ranges from 5×10^3 to 10^6. The molecular structures of amylopectin and amylose, as well as their respective fractions, are critical to the general physicochemical properties of starch. In addition to these factors, it has been experimentally proven that the degree of crystallinity equally affects important characteristics of starch such as thermal stability (Lemos et al., 2018). Structural characteristics like short and long chains of amylose and size of starch granules are primarily related to the starch source and therefore can be associated with the properties of the produced film. For example, the properties of starch isolated from various species of rice and corn present significant variations in amylose moisture content and solubility index even between species of the same cereal. A comparison between the physical and chemical properties of seven sweet potato varieties demonstrated a similar result, with significant differences in granule size, crystallinity, amylose content, and gelatinization properties among the different varieties (Contreras-Jiménez et al., 2019).

For starch solubility study, starch is soluble in some polar solvents like dimethyl sulfoxide (DMSO), dimethylformamide (DMF), pyridine, 90/10 v/v mixed solvent of DMSO/water, DMSO with lithium chloride (LiCl), DMF with LiCl, and N,N-dimethyl acetamide (DMAc). Generally, the more polar the solvent, the higher the solubility of starch will be in that solvent. DMSO appears to be a promising solvent media for starch modification owing to its high polarity and stability in neutral and alkaline pHs and fairly high temperatures. It is worth noting that the choice of solvent depends on the desired application (Ojogbo et al., 2020). Figure 6.2 presents the effect of two different types of starch content on physicochemical characteristics.

When starch is heated to a certain temperature, it becomes soluble in water, causing the grains to swell and burst. The semi-crystalline structure is also lost during this process, and the minor amylose particles begin percolating out of the granule and forming a network. As water is compressed, this network increases the viscosity of the mixture. This is referred to as starch gelatinization (Marichelvam et al., 2019). When starch is processed, its amylose content provides strength to the film (Niranjana Prabhu & Prashantha, 2018). Based on Tharanathan, when amylopectin is the predominant component of starch, the mechanical strength of films, such as tensile stress, decreases (2003). Table 6.1 shows the percentages of amylose and amylopectin in various starches (Marichelvam et al., 2019). Table 6.2 presents the tensile properties of certain starch-based films.

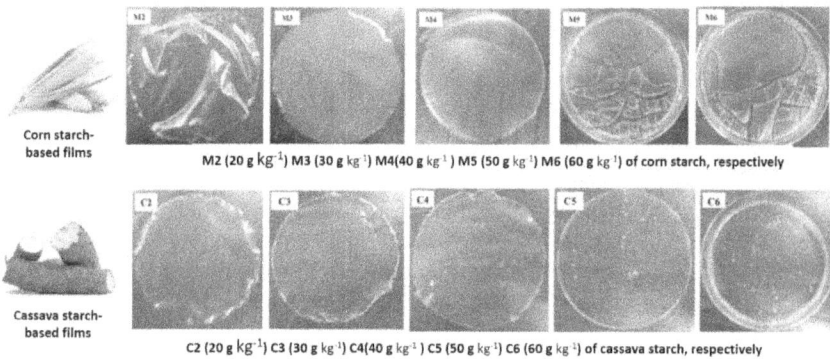

Corn starch-based films

M2 (20 g kg⁻¹) M3 (30 g kg⁻¹) M4(40 g kg⁻¹) M5 (50 g kg⁻¹) M6 (60 g kg⁻¹) of corn starch, respectively

Cassava starch-based films

C2 (20 g kg⁻¹) C3 (30 g kg⁻¹) C4(40 g kg⁻¹) C5 (50 g kg⁻¹) C6 (60 g kg⁻¹) of cassava starch, respectively

FIGURE 6.2 Physicochemical properties of starch-based films from different types of starch with their respective loading. Reproduced from Luchese et al. (2017).

TABLE 6.1
Percentage of Amylose and Amylopectin in Selected Starch Structures

Sources	Amylose (%)	Amylopectin (%)
Arrowroot	20.5	79.5
Banana	17	83
Cassava	18.6	81.4
Corn	28	72
Potato	17.8	82.2
Rice	35	65
Tapioca	16.7	83.3
Wheat	20	80

TABLE 6.2
Tensile Properties of Starch-Based Films

Type of Starch	Tensile Strength (MPa)	Elongation at Break (%)	Young's Modulus (MPa)	Ref.
Maize starch	1.49	51	14.2	(Żołek-Tryznowska &
Potato starch	3.05	70	14.5	Kałuża, 2021)
Oat starch	0.36	27	1.8	
Rice starch	1.80	49	9.6	
Tapioca starch	0.78	137	0.8	
Corn and wheat	15.50	30.00	–	(Song et al., 2018)
Wheat	3.29	15.21	0.12	(Basiak et al., 2016)

6.3   ESSENTIAL OILS

Essential oils are volatile liquids extracted from various parts of aromatic plants like barks, seeds, flowers, peels, fruit, roots, leaves, wood, and whole plants and named depending on from which plant they are obtained (El Sawi et al., 2019). According to the International Organization for Standardization (ISO), an essential oil is a "product obtained from a natural raw material of plant origin, by steam distillation, by mechanical processes from the epicarp of citrus fruits, or by dry distillation, after separation of the aqueous phase if any by physical processes", and it can also be treated physically without changing its composition (Baptista-Silva et al., 2020). Essential oils can be extracted by different methods, such as hydro-distillation, steam distillation, hydro-diffusion, and solvent extraction (Aziz et al., 2018).

The physical characteristics of essential oils include their high solubility in ether, alcohol, and fixed oils but low solubility in water, which is denser than oils. Essential oils are usually colorless and liquid at room temperature and are distinguished by their distinctive odor. These volatile liquids can be characterized by refractive index (RI) measurement and their high optical activity (Dhifi et al., 2016). These extracts of aromatic plants are composed of organic compounds such as carbon, hydrogen, and oxygen and in some cases nitrogen and sulfur derivatives. Carbon and hydrogen atoms tend to attract functional groups, resulting in a relatively inactive framework of atoms in the essential oils (Moghaddam & Mehdizadeh, 2017). These aromatic liquids are diverse due to the presence of different functional groups, and they exist in various forms, including aldehydes, alcohols, ethers, ketones, acids, amines, sulfides, epoxides, and others.

Essential oils can be divided into two types, terpenes and hydrocarbons. Terpenes are composed of a different number of isoprene units, and from the isoprene units, terpenes can be categorized into hemiterpenes (C_5H_8), monoterpenes (C_5H_8)$_2$, sesquiterpenes (C_5H_8)$_2$, diterpenes (C_5H_8)$_4$, and so on (Rubulotta & Quadrelli, 2019). Almost 90% of all essential oils are composed of monoterpenes.

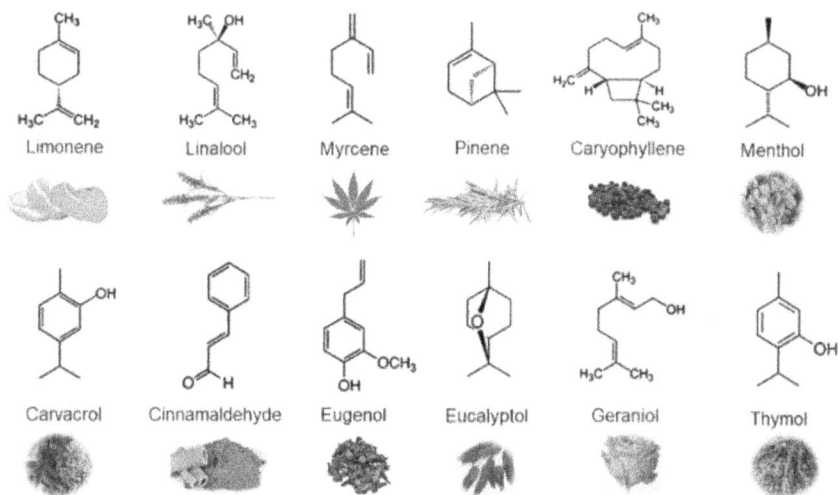

FIGURE 6.3   Main chemical structures of some typical compounds present in essential oils.

Some examples of monoterpene-structured essential oils are *Lavandula luisieri*, *Cymbopogon citratus*, and white and green tea (Dias et al., 2017). Hydrocarbon-based essential oils are made of carbon and hydrogen atoms. Depending on their structure, hydrocarbons are categorized into aliphatic, alkanes, and aromatic hydrocarbons. It is well known that citrus oil has a specific acid odor caused by aliphatic hydrocarbons that are composed of 8–10 carbon atoms connected linearly. Also, an aliphatic molecule with six carbon atoms provides a leafy green scent in floral oils, while octanal aldehydes are responsible for the smell in orange oil. Essential oils contain just a trace amount of aliphatic compounds that have oxygenated functional groups attached to them and are responsible for odor. On the other hand, alkanes are composed of carbon atoms liked together by single bonds, while alkynes contain carbon–carbon triple covalent bonds. Aromatic hydrocarbons are responsible for a pleasant odor due to the presence of benzene rings in their structure (Herman et al., 2019). Figure 6.3 presents the main chemical structures of some typical compounds present in essential oils such as phenols, aldehydes, ketones, terpenes, ethers, and alcohols, which are responsible for their outstanding biological activities. Among all of these components, terpenes are the most predominant. Terpenes are the secondary metabolites of plants formed by isoprene units and have high volatility. When biochemically modified, terpenes are called terpenoids (Wani et al., 2021).

6.3.1   Extraction Methods

Various methods have been used for the extraction of essential oil from plants, including steam distillation, hydrodistillation, supercritical fluid extraction, solvent extraction, microwave-assisted extraction, and ultrasound-assisted extraction

(Pateiro et al., 2018). The method used for the extraction of essential oil had a significant effect on the composition of the oil.

Khajeh et al. (2004) studied the composition of essential oil from *Carum copticum* extracted using both hydrodistillation and supercritical carbon dioxide methods. Apart from cost savings and shorter elution times, the supercritical carbon dioxide extraction method can manipulate the composition of the essential oil more accurately by changing the extraction parameters such as temperature and pressure. Latief et al. (2017) compared the extraction of *Melissa officinalis* using hydrodistillation and headspace solid-phase microextraction (HS-SPME) techniques. The compounds identified from *Melissa officinalis* using hydrodistillation and HS-SPME showed qualitative and quantitative differences. Chenni et al. (2016) studied and compared the essential oil extracted from Egyptian basil leaves using solvent-free microwave extraction (SFME) and hydrodistillation. The microwave extraction had a significant reduction in extraction time. There were no characteristic differences in composition or yields with both methods.

6.3.2 Antimicrobial Activity of Essential Oils

The presence of hydrophilic functional groups and/or lipophilicity in essential oil compounds imparts antimicrobial property, which is more predominant against Gram-positive than Gram-negative bacteria. There has been no single specific mechanism proposed for the antimicrobial activity of essential oil, as it depends on the constituents and concentration of essential oil, along with the nature of the microbe (Hasan et al., 2013). The activity of essential oil depends on its chemical constituents. The presence and location of functional groups on different compounds can affect the activity towards different pathogens (Kalemba & Kunicka, 2003). Five main proposed mechanisms of action of essential oil are degradation of the cell membrane, damage to membrane proteins, leakage of cell components, coagulation of cytoplasm, and collapse of the proton motive force (Gutiérrez-del-Río et al., 2018). Zhang et al. (2017) observed that the addition of black pepper oil broke down the cell membrane of *E. coli*, leading to cell deformation and death. The cell membrane responsible for functions including maintaining the energy of the cell, metabolic regulation, and solute transport is disrupted by the essential oil. The active compounds of essential oil attach themselves to the surface of the cell and penetrate to the phospholipid bilayer of the cell membrane, leading to the loss of structural integrity of the cell and subsequent cell death. Hydrophobic essential oils may also disrupt the microbial cell structure, resulting in increased permeability of the cell membrane and leading to loss of important intracellular contents such as proteins, reducing sugars, adenosine triphosphate (ATP), and DNA, leading to the destruction of the cell and leakage of electrolytes. Such cell structure damage was proposed when essential oil compounds carvacrol and thymol were used against Gram-negative bacteria (Nazzaro et al., 2013). Furthermore, the antimicrobial activity of essential oil cannot be attributed to a single cause but to a chain of reaction which involves many activities of bacterial cells (Macwan et al., 2016). Table 6.3 presents essential oil–incorporated starch-based film for antimicrobial applications.

TABLE 6.3

Types of Starch and Essential Oil Incorporated for Antimicrobial Film Applications

Type of Starch	Essential Oil	Type of Microorganism Activity	Ref.
Corn starch	Orange essential oil	*S. aureus* and *L. monocytogenes*	(do Evangelho et al., 2019)
Corn starch	*Zataria multiflora* Boiss.	*Escherichia coli* and *Staphylococcus aureus*	(Ghasemlou et al., 2013)
Corn/wheat starch	Lemon essential oil	*S. aureus* and *E. coli* suspensions	(Song et al., 2018)
Cassava starch	Cinnamon oil	Cinnamon and *Eurotium amstelodami*	(Souza et al., 2013)
Sweet potato starch	Octenylsuccination and oregano essential oil	*Escherichia coli* and *Staphylococcus aureus*	(Li et al., 2018)
Cassava starch/ chitosan	Oregano essential oil	*Salmonella enteritidis* and *Staphylococcus aureus*	(Pelissari et al., 2009)
Millet starch	Clove essential oil	*Syzygium aromaticum*	(G. Al-Hashimi et al., 2020)
Cassava starch/ gelatine	Cinnamon essential oil	*Fusarium oxysporum f.sp. gladiolo* and *Colletotrichum gloesporoides*	(Acosta et al., 2016)
Corn starch	Carvacrol essential oil and montmorillonite (MMT)	*Escherichia coli*	(de Souza et al., 2020)
Chitosan/ acetylated starch	Cinnamon and clove essential oils	*Escherichia coli*	(Choo et al., 2021)

6.4 PREPARATION OF STARCH-BASED FILMS WITH ESSENTIAL OILS

Starch-based films can be prepared using a variety of methods, including casting, extrusion, and solution-casting. The specific method used will depend on the desired properties of the film, as well as the equipment and resources available. In general, the preparation of starch-based films with essential oils starts with the preparation of the starch solution. Starch is typically dissolved in hot water to form a gel-like solution. The concentration of the starch solution will depend on the desired thickness and strength of the film. Essential oils are typically added to the starch solution after it has cooled to a temperature below 60°C, as high temperatures can cause degradation of the oils. The amount of essential oil added will depend on the desired aroma and antimicrobial properties of the film. The essential oils can be blended into the starch solution using a blender or mixer. Once the starch solution is prepared, it can be cast into a film using a casting machine or extruded into a film using an extrusion machine. The film is then typically dried at a low temperature to remove any excess water.

A study conducted by Wang et al. (2021) prepared corn starch-based films through the casting method with different concentrations of *Zanthoxylum bungeanum* essential oil (ZYO). The desired volume of corn starch was dissolved in distilled water, and then corn starch was completely gelatinized for 30 min at 90°C using a magnetic stirrer (800 rpm). Afterward, 25% glycerol (starch dry weight basis), v/v ZYO, and 1% Tween 80 (v/v) were added and mixed with the 90°C gelatinized solution. Then the solution was constantly dispersed with a homogenizer at the speed of 20,000 rpm for 10 min, and then the film solution was poured on a special polytetrafluoroethylene (PTFE) mold and dried in an oven set at 30°C for 48 h. After drying, the final film stripped easily from the PTFE molding and was stored at 25°C and 60% relative humidity for 48 h.

Another study used cassava-based starch and cinnamon essential oil as a antimicrobial agent in the prepared film (Iamareerat et al., 2018). Cassava starch–based film incorporated with cinnamon essential oil was made by mixing cassava starch, glycerol, and water and gradually heating on a hot plate with continuous stirring until the temperature reached 55°C. Different concentrations of essential oil were added gradually with simultaneous stirring at 1200 rpm. The mixture solutions were then heated to 70°C with continuous stirring and maintained for 20 min in a water bath. The solution was cast over a petri dish and placed into a vacuum-drying oven at 50°C for 30 min, followed by drying at 50°C for 24 h in a hot-air oven, and the films were then peeled off. All dried films were kept in plastic bags and stored in desiccators at 30–40% relative humidity (RH) for further study.

Syafiq et al. (2022) prepared sugar palm-based films using a conventional solution-casting method. The plasticizers used were glycerol, sorbitol, and their 1:1 combination ratio glycerol and glycerol to study the effect of each plasticizer on sugar palm starch (SPS) films. To begin with, a gelatinized sugar palm starch with 10% wt. aqueous dispersion was prepared via heating of the film-forming solution for 15 min at a temperature of 95 ± 2°C in a hot water bath with constant stirring, a crucial step to disintegrate starch granules to obtain a homogenous solution. Next, different plasticizers (w/w, starch basis) were added to the dispersion. Then the solution was further heated for 15 min at 95 ± 2°C. Next, cinnamon essential oil, sugar palm starch, and an emulsifier (Tween 80) were added into the dispersions. The solution was placed in a sonicator (VCX 500-W, Vibra-Cell, USA) for 30 min using 50% amplitude and 05 pulse to disperse the sugar palm starch. Then, the heating process continued for 15 min at 95°C. Before casting the film-forming solutions in glass Petri dishes, they were allowed to cool to room temperature. Glass Petri dishes were used as casting surfaces and provided a smooth and flat surface for the film formation. The newly cast films were dried in an oven at 40°C. Upon completion of drying for 24 h, the films were removed from the Petri dishes and placed at 53 ± 1% relative humidity in desiccators.

The influence of extrusion temperature (120, 130, and 140°C) and screw speed (25, 35, and 45 rpm.) on the properties of an active film formulated with starch, chitosan, and oregano essential oil was investigated by Pelissari et al. (2011). The proportion of starch (77%, w/w), chitosan (5%, w/w), and glycerol (18%, w/w) was selected because this formulation had the best mechanical properties and barrier results in a previous study aiming to develop films based on starch–chitosan blends. A concentration of 0.5% oregano essential oil (OEO) in relation to this total basic formulation was added due to effectiveness in inhibiting *Escherichia coli*, *Staphylococcus aureus*, *Salmonella enteritides*, and *Bacillus cereus* in another previous study (Pelissari et al., 2009). Starch, chitosan, and glycerol were mixed in a home mixer at the lowest speed, about 780 rpm, for 5 min. In the first stage of the extrusion process, the mixtures were extruded and pelletized with a temperature profile of 120/120/120/110°C and screw speed of 35 rpm. The cinnamon essential oil was added to the pellets, which were reprocessed. Next, the reprocessed pellets were extruded again for the formation of the blow-film.

6.5 PERFORMANCE OF ANTIMICROBIAL STARCH-BASED FILMS WITH ESSENTIAL OILS

Due to the potential health risks associated with synthetic antimicrobial agents, researchers are increasingly investigating alternative solutions, such as natural antimicrobial agents, for starch-based packaging films. Essential oils derived from plants are considered a safe source of antimicrobial compounds in the context of starch-based food packaging and have demonstrated their ability to effectively inhibit the growth of bacteria and molds in numerous studies. One such example is the development of an edible film using Nagara sweet potato, enhanced with garlic essential oil, for packaging *Kandangan Dodol* (Figure 6.4) (Sihombing et al., 2022). Synthesis of the edible film was carried out by prepared starch-derived Nagara sweet potato by extraction. Garlic essential oil was added with various concentrations (0%, 0.25%, 0.5%, 0.75% v/v total). The results showed an inhibition zone for the treatment of garlic essential oil extract against *E. coli* bacteria with a concentration of 0% of 32.00 mm, a concentration of 0.25% of 31.92 mm, a concentration of 0.5% of 31, 56 mm, and 0.75% concentration of 35.93 mm. This is in accordance with the idea that the higher the concentration of an antibacterial agent, the more inhibited bacterial

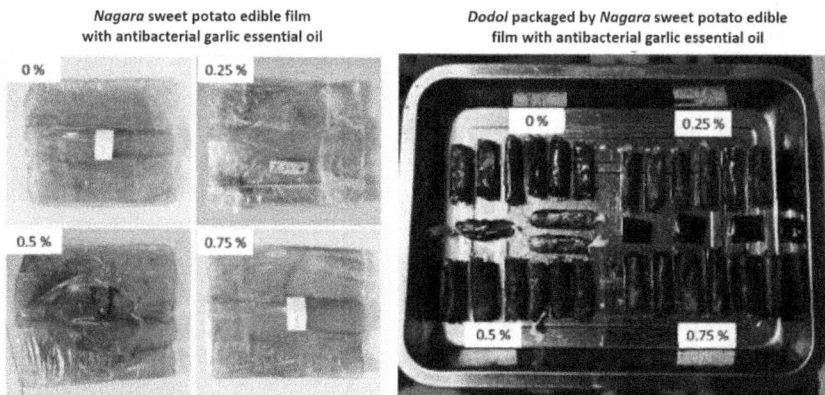

FIGURE 6.4 Edible film of Nagara sweet potato starch with the addition of garlic essential oil essential oil at 0.25%, 0.5%, and 0.75%. Reproduced from Sihombing et al. (2022).

growth will be. The larger inhibition zone formed is because the essential oil contains antibacterial active compounds such as phenol and its derivatives. A diameter of the inhibition zone of <10 mm was said to be non-existent to inhibit microbial growth, a diameter of 11 to 15 mm was categorized as weak, a diameter of 16 to 20 mm was categorized as moderate, and a diameter >20 mm was categorized as strong (Sari & Nasir, 2013). Thus, garlic essential oil has an inhibitory power extract against *E. coli* bacteria, which is indicated by the presence of a clear zone around the paper disc with a strong category at a concentration of 0.75%.

A hybrid of cinnamon oil and titanium dioxide (TiO$_2$) was incorporated into sago starch–based films as an active packaging film against growth of *Salmonella typhimurium, Escherichia coli*, and *Staphylococcus*. The results revealed that controlled sago starch films without TiO$_2$ and cinnamon oil showed no inhibition zones or antimicrobial activity. The inhibition zone of cinnamon oil and nano-incorporated sago starch films significantly increased by increasing both TiO$_2$ and cinnamon oil contents. The results revealed that the effects of sago starch films incorporating TiO$_2$ and cinnamon oil were more pronounced on Gram-positive bacteria than Gram-negative (Arezoo et al., 2020).

Another study developed chitosan with native glutinous rice starch film incorporating essential oils such as garlic, galangal, turmeric, and kaffir lime at fixed concentrations (0.312 mg/mL) against *Escherichia coli, Salmonella typhimurium, Listeria monocytogenes, Staphylococcus aureus*, and *Pseudomonas fluorescens* (Venkatachalam et al., 2023). The results show that the antimicrobial activity of the essential oils added to films was highly efficient against various Gram-positive and Gram-negative pathogens. Among the films with essential oils added, garlic and galangal essential oils exhibited superior inhibitory activity against *Escherichia coli, Salmonella typhimurium, Listeria monocytogenes, Staphylococcus aureus*, and *Pseudomonas fluorescens*, and turmeric and kaffir lime essential oils showed potential antimicrobial activity against *Lactobacillus plantarum* and *L. monocytogenes*. The authors found that the addition of garlic, galangal, and turmeric essential oil into the starch film could diminish the pathogenic bacteria completely. This study also found that chitosan and native glutinous rice starch are effective for creating edible films and, when combined with essential oils, can provide a long-term solution for storing dry and fresh foods without bacterial growth. Among the different essential oils used, this study recommended garlic and galangal essential oils as active ingredients to apply in polysaccharide-based composite film for overall performance as protective agents for preserved foods and storage.

The incorporation of different concentrations of *Zanthoxylum bungeanum* essential oil (0.5%, 1%, and 2%) in corn starch–based films formed by the casting method on antibacterial activity has been studied (Wang et al., 2021). The inclusion of *Zanthoxylum bungeanum* essential oil greatly improved the performance of corn starch–based films. Compared with oil-free films, corn starch–based films incorporating different concentrations of *Zanthoxylum bungeanum* essential oil showed good antibacterial activity against both Gram-negative and Gram-positive bacteria. The inhibition zone for Gram-positive bacteria (*S. aureus* and *L. monocytogenes*) was larger than that for Gram-negative bacteria (*E. coli*). Prepared composite films with good physical and antibacterial properties could be a promising approach for fabricating functional packaging materials. The major compounds that were identified were linalool (32.41%), glycerol tripelargonate (15.57%), tricapric glycerides (10.01%), glycerol tricaprylate (7.96%), and limonene (4.65%). The antibacterial effect of films incorporating *Zanthoxylum bungeanum* essential oil could be attributed to alcohols and ketones in *Zanthoxylum bungeanum* essential oil. Alcohols and ketones have strong antibacterial activities against *E. coli* and *S. aureus* via increasing membrane permeability. Zhou et al. (2020) also proved that alcohols in *Dalbergia pinnata* (Lour.) Prain essential oil extracts would have inhibitory effects on bacterial cells, such as *E. coli* and *S. aureus*.

Cassava starch–based films with geranium essential oil as an antimicrobial agent have been prepared by the casting method (Wang et al., 2022). In this experiment, the antibacterial activity of cassava starch-based films incorporating geranium essential oil (GEO) using the disk diffusion method was evaluated. The samples were initially inoculated with approximately 10^8 CFU/mL (equivalent to 0.5 McFarland) of *Staphylococcus aureus*, *Escherichia coli*, and *Listeria monocytogenes*, which had been previously cultured overnight. Subsequently, 100 µL of these cultures was seeded onto the samples. Finally, the samples were incubated at 37°C for 24 h, and the area of the zone of inhibition was measured with a digital caliper. The antibacterial properties of different percentages of geranium essential oil (0.5%, 1%, and 2%) in cassava starch–based films were compared based on the area of inhibition. The result shows prepared films with geranium essential oil had stronger antibacterial properties than films without GEO (*E. coli*: 0–1236.16 ± 11.61, *S. aureus*: 0–763.44 ± 10.47, *L. monocytogenes*: 0–963.91 ± 11.12 mm^2). When cassava starch–based films contact agar, they can slowly release geranium essential oil, thus playing an antibacterial role.

The results also reveal a more pronounced bacteriostatic effect on Gram-positive bacteria (*Staphylococcus aureus* and *Listeria monocytogenes*) compared to Gram-negative bacteria (*E. coli*). This discrepancy can be attributed to the presence of a peptidoglycan layer in Gram-positive bacteria, which has a specific obstructive impact, limiting the diffusion of hydrophobic compounds, such as essential oils, and consequently resulting in a less effective bacteriostatic effect. The essential oil itself has a specific bacteriostatic effect and has been widely used in the food processing industry. With an increase in geranium essential oil concentration from 0.5% to 2%, the antibacterial effect of cassava starch–based films becomes increasingly stronger. The essential oil can destroy the cell membrane and further deplete the cell's contents to achieve a bacteriostatic effect.

An intelligent packaging film from a starch/chitosan mixed base incorporating different amounts (0–2.5%, v/v) of *Zanthoxylum armatum* essential oil to against *S. aureus* and *E. coli* activities was prepared (Wang et al., 2022). With 2.5% v/v *Zanthoxylum armatum* DC essential oil, the film showed the best antibacterial activity with inhibition zones of up to 21.72 ± 1.25 and 18.73 ± 0.74 (cm) against *S. aureus* and *E. coli*, respectively. Starch/chitosan-based film as a control did not show any antibacterial activity against both microorganisms due to the release failure of water-soluble chitosan, an antibacterial agent, under neutral pH conditions. It has been shown that chitosan exhibits significant antimicrobial properties only in the form of gels or viscous acid solutions. On the contrary, the antibacterial ability of the *Zanthoxylum armatum* composite film was significantly higher ($P <$ 0.05). The diameter of the bacterial inhibition zone (8.76–21.72 mm) positively correlated with the *Zanthoxylum armatum* loading, which contains volatile antibacterial components where the volatile substances inhibit the growth of bacteria. In addition, the *Zanthoxylum armatum* film is more sensitive to Gram-positive than Gram-negative bacteria (Figure 6.5). The lipopolysaccharide component of the cell

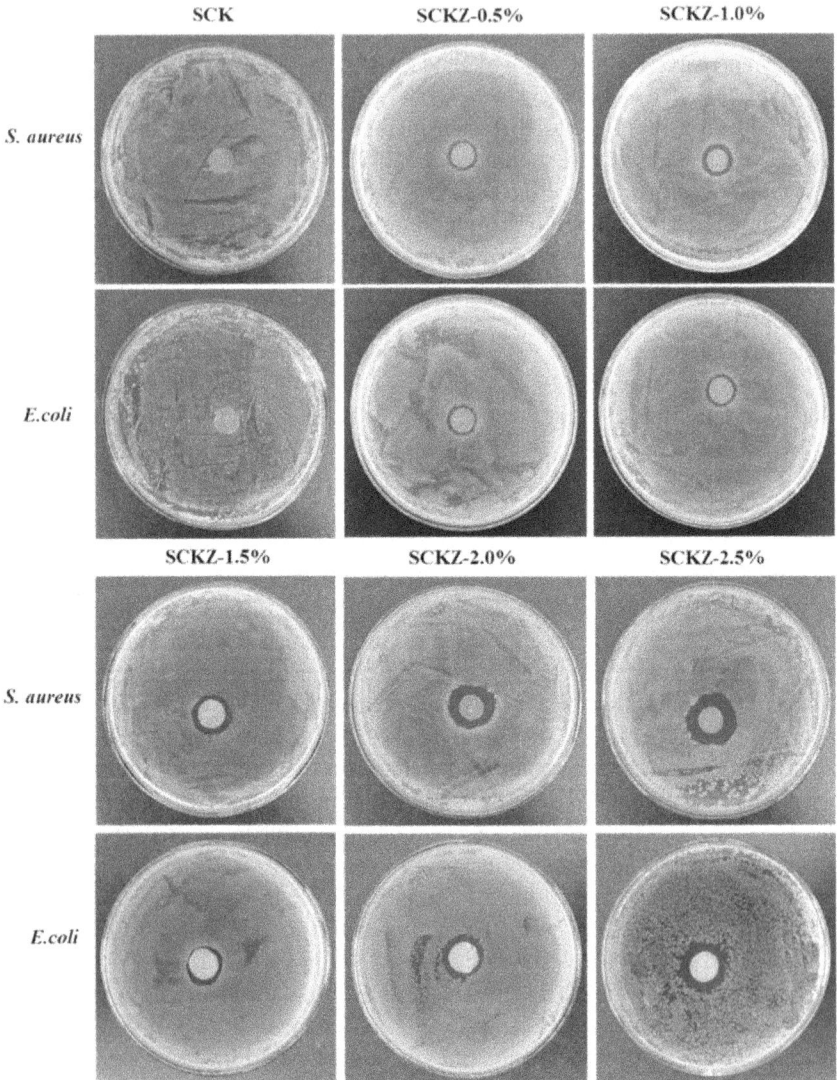

FIGURE 6.5 A significant observation on antibacterial activity of composite films against *S. aureus* and *E. coli.* in Gram-positive and Gram-negative bacteria. Reproduced from Wang et al. (2022).

wall of Gram-negative bacteria can prevent the entry of hydrophobic antibacterial compounds, which is not the case against Gram-positive bacteria.

A fixed 0.5% amount of rosemary, mint essential oil, nisin, and lactic acid were incorporated in chitosan pectin and starch to develop a novel functional packaging film from activity against all tested pathogenic strains (*Bacillus subtilis, Escherichia coli,* and *Listeria monocytogenes*) (Akhter et al., 2019). The presence of considerable amounts of polyphenols and antimicrobial agents in films were responsible for the exhibition of wide range of antioxidant and antimicrobial activity. However, addition of the essential oils increased film microstructure heterogeneity and hence reduced water barrier properties and tensile strength of the films while improving their flexibility. Thus, the developed active starch-based films incorporating essential oils and antimicrobial agents have novelty regarding biodegradability and could be a promising alternative to synthetic materials, potentially contributing to food shelf-life.

The results have shown considerable improvement in antimicrobial effects due to the presence of free amino groups in chitosan that bind to the cell surface, disturbing the cell membrane and thus causing cell death by inducing leakage of intracellular components. The inhibition of both Gram-positive and Gram-negative bacteria by starch-based films with added essential oils is mainly attributed to polyphenolic components, which are responsible for disrupting the biological activity of bacterial cell membranes. The results show the highest antioxidant activity of mint oil could be attributed to phenolic compounds with diverse structures like α-pinene, citronellol, and methyl eugenol that act as potent hydrogen donors and impart significant antioxidant activity to the films.

6.6 CONCLUSIONS

Obtaining essential oils from plants is a relatively simple process, and they offer significant benefits for food safety and preservation due to their antimicrobial and antioxidant properties. However, many food products are vulnerable to microbial spoilage during storage and require adequate protection. One way to protect foods for better shelf life is through the development of antibacterial packaging that includes antibacterial agents. Consumers are increasingly demanding natural and safe agents, which has led to a search for materials that can resist microbial growth while maintaining the food's sensory properties. Essential oils extracted from plants are a good option with good antimicrobial activity. Although essential oils have been successfully used in food applications, their volatility can cause the loss of the oils, which needs to be considered. Encapsulation and electrospinning methods may be helpful in preventing the loss of essential oils during the manufacturing process. Such encapsulation is necessary to ensure the long-term effectiveness of the essential oil. To fully benefit from these technologies, a thorough understanding of the essential oils and their release mechanism/kinetics is necessary. However, the use of essential oils in the food industry is currently limited due to their intense aroma and toxicity concerns when used beyond a specific limit. Therefore, it is essential to maintain a balance between the amount of essential oils used and their toxicity. More studies on the potential side effects of essential oils should be conducted before their widespread use in the industry. It is important to note that the antibacterial mechanism of essential oils is still not well understood.

REFERENCES

Abbas, M., Hussain, T., Arshad, M., Ansari, A. R., Irshad, A., Nisar, J., Hussain, F., Masood, N., Nazir, A., & Iqbal, M. (2019). Wound healing potential of curcumin cross-linked chitosan/polyvinyl alcohol. *International Journal of Biological Macromolecules*, *140*, 871–876.

Acosta, S., Chiralt, A., Santamarina, P., Rosello, J., González-Martínez, C., & Cháfer, M. (2016). Antifungal films based on starch-gelatin blend, containing essential oils. *Food Hydrocolloids*, *61*, 233–240.

Akhter, R., Masoodi, F. A., Wani, T. A., & Rather, S. A. (2019). Functional characterization of biopolymer based composite film: Incorporation of natural essential oils and antimicrobial agents. *International Journal of Biological Macromolecules*, *137*, 1245–1255.

Al-Hashimi, A. G., Ammar, A. B., Cacciola, F., & Lakhssassi, N. (2020). Development of a millet starch edible film containing clove essential oil. *Foods*, *9*(2), 184.

Ali, A., Wani, T. A., Wani, I. A., & Masoodi, F. A. (2016). Comparative study of the physicochemical properties of rice and corn starches grown in Indian temperate climate. *Journal of the Saudi Society of Agricultural Sciences*, *15*(1), 75–82.

Arezoo, E., Mohammadreza, E., Maryam, M., & Abdorreza, M. N. (2020). The synergistic effects of cinnamon essential oil and nano TiO2 on antimicrobial and functional properties of sago starch films. *International Journal of Biological Macromolecules*, *157*, 743–751.

Ayu Rafiqah, S., Khalina, A., Zaman, K., Tawakkal, I. S. M. A., Harmaen, A. S., & Nurrazi, N. M. (2021). Bioplastics: The future of sustainable biodegradable food packaging. *Bio-based Packaging: Material, Environmental and Economic Aspects*, 335–351.

Azadbakht, E., Maghsoudlou, Y., Khomiri, M., & Kashiri, M. (2018). Development and structural characterization of chitosan films containing Eucalyptus globulus essential oil: Potential as an antimicrobial carrier for packaging of sliced sausage. *Food Packaging and Shelf Life*, *17*, 65–72.

Aziz, Z. A. A., Ahmad, A., Setapar, S. H. M., Karakucuk, A., Azim, M. M., Lokhat, D., Rafatullah, M., Ganash, M., Kamal, M. A., & Ashraf, G. M. (2018). Essential oils: Extraction techniques, pharmaceutical and therapeutic potential-a review. *Current Drug Metabolism*, *19*(13), 1100–1110.

Baptista-Silva, S., Borges, S., Ramos, O. L., Pintado, M., & Sarmento, B. (2020). The progress of essential oils as potential therapeutic agents: A review. *Journal of Essential Oil Research*, *32*(4), 279–295.

Basiak, E., Debeaufort, F., & Lenart, A. (2016). Effect of oil lamination between plasticized starch layers on film properties. *Food Chemistry*, *195*, 56–63.

Bertoft, E. (2017). Understanding starch structure: Recent progress. *Agronomy*, *7*(3), 56.

Bertoft, E., & Nilsson, L. (2017). Starch: Analytical and structural aspects. In *Carbohydrates in food* (pp. 399–500). CRC Press.

Buleon, A., Colonna, P., Planchot, V., & Ball, S. (1998). Starch granules: Structure and biosynthesis. *International Journal of Biological Macromolecules*, *23*(2), 85–112.

Cheng, J., Lin, X., Wu, X., Liu, Q., Wan, S., & Zhang, Y. (2021). Preparation of a multifunctional silver nanoparticles polylactic acid food packaging film using mango peel extract. *International Journal of Biological Macromolecules*, *188*, 678–688.

Chenni, M., El Abed, D., Rakotomanomana, N., Fernandez, X., & Chemat, F. (2016). Comparative study of essential oils extracted from Egyptian basil leaves (Ocimum basilicum L.) using hydro-distillation and solvent-free microwave extraction. *Molecules*, *21*(1), 113.

Choo, K. W., Lin, M., & Mustapha, A. (2021). Chitosan/acetylated starch composite films incorporated with essential oils: Physiochemical and antimicrobial properties. *Food Bioscience*, *43*, 101287.

Contreras-Jiménez, B., Torres-Vargas, O. L., & Rodríguez-García, M. E. (2019). Physicochemical characterization of quinoa (Chenopodium quinoa) flour and isolated starch. *Food Chemistry*, *298*, 124982.

Copeland, L., Blazek, J., Salman, H., & Tang, M. C. (2009). Form and functionality of starch. *Food Hydrocolloids*, *23*(6), 1527–1534.

de Souza, A. G., Dos Santos, N. M. A., da Silva Torin, R. F., & dos Santos Rosa, D. (2020). Synergic antimicrobial properties of Carvacrol essential oil and montmorillonite in biodegradable starch films. *International Journal of Biological Macromolecules*, *164*, 1737–1747.

Dhifi, W., Bellili, S., Jazi, S., Bahloul, N., & Mnif, W. (2016). Essential oils' chemical characterization and investigation of some biological activities: A critical review. *Medicines*, *3*(4), 25.

Dias, N., Dias, M. C., Cavaleiro, C., Sousa, M. C., Lima, N., & Machado, M. (2017). Oxygenated monoterpenes-rich volatile oils as potential antifungal agents for dermatophytes. *Natural Product Research*, *31*(4), 460–464.

do Evangelho, J. A., da Silva Dannenberg, G., Biduski, B., El Halal, S. L. M., Kringel, D. H., Gularte, M. A., Fiorentini, A. M., & da Rosa Zavareze, E. (2019). Antibacterial activity, optical, mechanical, and barrier properties of corn starch films containing orange essential oil. *Carbohydrate Polymers*, *222*, 114981.

El Sawi, S. A., Ibrahim, M. E., El-Rokiek, K. G., & El-Din, S. A. S. (2019). Allelopathic potential of essential oils isolated from peels of three citrus species. *Annals of Agricultural Sciences*, *64*(1), 89–94.

Fernandes, E. M., Pires, R. A., Mano, J. F., & Reis, R. L. (2013). Bionanocomposites from lignocellulosic resources: Properties, applications and future trends for their use in the biomedical field. *Progress in Polymer Science*, *38*(10–11), 1415–1441.

Ghasemlou, M., Aliheidari, N., Fahmi, R., Shojaee-Aliabadi, S., Keshavarz, B., Cran, M. J., & Khaksar, R. (2013). Physical, mechanical and barrier properties of corn starch films incorporated with plant essential oils. *Carbohydrate Polymers*, *98*(1), 1117–1126.

Gutiérrez-del-Río, I., Fernández, J., & Lombó, F. (2018). Plant nutraceuticals as antimicrobial agents in food preservation: Terpenoids, polyphenols and thiols. *International Journal of Antimicrobial Agents*, *52*(3), 309–315.

Hasan, J., Crawford, R. J., & Ivanova, E. P. (2013). Antibacterial surfaces: The quest for a new generation of biomaterials. *TRENDS in Biotechnology*, *31*(5), 295–304.

Herman, R. A., Ayepa, E., Shittu, S., Fometu, S. S., & Wang, J. (2019). Essential oils and their applications-a mini review. *Advances in Nutrition & Food Science*, *4*(4).

Iamareerat, B., Singh, M., Sadiq, M. B., & Anal, A. K. (2018). Reinforced cassava starch based edible film incorporated with essential oil and sodium bentonite nanoclay as food packaging material. *Journal of Food Science and Technology*, *55*, 1953–1959.

Kalemba, D. A. A. K., & Kunicka, A. (2003). Antibacterial and antifungal properties of essential oils. *Current Medicinal Chemistry*, *10*(10), 813–829.

Kamarudin, S. H., Rayung, M., Abu, F., Ahmad, S., Fadil, F., Karim, A. A., Norizan, M. N., Sarifuddin, N., Mat Desa, M. S. Z., Mohd Basri, M. S., & Samsudin, H. (2022). A review on antimicrobial packaging from biodegradable polymer composites. *Polymers*, *14*(1), 174.

Khajeh, M., Yamini, Y., Sefidkon, F., & Bahramifar, N. (2004). Comparison of essential oil composition of Carum copticum obtained by supercritical carbon dioxide extraction and hydrodistillation methods. *Food Chemistry*, *86*(4), 587–591.

Khatami, M. H., Barber, W., & de Haan, H. W. (2021). Using geometric criteria to study helix-like structures produced in molecular dynamics simulations of single amylose chains in water. *RSC Advances*, *11*(20), 11992–12002.

Latief, R., Bhat, K. A., Khuroo, M. A., Shawl, A. S., & Chandra, S. (2017). Comparative analysis of the aroma chemicals of Melissa officinalis using hydrodistillation and HS-SPME techniques. *Arabian Journal of Chemistry*, *10*, S2485–S2490.

Lauer, M. K., & Smith, R. C. (2020). Recent advances in starch-based films toward food packaging applications: Physicochemical, mechanical, and functional properties. *Comprehensive Reviews in Food Science and Food Safety*, *19*(6), 3031–3083.

Lemos, P. V. F., Barbosa, L. S., Ramos, I. G., Coelho, R. E., & Druzian, J. I. (2018). The important role of crystallinity and amylose ratio in thermal stability of starches. *Journal of Thermal Analysis and Calorimetry*, *131*, 2555–2567.

Li, J., Ye, F., Lei, L., & Zhao, G. (2018). Combined effects of octenylsuccination and oregano essential oil on sweet potato starch films with an emphasis on water resistance. *International Journal of Biological Macromolecules*, *115*, 547–553.

Liu, Q., & Yang, H. (2019). Application of atomic force microscopy in food microorganisms. *Trends in Food Science & Technology*, *87*, 73–83.

Luchese, C. L., Spada, J. C., & Tessaro, I. C. (2017). Starch content affects physicochemical properties of corn and cassava starch-based films. *Industrial Crops and Products*, *109*, 619–626.

Macwan, S. R., Dabhi, B. K., Aparnathi, K. D., & Prajapati, J. B. (2016). Essential oils of herbs and spices: Their antimicrobial activity and application in preservation of food. *International Journal of Current Microbiology and Applied Sciences*, *5*(5), 885–901.

Manikandan, N. A., Pakshirajan, K., & Pugazhenthi, G. (2020). Preparation and characterization of environmentally safe and highly biodegradable microbial polyhydroxybutyrate (PHB) based graphene nanocomposites for potential food packaging applications. *International Journal of Biological Macromolecules*, *154*, 866–877.

Marichelvam, M. K., Jawaid, M., & Asim, M. (2019). Corn and rice starch-based bio-plastics as alternative packaging materials. *Fibers*, *7*(4), 32.

Maulana, M. I., Lubis, M. A. R., Febrianto, F., Hua, L. S., Iswanto, A. H., Antov, P., Kristak, L., Mardawati, E., Sari, R. K., Hidayat, W., Lo Giudice, V., Todaro, L., & Zaini, L. H. (2022). Environmentally friendly starch-based adhesives for bonding high-performance wood composites: A review. *Forests*, *13*(10), 1614.

McClements, D. J., & Öztürk, B. (2021). Utilization of nanotechnology to improve the application and bioavailability of phytochemicals derived from waste streams. *Journal of Agricultural and Food Chemistry*, *70*(23), 6884–6900.

Moghaddam, M., & Mehdizadeh, L. (2017). Chemistry of essential oils and factors influencing their constituents. In *Soft chemistry and food fermentation* (pp. 379–419). Elsevier.

Nazzaro, F., Fratianni, F., De Martino, L., Coppola, R., & De Feo, V. (2013). Effect of essential oils on pathogenic bacteria. *Pharmaceuticals*, *6*(12), 1451–1474.

Niranjana Prabhu, T., & Prashantha, K. (2018). A review on present status and future challenges of starch based polymer films and their composites in food packaging applications. *Polymer Composites*, *39*(7), 2499–2522.

Norrrahim, M. N. F., Nurazzi, N. M., Jenol, M. A., Farid, M. A. A., Janudin, N., Ujang, F. A., Yasim-Anuar, T. A. T., Syed Najmuddin, S. U. F., & Ilyas, R. A. (2021). Emerging development of nanocellulose as an antimicrobial material: An overview. *Materials Advances*, *2*(11), 3538–3551.

Ojogbo, E., Ogunsona, E. O., & Mekonnen, T. H. (2020). Chemical and physical modifications of starch for renewable polymeric materials. *Materials Today Sustainability*, *7*, 100028.

Pateiro, M., Barba, F. J., Domínguez, R., Sant'Ana, A. S., Khaneghah, A. M., Gavahian, M., Gómez, B., & Lorenzo, J. M. (2018). Essential oils as natural additives to prevent oxidation reactions in meat and meat products: A review. *Food Research International, 113,* 156–166.

Pelissari, F. M., Grossmann, M. V. E., Yamashita, F., & Pineda, E. A. G. (2009). Antimicrobial, mechanical, and barrier properties of cassava starch– chitosan films incorporated with oregano essential oil. *Journal of Agricultural and Food Chemistry, 57*(16), 7499–7504.

Pelissari, F. M., Yamashita, F., & Grossmann, M. V. E. (2011). Extrusion parameters related to starch/chitosan active films properties. *International Journal of Food Science & Technology, 46*(4), 702–710.

Rodrigues, S. C. S., da Silva, A. S., de Carvalho, L. H., Alves, T. S., & Barbosa, R. (2020). Morphological, structural, thermal properties of a native starch obtained from babassu mesocarp for food packaging application. *Journal of Materials Research and Technology, 9*(6), 15670–15678.

Rubulotta, G., & Quadrelli, E. A. (2019). Terpenes: A valuable family of compounds for the production of fine chemicals. In *Studies in surface science and catalysis* (Vol. 178, pp. 215–229). Elsevier.

Sari, K. I. P., & Nasir, N. (2013). Uji antimikroba ekstrak segar jahe-jahean (Zingiberaceae) terhadap Staphylococcus aureus, Escherichia coli dan Candida albicans. *Jurnal Biologi UNAND, 2*(1).

Schaefer, D., & Cheung, W. M. (2018). Smart packaging: Opportunities and challenges. *Procedia Cirp, 72,* 1022–1027.

Sihombing, N., Elma, M., Thala'Ah, R. N., Simatupang, F. A., Pradana, E. A., & Rahma, A. (2022, March). Garlic essential oil as an edible film antibacterial agent derived from Nagara sweet potato starch applied for packaging of Indonesian traditional food— dodol. In *IOP Conference Series: Earth and Environmental Science* (Vol. 999, No. 1, p. 012026). IOP Publishing.

Song, X., Zuo, G., & Chen, F. (2018). Effect of essential oil and surfactant on the physical and antimicrobial properties of corn and wheat starch films. *International Journal of Biological Macromolecules, 107,* 1302–1309.

Souza, A. C., Goto, G. E. O., Mainardi, J. A., Coelho, A. C. V., & Tadini, C. C. (2013). Cassava starch composite films incorporated with cinnamon essential oil: Antimicrobial activity, microstructure, mechanical and barrier properties. *LWT-Food Science and Technology, 54*(2), 346–352.

Syafiq, R. M. O., Sapuan, S. M., Zuhri, M. Y. M., Othman, S. H., & Ilyas, R. A. (2022). Effect of plasticizers on the properties of sugar palm nanocellulose/cinnamon essential oil reinforced starch bionanocomposite films. *Nanotechnology Reviews, 11*(1), 423–437.

Tharanathan, R. N. (2003). Biodegradable films and composite coatings: Past, present and future. *Trends in Food Science & Technology, 14*(3), 71–78.

Venkatachalam, K., Rakkapao, N., & Lekjing, S. (2023). Physicochemical and antimicrobial characterization of chitosan and native glutinous rice starch-based composite edible films: Influence of different essential oils incorporation. *Membranes, 13*(2), 161.

Vianna, T. C., Marinho, C. O., Júnior, L. M., Ibrahim, S. A., & Vieira, R. P. (2021). Essential oils as additives in active starch-based food packaging films: A review. *International Journal of Biological Macromolecules, 182,* 1803–1819.

Wang, B., Sui, J., Yu, B., Yuan, C., Guo, L., Abd El-Aty, A. M., & Cui, B. (2021). Physicochemical properties and antibacterial activity of corn starch-based films incorporated with Zanthoxylum bungeanum essential oil. *Carbohydrate Polymers, 254,* 117314.

Wang, B., Yan, S., Qiu, L., Gao, W., Kang, X., Yu, B., Cui, B., & Abd El-Aty, A. M. (2022). Antimicrobial activity, microstructure, mechanical, and barrier properties of cassava starch composite films supplemented with geranium essential oil. *Frontiers in Nutrition*, *9*, 963.

Wang, Y., Luo, J., Hou, X., Wu, H., Li, Q., Li, S., Li, M., Liu, X., Cheng, A., Zhang, Z., & Shen, G. (2022). Physicochemical, antibacterial, and biodegradability properties of green Sichuan pepper (Zanthoxylum armatum DC.) essential oil incorporated starch films. *LWT*, *161*, 113392.

Wang, Y., & Zhang, G. (2021). The preparation of modified nano-starch and its application in food industry. *Food Research International*, *140*, 110009.

Wani, A. R., Yadav, K., Khursheed, A., & Rather, M. A. (2021). An updated and comprehensive review of the antiviral potential of essential oils and their chemical constituents with special focus on their mechanism of action against various influenza and coronaviruses. *Microbial Pathogenesis*, *152*, 104620.

Zhang, J., Ye, K. P., Zhang, X., Pan, D. D., Sun, Y. Y., & Cao, J. X. (2017). Antibacterial activity and mechanism of action of black pepper essential oil on meat-borne Escherichia coli. *Frontiers in Microbiology*, *7*, 2094.

Zhou, W., He, Y., Lei, X., Liao, L., Fu, T., Yuan, Y., Huang, X., Zou, L., Liu, Y., Li, J., & Ruan, R. (2020). Chemical composition and evaluation of antioxidant activities, antimicrobial, and anti-melanogenesis effect of the essential oils extracted from Dalbergia pinnata (Lour.) Prain. *Journal of Ethnopharmacology*, *254*, 112731.

Żołek-Tryznowska, Z., & Kałuża, A. (2021). The influence of starch origin on the properties of starch films: Packaging performance. *Materials*, *14*(5), 1146.

7 Chitosan-Based Chemical Sensors
Sensing Mechanism and Detection Capacity

7.1 INTRODUCTION

Chitin is the second most abundant natural biopolymer and the precursor of chitosan. This polysaccharide was principally extracted from the exoskeleton of certain species such as crustaceans, mollusks, insects, and even certain fungi. For the first time, Braconnot separated chitin from fungi in 1811 and named it "chitin", taken from the Greek word that means tunic or cover. Further, it was extracted from elytrum of the cock-chafer beetle by Odier in 1923 via treatment of warm alkaline solutions. The exoskeletons of crustaceans and arthropods are the most significant source of chitin, which is identified as a non-toxic copolymer comprising β-(1–4)-2-acetamido-d-glucose and β-(1–4)-2-amino-d-glucose units (Hasan et al., 2020; Kazi et al., 2019). Chitin comes about as structured crystalline microfibrils that outline structural components in the exoskeleton of arthropods or in the cell wall of fungi and yeast. In recent years, chitin has been of major interest as a material identified as an obstinate polymer due to its insoluble nature in various common solvents. Derivatives of chitin are obtained by the alkaline deacetylation reaction, using preferably 40–50% sodium hydroxide (NaOH), but this reaction is found to be incomplete most of the time (Figure 7.1) (Kaur & Dhillon, 2014; Lizardi-Mendoza et al., 2016).

Amino groups (NH_2) present in chitosan can be effectively harnessed for surface functionalization, allowing for the selective detection of specific analytes via chemical interactions with these functional groups. In addition, chitosan-based sensors have shown high sensitivity, selectivity, and reproducibility, making them attractive for a wide range of sensing applications. This chapter will discuss the properties of chitosan, its synthesis, various functionalization techniques, detection mechanisms, and capacity behavior of chitosan-based chemical sensors and their potential applications in environmental monitoring. Other properties like low methanol permeability, amphiphilicity, applicability in high temperature, and a low relative humidity environment make chitosan an excellent candidate for solid polymer electrolyte– or polymer electrolyte–based fuel cells. Similarly, the high adsorption capacity for contaminants, ease of physical and chemical modifications, resistance to microbiological degradation, inherent availability of NH_2 and abundant hydroxyl (OH) groups, and affinity towards heavy metal make chitin and chitosan ideal candidates for environmental applications (Musarurwa & Tavengwa, 2020). The recent advancements

DOI: 10.1201/9781003416043-7

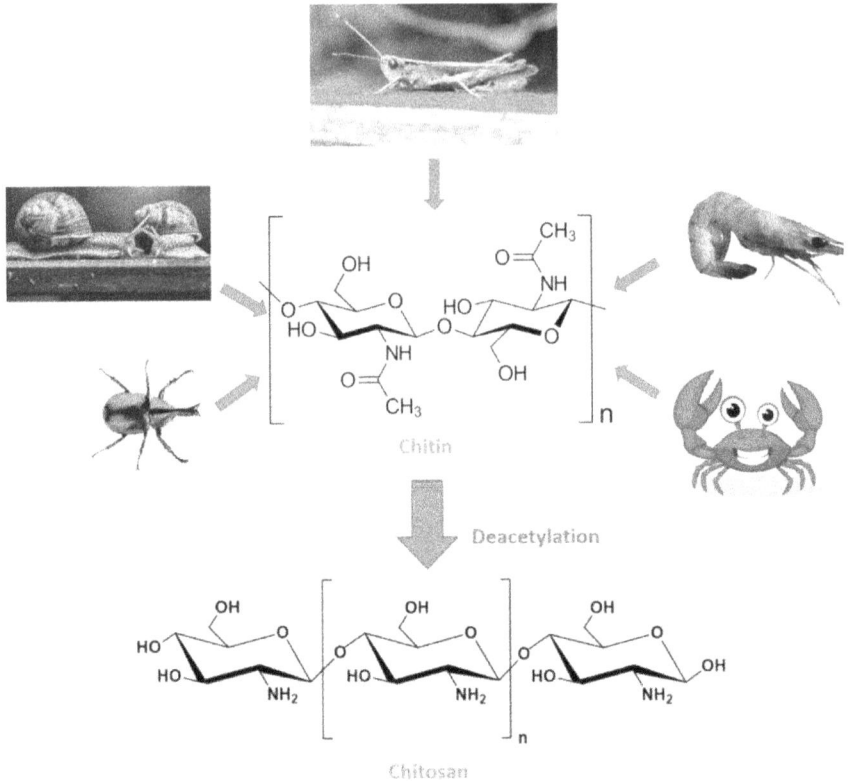

FIGURE 7.1   Schematic diagram of preparation of chitosan.

in chitosan-based composites have resulted in the evolution of various studies to further enhance its applicability in varying technological applications. This exploration plays a significant role in fabricating sensitive sensors.

7.2   CHITIN AND CHITOSAN: STRUCTURE AND CHARACTERISTICS

Chitosan is structurally similar to cellulose, with the key distinction being the replacement of a secondary hydroxyl on the second carbon atom of the hexose repeat unit by an acetamide group. It is derived from chitin through a deacetylation process in an alkaline medium. In fact, chitosan is a copolymer composed of β-(1–4)-2-acetamido-d-glucose and β-(1–4)-2-amino-d-glucose units, typically with the latter exceeding 60% in composition. The significance of chitosan lies in its antimicrobial properties, supported by its cationic nature and its ability to form films. It is often characterized by its degree of deacetylation and average molecular weight. As a rule, a polymer with 2-amino-2-deoxy-d-glucose content concentration beyond 50% is considered chitosan, and that with a concentration lower than 50% is considered chitin. A typical monomer has two hydroxyl groups, one primary hydroxyl at C-6, one secondary hydroxyl at C-3, and amino group or N-acetyl group (at C-2) positions (Anitha et al., 2014). The N-acetyl group can form linear inter- and intramolecular hydrogen bonds, resulting in higher crystallinity, increase in molecular weight, and insolubility in water. The difference in molecular weight can be as much as 10° Da. Chitosan is also the only commercially available water-soluble cationic polymer due to the positive charges by its amino groups. It can be differentiated from chitin by molecular mass, degree of deacetylation (DA), crystallinity, moisture, and protein content. Among the different parameters, the degree of DA is the most influential factor and can determine the final properties of chitosan. For instance, deacetylation can influence the solubility, flexibility, polymer conformation, surface area, porosity, conductivity, tensile strength, crystallinity, biodegradability, biocompatibility, mucoadhesion, hemostatic, analgesic, adsorption enhancing, anti-microbial, and antioxidant properties (Singh & Nagendran, 2016).

Chitin is present in nature mainly in three polymorphic forms, α, β, and γ. Different forms have different properties depending upon the various microstructures. The α structure has the antiparallel alignment with the highest crystallinity and is the most abundant of the three. It is mainly obtained from the exoskeletons of crabs, lobsters, and shrimp; insect cuticles; fungal and yeast cell walls; and marine sponges. The β structure has a parallel alignment and is mainly obtained from squid, the extracellular fibers of diatoms, and the spines and chaetae of certain annelids. The γ structure has a mixed alignment where two parallel chains are followed by one anti-parallel chain in the common source. Chitosan, on the other hand, has a noticeable heteropolysaccharide structure with linear β-1,4-linkage between the units (El Knidri et al., 2018). The quantity and arrangement order significantly impact the material's properties in different conditions. For example, in heterogeneous conditions, the polymer displays an uneven distribution of N-acetyl-D-glucosamine and D-glucosamine units, primarily occurring in a random manner with a limited extent of blockwise distribution. Consequently, the physicochemical characteristics of chitosan vary from those of randomly acetylated chitosan obtained under homogeneous conditions. Chitosan is the deacetylated form of chitin and has a degree of deacetylation beyond 50% in comparison with its precursor, which results in some characteristic changes in the behavior. In contrast to chitin, some of the characteristics are unique to chitosan, such as the distribution of acetyl groups along the chains, average DA, solubility, and molecular weight (Abdou et al., 2008). The characteristic differences between chitin and chitosan are mainly derived from the compositional difference of N-acetyl-2-amino-2-deoxy-d-glucose and d-glucosamine monomer units. Hence, to understand the characteristics of chitosan, determination of its average DA is of prime importance.

The process of converting chitin into chitosan is a pivotal factor in the distribution of acetyl groups within the chitosan polymer chain. Typically, a random distribution of acetyl groups is achieved under homogeneous conditions, whereas a blockwise distribution of acetyl groups occurs under heterogeneous conditions. Understanding this distribution is of paramount significance, as it governs inter-chain interactions and imparts hydrophobic characteristics, with hydrogen bonds and acetyl groups playing distinct roles in this context. Another important characteristic that the DA directly influences is the solubility of chitosan. This is particularly important since, per some researchers, chitosan is the deacetylated form of chitin and can be considered chitosan only when it becomes soluble in aqueous acidic media. Solubility in chitosan is achieved through the establishment of inter-chain interactions facilitated by hydrogen bonds. Additionally, deacetylation leads to the substitution of hydrophobic groups with hydrophilic amino groups. This enhanced comprehension of the solubility mechanism has driven numerous research endeavors focused on the modification of chitosan to enhance its solubility and mitigate insolubility issues (Ding et al., 2015). Another important characteristic exhibited by chitosan is the importance of the size of the polymer chain in determining its physicochemical properties.

7.3 PREPARATION OF CHITOSAN

Previously, conventional techniques for extracting chitin and chitosan had to make a trade-off between achieving large-scale production and minimizing economic impact. However, new techniques have emerged that strike a better balance between scalability and affordability. By using microwave extraction instead of traditional methods, it is possible to reduce processing time by up to 50 minutes with only moderate substitution. For example, an alkaline-based environment that is assisted by microwave irradiation has proved highly efficient for the deacetylation of chitin, with degrees of deacetylation for chitosan from different sources (fish shells, shrimp, and crab) of 75%, 78%, and 70%, respectively. Microwave synthesis also offers substantial benefits over conventional extractions, such as significantly reducing reaction time, increasing yield and purity, and minimizing unwanted side reactions (Safavy et al., 2007). Another promising method involves using both acidic and basic environments for chitosan isolation, which can reduce overall processing time to just a few minutes while still achieving around 83% deacetylation. Microwave technology has thus gained considerable attention as a means of extracting chitosan, with some authors claiming to have reduced processing time by a factor of 16. Combining chemicals with microwave technology has provided better control over the extraction process, allowing for adjustment of microwave power based on the specific extraction stage (El Knidri et al., 2016).

7.4 SURFACE MODIFICATION OF CHITOSAN

Chitosan are interesting bioactive polymers with primary hydroxyl, secondary hydroxyl (–OH), and amino (–NH$_2$) or acetyl functional groups, which could facilitate the possibilities of diverse chemical modifications. Chemical modification result in various types of derivatives with modified properties for specific applications such as pharmaceutical, biomedical, biotechnological, cosmetic, agricultural, food and non-food industries, water treatment, paper, and textiles. The ability of chitosan to undergo a wide range of surface modifications suggests its huge potential to the scientific community and industry in the current scenario. A schematic diagram of chitosan-based surface modification is shown in Figure 7.2. The functionalization of polysaccharides like chitin and chitosan has been of great interest for many years, since it is a convenient way to improve and introduce properties and helps to prepare

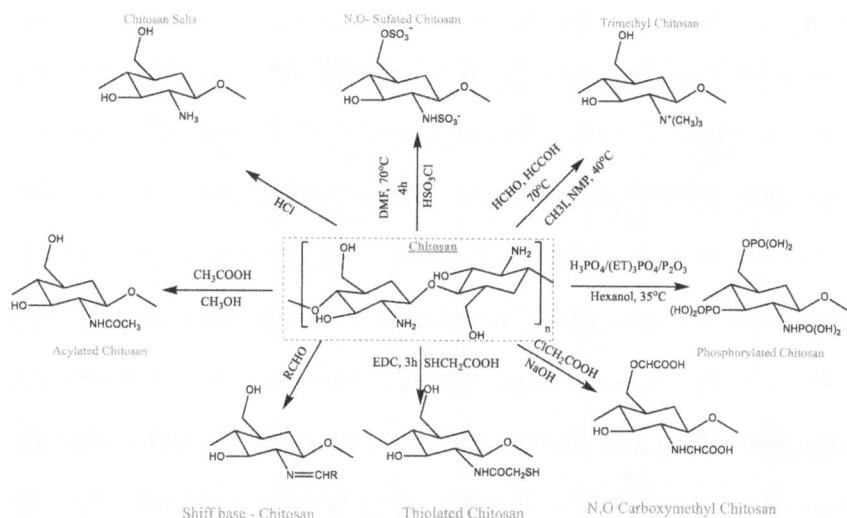

FIGURE 7.2 Schematic diagram of chitosan-based surface modification. Reproduced from Ibrahim et al. (2023).

new materials with specialized characteristics for developing more advanced and innovative materials, especially for highly sensitive chemical sensors. The most commonly employed functionalization includes carboxymethylation, alkylation, sulfonation, succinylation, and phosphorylation, and crosslinking and graft polymerization have been discussed. Even though many techniques have been adopted, carboxymethylation and alkylation drawn more attention, as carboxymethylation is suitable for a wide range of applications since it offers relatively easy chemical modification, whereas alkylation helps to develop both hydrophilic and hydrophobic derivatives (Peter et al., 2021).

7.5 ROLE OF CHITOSAN-BASED CHEMICAL SENSORS

Chitosan has gained significant attention as a sensing material in chemical sensors due to its ability to interact with various analytes and convert chemical information into an electrical signal. The role of chitosan-based chemical sensors is to detect and quantify specific chemical species based on the interaction between the chitosan sensing material and the analyte of interest. When the analyte binds to the chitosan, it causes a change in the physical or electrical properties of the chitosan, which can be detected and quantified using various sensing techniques. The use of chitosan as a sensing material offers several advantages, such as its natural polymer properties, which make it a cleaner, non-toxic, and environmentally friendly material. Its hydrophilicity, gel-forming ability, doping practicability, enhanced mechanical reliability, high permeability, and ease of use of reactive functional groups for chemical changes make it an interesting sensing material. Due to the abundance of NH_2 and OH functional groups, chitosan can easily react with bioactive molecules, making it an excellent substrate for covalent immobilization through covalent binding or electrostatic interactions (Baranwal et al., 2018). Chitosan is also an excellent immobilization matrix for electroactive materials on the electrode surface and demonstrates good adhesion characteristics. Research has shown that chitosan can accommodate silver nanoparticles and make it easier to form stable and aggregation-free nanocomposites by facilitating the interfaces between multiwalled carbon nanotubes (MWCNTs) and silver nanoparticles (Kangkamano et al., 2017; Park et al., 2018).

Chitosan's natural or chemically fabricated structure can also chelate fundamental oxygen and nitrogen atoms through free electron pairs, making it a proficient sorbent substance for heavy metals, pesticides, and dyes (Cai et al., 2021). The reactive binding sites can be covalently bound to mediators, enabling the production of electrode surfaces that efficiently transport electrons from chitosan. These surfaces can be used as a support for immobilizing graphene oxide (GO) during the fabrication of bio-anodes for biofuel cells. Chitosan's NH_2 and OH functional groups provide a hydrophilic surface for carbon nanostructures, allowing for covalent, absorptive, or ionic linkages with bioactive molecules. These linkages have also been used to create superior sensors for target metal ions. Chitosan's distinctive physiochemical characteristics allow it to maintain its innovative properties when firmly captured on electrode surfaces (Bakhsh et al., 2019; Zhou et al., 2018).

7.6 CHITOSAN-BASED CHEMICAL SENSOR SENSING MECHANISMS

The key aspects expected for the development of a chemical sensor include sensitivity in the parts per million (ppm) to billion (ppb) range where trace levels are involved, absolute discrimination, mild operation temperature, low power consumption, practical size, volume and mass, and low cost for large-scale applications (Norizan et al., 2020; Norizan et al., 2021; Nurazzi, Harussani et al., 2021). Figure 7.3 summarizes the four major important aspects for chemical sensors.

In a study by Nasution et al. on chitosan-based sensors for acetone detection, stability was determined by a continuous increase in response without significant fluctuations in the graph (2013). In some cases, small variations in humidity and operating temperatures may cause this, as supported by relative humidity values ranging from 60 to 68% and temperatures of 25 to 30°C. Additionally, chitosan-based sensors showed excellent repeatability, as demonstrated by almost identical maximum response values for each exposure time of 5 min/measurement in five uninterrupted measurements. The good response, recovery, stability, and repeatability are indicative of a practical and effective sensor. Although the maximum response values for all different concentrations tended to decrease over time, the chitosan-based sensor still functions well because the decrease in response is minor. This suggests that oxidation on the aluminum layer from fabrication and swelling on chitosan have no significant effect on chitosan-based sensor response. Several reports have noted that the chemisorption of oxygen species on the sensor surface greatly affects sensor resistance (Mbarek et al., 2007).

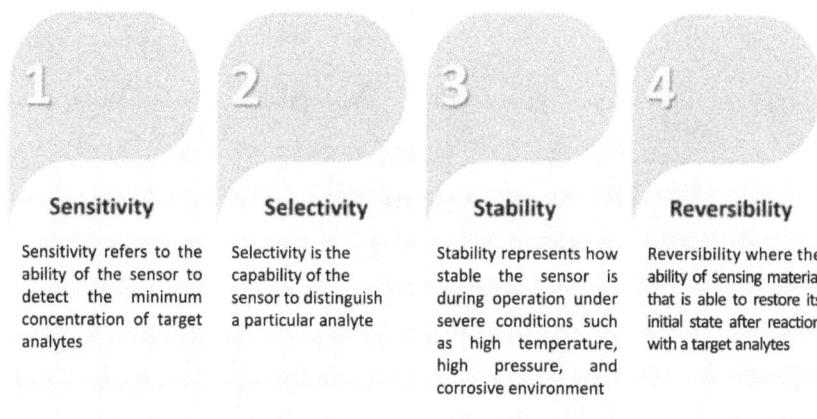

FIGURE 7.3   Four major important aspects for chemical sensor capacity. Reproduced from Nurazzi et al. (2021).

Considering this, changes in output voltage values of chitosan film sensors in normal air were analyzed. It was found that when an input voltage of 5 V was applied to the sensor in normal air, the output voltage value was initially 102 mV. However, the output voltage decreased over time until it reached a steady state where no significant changes in output voltage occurred when the input voltage was kept at 5 V. Under normal air, different voltages were tested, and it took each around 45 to 60 min to give steady output voltage values. This can be explained by the transfer of electrons from the valence band to the conduction band, which form ionic species such as O_2^- or O^- as a barrier for electron movements and are trapped by the chemisorbed oxygen atoms, as described by Equations 7.1, 7.2, and 7.3.

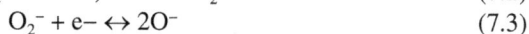

$$O_2 \text{ (gas)} \leftrightarrow O_2 \text{ (absorbed)} \tag{7.1}$$
$$O_2 \text{ (absorbed)} + e^- \leftrightarrow O_2^- \tag{7.2}$$
$$O_2^- + e- \leftrightarrow 2O^- \tag{7.3}$$

When the chitosan sensor is exposed to normal air, oxygen chemisorbs onto the surface of the chitosan particles per Equation 7.1. At a certain input voltage, the number of free electrons in the conduction band that move randomly decreases as the chemisorbed oxygen traps and transfers them from one particle to another, resulting in increased resistance and decreased output voltage of the chitosan film. This continues until the oxygen species become saturated, and according to Equations (7.2) and (7.3), no further oxygen species are formed. On the other hand, when the sensor is exposed to water molecules (H_2O), the molecules sticking to the surface of the chitosan particles react with O^- ions to produce O_2 gases and release electrons, as described in Equation (7.4).

$$\text{Chitosan-NH}_2 + 2O^- + H_2O\text{-Chitosan-NH}_2 + O_2 + H_2O + 2e^- \tag{7.4}$$

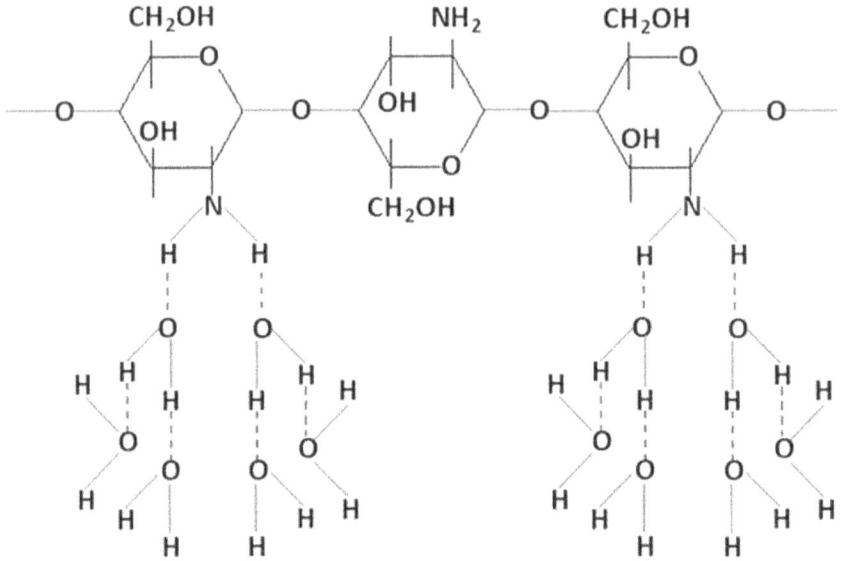

FIGURE 7.4 Hydrogen bonds formed when amino groups react with water molecules. Reproduced from Nasution et al. (2013).

When the chitosan film sensor is exposed to moist air, the water molecules create surface tension on the film, which, along with reacting with O^- ions, aids the movement of electrons to escape from the oxygen species. As a result, the electrical conductivity of the film increases. However, in the presence of acetone vapor molecules, the inherent molecular vibration of the acetone vapor molecules breaks the surface tension of the water molecules, making it easier for them to vaporize and reducing their interaction with the oxygen species. The conversion of ionic O^- to O_2 gas reduces the barrier width and enables electron transfer from one particle to another. But the free electrons may not easily move across the particles due to gaps between them. In chitosan films, the hydrogen bonds formed between the hydrogen of the amino groups and the oxygen of water molecules act as a pathway or electrical bridge for electron transfer between the particles (Figure 7.4). The membrane form of the chitosan film layer, with a uniform particle size distribution and close particle distance, supports the possibility of a pathway or electrical bridge, promoting electron transfer across the available particles. This is indicated by the rapid and significant increase in the electrical response.

7.7 CONCLUSIONS

Chitin, the second most prevalent natural polysaccharide after cellulose, is an amino polysaccharide that can be modified chemically and mechanically to produce novel properties and functions. Chitin and its derivative chitosan are biopolymers that have gained significant attention in recent years due to their abundant nature, biorenewability, and potential applications in various fields, particularly in chemical sensors. Chitosan, in particular, is a versatile material with a functionalized surface that enables easy blending with other materials to improve physicochemical properties. The unique properties of chitin and its derivatives, such as biodegradability, renewability, and antibacterial and fungistatic properties, offer promising opportunities for the development of sustainable products. However, the intractability and insolubility of chitin have limited its widespread use, and the high cost and complex chemical modifications have also restricted its applications to the biomedical, pharmaceutical, and cosmetic industries. Chitosan, which is water soluble under specified conditions, has emerged as the most important derivative of chitin due to its chemical modifiability and potential uses in various fields. Despite some limitations, chitin and chitosan hold great promise for a green and sustainable future.

Chitosan-based chemical sensors have shown promising results in various applications due to their unique properties such as biocompatibility, biodegradability, and sensitivity to various analytes. The sensing mechanism involves the interaction of analyte molecules with chitosan and the subsequent changes in electrical, optical, or mechanical properties of the sensor. Chitosan-based sensors have demonstrated excellent detection capacity for various analytes such as heavy metals, organic compounds, gases, and biological molecules. The sensitivity and selectivity of chitosan-based sensors can be further improved by modifying the chitosan structure, introducing functional groups, or incorporating additional materials. Based on the literature, molecular imprinted polymers (MIPs) have been widely used in various fields and have the potential to replace biological entities in different applications due to their unique properties. Chitosan is an attractive biopolymer for the development of MIPs due to its excellent characteristics. Chitosan-MIPs have been successfully integrated into the transducer area to develop electrochemical sensors. Various methods have been employed to produce chitosan-based membranes, thin films, and three-dimensional structures by exploiting its physicochemical properties. Electrodeposition is a cost-effective, simple, and fast method that allows for the controlled deposition of highly porous chitosan films. Although small molecules such as pesticides and pharmaceuticals have been successfully imprinted using MIPs, imprinting larger structures remains a challenge. The sensitivity of the sensors developed using chitosan is influenced by various parameters that affect its behavior. Therefore, introducing chemical modifications to chitosan could potentially improve the performance of sensors. In the future, the multi-imprinting of several templates on chitosan could lead to a wide range of fundamental experiments. Overall, chitosan-MIPs show promising potential for the development of advanced sensor technologies in various fields.

REFERENCES

Abdou, E. S., Nagy, K. S. A., & Elsabee, M. Z. (2008). Extraction and characterization of chitin and chitosan from local sources. *Bioresource Technology*, *99*(5), 1359–1367.

Anitha, A., Sowmya, S., Kumar, P. T. S., Deepthi, S., Chennazhi, K. P., Ehrlich, H., Tsurkan, M., & Jayakumar, R. (2014). Chitin and chitosan in selected biomedical applications. *Progress in Polymer Science*, *39*(9), 1644–1667.

Bakhsh, E. M., Ali, F., Khan, S. B., Marwani, H. M., Danish, E. Y., & Asiri, A. M. (2019). Copper nanoparticles embedded chitosan for efficient detection and reduction of nitroaniline. *International Journal of Biological Macromolecules, 131*, 666–675.

Baranwal, A., Kumar, A., Priyadharshini, A., Oggu, G. S., Bhatnagar, I., Srivastava, A., & Chandra, P. (2018). Chitosan: An undisputed bio-fabrication material for tissue engineering and bio-sensing applications. *International Journal of Biological Macromolecules, 110*, 110–123.

Cai, L., Ying, D., Liang, X., Zhu, M., Lin, X., Xu, Q., Cai, Z., Xu, X., & Zhang, L. (2021). A novel cationic polyelectrolyte microsphere for ultrafast and ultra-efficient removal of heavy metal ions and dyes. *Chemical Engineering Journal, 410*, 128404.

Ding, F., Qian, X., Zhang, Q., Wu, H., Liu, Y., Xiao, L., Deng, H., Du, Y., & Shi, X. (2015). Electrochemically induced reversible formation of carboxymethyl chitin hydrogel and tunable protein release. *New Journal of Chemistry, 39*(2), 1253–1259.

El Knidri, H., Belaabed, R., Addaou, A., Laajeb, A., & Lahsini, A. (2018). Extraction, chemical modification and characterization of chitin and chitosan. *International Journal of Biological Macromolecules, 120*, 1181–1189.

El Knidri, H., El Khalfaouy, R., Laajeb, A., Addaou, A., & Lahsini, A. (2016). Eco-friendly extraction and characterization of chitin and chitosan from the shrimp shell waste via microwave irradiation. *Process Safety and Environmental Protection, 104*, 395–405.

Hasan, M., Gopakumar, D. A., Olaiya, N. G., Zarlaida, F., Alfian, A., Aprinasari, C., Alfatah, T., Rizal, S., & Khalil, H. A. (2020). Evaluation of the thermomechanical properties and biodegradation of brown rice starch-based chitosan biodegradable composite films. *International Journal of Biological Macromolecules, 156*, 896–905.

Ibrahim, M. A., Alhalafi, M. H., Emam, E. A. M., Ibrahim, H., & Mosaad, R. M. (2023). A review of chitosan and chitosan nanofiber: Preparation, characterization, and its potential applications. *Polymers, 15*(13), 2820.

Kangkamano, T., Numnuam, A., Limbut, W., Kanatharana, P., & Thavarungkul, P. (2017). Chitosan cryogel with embedded gold nanoparticles decorated multiwalled carbon nanotubes modified electrode for highly sensitive flow based non-enzymatic glucose sensor. *Sensors and Actuators B: Chemical, 246*, 854–863.

Kaur, S., & Dhillon, G. S. (2014). The versatile biopolymer chitosan: Potential sources, evaluation of extraction methods and applications. *Critical Reviews in Microbiology, 40*(2), 155–175.

Kazi, G. A. S., Yamanaka, T., & Osamu, Y. (2019). Chitosan coating an efficient approach to improve the substrate surface for in vitro culture system. *Journal of the Electrochemical Society, 166*(9), B3025.

Lizardi-Mendoza, J., Monal, W. M. A., & Valencia, F. M. G. (2016). Chemical characteristics and functional properties of chitosan. In *Chitosan in the preservation of agricultural commodities* (pp. 3–31). Elsevier.

Mbarek, H., Saadoun, M., & Bessaïs, B. (2007). Porous screen printed indium tin oxide (ITO) for NOx gas sensing. *Physica Status Solidi C, 4*(6), 1903–1907.

Musarurwa, H., & Tavengwa, N. T. (2020). Application of carboxymethyl polysaccharides as bio-sorbents for the sequestration of heavy metals in aquatic environments. *Carbohydrate Polymers, 237*, 116142.

Nasution, T. I., Nainggolan, I., Hutagalung, S. D., Ahmad, K. R., & Ahmad, Z. A. (2013). The sensing mechanism and detection of low concentration acetone using chitosan-based sensors. *Sensors and Actuators B: Chemical, 177*, 522–528.

Norizan, M. N., Moklis, M. H., Demon, S. Z. N., Halim, N. A., Samsuri, A., Mohamad, I. S., Knight, V. F., & Abdullah, N. (2020). Carbon nanotubes: Functionalisation and their application in chemical sensors. *RSC Advances, 10*(71), 43704–43732.

Norizan, M. N., Zulaikha, N. D. S., Norhana, A. B., Syakir, M. I., & Norli, A. (2021). Carbon nanotubes-based sensor for ammonia gas detection: An overview. *POLIMERY, 66*(3), 175–186.

Nurazzi, N. M., Abdullah, N., Demon, S. Z. N., Halim, N. A., Azmi, A. F. M., Knight, V. F., & Mohamad, I. S. (2021). The frontiers of functionalized graphene-based nanocomposites as chemical sensors. *Nanotechnology Reviews, 10*(1), 330–369.

Nurazzi, N. M., Harussani, M. M., Zulaikha, N. D. S., Norhana, A. H., Syakir, M. I., & Norli, A. (2021). Composites based on conductive polymer with carbon nanotubes in DMMP gas sensors: An overview. *POLIMERY, 66*(2), 85–97.

Park, I., Nguyen, T., Park, J., Yoo, A. Y., Park, J. K., & Cho, S. (2018). Impedance characterization of chitosan cytotoxicity to MCF-7 breast cancer cells using a multidisc indium tin oxide microelectrode array. *Journal of the Electrochemical Society, 165*(2), B55.

Peter, S., Lyczko, N., Gopakumar, D., Maria, H. J., Nzihou, A., & Thomas, S. (2021). Chitin and chitosan based composites for energy and environmental applications: A review. *Waste and Biomass Valorization, 12*, 4777–4804.

Safavy, A., Raisch, K. P., Mantena, S., Sanford, L. L., Sham, S. W., Krishna, N. R., & Bonner, J. A. (2007). Design and development of water-soluble curcumin conjugates as potential anticancer agents. *Journal of Medicinal Chemistry, 50*(24), 6284–6288.

Singh, P., & Nagendran, R. (2016). A comparative study of sorption of chromium (III) onto chitin and chitosan. *Applied Water Science, 6*, 199–204.

Zhou, G., Sun, J., Yaseen, M., Zhang, H., He, H., Wang, Y., Yang, H., Liang, J., Zhang, M., Liqin, Z., & Zhou, L. (2018). Synthesis of highly selective magnetite (Fe3O4) and tyrosinase immobilized on chitosan microspheres as low potential electrochemical biosensor. *Journal of the Electrochemical Society, 165*(2), G11.

8 Seaweed-Based Biopolymers for Sustainable Applications

8.1 INTRODUCTION

The increasing global demand for materials and the adverse environmental impacts of conventional plastics and synthetic polymers have driven researchers and industries to seek sustainable alternatives. Plastic waste primarily comes from packaging. The difficulty of plastics to biodegrade and their reliance on non-renewable natural resources are two major barriers to their increased use. The accumulation of plastic items in the environment, which has a negative impact on wildlife, wildlife habitats, and humans, has put a great deal of strain on the environment (Kumar et al., 2021; Zolotova et al., 2022). Furthermore, plastic waste will cause drain plugging and dead aquatic life and pollute the river and sea. Biopolymers are the subject of ongoing research and exploration by scientists and researchers worldwide. Their utilization has the potential to promote sustainability and minimize the environmental consequences associated with the disposal of petroleum-based polymers. By being biodegradable in nature, these polymers can naturally degrade, thereby reducing the need for labor-intensive plastic removal efforts (Abdul Khalil et al., 2018a). The advent of bioplastics or biodegradable polymers as alternatives to non-biodegradable or petroleum-based polymers offers a means to mitigate environmental harm and serve as an example for other industries to adopt eco-friendly materials.

Mangaraj et al. (2019) provide a definition for biodegradable polymers as polymers that undergo degradation through microbial metabolism. This degradation process leads to the production of inorganic compounds, carbon dioxide, and water (Wróblewska-Krepsztul et al., 2018). Biodegradable polymers can be derived from various sources, such as biomass materials and microorganisms (Abdul Khalil et al., 2017), and they offer reduced environmental pollution compared to petroleum-based polymers (Atef et al., 2014). Biopolymers, which are a type of natural polymer obtained from natural sources, are particularly attractive due to their environmental advantages (Matei et al., 2022). These biopolymers exhibit variations in terms of size, type, and functional properties. The two commonly used types of biodegradable polymers in various applications are natural polymers (biopolymers) and synthetic polymers.

The growing concern about environmental sustainability has led to the exploration of alternative materials for various applications. Biopolymers including starch, gluten, seaweed, and guar gum are viable substitutes to create packaging material due to their nontoxicity, biodegradability, and ability to be derived from renewable natural resources (Baranwal et al., 2022; Das et al., 2022). Seaweed, a versatile

DOI: 10.1201/9781003416043-8

marine resource, has gained significant attention due to its abundant availability, rapid growth, and numerous ecological benefits (Farghali et al., 2023). This chapter discusses the potential of seaweed-based biopolymers as a sustainable alternative to conventional materials. The unique properties of seaweed and its potential applications in various industries, such as packaging, agriculture, textiles, and biomedical engineering, will be explored. Additionally, the challenges and future prospects of seaweed-based biopolymers for a greener and more sustainable future are highlighted.

Seaweed, also known as marine macroalgae, offers a promising solution due to its renewable nature, high growth rate, and low environmental footprint (Jayakumar et al., 2023). Seaweed, a fascinating and diverse group of marine plants, is gaining increasing attention as a valuable source of biopolymers. Biopolymers are natural polymers derived from living organisms that have garnered significant interest due to their renewable and sustainable nature, as well as their potential to replace synthetic polymers derived from fossil fuels (Udayakumar et al., 2021). Seaweed biopolymers offer a multitude of advantages, including their abundance, biodegradability, and versatile applications across various industries. Seaweeds, also known as macroalgae, are found abundantly in oceans and other bodies of water around the world. They encompass a wide range of species, such as brown algae (Phaeophyceae), red algae (Rhodophyceae), and green algae (Chlorophyceae). These marine plants are known for their rapid growth rates and high biomass production, making them an attractive and sustainable resource for the extraction of biopolymers (Yang et al., 2021).

Seaweeds are plant-like organisms that thrive in coastal waters, ranging from intertidal areas to deep regions. They require sunlight and a solid surface to grow, as they lack roots for nutrient delivery. Instead, the entire organism is responsible for absorbing nutrients and performing photosynthesis (Jagtap & Meena, 2021). Seaweeds are highly successful photosynthetic organisms, exhibiting rapid growth and yielding 6 to 40 times more biomass compared to terrestrial plants. Their natural characteristics enable them to thrive in diverse environments without the need for pesticides (Ditchburn & Carballeira, 2019). In the mangrove habitat, which is known for its challenging conditions, 206 seaweed species have been identified. However, the growth of seaweeds in mangroves is hindered due to their reliance on the aerial roots of the trees for support.

Seaweed, similar to other polysaccharide materials, is an abundant and inexpensive green source of polysaccharides, distinguished by its extraction from the sea. It is a renewable natural resource with substantial untapped industrial potential. Notable seaweed-derived products, such as alginate, carrageenan, and agar, exhibit excellent film-forming properties (Abdou & Sorour, 2014; Benavides et al., 2012; Huq et al., 2012; Kulig et al., 2017; Mostafavi & Zaeim, 2020a; Parreidt et al., 2018; Pranoto et al., 2005; Sedayu et al., 2019; Tavassoli-Kafrani et al., 2016). However, when compared to traditional non-renewable polymers, seaweed films have relatively low water vapor barrier characteristics and mechanical properties (Sedayu et al., 2020). Consequently, seaweed is often combined with other materials to enhance the properties of seaweed films.

The utilization of seaweed biopolymers offers numerous advantages over synthetic polymers. First, seaweed is a renewable resource, with many species capable of fast growth and regeneration, making it a sustainable alternative to petroleum-based polymers. Furthermore, seaweed biopolymers are biodegradable, meaning they can be broken down naturally by microorganisms, reducing environmental pollution and waste accumulation. This characteristic is particularly crucial in combating the growing concerns surrounding plastic pollution and the need for more environmentally friendly materials. Moreover, seaweed biopolymers exhibit a wide range of functional properties, making them suitable for diverse applications. They possess excellent film-forming abilities, moisture retention, and emulsifying properties, making them valuable in food products, pharmaceutical formulations, and cosmetics. The potential applications of seaweed biopolymers also extend to the agricultural sector, where they can be used as biofertilizers, soil conditioners, and plant growth promoters.

8.2 SEAWEED POLYSACCHARIDES

Over the past few decades, the significance of seaweed as a marine resource for humans has grown considerably, particularly in the context of multibillion-dollar industries that heavily rely on seaweed-derived polysaccharides. Seaweeds can be classified into three primary groups based on the pigments they possess: green (Chlorophyta), red (Rhodophyta), and brown (Heterokontophyta) seaweeds (Aryee et al., 2018; Ferrara, 2020; Abdul Khalil et al., 2018b; Rajauria et al., 2015), as shown in Figure 8.1.

Red seaweed is characterized by its unique red color, which is primarily attributed to the presence of a pigment called phycoerythrin (Vilar et al., 2021). This pigment is responsible for absorbing blue light and reflecting red light, giving the seaweed its characteristic reddish hue. The abundance of phycoerythrin in red seaweed distinguishes it from other types of seaweed and contributes to its vibrant coloration (Freitas et al., 2021). Green seaweeds contain significant levels of chlorophyll a and b, along with lesser quantities of β-carotene and xanthophylls. These pigments are essential for photosynthesis, enabling the seaweeds to capture light

FIGURE 8.1 Marine seaweed samples: (a) (Phaeophyceae) brown variant (*Padina gymnospora*), (b) (*Rhodophyta*) red variant (*Kappaphycus alvarezii*), and (c) green variant (*Kappaphycus striatus*). Reproduced from Bhuyar et al. (2021).

energy effectively. β-carotene contributes to their yellow to orange color, while xanthophylls provide a range of hues, such as yellow and brown. These pigments not only aid in light absorption but also protect the seaweeds from excessive light and oxidative damage (Wang et al., 2021). Meanwhile, brown seaweeds, with a color spectrum ranging from yellow and olive-green to brown, possess elevated levels of the xanthophyll fucoxanthin while having lower quantities of other xanthophylls, chlorophyll a and c, and β-carotene (Din et al., 2022). The distinctive brown color of brown seaweed is primarily attributed to the presence of the pigment fucoxanthin, which effectively masks the green color of chlorophyll. However, when red or brown seaweed is subjected to heat, the pigments undergo denaturation, causing the reemergence of the green color of chlorophyll (Bonanno & Orlando-Bonaca, 2018).

While marine environments are the primary habitats for red and brown seaweeds, it is worth noting that green seaweeds have a broader distribution. They can be found not only in marine ecosystems but also in a variety of other environments (Salehi et al., 2019). This wider distribution allows green seaweeds to thrive in freshwater locations such as rivers and lakes, as well as unexpected terrestrial settings like houses, rocks, and damp areas. The adaptability of green seaweeds to different habitats contributes to their ecological success and diverse ecological roles (Abdul Khalil et al., 2017). These seaweeds are highly nutritious and rich in minerals and vitamins. The vitamin composition of seaweeds varies depending on factors such as location, season, and species. However, what remains consistent is their exposure to ample sunlight in their aquatic environment. This sunlight exposure, along with the nutrient-rich water they inhabit, contributes to the high vitamin content of seaweeds. Incorporating seaweeds into a balanced diet can provide essential vitamins and minerals, offering significant health benefits.

In addition to their nutritional value, seaweeds have gained significant recognition for their versatile applications in both herbal medicine and human cuisine. With their edible nature and rich content of essential elements, seaweeds are widely used in various culinary preparations. They can be incorporated into raw salads, soups, pastries, dinners, and sauces, offering a unique flavor and texture to these dishes (Raja et al., 2022). While all seaweed species share similar chemical components, there is variation in the proportion of protein content among different species (El-Said & El-Sikaily, 2013; Fleurence et al., 2017). This variation in protein content contributes to the diversity of nutritional profiles found in different types of seaweeds. Table 8.1 provides the chemical compositions of different seaweeds (Abdul Khalil et al., 2017). It shows that water, carbohydrates, minerals, and lipids are consistent across seaweed types. However, there is significant variation in protein content. Brown seaweed generally contains 3% to 15% protein, while red and green seaweeds have a broader range of 10% to 47% protein content.

Seaweed possesses immense value beyond its nutritional benefits, owing to its substantial carbohydrate concentration that makes it a sought-after resource across multiple industries (Sultana et al., 2023). Various types of seaweed serve as abundant natural sources of essential polysaccharides such as alginate, carrageenan, and agar. Agar, derived from a polysaccharide present in the cell walls of red algae, plays a crucial role in providing structural support. On the other hand, alginate, predominantly obtained from large brown seaweeds, is highly valuable for its ability to react

TABLE 8.1
Chemical Compositions of Seaweeds

Components	Compositions
Water	80–90%
Carbohydrates	50% dry weight
Proteins	Brown seaweed: 3–15% dry weight
	Red or green seaweeds: 10–47% dry weight
Minerals	7–38% dry weight
Lipids	1–3% dry weight

with divalent and trivalent cations, making it indispensable in the formation of alginate films. Meanwhile, carrageenans, linear sulfated polysaccharides extracted from edible red seaweeds, find extensive applications in the food packaging and pharmaceutical industries due to their hydrophilic and anionic properties.

Furthermore, these derivatives are widely utilized in the food industry as hydrocolloids, serving various functions such as thickening, gelling, and stabilizing agents (Liao et al., 2021). Moreover, these hydrocolloids have extensive applications in biotechnology, microbiology, biomedical fields, and the plastics industry (Abdul Khalil et al., 2018; Gade et al., 2013). When combined with water, these hydrocolloids have the capability to form viscous dispersions or gels (Gurram, 2022). As high molecular weight hydrophilic polymers (polysaccharides), they possess a strong affinity for water molecules due to the presence of multiple hydroxyl (-OH) groups, which renders them hydrophilic compounds (Gunathilake et al., 2022). They are commonly employed as thickening or gelling agents to modify the functional properties of aqueous solutions.

Table 8.2 provides a comprehensive overview of the polysaccharides found in different types of seaweeds (Abdul Khalil et al., 2017). According to the table, red seaweed contains a wide range of polysaccharides, including agar, carrageenan, cellulose, floridean starch, mannan, porphyrin, sulphated galactans, and xylans. Brown seaweed, on the other hand, contains polysaccharides such as alginate, cellulose, fucoidan, laminarin, mannitol, and sargassan. Green seaweed is characterized by the presence of cellulose, sulphated galactans, sulfuric acid, and xylans. This diversity of polysaccharides underscores the versatility and potential applications of seaweed in various industries.

8.3 TYPES OF SEAWEEDS

8.3.1 RED SEAWEED

Red seaweed, scientifically referred to as Rhodophyta, represents a variety of marine algae within the phylum Rhodophyta. This group is among the largest assemblages of seaweed, encompassing over 7,000 distinct species inhabiting tropical and temperate waters across the globe (Ismail et al., 2020). The distinguishing characteristic of red

TABLE 8.2

Polysaccharides from Red, Brown, and Green Seaweeds

Polysaccharides	Red Seaweed	Brown Seaweed	Green Seaweed
Agar	✓	–	–
Alginate	–	✓	–
Carrageenan	✓	–	–
Cellulose	✓	✓	✓
Floridean starch (α-1,4-bindingglucan)	✓	–	–
Fucoidan (sulphatedfucose)	–	✓	–
Laminarin (β-1, 3 glucan)	–	✓	–
Mannan	✓	–	–
Mannitol	–	✓	–
Porphyran	✓	–	–
Sargassan	–	✓	–
Sulfated galactans	✓	–	✓
Sulfuric acid polysaccharides	–	–	✓
Xylans	✓	–	✓

seaweed is its remarkable red or purplish hue, although certain species may exhibit shades of green or brown owing to variations in pigmentation. Structurally, red seaweed possesses a complex organization comprising a holdfast, stipe, and fronds. The holdfast functions akin to a root-like structure, ensuring the seaweed's attachment to the substrate. Linking the holdfast to the fronds is the stipe, resembling a stem-like component. The fronds, which resemble leaf-like blades, furnish the seaweed with a surface area essential for photosynthesis. Notably, red seaweed exhibits a noteworthy adaptation enabling its proliferation across a broad spectrum of water depths. Some species thrive in shallow intertidal zones, while others are capable of thriving in depths reaching several hundred meters. This adaptability can be attributed to the presence of specialized pigments, such as phycoerythrins and phycocyanins, which enable red seaweed to effectively absorb light across varying depths.

Red seaweed holds a vital position within marine ecosystems, playing multiple crucial roles. It serves as a habitat and refuge for a diverse array of marine organisms, encompassing fish, invertebrates, and microorganisms. The intricate and three-dimensional structure of red seaweed offers shelter and a substrate for attachment, fostering biodiversity and enhancing the productivity of marine ecosystems. Moreover, red seaweed serves as a significant food source for numerous marine animals. Various herbivorous species, including sea urchins, snails, and certain fish, heavily rely on the nutritional richness of red seaweed. Humans also utilize it in different forms, such as nori (sushi wraps) and as an ingredient in various processed foods.

The economic value of red seaweed is substantial, finding applications in diverse industries. One of its primary uses lies in the production of hydrocolloids, particularly carrageenan and agar. These valuable substances are extracted from red seaweed and employed as gelling agents, stabilizers, and thickeners in an extensive range of products, spanning food, cosmetics, and pharmaceuticals. Additionally, red seaweed is being actively explored as a potential source of biofuels due to its high carbohydrate content. Researchers are investigating methods to extract and convert the sugars present in red seaweed into bioethanol and other forms of renewable energy.

8.3.2 GREEN SEAWEED

Green seaweed, scientifically referred to as Chlorophyta or green algae, encompasses a diverse group of marine algae within the phylum Chlorophyta. It stands as one of the most abundant and varied assemblages of algae, with numerous species identified in marine, freshwater, and terrestrial environments. Green seaweeds exhibit a distinctive green coloration, owing to the presence of chlorophyll pigments, enabling them to engage in photosynthesis and convert sunlight into energy. These algae predominantly inhabit shallow coastal waters and intertidal zones, where they firmly attach themselves to rocks, shells, or other substrates through specialized structures called holdfasts. In terms of structure, green seaweeds exhibit considerable variation in size and shape. Some species are microscopic, forming a slimy film over rocks or surfaces, while others can develop into large and intricate structures reminiscent of plants. Their forms span a wide spectrum, including filamentous, sheet-like, and branched arrangements.

Green seaweeds play a vital ecological role within marine ecosystems, offering both habitats and nourishment for diverse marine organisms, including invertebrates and small fish. Additionally, they actively contribute to oxygen production and nutrient cycling processes, thus playing a crucial part in maintaining the overall balance of marine ecosystems. Beyond their ecological significance, green seaweeds possess practical applications across various industries. Culturally, green seaweeds serve as important food sources, valued for their nutritional richness characterized by high protein and vitamin content. Moreover, certain species of green seaweed are harvested specifically for the extraction of commercially valuable compounds. These compounds include agar, carrageenan, and alginates, which find application in the food industry, pharmaceuticals, and other industrial sectors. The scientific community also demonstrates considerable interest in green seaweeds due to their potential in biofuel production. These algae possess the capacity to accumulate substantial levels of lipids (oils), which can be converted into biofuels such as biodiesel, thereby contributing to sustainable energy solutions.

8.3.3 BROWN SEAWEED

Brown seaweed, commonly known as kelp, is a type of marine algae that belongs to the Phaeophyceae class. It is found in cool coastal waters worldwide, particularly in the northern hemisphere. This seaweed is characterized by its distinctive brown color, which is due to the presence of pigments such as fucoxanthin and chlorophyll (Sheath & Wehr, 2015). It is one of the largest and most complex forms of seaweed, often forming dense underwater forests known as kelp forests. These forests serve

as crucial habitats and shelters for various marine organisms, including fish, invertebrates, and other types of algae. The structure of brown seaweed consists of a holdfast, which anchors it to rocks or other surfaces, a stem-like structure called a stipe
that provides support and transports nutrients, and leaf-like structures called blades.
The blades are typically large and flat, allowing them to capture sunlight for photosynthesis. Brown seaweed is known for its rapid growth rate. In favorable conditions,
it can grow several centimeters per day and reach lengths of over 30 meters. This fast
growth is facilitated by the absorption of nutrients from the water, such as nitrogen
and phosphorus (Saunders, 2001; Reed et al., 2006; Minhas et al., 2020).

Brown seaweed plays a crucial ecological role in marine ecosystems. It serves as a
source of food and habitat for a diverse array of organisms, contributing to increased
biodiversity and supporting productive fisheries. Additionally, brown seaweed helps
mitigate climate change by sequestering carbon dioxide from the atmosphere. It
has the ability to absorb and store carbon, making it an important factor in carbon
sequestration. Apart from its ecological significance, brown seaweed has practical
applications in various industries. It is used as a food source in many cultures, particularly in East Asian countries, where it is a staple ingredient in dishes like soups,
salads, and sushi. Brown seaweed is also utilized in the production of fertilizers,
animal feed, cosmetics, and pharmaceuticals (Jayathilake & Costello, 2020).

Moreover, brown seaweed has gained attention for its potential health benefits.
It is rich in essential minerals, vitamins, antioxidants, and bioactive compounds,
which are believed to possess anti-inflammatory, antiviral, and anticancer properties. Consequently, researchers are investigating the potential therapeutic applications of brown seaweed and its extracts in medicine and functional foods.

8.4   SEAWEED DERIVATIVES

8.4.1   ALGINATE

Alginate is a polysaccharide that occurs naturally and is derived from brown seaweed. It consists of long chains made up of repeated units of guluronic acid (G units)
and mannuronic acid (M units). The ratio of G to M units determines the physical
and chemical characteristics of alginate (Kim et al., 2011) Figure 8.2 illustrates the
chemical structure of alginate. This biomaterial is highly versatile and finds applications in a wide range of industries, including food, biomedical, and pharmaceutical.
Alginate possesses several important properties that contribute to its usefulness in
various applications. Its ability to form gels, biocompatibility, and controlled-release
capabilities make it a valuable ingredient in numerous products and technologies
(Mikkonen, 2013).

Alginate exhibits gel-forming properties when exposed to divalent cations, such
as calcium. The presence of calcium ions triggers the gelation process, leading to the
formation of a three-dimensional network structure (Tavassoli-Kafrani et al., 2016).
This characteristic finds practical applications in various fields, including food and
pharmaceuticals, where alginate gels serve as thickeners, stabilizers, or matrices for
encapsulation. Alginate is known for its biocompatibility and is extensively used in
biomedical and tissue engineering applications. It can function as a scaffold material

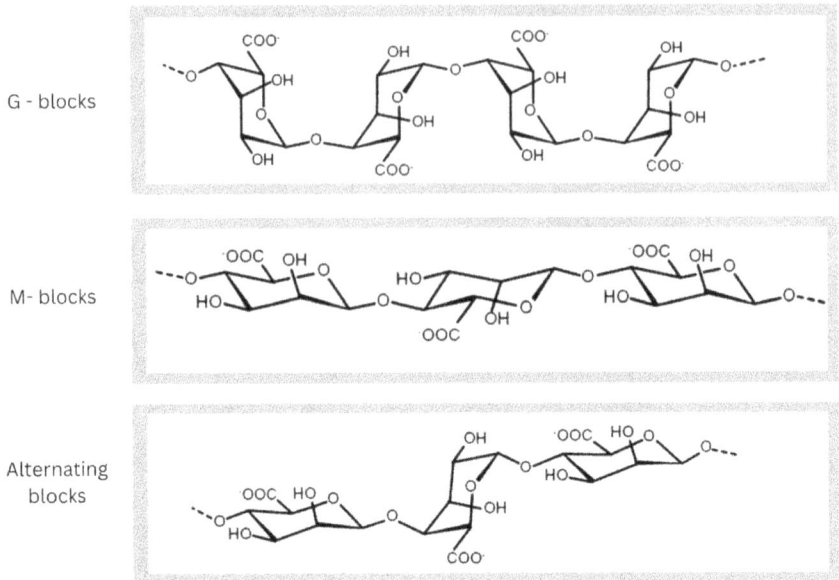

FIGURE 8.2 Chemical structure of alginates. Reproduced from Lee and Mooney (2012).

for cell encapsulation and tissue regeneration. The gel-forming nature of alginate enables the encapsulation of cells, safeguarding them while providing an appropriate microenvironment for their growth and function (Lee & Mooney, 2012).

Alginate gels are also employed for controlled release purposes, such as delivering drugs or nutrients. The porous structure of the gel allows for the diffusion of the encapsulated substance, facilitating a controlled and sustained release over time. In the food industry, alginate is commonly utilized as a thickening agent, stabilizer, or gelling agent. It can create diverse textures in foods, including gels, foams, or films. Ice cream, salad dressings, sauces, and processed meats often incorporate alginate. Moreover, alginate dressings are frequently used in wound healing due to their ability to absorb exudate and maintain a moist environment, promoting faster recovery. In pharmaceutical formulations, alginate finds applications in drug delivery systems and as a binder in tablet formulations.

8.4.2 CARRAGEENAN

Carrageenan, derived from specific species of red seaweed, is a natural ingredient used in a wide range of culinary and industrial applications for many centuries. Its exceptional properties in gelling, thickening, and stabilizing make it highly valuable. The name "carrageenan" originates from the Irish term "carragheen", which translates to "little rock" (Necas & Bartosikova, 2013). There are three primary types of carrageenan: kappa, iota, and lambda. The chemical structure of carrageenan can be observed in Figure 8.3. Each type exhibits distinct gel-forming capabilities and finds diverse applications. Kappa carrageenan forms firm and brittle gels when

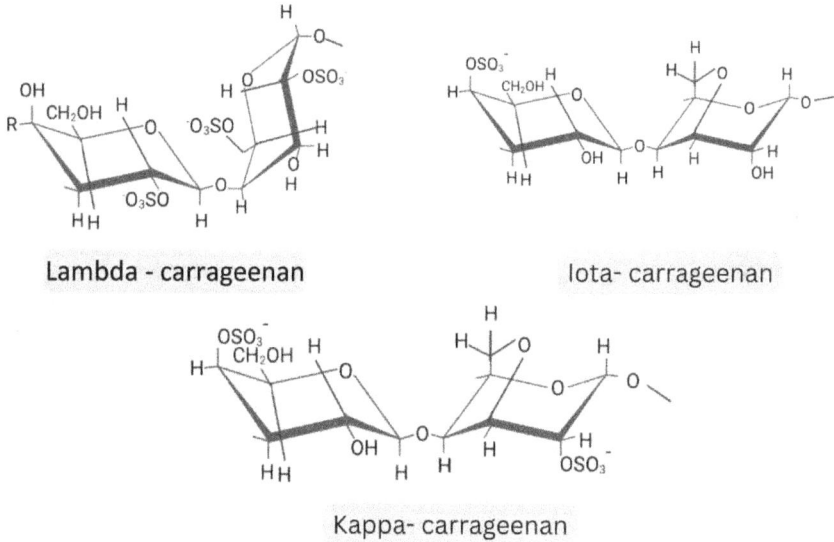

FIGURE 8.3 Chemical structures of carrageenans. Reproduced from Tavassoli-Kafrani et al. (2016).

combined with potassium ions, while iota carrageenan creates soft and elastic gels in the presence of calcium ions. Lambda carrageenan, however, does not form gels but functions as a thickener and stabilizer (Mugdha Bhat et al., 2020; Tasende & Manríquez-Hernández, 2016; Vasconcelos & Pomin, 2018).

Carrageenan is a popular food additive in the food industry, particularly in dairy products like ice cream, chocolate milk, yogurt, and cottage cheese. It serves to enhance texture, prevent separation, and improve mouthfeel. Additionally, carrageenan functions as a thickener and emulsifier in meat products, salad dressings, sauces, and processed foods. Beyond its culinary applications, carrageenan finds uses in various industries. It is utilized in pharmaceuticals, cosmetics, toothpaste, and even acts as a clarifying agent in beer and wine production. In the pharmaceutical sector, carrageenan can be found in cough syrups, capsules, and creams due to its binding and thickening properties.

Regulatory bodies such as the U.S. Food and Drug Administration (FDA), the European Food Safety Authority (EFSA), and the Joint Expert Committee on Food Additives (JECFA) consider carrageenan safe for consumption. However, there has been some controversy surrounding its safety, particularly regarding potential digestive issues. Certain studies have suggested that specific forms of carrageenan may induce inflammation and gastrointestinal problems in animals. Nonetheless, the general consensus is that the levels of carrageenan used in food products are not harmful to humans. It is important to note that there is a distinction between food-grade carrageenan and degraded carrageenan, also known as poligeenan. Poligeenan is a different substance produced through the acid treatment of carrageenan. Poligeenan is not approved for food use and is considered potentially harmful due to its capacity to cause intestinal damage. To summarize, carrageenan is a natural ingredient derived from red seaweed that is widely employed in the food industry due to its gelling, thickening, and stabilizing properties. While it is generally regarded as safe for consumption, concerns have been raised regarding potential digestive issues associated with specific forms of carrageenan.

8.4.3 AGAR

Agar, also known as agar-agar, is a gelatinous substance derived from red algae (Mostafavi & Zaeim, 2020) The primary sources of agar extraction are two types of algae, *Gelidium* sp. and *Gracilaria* sp. (Qin, 2018). In seaweed, agar serves as a supportive structure in cell walls, analogous to hemicellulose in terrestrial plants. However, agar exhibits greater flexibility and can withstand strong ocean currents and waves (Lee et al., 2017). Chemically, agar primarily consists of repeating units of D-galactose and 3,6-anhydro-L-galactose, with minimal variations and a low concentration of sulfate esters (Figure 8.4) (Wang et al., 2020). The composition of agar is predominantly polysaccharides, specifically agarose and agaropectin, which impart its characteristic gel-forming properties (El-Hefian et al., 2012). Throughout history, agar has found applications in both culinary and scientific fields.

Agar has gained popularity for its versatility, particularly as a vegetarian alternative to gelatin. With its unique gel-forming properties, wide range of applications, and natural origin, agar has become a valuable ingredient in diverse industries. In the food industry, agar serves as a vegetarian substitute for gelatin in desserts, confectionery, and other food preparations. It can be used to create jellies, custards, and puddings and also functions as a thickening agent for soups and sauces. Agar sets at a higher temperature than gelatin, resulting in a firmer texture. By dissolving agar in hot water and then cooling it, a strong and stable gel can be formed even at low concentrations of 0.2–1.0%. This property makes it suitable for a wide range of applications where a solid or semi-solid gel is desired. Agar finds extensive use in laboratories and scientific research. It is commonly employed as a solidifying agent in microbiology to cultivate bacteria, fungi, and other microorganisms. Petri dishes and agar plates are utilized for the growth and isolation of microbial colonies, enabling the study of their growth, species identification, and testing the efficacy of antibiotics. In the medical and pharmaceutical industries, agar-based gels play a significant role. Agar hydrogels can serve as a medium for drug delivery, wound dressings, and tissue engineering. They offer a biocompatible and stable matrix that can release

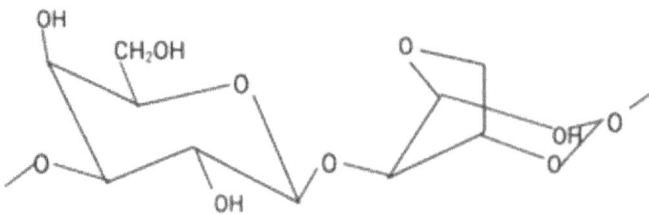

FIGURE 8.4   Chemical structure of agars.

drugs or support cell growth. Agar also has various industrial applications. It is used in cosmetic production, providing texture and stability to creams, lotions, and hair care products. Additionally, agar finds use in the production of photographic emulsions, textile printing as a thickening agent, and paper manufacturing as a binder.

8.5 PROPERTIES OF SEAWEED-BASED BIOPOLYMERS

Seaweed is abundant in polysaccharides such as alginate, carrageenan, and agar, which can be extracted and further processed to create biopolymers. These biopolymers exhibit exceptional characteristics, including biodegradability, biocompatibility, and the ability to form films. Due to their distinctive chemical composition and physical properties, seaweed-based biopolymers offer a versatile solution for numerous applications. Figure 8.5 shows the notable merits and unique structural and functional entities of marine-based seaweed polysaccharides.

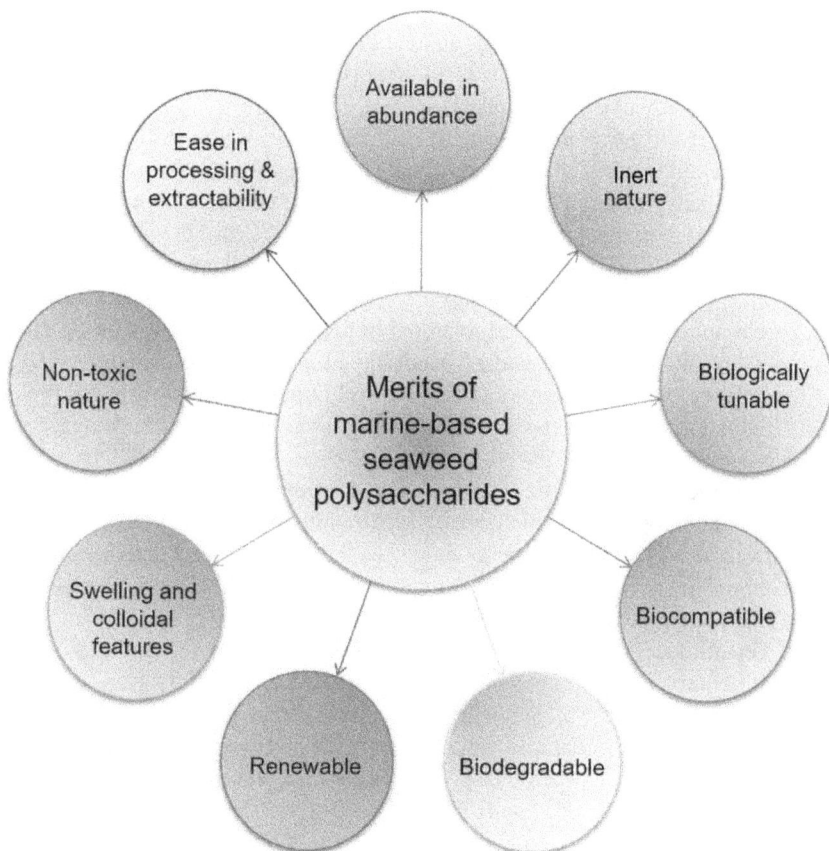

FIGURE 8.5   Schematic illustration of notable merits, unique structural and functional entities of marine-based seaweed polysaccharides. Reproduced from Bilal and Iqbal (2020).

Seaweed-based biopolymers are derived from various species of seaweed and offer a range of properties that make them attractive for various applications. Here are some key properties of seaweed-based biopolymers:

(1) Biodegradability: Seaweed-based biopolymers are typically biodegradable, meaning they can be naturally broken down by microbial activity or enzymatic degradation. This property makes them environmentally friendly, as they can reduce waste accumulation and minimize their impact on ecosystems compared to conventional synthetic polymers that persist in the environment for long periods.

(2) Renewable source: Seaweed is a renewable resource that can be sustainably harvested without depleting natural reserves. Unlike petroleum-based polymers that rely on finite fossil fuel reserves, seaweed-based biopolymers offer a more sustainable alternative, as seaweed can be cultivated and harvested in large quantities without harming the environment.

(3) High strength and flexibility: Seaweed-based biopolymers possess good mechanical properties, including high tensile strength and flexibility. This makes them suitable for applications that require durable materials capable of withstanding stress and deformation, such as packaging, textiles, and structural materials.

(4) Water absorbency: Seaweed-based biopolymers have the ability to absorb and retain water. This property is advantageous for applications such as hydrogels, wound dressings, and agricultural products. The water absorbency of these biopolymers allows for controlled release of water or active substances, contributing to the efficacy of these applications.

(5) Biocompatibility: Seaweed-based biopolymers are generally biocompatible, meaning they are well tolerated by living organisms. This property is crucial for medical and pharmaceutical applications, including drug delivery systems, tissue engineering, and implants. Seaweed-based biopolymers can interact with biological systems without causing adverse reactions or toxicity.

(6) Antimicrobial properties: Some seaweed species possess natural antimicrobial compounds, which can be incorporated into biopolymers derived from seaweed. This property enhances the antimicrobial activity of the biopolymer, making it useful for applications in healthcare, food packaging, and personal care products.

(7) Versatility: Seaweed-based biopolymers can be processed and modified to achieve a wide range of properties and functionalities. They can be blended with other biopolymers or additives to enhance specific characteristics, such as mechanical strength, thermal stability, or barrier properties.

(8) Low toxicity: Seaweed-based biopolymers are generally non-toxic and safe for use in various applications. As they are often derived from edible seaweed species, which have a long history of consumption by humans, the risk of adverse health effects is minimized. This makes them suitable for applications in contact with food, pharmaceuticals, and other consumer products.

(9) UV resistance: Some seaweed-based biopolymers exhibit good resistance to UV radiation. This property is advantageous for applications that require protection against sunlight-induced degradation, such as coatings, films, and outdoor materials. Seaweed-based biopolymers can help maintain the integrity and longevity of products exposed to sunlight.

(10) Cost-effectiveness: Seaweed is abundant in many coastal areas, and its cultivation and processing can be relatively cost-effective compared to other biopolymers. The availability and low production costs make seaweed-based biopolymers economically viable for large-scale production, which contributes to their potential for widespread adoption in various industries. Overall, seaweed-based biopolymers offer a promising and sustainable alternative to conventional synthetic polymers, with a wide range of applications in various industries.

8.6   APPLICATIONS OF SEAWEED

Seaweed biopolymers offer a promising and environmentally friendly substitute for synthetic polymers, presenting a sustainable option across various industries. The abundant and renewable nature of seaweed, coupled with its biodegradability and versatile functional properties, makes it an appealing source for extracting biopolymers such as alginate, carrageenan, and agar. As sustainability becomes increasingly important and the demand for eco-friendly materials rises, seaweed biopolymers have the potential to bring about a transformative shift, significantly reducing the environmental impact associated with conventional polymer usage. In recent times, seaweed has garnered significant attention due to its polysaccharide composition, leading to its application in a wide range of fields. These include food production (Holdt & Kraan, 2011; Kharkwal, 2009; Lafarga et al., 2020), tissue engineering (Carvalho et al., 2020; Laurienzo, 2010), biosensors (Daryaii et al., 2020), biostimulants (Crouch & Van Staden, 1993; Kavipriya et al., 2011; Nanda et al., 2022), biofuel production (Alvarado-Morales et al., 2013; del Río et al., 2020; Jiang et al., 2016), and even drug delivery systems (Pudjiastuti et al., 2017; Venkatesan et al., 2016). Figure 8.6 shows applications of seaweed.

Seaweed-based biopolymers have gained recognition for their sustainable applications in various industries, and here are some specific examples:

(1) Water Treatment: Seaweed-based biopolymers exhibit exceptional adsorption properties, making them highly effective in water treatment processes. They have the ability to remove pollutants, heavy metals, and organic contaminants from wastewater. The high adsorption capacity of seaweed-based materials contributes to efficient water purification and aids in environmental remediation.

(2) Cosmetics and Personal Care Products: The cosmetics and personal care industry has increasingly embraced the use of seaweed-based biopolymers. These biopolymers can be found in a range of products such as moisturizers, shampoos, and facial masks. Seaweed extracts are rich in beneficial compounds like vitamins, minerals, and antioxidants,

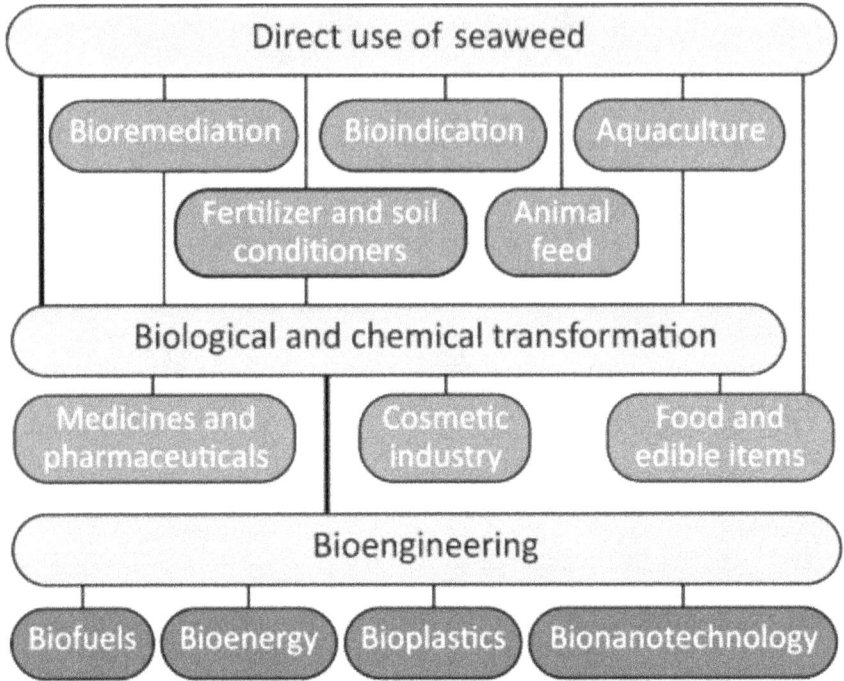

FIGURE 8.6 An overview of seaweed application. Reproduced from Ditchburn and Carballeira (2019).

which provide nourishment and protection for the skin, hair, and nails. Incorporating seaweed-based biopolymers into personal care products offers a natural and sustainable alternative (Lopez-Hortas et al., 2021).

(3) Energy Production: Seaweed serves as a promising feedstock for biofuel production. The carbohydrates present in seaweed can be converted into biofuels, including bioethanol and biogas, through various processes. This utilization of seaweed as a renewable resource offers a sustainable alternative to fossil fuels, contributing to the reduction of greenhouse gas emissions and dependence on finite resources.

(4) Seaweed-Based Biopolymers in Packaging: The packaging industry faces significant challenges related to plastic waste. Seaweed-based biopolymers provide an eco-friendly alternative to conventional plastic packaging materials. These biopolymers possess desirable properties such as barrier properties, mechanical strength, and flexibility, making them suitable for food packaging applications. Additionally, they are biodegradable and renewable and have a lower environmental impact compared to petroleum-based plastics. Incorporating seaweed-based films and coatings into packaging can help reduce plastic waste and promote a more sustainable approach to packaging materials. Figure 8.7 shows the use of seaweed as an active component for important applications.

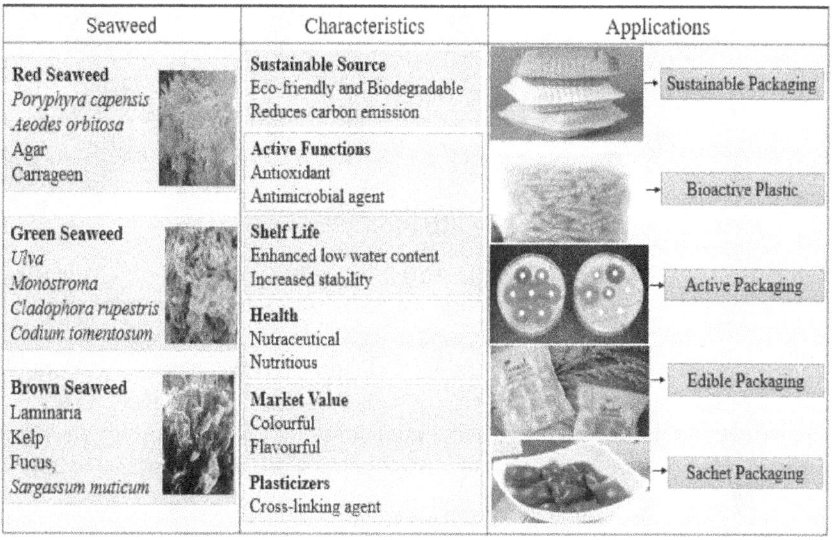

Seaweed	Characteristics	Applications
Red Seaweed *Poryphyra capensis* *Aeodes orbitosa* Agar Carrageen	**Sustainable Source** Eco-friendly and Biodegradable Reduces carbon emission	Sustainable Packaging
	Active Functions Antioxidant Antimicrobial agent	Bioactive Plastic
Green Seaweed *Ulva* *Monostroma* *Cladophora rupestris* *Codium tomentosum*	**Shelf Life** Enhanced low water content Increased stability	Active Packaging
	Health Nutraceutical Nutritious	Edible Packaging
Brown Seaweed Laminaria Kelp Fucus, *Sargassum muticum*	**Market Value** Colourful Flavourful	Sachet Packaging
	Plasticizers Cross-linking agent	

FIGURE 8.7 Seaweed as an active component and its potential applications. Reproduced from Carina et al. (2021).

(5) Seaweed-Based Biopolymers in Agriculture: Seaweed extracts have long been utilized in agriculture for their positive impact on plant growth, disease resistance, and nutrient absorption. However, recent studies have demonstrated the potential of seaweed-based biopolymers as biodegradable mulches, seed coatings, and soil conditioners. These applications not only improve soil health and promote sustainable agricultural practices but also reduce plastic waste. Seaweed-based biopolymers serve as effective soil conditioners and fertilizers, containing natural compounds, trace elements, and minerals that enhance soil fertility and plant growth. Furthermore, they enhance crop resilience against diseases and pests, reducing the need for chemical pesticides.

(6) Seaweed-Based Biopolymers in Textiles: The textile industry heavily relies on synthetic fibers derived from non-renewable resources, contributing to environmental pollution. Seaweed-based biopolymers offer an eco-friendly alternative to synthetic materials. They can be processed into fibers and fabrics with desirable properties such as antimicrobial activity, moisture wicking, and biodegradability. By incorporating seaweed-based biopolymers into textiles, we can mitigate environmental impact and support the transition to a circular economy.

(7) Seaweed-Based Biopolymers in Biomedical Engineering: Seaweed-based biopolymers have gained recognition in the biomedical field for their biocompatibility and unique properties. These materials exhibit antimicrobial and anticoagulant properties, controlled release capabilities, and excellent wound healing characteristics, making them suitable for various

biomedical applications. They can be employed in tissue engineering scaffolds, drug delivery systems, and wound dressings. Seaweed-based biopolymers provide a sustainable and biologically active platform for a range of biomedical applications, supporting advancements in wound care, tissue regeneration, and drug therapies (Bilal & Iqbal, 2020).

8.7   CHALLENGES AND FUTURE PROSPECTS

Despite their numerous advantages, seaweed-based biopolymers face challenges in terms of scalability, processing techniques, and cost-effectiveness that need to be addressed for their widespread adoption and future prospects in sustainable applications. However, ongoing research and technological advancements are addressing these obstacles. The future prospects for seaweed-based biopolymers look promising, with the potential for large-scale production and wider adoption in various industries. Several challenges for the utilization of seaweed in sustainable applications are as follows:

(1) Supply Chain and Cultivation: One of the main challenges is establishing a reliable and scalable supply chain for seaweed biomass. Large-scale cultivation of seaweed requires suitable marine environments, appropriate farming techniques, and efficient harvesting methods. Ensuring a consistent and sustainable supply of seaweed biomass is crucial for the production of seaweed-based biopolymers.

(2) Processing and Extraction: Efficient processing and extraction methods are necessary to obtain high-quality biopolymers from seaweed biomass. This involves removing impurities, extracting the desired components, and converting them into usable forms. Development of cost-effective and energy-efficient extraction techniques is essential to make seaweed-based biopolymers economically viable.

(3) Material Properties and Performance: Seaweed-based biopolymers need to exhibit desirable material properties, such as mechanical strength, flexibility, and thermal stability, to compete with conventional plastics. Researchers are working on modifying the chemical and physical properties of these biopolymers to enhance their performance and expand their range of applications.

(4) Standardization and Regulation: The development of standardized testing methods, certification processes, and regulations for seaweed-based biopolymers is essential to ensure their quality, safety, and compatibility with existing industrial processes. Standardization efforts will facilitate the adoption of these biopolymers in various industries and promote market acceptance.

(5) Market Acceptance and Consumer Awareness: Creating awareness among consumers about the environmental benefits of seaweed-based biopolymers is crucial for their market acceptance. Education and outreach programs can help promote the use of these sustainable materials and encourage consumers to choose products made from seaweed-based biopolymers.

Despite these challenges, seaweed-based biopolymers hold great promise for sustainable applications in various industries. Some of the future prospects include:

(1) Packaging: Seaweed-based biopolymers can be used as an alternative to conventional plastic packaging materials. They offer advantages such as biodegradability and a reduced carbon footprint, making them attractive for sustainable packaging solutions.

(2) Biomedical Applications: Seaweed-based biopolymers have potential applications in the biomedical field, including drug delivery systems, wound dressings, and tissue engineering scaffolds. These biopolymers are biocompatible, and their natural properties make them suitable for various biomedical applications.

(3) Agriculture and Horticulture: Seaweed-based biopolymers can be utilized in agriculture as biodegradable mulches, soil conditioners, and biofertilizers. They can enhance crop growth, improve soil health, and reduce the environmental impact of conventional agricultural practices.

(4) Textiles and Fibers: Seaweed-based biopolymers can be used in the production of sustainable textiles and fibers. They offer potential as a renewable and biodegradable alternative to synthetic fibers, contributing to the reduction of microplastic pollution associated with the textile industry.

(5) Water Treatment and Filtration: Seaweed-based biopolymers have shown promise in water treatment and filtration applications. They can be used for the removal of heavy metals, dyes, and other contaminants from water, providing an eco-friendly approach to address water pollution challenges.

8.8   CONCLUSION

Seaweed-based biopolymers present a sustainable and renewable option for various industries like packaging, agriculture, textiles, and biomedical engineering, offering an eco-friendly alternative to conventional materials. These biopolymers possess unique properties that make them suitable for a wide range of applications while also minimizing the environmental impact. To fully harness the potential of seaweed-based biopolymers and transition towards a greener and more sustainable future, it is essential to continue research, foster technological innovation, and promote collaboration among academia, industry, and policymakers. Although seaweed has found extensive use in industrial applications and functional foods, there are still several challenges that need to be addressed. One limitation is the low solubility of certain seaweed types under neutral pH conditions, which restricts their applicability. Therefore, chemical modification of the seaweed structure has become an important approach to enhance its properties and expand potential applications.

In recent years, enzymatic modification has emerged as a promising alternative to hazardous chemical methods. Enzymatic processes encompass various techniques such as reducing polymer molecular weight, transesterification, oxidation, glycosylation, and ester formation. However, the cost of certain enzymes remains a concern. Another strategy to enhance the biological properties and application potential of seaweed is to form complexes with other molecules, such as polyphenols. This review focuses on highlighting the structural differences among prominent seaweed derived from algae, exploring available modification techniques, and providing an overview of their biological activities. Additionally, novel methods are being developed to streamline and improve the extraction processes of seaweed-based biopolymers. It is important to address the reduction of chemical reagents and optimize operations to enhance yields, which presents a significant challenge. Therefore, further research is crucial to refine the extraction and purification processes of seaweed.

REFERENCES

Abdou, E. S., & Sorour, M. A. (2014). Preparation and characterization of starch/carrageenan edible films. *International Food Research Journal, 21*(1).

Abdul Khalil, H. P. S., Lai, T. K., Tye, Y. Y., Rizal, S., Chong, E. W. N., Yap, S. W., Hamzah, A. A., Nurul Fazita, M. R., & Paridah, M. T. (2018a). A review of extractions of seaweed hydrocolloids: Properties and applications. *Express Polymer Letters, 12*(4), 296–317. https://doi.org/10.3144/expresspolymlett.2018.27

Abdul Khalil, H. P. S., Yap, S. W., Tye, Y. Y., Tahir, P. M., Rizal, S., & Fazita, M. N. (2018b). Effects of corn starch and *Kappaphycus alvarezii* seaweed blend concentration on the optical, mechanical, and water vapor barrier properties of composite films. *BioResources, 13*(1), 1157–1173.

Abdul Khalil, H. P. S., Tye, Y. Y., Saurabh, C. K., Leh, C. P., Lai, T. K., Chong, E. W. N., Nurul Fazita, M. R., Hafiidz, J. M., Banerjee, A., & Syakir, M. I. (2017). Biodegradable polymer films from seaweed polysaccharides: A review on cellulose as a reinforcement material. *Express Polymer Letters, 11*(4), 244–265. https://doi.org/10.3144/expresspolymlett.2017.26

Alvarado-Morales, M., Boldrin, A., Karakashev, D. B., Holdt, S. L., Angelidaki, I., & Astrup, T. (2013). Life cycle assessment of biofuel production from brown seaweed in Nordic conditions. *Bioresource Technology, 129*, 92–99. https://doi.org/10.1016/j.biortech.2012.11.029

Aryee, A. N., Agyei, D., & Akanbi, T. O. (2018). Recovery and utilization of seaweed pigments in food processing. *Current Opinion in Food Science, 19*(2), 113–119. https://doi.org/10.1016/j.cofs.2018.03.013

Atef, M., Rezaei, M., & Behrooz, R. (2014). Preparation and characterization agar-based nanocomposite film reinforced by nanocrystalline cellulose. *International Journal of Biological Macromolecules, 70*, 537–544. https://doi.org/10.1016/j.ijbiomac.2014.07.013

Baranwal, J., Barse, B., Fais, A., Delogu, G. L., & Kumar, A. (2022). Biopolymer: A sustainable material for food and medical applications. *Polymers, 14*(5). https://doi.org/10.3390/polym14050983

Benavides, S., Villalobos-Carvajal, R., & Reyes, J. E. (2012). Physical, mechanical and antibacterial properties of alginate film: Effect of the crosslinking degree and oregano essential oil concentration. *Journal of Food Engineering, 110*(2), 232–239. https://doi.org/10.1016/j.jfoodeng.2011.05.023

Bhuyar, P., Sundararaju, S., Rahim, M. H. A., Unpaprom, Y., Maniam, G. P., & Govindan, N. (2021). Antioxidative study of polysaccharides extracted from red (Kappaphycus alvarezii), green (Kappaphycus striatus) and brown (Padina gymnospora) marine macroalgae/seaweed. *SN Applied Sciences, 3*(4). https://doi.org/10.1007/s42452-021-04477-9

Bilal, M., & Iqbal, H. M. N. (2020). Marine seaweed polysaccharides-based engineered cues for the modern biomedical sector. *Marine Drugs, 18*(1). https://doi.org/10.3390/md18010007

Bonanno, G., & Orlando-Bonaca, M. (2018). Chemical elements in Mediterranean macroalgae. A review. *Ecotoxicology and Environmental Safety, 148*(July 2017), 44–71. https://doi.org/10.1016/j.ecoenv.2017.10.013

Carina, D., Sharma, S., Jaiswal, A. K., & Jaiswal, S. (2021). Seaweeds polysaccharides in active food packaging: A review of recent progress. *Trends in Food Science and Technology, 110*, 559–572. https://doi.org/10.1016/j.tifs.2021.02.022

Carvalho, D. N., Inácio, A. R., Sousa, R. O., Reis, R. L., & Silva, T. H. (2020). Seaweed polysaccharides as sustainable building blocks for biomaterials in tissue engineering. In *Sustainable seaweed technologies: Cultivation, biorefinery, and applications.* https://doi.org/10.1016/B978-0-12-817943-7.00019-6

Crouch, I. J., & Van Staden, J. (1993). Commercial seaweed products as biostimulants in horticulture. *Journal of Home & Consumer Horticulture, 1*(1), 19–76. https://doi.org/10.1300/j280v01n01_03

Daryaii, L. B., Samsampour, D., Bagheri, A., & Sohrabipour, J. (2020). High content of heavy metals in seaweed species: A case study in the Persian Gulf and the Gulf of Oman in the southern coast of Iran. *Journal of Phycological Research, 4*(2), 544–560.

Das, A., Ringu, T., Ghosh, S., & Pramanik, N. (2022). A comprehensive review on recent advances in preparation, physicochemical characterization, and bioengineering applications of biopolymers. In *Polymer bulletin.* Springer Science and Business Media Deutschland GmbH. https://doi.org/10.1007/s00289-022-04443-4

del Río, P. G., Gomes-Dias, J. S., Rocha, C. M. R., Romaní, A., Garrote, G., & Domingues, L. (2020). Recent trends on seaweed fractionation for liquid biofuels production. *Bioresource Technology, 299*, 122613. https://doi.org/10.1016/j.biortech.2019.122613

Din, N. A. S., Mohd Alayudin, 'Ain Sajda, Sofian-Seng, N. S., Rahman, H. A., Mohd Razali, N. S., Lim, S. J., & Wan Mustapha, W. A. (2022). Brown algae as functional food source of fucoxanthin: A review. *Foods, 11*(15), 1–35. https://doi.org/10.3390/foods11152235

Ditchburn, J. L., & Carballeira, C. B. (2019). Versatility of the humble seaweed in biomanufacturing. *Procedia Manufacturing, 32*, 87–94. https://doi.org/10.1016/j.promfg.2019.02.187

El-Hefian, E. A., Nasef, M. M., & Yahaya, A. H. (2012). Preparation and characterization of chitosan/agar blended films: Part 1. Chemical structure and morphology. *E-Journal of Chemistry, 9*(3), 1431–1439. https://doi.org/10.1155/2012/781206

El-Said, G. F., & El-Sikaily, A. (2013). Chemical composition of some seaweed from Mediterranean Sea coast, Egypt. *Environmental Monitoring and Assessment, 185*(7), 6089–6099. https://doi.org/10.1007/s10661-012-3009-y

Farghali, M., Mohamed, I. M. A., Osman, A. I., & Rooney, D. W. (2023). Seaweed for climate mitigation, wastewater treatment, bioenergy, bioplastic, biochar, food, pharmaceuticals, and cosmetics: A review. *Environmental Chemistry Letters, 21*(1), 97–152. https://doi.org/10.1007/s10311-022-01520-y

Ferrara, L. (2020). Seaweeds: A food for our future. *Journal of Food Chemistry and Nanotechnology, 6*(2), 56–64. https://doi.org/10.17756/jfcn.2020-084

Fleurence, J., Morançais, M., & Dumay, J. (2017). Seaweed proteins. In *Proteins in food processing* (2nd ed.). https://doi.org/10.1016/B978-0-08-100722-8.00010-3

Freitas, M. V., Pacheco, D., Cotas, J., Mouga, T., Afonso, C., & Pereira, L. (2021). Red seaweed pigments from a biotechnological perspective. *Phycology, 2*(1), 1–29. https://doi.org/10.3390/phycology2010001

Gade, R., Siva Tulasi, M., & Aruna Bhai, V. (2013). Seaweeds: A novel biomaterial. *International Journal of Pharmacy and Pharmaceutical Sciences*, 40–44.

Gunathilake, T., Akanbi, T. O., Suleria, H. A. R., Nalder, T. D., Francis, D. S., & Barrow, C. J. (2022). Seaweed phenolics as natural antioxidants, aquafeed additives, veterinary treatments and cross-linkers for microencapsulation. *Marine Drugs*, *20*(7). https://doi. org/10.3390/md20070445

Gurram, S. (2022). Role of hydrocolloids in food systems. *The Pharma Innovation Journal*, *11*(8), 1748–1755.

Holdt, S. L., & Kraan, S. (2011). Bioactive compounds in seaweed: Functional food applications and legislation. *Journal of Applied Phycology*, *23*(3), 543–597. https://doi. org/10.1007/s10811-010-9632-5

Huq, T., Salmieri, S., Khan, A., Khan, R. A., Le Tien, C., Riedl, B., Fraschini, C., Bouchard, J., Uribe-Calderon, J., Kamal, M. R., & Lacroix, M. (2012). Nanocrystalline cellulose (NCC) reinforced alginate based biodegradable nanocomposite film. *Carbohydrate Polymers*, *90*(4), 1757–1763. https://doi.org/10.1016/j.carbpol.2012.07.065

Ismail, M. M., Alotaibi, B. S., & El-Sheekh, M. M. (2020). Therapeutic uses of red macroalgae. *Molecules*, *25*(19), 1–14. https://doi.org/10.3390/molecules25194411

Jagtap, A. S., & Meena, S. N. (2021). Seaweed farming: A perspective of sustainable agriculture and socio-economic development. In *Natural resources conservation and advances for sustainability* (pp. 493–501). https://doi.org/10.1016/B978-0-12-822976-7.00022-3

Jayakumar, A., Radoor, S., Siengchin, S., Shin, G. H., & Kim, J. T. (2023). Recent progress of bioplastics in their properties, standards, certifications and regulations: A review. *Science of the Total Environment*, *878*. https://doi.org/10.1016/j.scitotenv.2023.163156

Jayathilake, D. R. M., & Costello, M. J. (2020). The kelp biome. In *Encyclopedia of the world's biomes* (pp. 4–5). Elsevier Inc. https://doi.org/10.1016/B978-0-12-409548-9.11768-3

Jiang, R., Ingle, K. N., & Golberg, A. (2016). Macroalgae (seaweed) for liquid transportation biofuel production: What is next? *Algal Research*, *14*, 48–57. https://doi.org/10.1016/j. algal.2016.01.001

Kavipriya, R., Dhanalakshmi, P. K., Jayashree, S., & Thangaraju, N. (2011). Seaweed extract as a biostimulant for legume crop, green gram. *Journal of Ecobiotechnology*, *3*(8), 16–19.

Kharkwal, A. C., Joshi, D. D., Bahuguna, P. P., & Kharkwal, A. (2009). Algae as future drugs. *Asian Journal of Pharmaceutical and Clinical Research*, *5*(3), 3–6.

Kim, H. S., Lee, C. G., & Lee, E. Y. (2011). Alginate lyase: Structure, property, and application. *Biotechnology and Bioprocess Engineering*, *16*(5), 843–851. https://doi. org/10.1007/s12257-011-0352-8

Kulig, D., Zimoch-Korzycka, A., Kró, Z., Oziembłowski, M., & Jarmoluk, A. (2017). Effect of film-forming alginate/chitosan polyelectrolyte complex on the storage quality of pork. *Molecules*, *22*(1). https://doi.org/10.3390/molecules22010098

Kumar, R., Verma, A., Shome, A., Sinha, R., Sinha, S., Sharma, P., & Prasad, P. V. V. (2021). Impacts of plastic pollution on ecosystem services, sustainable development goals, and need to focus on circular economy and policy interventions. *Sustainability*, *13*(17), 1–40.

Lafarga, T., Acién-Fernández, F. G., & Garcia-Vaquero, M. (2020). Bioactive peptides and carbohydrates from seaweed for food applications: Natural occurrence, isolation, purification, and identification. *Algal Research*, *48*(April). https://doi.org/10.1016/j. algal.2020.101909

Laurienzo, P. (2010). Marine polysaccharides in pharmaceutical applications: An overview. *Marine Drugs*, *8*(9), 2435–2465. https://doi.org/10.3390/md8092435

Lee, K. Y., & Mooney, D. J. (2012). Alginate: Properties and biomedical applications. *Progress in Polymer Science (Oxford)*, *37*(1), 106–126. https://doi.org/10.1016/j. progpolymsci.2011.06.003

Lee, W. K., Lim, Y. Y., Leow, A. T. C., Namasivayam, P., Ong Abdullah, J., & Ho, C. L. (2017). Biosynthesis of agar in red seaweeds: A review. *Carbohydrate Polymers*, *164*, 23–30. https://doi.org/10.1016/j.carbpol.2017.01.078

Liao, Y. C., Chang, C. C., Nagarajan, D., Chen, C. Y., & Chang, J. S. (2021). Algae-derived hydrocolloids in foods: Applications and health-related issues. *Bioengineered*, *12*(1), 3787–3801. https://doi.org/10.1080/21655979.2021.1946359

Lopez-Hortas, L., Florez-Fernandez, N., Torres, M. D., Ferreira-Anta, T., Casas, M. P., Balboa, E. M., Falque, E., & Domínguez, H. (2021). Applying seaweed compounds in cosmetics, cosmeceuticals and nutricosmetics. *Marine Drugs*, *19*(10). MDPI. https://doi.org/10.3390/md19100552

MacArtain, P., Gill, C. I. R., Brooks, M., Campbell, R., & Rowland, I. R. (2007). Nutritional value of edible seaweeds. *Nutrition Reviews*, *65*(12), 535–543. https://doi.org/10.1301/nr.2007.dec.535-543

Mangaraj, S., Yadav, A., Bal, L. M., Dash, S. K., & Mahanti, N. K. (2019). Application of biodegradable polymers in food packaging industry: A comprehensive review. *Journal of Packaging Technology and Research*, *3*(1), 77–96. https://doi.org/10.1007/s41783-018-0049-y

Matei, E., Predescu, A. M., Râpă, M., Țurcanu, A. A., Mateș, I., Constantin, N., & Predescu, C. (2022). Natural polymers and their nanocomposites used for environmental applications. *Nanomaterials*, *12*(10). https://doi.org/10.3390/nano12101707

Mikkonen, K. S. (2013). Recent studies on hemicellulose-based blends, composites and nanocomposites. *Advanced Structured Materials*, *18*, 313–336. https://doi.org/10.1007/978-3-642-20940-6_9

Minhas, A., Kaur, B., & Kaur, J. (2020). Genomics of algae: Its challenges and applications. In *Pan-genomics: Applications, challenges, and future prospects* (Issue 1989). Elsevier Inc. https://doi.org/10.1016/b978-0-12-817076-2.00013-5

Mostafavi, F. S., & Zaeim, D. (2020a). Agar-based edible films for food packaging applications—A review. *International Journal of Biological Macromolecules*, *159*, 1165–1176. Elsevier B.V. https://doi.org/10.1016/j.ijbiomac.2020.05.123

Mugdha Bhat, K., Sharma, A., Rao, N. N., & Biotechnology, B. (2020). Carrageenan-based edible biodegradable food packaging: A review. *International Journal of Food Science and Nutrition*, *5*(September), 2455–4898.

Nanda, S., Kumar, G., & Hussain, S. (2022). Utilization of seaweed-based biostimulants in improving plant and soil health: Current updates and future prospective. *International Journal of Environmental Science and Technology*, *19*(12), 12839–12852. https://doi.org/10.1007/s13762-021-03568-9

Necas, J., & Bartosikova, L. (2013). Carrageenan: A review. *Veterinarni Medicina*, *58*(4), 187–205. https://doi.org/10.17221/6758-VETMED

Parreidt, T. S., Müller, K., & Schmid, M. (2018). Alginate-based edible films and coatings for food packaging applications. *Foods*, *7*(10). https://doi.org/10.3390/foods7100170

Pranoto, Y., Salokhe, V. M., & Rakshit, S. K. (2005). Physical and antibacterial properties of alginate-based edible film incorporated with garlic oil. *Food Research International*, *38*(3), 267–272. https://doi.org/10.1016/j.foodres.2004.04.009

Pudjiastuti, P., Fauzi, M. A. R. D., & Darmokoesoemo, H. (2017). Drug delivery hard shell capsules from seaweed extracts. *Journal of Chemical Technology and Metallurgy*, *53*(6), 1140–1144.

Qin, Y. (2018). Seaweed bioresources. In *Bioactive seaweeds for food applications: Natural ingredients for healthy diets*. https://doi.org/10.1016/B978-0-12-813312-5.00001-7

Raja, K., Kadirvel, V., & Subramaniyan, T. (2022). Seaweeds, an aquatic plant-based protein for sustainable nutrition: A review. *Future Foods*, *5*(December 2021), 100142. https://doi.org/10.1016/j.fufo.2022.100142

Rajauria, G., Cornish, L., Ometto, F., Msuya, F. E., & Villa, R. (2015). Identification and selection of algae for food, feed, and fuel applications. In Brijesh K. Tiwari, Declan J. Troy (Eds.), *Seaweed sustainability* (pp. 315–345). Academic Press.

Reed, D. C., Kinlan, B. P., Raimondi, P. T., Washburn, L., Gaylord, B., & Drake, P. T. (2006). A metapopulation perspective on the patch dynamics of giant kelp in Southern California. *Marine Metapopulations*, 353–386. https://doi.org/10.1016/B978-012088781-1/50013-3

Salehi, B., Sharifi-rad, J., Seca, A. M. L., & Pinto, D. C. G. A. (2019). Current trends on seaweeds: Looking at chemical. *Molecules*, *24*(22), 4182.

Saunders, G. W. (2001). Brown algae. *ELS*. https://doi.org/10.1038/npg.els.0000329

Sedayu, B. B., Cran, M. J., & Bigger, S. W. (2019). A review of property enhancement techniques for carrageenan-based films and coatings. *Carbohydrate Polymers*, *216*, 287–302. Elsevier Ltd. https://doi.org/10.1016/j.carbpol.2019.04.021

Sedayu, B. B., Cran, M. J., & Bigger, S. W. (2020). Improving the moisture barrier and mechanical properties of semi-refined carrageenan films. *Journal of Applied Polymer Science*, *137*(41). https://doi.org/10.1002/app.49238

Sheath, R. G., & Wehr, J. D. (2015). Introduction to the freshwater algae. *Freshwater Algae of North America: Ecology and Classification*, 1–11. https://doi.org/10.1016/B978-0-12-385876-4.00001-3

Sultana, F., Wahab, M. A., Nahiduzzaman, M., Mohiuddin, M., Iqbal, M. Z., Shakil, A., Mamun, A. Al, Khan, M. S. R., Wong, L. L., & Asaduzzaman, M. (2023). Seaweed farming for food and nutritional security, climate change mitigation and adaptation, and women empowerment: A review. *Aquaculture and Fisheries*, *8*(5), 463–480. https://doi.org/10.1016/j.aaf.2022.09.001

Tasende, M. G., & Manríquez-Hernández, J. A. (2016, May). Carrageenan properties and applications: A review. In Leonel Pereira (Ed.), *Carrageenans: Sources and extraction methods, molecular structure, bioactive properties and health effects* (pp. 17–49). Nova Science Publishers.

Tavassoli-Kafrani, E., Shekarchizadeh, H., & Masoudpour-Behabadi, M. (2016). Development of edible films and coatings from alginates and carrageenans. *Carbohydrate Polymers*, *137*, 360–374. Elsevier Ltd. https://doi.org/10.1016/j.carbpol.2015.10.074

Udayakumar, G. P., Muthusamy, S., Selvaganesh, B., Sivarajasekar, N., Rambabu, K., Banat, F., Sivamani, S., Sivakumar, N., Hosseini-Bandegharaei, A., & Show, P. L. (2021). Biopolymers and composites: Properties, characterization and their applications in food, medical and pharmaceutical industries. *Journal of Environmental Chemical Engineering*, *9*(4). Elsevier Ltd. https://doi.org/10.1016/j.jece.2021.105322

Vasconcelos, A. A., & Pomin, V. H. (2018). Marine carbohydrate-based compounds with medicinal properties. *Marine Drugs*, *16*(7). https://doi.org/10.3390/md16070233

Venkatesan, J., Anil, S., Kim, S. K., & Shim, M. S. (2016). Seaweed polysaccharide-based nanoparticles: Preparation and applications for drug delivery. *Polymers*, *8*(2), 1–25. https://doi.org/10.3390/polym8020030

Vilar, E. G., O'Sullivan, M. G., Kerry, J. P., & Kilcawley, K. N. (2021). A chemometric approach to characterize the aroma of selected brown and red edible seaweeds/ extracts. *Journal of the Science of Food and Agriculture*, *101*(3), 1228–1238. https://doi.org/10.1002/jsfa.10735

Wang, L., Liu, Z., Jiang, H., & Mao, X. (2021). Biotechnology advances in β-carotene production by microorganisms. *Trends in Food Science and Technology*, *111*(February), 322–332. https://doi.org/10.1016/j.tifs.2021.02.077

Wang, X., Jiang, H., Zhang, N., Cai, C., Li, G., Hao, J., & Yu, G. (2020). Anti-diabetic activities of agaropectin-derived oligosaccharides from *Gloiopeltis furcata* via regulation of mitochondrial function. *Carbohydrate Polymers*, *229*(July), 115482. https://doi.org/10.1016/j.carbpol.2019.115482

Wróblewska-Krepsztul, J., Rydzkowski, T., Borowski, G., Szczypiński, M., Klepka, T., & Thakur, V. K. (2018). Recent progress in biodegradable polymers and nanocomposite-based packaging materials for sustainable environment. *International Journal of*

Polymer Analysis and Characterization, *23*(4), 383–395. https://doi.org/10.1080/1023 666X.2018.1455382

Yang, Y., Zhang, M., Alalawy, A. I., Almutairi, F. M., Al-Duais, M. A., Wang, J., & Salama, E. S. (2021). Identification and characterization of marine seaweeds for biocompounds production. *Environmental Technology and Innovation*, *24*. https://doi.org/10.1016/j. eti.2021.101848

Zolotova, N., Kosyreva, A., Dzhalilova, D., Fokichev, N., & Makarova, O. (2022). Harmful effects of the microplastic pollution on animal health: A literature review. *PeerJ*, *10*(June). https://doi.org/10.7717/peerj.13503

9 Characteristics and Performance of Emerging Biopolymers from Sugar Palm Starch for Packaging

9.1 INTRODUCTION

The primary objective of food packaging is to ensure the preservation of the safety and quality of food products during transportation and storage, as well as to prevent unfavorable conditions like moisture, mechanical shock, vibration, odor, gases, oxygen, and chemical contamination, which can lead to spoilage caused by microorganisms, among other factors. Additionally, packaging serves to provide crucial product information to customers and promote the product. In contemporary society, food producers are striving to improve packaging systems by incorporating user-friendly features to accommodate the active lifestyle of modern consumers. Packaging materials not only serve as protective barriers but also possess other established functions. Compared to sturdy goods such as furniture, home appliances, and electronics, food packaging materials are designed for temporary use, with a focus on safety.

The increasing demands on global resources due to technological advancements and consumer expectations have led to significant environmental sustainability and material accessibility issues. Food, beverage, and related commodity packaging account for more than 50% of the global packaging market, which results in a considerable amount of waste. The techniques used in recycling programs are inadequate for handling the excess waste, which contributes to environmental devastation caused by non-biodegradable plastics. Therefore, natural biopolymers such as starch-based plastics are being investigated as potential substitutes for conventional plastics. Although starch-based polymers have the potential for large-scale production of bio-plastic film, they have poor physical properties that limit their usage in numerous applications, particularly for food packaging. The solution to this problem lies in understanding the rheological and thermal properties of starch and developing starch formulations. Recently, attention has been focused on the use of cellulose nanofiller reinforcement to produce nanocomposites and improve the properties of starch-based materials.

Starch, a biopolymer, can be degraded by microorganisms such as fungi, bacteria, and enzymes (Ilyas et al., 2020). It is naturally occurring and can be used as a raw material for biodegradable materials. In certain cases, starch-based films might be a less costly alternative to polylactic acid (PLA) films for food packaging. Since the 1980s, the filmogenic properties of starch materials have been extensively studied

 DOI: 10.1201/9781003416043-9

(Basiak et al., 2018). Unlike other biopolymers, starch is abundant, low-cost, biodegradable, and edible, with exceptional filmogenic capability (Sanyang et al., 2018). Thus, starch is one of the most promising options for replacing petroleum-based plastics. Starch-based films and composites have great potential as ecologically suitable materials for food packaging. Starch-based films may be used to create packaging with antibacterial and antioxidant properties, and they find application in several industries, primarily in the food, pharmaceutical, cosmetic, and paper sectors, as a matrix, binder, or filler (Chen et al., 2023; Malik et al., 2023; Ramakrishnan et al., 2023). The reinforcement of the starch matrix and nanofibers results in a 3D hydrogen bonding network, which enhances the performance of the nanocomposite, making it a promising material for food packaging purposes (Sanyang et al., 2018). However, some unfavorable properties of natural starch, such as insolubility in cold water, have hindered its wider use. Yet starch processing, such as modification by chemical, enzymatic, and physical methods, can change these properties, making them attractive for commercialization and marketable. This processing requires specialized operations of the processing unit, such as ultrasound treatment (Hu et al., 2014; Mohamed et al., 2017).

Starch is primarily sourced from agricultural plants such as potato, rice, and corn and is a polysaccharide made of glucose units bound by α-glycosidic bonds. In plants, starch is formed as hydrophilic granules composed of different types of polysaccharides: a linear and crystalline amylose (poly-α-1,4-d-glucopyranoside), branched and amorphous amylopectin (poly-α-1,4-d-glucopyranoside), and α-1,6-d-glucopyranoside (Żołek-Tryznowska & Kałuża, 2021). The original chemical composition of starch is based on two macromolecular components: amylose and amylopectin, and the structures of starches vary depending on the amylose and amylopectin content and the size of the starch granules. Maize, wheat, potato, and rice starches dominate the world market, with 84%, 7%, 4%, and 1%, respectively (Basiak et al., 2017). Research has been conducted to produce starch from natural fibers, including kenaf (Sulaiman et al., 2017), sugarcane bagasse (Shahi et al., 2020), cotton (Soni & Mahmoud, 2015), and oil palm empty fruit bunches (Supian et al., 2020). The significance of using biobased plastics, such as thermoplastic starch, in place of traditional packaging plastics is emphasized, and this chapter introduces sugar palm starch as a new and innovative option for developing biobased packaging materials.

9.2 TYPES OF PACKAGING MATERIALS

In the past, various materials such as glass, plastic, metal, and paper were commonly used for packaging. Among these materials, petroleum-based plastics have been widely utilized due to their desirable processing properties, physicochemical characteristics, and aesthetic appeal. Plastic is a preferred material for food packaging due to its flexibility, lightweight, portability, and affordability (Saha et al., 2020). While glass and metal are easier to reuse, they are expensive to recycle. In contrast, plastics can be completely recycled at a lower cost. Paper packaging is mostly made from wood from different tree species. However, the increasing use of paper packaging has led to massive deforestation. Additionally, paper packaging is less durable and is sensitive to water, and recycling it is more difficult and inefficient compared to plastics, which consume less energy and water. Therefore, plastic packaging continues to dominate the packaging market over paper. According to Fang and Fowler (2003), more than 30 million tons of petroleum-based plastics are used for packaging annually, including gardening supplies, blood platelet bags, food packaging, agricultural mulch films, waste disposal bags, and woven plastic fibers bags used for sand or shopping bags, equivalent to approximately 25% of plastic produced worldwide, with consumption continuing to rise (Fang & Fowler, 2003). It is projected that about 30 types of the plastics are known that are used on a commercial scale. The problem of using plastic is that it is not degradable, cost effective, or environment friendly. The production of waste plastics is projected to be around 300 million tons per year, which is a massive amount (Aslam et al., 2023).

Although petroleum-based plastics have many advantages, such as resistance to microbial attack, they are difficult to degrade and cause serious environmental problems due to the accumulation of plastic waste in the environment. As a result, there has been an increase in research on developing new biobased plastics from renewable sources that have controllable lifetimes and can be used for various applications, including biomedical implants, wildlife conservation, drug release, agriculture, forestry, and waste management (Sanyang et al., 2018). Starch-based packaging materials are one of the alternatives to synthetic plastics and are expected to surpass an annual utilization of thousands of tons in the coming years. The use of biobased packaging is on the rise, especially in countries where landfills are the primary waste management method.

9.2.1 Production of Biobased Plastics

Referring to Storz and Vorlop (2013), there are three primary methods for obtaining biobased plastics. The first approach involves modifying natural polymers while keeping the polymer backbone mainly intact. This is the most commonly used method for producing biobased polymers and fibers, such as starch and cellulose-based plastics, which are used in non-food and non-plastic applications. The second method involves a two-step biomass conversion process. The first step involves the production of biobased precursors or monomers through biochemical and/or chemical transformation. The second step involves the polymerization of the monomers to produce the final product. This approach is complex and can be divided into several sub-steps. If the monomers obtained are biobased versions of conventional monomers, they are referred to as drop-in replacements. Plastics made from drop-in monomers are advantageous because they can be processed and recycled similarly to their petrochemical counterparts. However, novel biobased plastics that use new structures or have not been used before require the development and implementation of new recycling systems. This second approach to biobased plastics is becoming increasingly important due to advancements in chemical and bio-technological production of monomers. Compared to the conventional plastics which they could replace, they often show an improved functionality and thus additional markets and applications. However, in contrast to drop-in plastics, novel biobased plastics

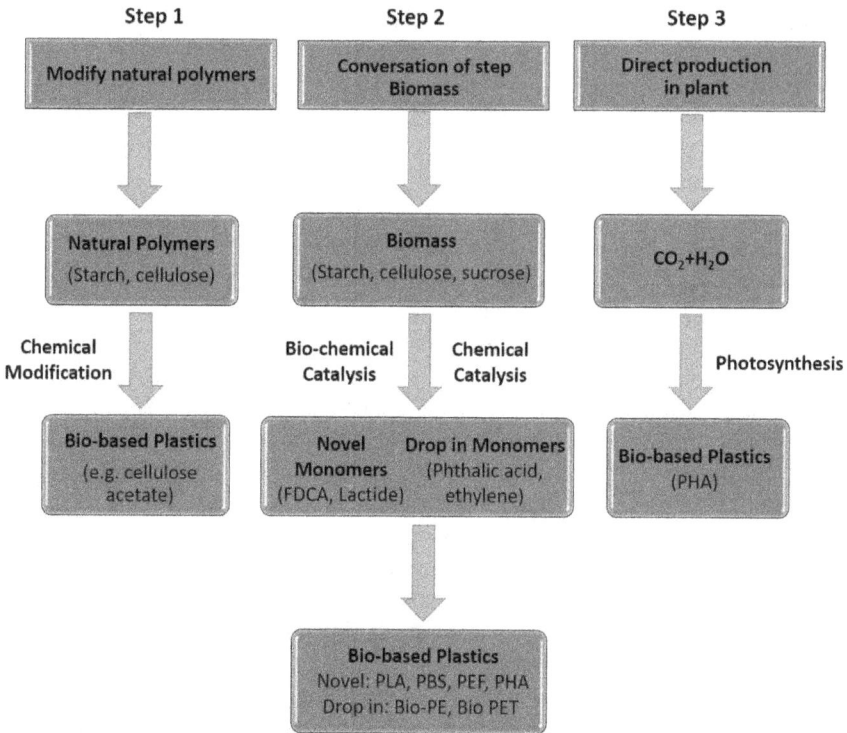

FIGURE 9.1 Three primary methods for obtaining biobased plastics.

require the development and implementation of new recycling systems. One example is biopolymers from biomass conversion process like PLA, polybutylene succinate (PBS), and biobased polyethylene (PE) prepared from bio-ethanol–derived ethylene. The third method involves the production of a polymeric material that can be used directly as a plastic without further modification, directly in microorganisms or plants. This route is becoming more feasible due to progress in genetic engineering and biotechnology. Because of the progress in genetic engineering and biotechnology which enables researchers to move genes responsible for the production of a polymer like poly(hydroxyalkanoate) (PHA) from bacteria into crops, this route is becoming more feasible. However, environmental and regulatory issues make the direct production of biobased plastics via photosynthesis complex. Despite intensive research, no significant quantities of biobased plastics have been produced using this method (Storz & Vorlop, 2013). Figure 9.1 presents an example route of three primary methods for obtaining biobased plastics.

9.2.2 BIOPOLYMERS FROM STARCH

Starch, a complex carbohydrate, is synthesized by plants to store energy. However, its natural form cannot be used in thermoplastic processing due to its semi-crystalline structure, which degrades before reaching its melting point. To address this, a bio-based plastic called thermoplastic starch has been developed. Thermoplastic starch is created by mixing and heating starch granules with plasticizers, typically water and glycerol, in a process called destructurization. It is an appealing material, as it is inexpensive, biodegradable, and readily available in large quantities. Moreover, thermoplastic starch can be processed using standard equipment. However, its high hydrophilicity makes it unsuitable for use in humid environments (Liu et al., 2009), which limits its application to niche products such as fast-dissolving dishwasher tabs or adhesive tapes. To enhance the properties of thermoplastic starch, various solutions have been proposed. These include the use of less volatile and water-sensitive plasticizers like sorbitol or xylitol, as well as the addition of hydrophobic fillers such as lignin (Mouren & Avérous, 2023).

Blending thermoplastic starch with hydrophobic plastics has expanded the range of applications for these plastic materials. By incorporating a hydrophobic component, the direct absorption of water is prevented, significantly enhancing water resistance. Also, by adjusting the content of plasticizer, the properties of the material can be modified. However, the maximum amount of starch that can be included in blends is typically limited to around 25 to 30% due to the incompatibility of most hydrophobic polymers with starch. To achieve higher starch contents, reactive blending with compatibilizers is employed, which chemically links the components (Kalambur & Rizvi, 2006). Extensive research has been conducted on blends of starch with both conventional and biobased polymers (Florencia et al., 2020; Li et al., 2023; Tian et al., 2021). Currently, various commercial grades of starch blends are available. These primarily consist of polyester-based blends used for short-lived biodegradable products such as bags, packaging films, agricultural mulch films, and protection foams. Additionally, non-biodegradable starch-polyolefin blends have also entered the market.

For example, Cereplast, a US company, offers thermoplastic starch blends with polyethylene and polypropylene (PP) specifically designed for long-term applications in automobiles, consumer goods, and construction. These blends have significantly lower carbon footprints compared to pure polyolefins. Another starch-based product blend with polymers is produced by Easygreen from Guangdong, China. These blends, derived from corn-based starches, exhibit rapid decomposition in a controlled composting environment, typically breaking down within a matter of months. Moreover, their production demands 65% less energy than that of traditional plastics, resulting in competitive production costs. Notably, corn starch–based containers are entirely free from bisphenol A (BPA) and phthalates, alleviating any concerns related to endocrine disruption. Apart from several companies operating pilot/demo scale plants (<10 ktpa), the major producers of starch plastics are Novamont (Italy, 120 ktpa) and Rodenberg (Netherlands, 47 ktpa). Despite the widespread use of starch-based plastics, the production capacities are not projected to grow as rapidly as those for other biobased plastics due to challenges associated with incorporating high amounts of starch (>30%) while maintaining desirable material properties, even with reactive blends.

9.3 SUGAR PALM FIBER

The sugar palm tree, *Arenga pinnata*, comes from a forest plant that can be found abundantly in Southeast Asian countries such as Malaysia and Indonesia. Specifically, in peninsular Malaysia, at Kampung Kuala Jempol, Negeri Sembilan, there is activity to generate income by the local villagers from the yield of sugar palm trees (Nurazzi et al., 2020). Palm sap tapping has been historically popular, as the sap was commonly used as the base material for making traditional sugar blocks and could also be processed to become crystal and brown sugar as an alternative to commercialized sugarcane granular sugar, which was locally known as *gula kabung* or *gula enau*. Its fruits may also be processed for making pickles, juices, and desserts, and were usually canned for the food industry. The most important part after the palm sugar and fruits is the black fiber, called *ijuk* in Malay. This black fiber has many applications and uses, such as the manufacture of brooms, paint brushes, septic tank base filters, clear water filters, door mats, carpets, and ropes for sea cordage. It is like other natural fibers like kenaf, banana pseudo-stems, bagasse, pineapple leaves, oil palm fiber, and sugar palm fiber in that it does not require secondary processing such as water retting or a mechanical decorticating process to yield fibers (Mohd Nurazzi et al., 2020).

Recent research has demonstrated the advancing role of sugar palm fiber in diverse engineering applications. Notably, it has found utility in road construction for enhancing soil stability and replacing traditional geotextiles with fiberglass reinforcement. Furthermore, in specific scenarios, it has proven valuable for applications involving underwater and underground cables. Nurazzi et al. (2018) successfully developed a boat using hybrid woven sugar palm fiber and glass fiber with unsaturated polyester as a matrix. In the field of material engineering, it is used as a reinforcement in polymer matrix composites (Nurazzi et al., 2018).

Recent developments in palm sugar have advanced since palm sugar can now be fermented with yeast to produce alcoholic beverages. Bio-ethanol is used as a raw material for goods such as chemical products, solvents, pharmaceutical, cosmetics, medicines, and beverages. Palm sugar can also be used for the production of bio-fuel as a renewable source of energy just like other bio-ethanol plant sources. It is interesting to note that sugar palm can yield the highest production of bio-ethanol (20,160 l/ha/year) compared to other sources such as cassava (4,500 l/ha/year), sugarcane (5,025 l/ha/year), sago (4,133 l/ha/year), and sweet sorghum (6,000 l/ha/year) (Norizan et al., 2017).

9.4 SUGAR PALM STARCH

Commercially, starches are mainly sourced from tubers (such as potatoes and sweet potatoes), cereals (such as rice and wheat), roots (such as cassava and yam), and legumes (such as beans and green peas). These starches often serve as food sources in impoverished regions. Consequently, the use of such carbohydrates as matrices in polymer composites has faced significant criticism and controversy. To address

FIGURE 9.2   (a) Sugar palm tree and (b) collected sugar palm fiber.

these concerns, recent research efforts have focused on developing biopolymers (such as starches and PLA) from non-food sources, aiming to alleviate the debate surrounding the utilization of food sources as polymeric matrices. In this context, sugar palm starch has emerged as a potential alternative. Similar to commercial sago, sugar palm starch is derived from the core of the sugar palm tree's stem (Sahari et al., 2014). Not all sugar palm trees produce sap rich in sugar from their flower bunches. The unproductive trees can account for a significant portion of a plantation, sometimes reaching up to half of the total trees (Elbersen & Oyen, 2010). Starch is typically extracted from these unproductive trees using similar procedures to those employed in sago starch production. It has been reported that a single sugar palm tree can yield 50–100 kg of starch (Sahari et al., 2012).

9.5   PREPARATION OF SUGAR PALM STARCH

Figure 9.3 demonstrates the process of extracting sugar palm starch from the trunk of the sugar palm tree. To initiate the sugar palm starch extraction process, the sugar palm tree is felled just before its initial flowering stage. The trunk is then longitudinally split, allowing the removal of the woody fiber that is mixed with the starch powder found in the soft inner core of the sugar palm trunk. Subsequently, the washing process takes place. Water is gradually introduced into the fiber and starch mixture, and thorough manual kneading is performed. The mixture is filtered to enable the water to pass through the sieve while suspending the starch granules. Ample time

Cutting the tree

Mixture

Stem

Sugar palm starch

Washing process

FIGURE 9.3 Extraction process to obtain sugar palm starch from sugar palm trunk. Reproduced from Sanyang et al. (2016b).

is provided for the starch to settle at the bottom of the container, and then the water is decanted. The resulting white powdered starch is initially exposed to the open air for a period of time, followed by drying in an air circulating oven at a temperature of 120°C for 24 hours (Sahari et al., 2014).

9.6 CHARACTERISTICS OF SUGAR PALM STARCH–BASED BIOPOLYMERS

Sahari et al. (2014) conducted a study to examine the properties of sugar palm starch and assess its potential as a new alternative polymer. Compared to other starches like tapioca (17%), sago (24–27%), potato (20–25%), wheat (26–27%), and maize (26–28%) (Lu et al., 2009), sugar palm starch exhibited superior amylose content (37.60%). Amylose is a component of starch characterized by a highly branched structure consisting of α-D-glucopyranosyl residues with (1→4) linkages, connected by (1→6)-α-linkages. About 5–6% of these linkages occur at branch points. As a result, amylopectin, the highly branched component of starch, has a high molecular weight (107–109 Da) but low intrinsic viscosity (120–190 ml/g) due to its extensively branched structure (Japar & Sapaun Salit, 2012). Sugar palm starch has a low protein content of 0.10% and fat content of 0.27% (w/w) (Adawiyah et al., 2013).

The density of sugar palm starch is comparable to other biopolymers, with a value of 1.54 g/cm^3. Its ash content is similar to tapioca, sago, and wheat (0.2%), while potato has a higher ash content (0.4%). Like all starches, sugar palm starch has a moisture content ranging from 10% to 20% under normal atmospheric conditions (Sahari et al., 2014). Consequently, sugar palm starch is sensitive to moisture due to the presence of hydroxyl functional groups, as indicated by the strong peak at 3200–3500 cm^{-1}. Adawiyah et al. (2013) conducted a comparative study characterizing

TABLE 9.1

Characteristics of Sugar Palm Starch

Characterization	Parameters	Sugar Palm Starch
Chemical composition	Amylose (%w/w)	37.0 ± 1.46
	Fat (%w/w)	0.27 ± 0.00
	Protein (%w/w)	0.10 ± 0.00
	Moisture (%w/w)	9.03 ± 0.00
	Ash (%w/w)	0.20 ± 0.00
Gelatinization properties	Onset temp. (T_O) (°C)	63.0 ± 0.12
	Peak temp. (T_P) (°C)	67.7 ± 0.07
	Conclusion temp. (T_C) (°C)	74.6 ± 0.42
	Range (T_C–T_O) (°C)	11.6 ± 0.49
	ΔH (J/g)	15.4 ± 0.25
Mechanical properties	Stress at 10% strain (kPa)	0.61 ± 0.10
	Stress at shoulder point (kPa)	23.0 ± 3.65
	Strain at shoulder point (%)	54.4 ± 3.54
	Working until shoulder point (N mm)	20.2 ± 1.80
	Breaking stress (kPa)	29.8 ± 2.64
	Breaking strain (%)	60.1 ± 2.61
	Work until breaking point (N mm)	29.6 ± 2.45
	Compressive force after breaking at 70% strain (N)	8.77 ± 0.59
	Compressive force at 90% strain (N)	43.8 ± 2.34
	Working until 90% strain (N mm)	108 ± 6.11
	Adhesive force (N)	-3.64 ± 0.96

sugar palm (*Arenga pinnata*) starch and sago starch, which share similar properties and are often sold commercially under the same name. Table 9.1 provides a detailed comparison between sugar palm starch (SPS) and sago starch (Adawiyah et al., 2013).

9.7 PROCESS OF MODIFYING SUGAR PALM STARCH AND ITS PURPOSES

Sugar palm starch, like many other biopolymers, displays hydrophilic characteristics due to the presence of hydroxyl or polar groups. The development of starches for packaging films faces significant challenges, including brittleness, processability issues, high sensitivity to moisture, rapid retrogradation, and unsatisfactory mechanical and barrier properties. To create high-performance thermoplastic starch suitable for packaging purposes from native sugar palm starch, it is crucial to address these limitations. Several approaches have been employed to overcome these obstacles, including by incorporating various types and concentrations of plasticizers into the starch matrix, then combining sugar palm starch with other polymers that offer better functional performance and adding natural fibers, cellulose, and nanocellulose into the starch composition. By implementing these modifications to sugar palm starch, its functional properties can be significantly improved, ultimately optimizing its potential as an effective material for food packaging and any desired applications.

9.7.1 MECHANISM OF PLASTICIZATION FOR STARCH

In general, biopolymers mostly demonstrate inferior properties compared to petrochemical-derived polymers. Modification techniques such as plasticization, blending, and incorporation of fillers and reinforcements are common effective methods to improve the properties of biopolymers (Imre & Pukánszky, 2013). Polymer science defines two types of plasticizers: internal and external (Figure 9.4). Internal plasticizers are integrated into the polymer structure through copolymerization, grafting, or reaction with the original polymer. They make the polymer chains less compact and rigid, leading to a reduction in the glass transition temperature (T_g) and elastic modulus, thereby softening the polymers. On the other hand, external plasticizers are non-volatile molecules added to interact with polymers without undergoing any chemical reaction. In external plasticization, important molecular forces such as dispersion forces, induction forces, dipole–dipole interactions, and hydrogen bonds play a role (Mekonnen et al., 2013). A polymer can be internally plasticized by chemically modifying the polymer or monomer so that the flexibility is increased. This involves copolymerization of the monomers of the desired polymer (with high T_g) and that of the plasticizer (with low T_g) so that the plasticizer is an integral part of the polymer chain.

The most widely used internal plasticizer monomers are vinyl acetate and vinylidene chloride, while an external plasticizer is the most commonly used method of plasticization because low-cost liquid plasticizers give the formulator freedom in

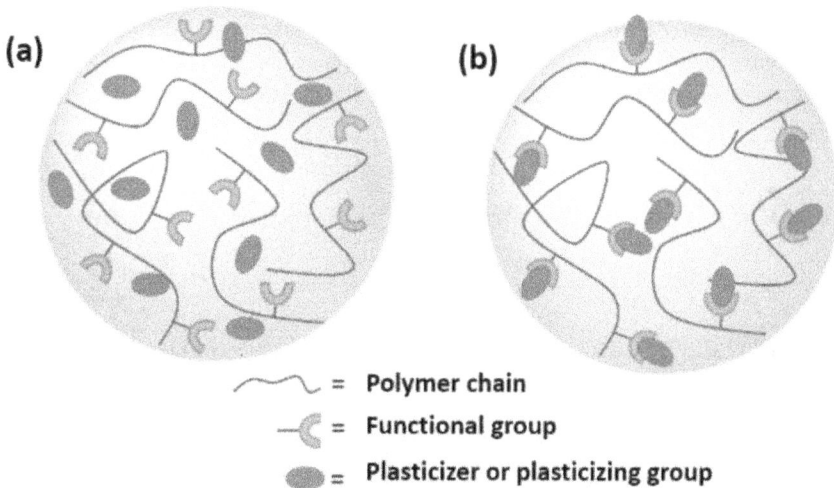

$\sim\!\sim$ = Polymer chain

$-\!\!\!\zeta$ = Functional group

⬤ = Plasticizer or plasticizing group

FIGURE 9.4 External (plasticizers are free) and internal (plasticizing groups are bound to the polymer) plasticizers. Reproduced from Klähn et al. (2019).

developing formulations for a range of products (from semi-rigid to highly flexible depending on the quantity). The most widely used external plasticizers include esters formed from the reaction of acids or acid anhydrides with alcohols. There are two main groups of external plasticizers. Primary plasticizers that enhance elongation, softness, and flexibility of polymer. They are highly compatible with polymers and can be added in large quantities. For example: up to 50% of vinyl gloves are made up of plasticizers, which make the polyvinyl chloride (PVC) flexible and soft enough to wear. A secondary plasticizer is one that typically cannot be used as the sole plasticizer in a plasticized polymer. Secondary plasticizers may have limited compatibility with the polymer and/or high volatility. They may or may not contain functional groups that allow them to solvate the polymer at processing temperatures. Extenders are a subset of secondary plasticizers. They are commonly employed with primary plasticizers to reduce costs in general-purpose flexible PVC. They are mostly low-cost oils with limited compatibility in PVC. They are added to reduce cost and in some cases to improve fire resistance. Examples of extenders include naphthenic hydrocarbons, aliphatic hydrocarbons, chlorinated paraffins (fire resistance), and others.

Various theories have been proposed to explain the mechanism and action of plasticizers on polymers, for example, lubricity theory, gel theory, and free volume theory (Figure 9.5). The lubricity theory suggests that plasticizers act as lubricants, reducing friction and facilitating the mobility of polymer chains, thereby decreasing deformation. The gel theory builds upon the lubricity theory and proposes that plasticizers disrupt and replace polymer–polymer interactions (such as hydrogen bonds and van der Waals or ionic forces) that hold polymer chains together. This disruption leads to a reduction in the polymer gel structure and increased flexibility. The free volume theory defines the internal space available in a polymer for chain movement as the "free volume". Rigid resins have limited free volume, while flexible resins have relatively large amounts. Plasticizers increase the free volume of resins and maintain it even after the polymer–plasticizer mixture is cooled down. The free volume theory explains the role of plasticizers in lowering the T_g. Referring to Shtarkman and Razinskaya, these theories are widely used in selecting plasticizers

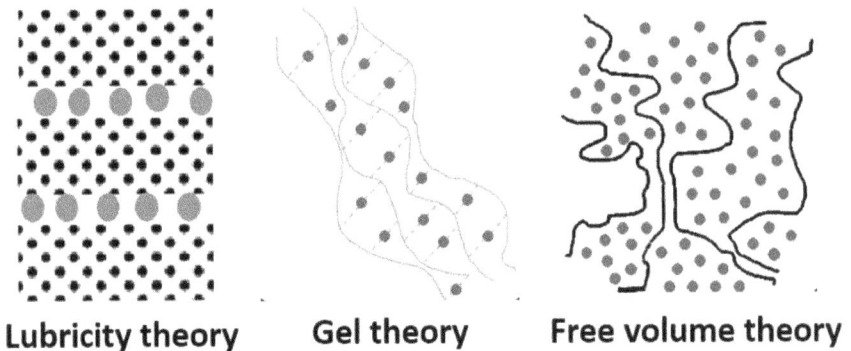

Lubricity theory   Gel theory   Free volume theory

FIGURE 9.5   Theories of plasticization.

for polymers (1983). They argue that the current plasticization theories lack direct studies on the plasticization mechanism and have limited predictive capabilities, being applicable only in specific cases. Therefore, the authors suggest the need for compatibility, efficiency, and property studies that consider the polymeric system's structure to select specific plasticizers rather than relying solely on existing theories.

It is worth noting that the aforementioned plasticization theories were primarily developed for synthetic plastics, particularly PVC. Limited attention has been given to developing new theories or improving existing ones to explain the plasticization mechanism in newly developed biobased plastics. These biobased plastics differ significantly from traditional synthetic polymers due to the complex nature of their biological feedstock macromolecules. Thus, further research is required to explore alternative and more comprehensive plasticization possibilities and theories.

9.7.2 PLASTICIZATION OF SUGAR PALM STARCH

Starch, in its pure form, lacks the ability to melt and cannot be processed as a thermoplastic (Imre & Pukánszky, 2013). Although neat starch exhibits a relatively high modulus and strength, it possesses limited deformability and impact resistance due to the rigid nature of its chains (Jiang & Zhang, 2017). Consequently, the process of plasticization is commonly employed to improve the processability and other properties of starch. Plasticization involves the use of low molecular weight polar compounds such as water, glycerol, urea, and formamide (Imre & Pukánszky, 2013; Whistler & Daniel, 1984) to replace the intermolecular bonds between polymer chains with bonds between the macromolecules and the plasticizer. This disruption of starch granules and disorientation of the crystalline structure is caused by the influence of plasticizers, heat, and shear. Plasticization leads to a decrease in the glass transition temperature and processing temperature of starch. As a result, the material becomes melt-processable, exhibiting increased flexibility, workability, distensibility, and deformability (Jiang & Zhang, 2017; Whistler & Daniel, 1984). Plasticizing starch enhances the flexibility of polymer chains and improves fracture resistance. It also reduces tension during deformation, hardness, density, and viscosity and facilitates easier incorporation and dispersion of fillers or reinforcements. Additionally, plasticization affects properties such as degree of crystallinity, optical clarity, and resistance to biological degradation, among others (Vieira et al., 2011). The resulting plasticized starch is commonly referred to as thermoplastic starch.

Syafiq et al. (2022) conducted a study to examine how various types and amounts of plasticizers affect the physical, mechanical, and antibacterial properties of bionanocomposite films made from sugar palm nanocellulose, sugar palm starch, and cinnamon essential oil. The researchers found that without a plasticizer, sugar palm starch films were brittle, had visible cracks, and were difficult to remove from casting surfaces. However, the addition of plasticizers helped to overcome brittleness and improve the flexibility and peelability of the starch films. The results indicated that both the concentration and type of plasticizer influenced the density, thickness, tensile strength, moisture content, surface structure, and elongation at breaks of the films. The films were prepared using glycerol, sorbitol, and a combination of both as plasticizers at ratios of 1.5, 3.0, and 4.5 wt.%. The antibacterial activity of the bionanocomposite films was not significantly affected by the type of plasticizer, indicating that inhibition zone effectiveness was maintained. As the concentration of the plasticizer increased from 1.5 to 4.5 wt.%, the density of the films decreased, while the moisture content and thickness increased, regardless of the plasticizer type used. Overall, films plasticized with a blend of both glycerol and sorbitol demonstrated the best physical and mechanical properties. The plasticizing effect of different concentrations of the plasticizer was attributed to the weakening of hydrogen bonds between the starch intermolecular chains, caused by the formation of starch–plasticizer complexes. At a concentration of 4.5 wt.% plasticizer, an antiplasticization effect was observed for the glycerol- and glycerol/sorbitol–plasticized films, whereas plasticization behavior was observed at lower concentrations. Interestingly, films plasticized with a combination of glycerol and sorbitol improved the tensile strength compared to films plasticized with glycerol alone while reducing the brittleness compared to films plasticized with sorbitol alone. Optimum physical, mechanical, and antibacterial performance was achieved by plasticizing the sugar palm nanocellulose/sugar palm starch/cinnamon essential oil bionanocomposite films using a blend of glycerol and sorbitol. However, further research is needed to investigate the effects of different plasticizer types and concentrations on the solubility, water absorption, water vapor permeability, and barrier properties of films based on sugar palm starch in order to discover the optimal combination for developing biodegradable food packaging films.

Sahari et al. (2013) conducted a study to investigate the impact of plasticization on the properties of sugar palm starch. Plasticized sugar palm starch samples were prepared using varying concentrations of glycerol (15%, 20%, 30%, and 40% w/w). The results indicated that as the amount of glycerol increased, the densities (g/cm^3), moisture content (%), and water absorption (%) of the sugar palm starch decreased. This can be attributed to the ability of glycerol, with its low molecular weight, to replace the intermolecular bonds between the polymer chains of sugar palm starch, thereby reducing the secondary forces among them. The increased concentration of glycerol resulted in stronger hydrogen bonding between glycerol and sugar palm starch, leading to a decrease in the water absorption capacity of the plasticized sugar palm starch. As a result, water molecules found it more difficult to penetrate into the plasticized sugar palm starch since the stronger hydrogen bonds hindered their interaction with the plasticizer and with the sugar palm starch. Consequently, the entry of water molecules into the plasticized sugar palm starch was impeded.

Moreover, the addition of 15%, 20%, and 30% w/w glycerol consistently led to an increase in the tensile strength and elongation at break of sugar palm starch. The highest values were observed for sugar palm starch with 30% w/w glycerol, with a tensile strength of 2.42 MPa and elongation at break of 8.03%. However, when the glycerol concentration was further increased to 40% w/w, there was a significant decrease in both tensile strength and elongation at break of sugar palm starch with 30% w/w glycerol by 79.34% and 31.26%, respectively. The reduced tensile strength in sugar palm starch with 40% w/w glycerol was attributed to the excessive amount of glycerol, leading to poor adhesion with SPS. Conversely, the tensile modulus of sugar palm starch decreased as the plasticizer concentration increased from 15% to

40% w/w. These findings suggest that plasticized sugar palm starch exhibits greater flexibility when subjected to tension or mechanical stress.

The introduction of more glycerol to sugar palm starch resulted in an increase in surface smoothness, rendering it softer and non-brittle compared to pure sugar palm starch. The plasticizer effectively reduced internal hydrogen bonding while increasing intermolecular spacing, thereby reducing brittleness (Bibers et al., 1999). In the Fourier-transform infrared (FTIR) spectra of plasticized sugar palm starch with varying glycerol concentrations, strong peaks at 3,200 to 3,500 cm^{-1} were observed, indicating the presence of O–H groups. This suggests that the hydroxyl groups of plasticized sugar palm starch decreased as the glycerol concentration increased (Sahari et al., 2013).

9.8   PERFORMANCE OF SUGAR PALM STARCH COMPOSITES

Sugar palm nanocrystalline cellulose (SPNCC) was obtained by subjecting sugar palm fibers to hydrolysis treatment (Ilyas et al., 2018). These SPNCCs were then incorporated into the sugar palm starch matrix at concentrations ranging from 0.1 to 1.0 wt.% using a solution-casting method to create bionanocomposite films. Examination of the field emission scanning electron microscopy (FESEM) micrographs demonstrated excellent dispersion of SPNCC nanofibers within the SPS matrix, resulting in enhanced mechanical, thermal, and water barrier properties of the SPS-based films. Notably, the addition of SPNCCs as fillers in the SPS nanocomposite films led to a remarkable increase in both the Young's modulus and tensile strength, with values rising from 54 to 178.83 MPa and 4.80 to 11.47 MPa, respectively, as the concentration of nanofillers increased from 0 to 1.0 wt.%. Moreover, the presence of SPNCCs led to an elevation in the T_g value associated with the starch-rich phase, indicating a reduction in the flexibility of the starch molecular chains, and exhibited superior water resistance. The observed performance enhancements in these SPS/SPNCC bionanocomposite films can be attributed to the high compatibility and intermolecular hydrogen bonding interaction between the two components resulting from their chemical similarities. Consequently, this study highlights the significant potential of SPS/SPNCC nanocomposite films for packaging applications.

Biodegradable bilayer films using SPS and PLA have been characterized (Sanyang et al., 2016a). SPS-PLA bilayer films were successfully created without the need for any compatibilizers, adhesives, or chemical modifications of the film surfaces. The experimental findings demonstrated that the physical, mechanical, and water barrier properties of the SPS-based films were enhanced by incorporating a 50% PLA layer in the preparation of 50% PLA layer onto 50% SPS layer (SPS50-PLA50) bilayer films. Comparing SPS50-PLA50 with 100% SPS (SPS100), it was observed that SPS50-PLA50 exhibited a 76.36% increase in tensile strength and a 96.70% decrease in water vapor permeability (WVP). Additionally, the water absorption value of SPS50-PLA50 decreased by 65.89% compared to SPS100. Fourier-transform infrared spectroscopy results of the bilayer films indicated no specific interaction between SPS and PLA, suggesting that the two components were immiscible. Consequently, scanning electron microscopy image of the cross-section of SPS50-PLA50 revealed a gap between the interface of the two layers, indicating poor interfacial adhesion between SPS and PLA. Overall, the incorporation of a PLA layer onto SPS films improved the mechanical strength and water barrier properties of the SPS-based films, thereby enhancing their suitability for food packaging applications.

The degradation and physical properties of sugar palm starch/sugar palm nanofibrillated cellulose (SPNFC) bionanocomposites were studied by Atikah et al. (2019). SPNFCs were isolated from sugar palm fiber, while SPS was extracted from sugar palm trunks. The SPNFCs were reinforced with SPS biopolymer as biodegradable reinforcement materials of different diameter/length based on the number of passes of the high-pressurize homogenization process (5, 10, and 15 passes represented by SPS/SPNFC-5, SPS/SPNFC-10, and SPS/SPNFC-15). These SPNFCs were incorporated into SPS and plasticized with glycerol and sorbitol via the solution casting method. From the biodegradation study, it was found that the neat biopolymer SPS degraded faster in SPS/SPNFC bionanocomposites, which lost 85.8% of their weight at day 9 compared to 69.9% by the SPS/SPNFC-15 bionanocomposite. The performance improvement of the SPS/SPNFC bionanocomposites might be due to the high compatibility derived from intermolecular hydrogen bonding interaction between these two biomaterials as a result of their chemical similarities, good dispersion, and adhesion of the SPNFCs within the SPS biopolymer matrix. The SPNFCs established in this current work are intended to be utilized in SPNFC/starch-based nanocomposites for potential packaging applications.

In their study, Hasan et al. (2020) developed active edible films using sugar palm starch and chitosan, incorporating extra virgin olive oil as antioxidants. The film's tensile strength, elongation at break, thermal stability, barrier properties, antioxidant activity, and antimicrobial activity were investigated. The results indicated that the chitosan/SPS blend films with varying amounts of extra virgin olive oil exhibited improved surface roughness, mechanical properties (tensile strength and elongation at break), thermal stability, barrier properties, antimicrobial activity, and antioxidant activity. X-ray diffraction analysis (XRD) confirmed that the addition of extra virgin olive oil did not affect the structure of the chitosan/SPS matrix. The optimal content of extra virgin olive oil for producing a strong and flexible film was found to be 2% w/w. The film containing 2% extra virgin olive oil exhibited the highest tensile strength, elongation at break, and antioxidant activity against the 2,2-diphenyl-1-picrylhydrazyl (DPPH) radical scavenger.

On the other hand, incorporating 1.0% w/w of extra virgin olive oil did not significantly increase the tensile strength and elongation at break values ($p < 0.05$) due to uneven distribution throughout the matrix. However, the addition of 2% w/w extra virgin olive oil significantly increased these values ($p < 0.05$). Similarly, adding 5% w/w extra virgin olive oil significantly increased the tensile strength value but was lower than that at 2% w/w. This discrepancy was attributed to the uniform distribution of extra virgin olive oil within the film matrix, leading to stronger interactions between the compounds of extra virgin olive oil and chitosan/SPS, thus enhancing the tensile strength and elongation at break values. The increase in film thickness resulting from the addition of extra virgin olive oil was consistent with the data. The hydroxyl groups of extra virgin olive oil interacted with chitosan/SPS, contributing to the increase in film thickness. The enhanced tensile strength and elongation at break values were likely due to the presence of mono-unsaturated lipids in extra virgin olive oil, which formed a flexible complex with the chitosan/SPS matrix. This effect was achieved through both hydrophilic and hydrophobic mechanisms, as the free fatty acids (oleic acid) from extra virgin olive oil interacted electrostatically with the amino group of chitosan, forming hydrogen bonds, while the lipid tail of fatty acids interacted hydrophobically via van der Waals relaxation. Overall, the findings demonstrate the promising potential of chitosan/SPS-extra virgin olive oil blend films as active edible films, providing sustainable alternatives to pure chitosan/SPS films for various packaging applications.

The effect of sugar palm nanocrystalline cellulose loading (0.00–0.10 wt.%) in sugar palm starch on the electrical resistance, resistivity, and conductivity of SPS/SPNCC nanocomposite films was investigated (Hazrol et al., 2020). The experiments were conducted using the four-probe method and Ohm's law, and resistivity and conductivity equations were utilized to obtain the electrical properties. The conductivity was determined with different amounts of current ranging from 0.0–0.20 µA and a concentration of SPNCCs in the film from 0.00–0.10 wt.%. From 0.00 to 0.05 µA, the conductivity of SPS/SPNCC films show a huge spike that reduced and remained stable after 0.1 µA. It was found that the peak conductivity occurred with films carrying the concentration of 0.00 wt.% and current value at 0.02 µA, where the fiber holds more charge before it is discharged and becomes stable. It was also discovered that 0.10 wt.% concentration of SPNCC gave the least electrical conductivity value, 6.813×10^{-5} S/mm. The trend observed from the 0.10 wt.% concentration trace showed good results with an increase in current value. For the layered effect of electrical resistivity of various concentration of SPS/SPNCC films at room temperature, it can be summarized that for a single layer of SPS/SPNCC films, the resistivity increased from 0.37×10^3 ($\Omega \cdot$ mm) at 0.05 µA to 4.8×10^3 ($\Omega \cdot$ mm) at 0.50 µA. This can be compared to SPS/SPNCC concentrations of SPNCCs at 1.0 wt.% for three-layer films that had a lower resistivity of 0.11×10^3 ($\Omega \cdot$ mm) at 0.05 µA and increased resistivity value to 5.95×103 ($\Omega \cdot$ mm) at 0.50 µA. There were also no changes to the physical appearance of the film due to the small amounts of current tested. The resistivity increased uniformly by increasing the current supply; thus this hypothesis applies to conductivity, which will decrease in value with the increase in the current being supplied.

A study conducted by Jumaidin et al. (2017) aimed to assess how seaweed affects the mechanical, thermal, and biodegradation properties of thermoplastic sugar palm starch/agar (TPSA) composites. The researchers examined TPSA blends containing *Eucheuma cottonii* seaweed waste as a biofiller. The composites were created by melt-mixing and hot pressing at a temperature of 140°C for 10 minutes. Incorporating seaweed in amounts ranging from 0 to 40 wt.% significantly enhanced the tensile, flexural, and impact properties of the TPSA/seaweed composites. Analysis of the tensile fracture surface using scanning electron microscopy revealed a uniform surface with the formation of a cleavage plane. Furthermore, the inclusion of seaweed promoted the biodegradation of the composites when buried in soil for 2 and 4 weeks, demonstrating that adding seaweed increased the weight loss of the materials, indicating a faster biodegradation process for the composites. This can be attributed to the similar hydrophilic nature of seaweed and TPSA, leading to strong adhesion between the filler and matrix. Fourier-transform infrared spectroscopy results indicated a higher number of intermolecular hydrogen bonds, suggesting good compatibility between seaweed and TPSA. These findings underscore the potential of *Eucheuma cottonii* seaweed waste as an excellent filler in biocomposites, offering various characteristics ranging from strength reinforcement to promoting biodegradation. Overall, integrating seaweed into TPSA enhances its properties, making it suitable for manufacturing short-life products like trays and plates.

The characteristics of a thermoplastic sugar palm starch/agar blend on thermal, tensile, and physical properties were studied by Jumaidin et al. (2016). The investigation on the behavior of a thermoplastic composed of biodegradable sugar palm starch and agar, in the 10–40 wt.% range, revealed through FTIR analysis that SPS and agar exhibited compatibility and formed inter-molecular hydrogen bonds. The inclusion of agar in the composition led to enhanced tensile properties, including Young's modulus and tensile strength, of the thermoplastic starch. Moreover, the thermal stability and moisture absorption capabilities increased as the agar content rose. Notably, the study identified that thermoplastic SPS with 30 wt.% agar exhibited the highest tensile strength. However, a higher agar content (40 wt.%) resulted in slightly diminished tensile strength accompanied by rough cleavage fracture. In conclusion, the incorporation of agar proved beneficial for improving the thermal and tensile properties of SPS-based thermoplastics, thereby expanding the potential applications of this environmentally friendly material. Particularly, this eco-friendly material shows great promise for applications requiring short-term usage, such as packaging, containers, trays, and similar products.

An antibacterial biopolymer composite film based on sugar palm starch, derived from renewable sources and incorporating inorganic silver nanoparticles (AgNPs) as the primary antibacterial component, was successfully created (Rozilah et al., 2020). The composite films were fabricated using the solution casting method, and their mechanical and physicochemical properties were assessed through tensile testing, FTIR analysis, thermal gravimetric analysis (TGA), antibacterial screening tests, and FESEM imaging. The addition of AgNPs to the composite films resulted in improvements in both mechanical and antibacterial properties compared to films without active metals. The inclusion of inorganic AgNPs as nanofillers in the film's matrix addressed the weaknesses observed in the neat composite films, effectively preventing bacterial growth. Tensile strength values ranged from 8 to 408 kPa, while the elasticity modulus varied between 5.72 and 9.86 kPa. The FTIR analysis demonstrated a decrease in transmittance values upon the addition of AgNPs, leading to minor changes in the chemical structure, slight variations in intensity peaks, and longer wavelengths because there were some modifications on the functional groups with the addition of Ag^+ into the formulation that caused slight changes to the intensity peaks by lowering the transmittance and shortening the frequency. The incorporation of these active films also increased the weight of degradation and elevated the decomposition temperature due to the enhanced thermal stability of the AgNPs. In terms of antibacterial efficacy, the inhibited areas for *Escherichia coli* measured between 7.66 and 7.83 mm, for *Salmonella cholerasuis* between 7.5 and 8.0 mm, and for *Staphylococcus aureus* between 0.1 and 0.5 mm. Microscopic analysis revealed that the average size of all microbes ranged from 0.57 to 2.90 mm for films containing 1 and 4 wt.% AgNPs. Based on comprehensive evaluation, the optimal composition for the biopolymer composite films was determined to be 3 wt.% AgNPs. This composition successfully met all the required mechanical properties while exhibiting superior antimicrobial properties. Consequently, the development of organic-inorganic hybrid antibacterial biopolymer composite films proved a suitable approach for creating antibacterial coatings.

According to Nazrin et al. (2021), optimizing the reinforcement of SPCNCs in blend bionanocomposites depends on the effective dispersion of thermoplastic starch (TPS) within the PLA phase. To achieve this, a balanced ratio between PLA and TPS needs to be determined. Blending different ratios of TPS with PLA affected the burning rate and behavior of the resulting blended bionanocomposites. PLA60TPS40 (40% TPS) and PLA40TPS60 (60% TPS) showed increased flammability compared to the other ratios, with sustained burning up to the 100 mm reference line. This heightened flammability can be attributed to the plasticizers (glycerol and sorbitol) incorporated into TPS during fabrication. Glycerol and sorbitol are known to be flammable substances, with flash points of 176°C and above 300°C, respectively. The higher concentration of plasticizers in TPS significantly influenced the burning rate, as observed in the faster burning rate of PLA40TPS60 (16.39 mm/min) compared to PLA60TPS40 (15.29 mm/min). This phenomenon can be attributed to the migration of glycerol from TPS to PLA due to its lower molecular weight, thereby activating flammability in the PLA phase. Other studies (Esmaeili et al., 2019; H. Li & Huneault, 2011) reported that during the melt mixing process, glycerol with lower molecular weight tends to migrate from the TPS to PLA matrix, while sorbitol with a higher molecular weight remains within TPS. Sorbitol contributed to reducing particle size and promoting a fine dispersion of TPS within the PLA phase, resulting in a homogeneous composition and uniform distribution of TPS.

This uniform distribution of TPS within PLA triggered continuous combustion of the samples. However, the burning rate of PLA40TPS60 and PLA60TPS40 remained below 40 mm/min, classifying them as horizontal burning (HB) in UL94 ratings. In other words, they were considered the least flame-retardant materials that required improvement in their flammability characteristics To improve the flame-retardant properties of starch-based polymer blends, glycerol was replaced with glycerol phosphate. Scanning electron microscope (SEM) images revealed that PLA60TPS40 exhibited a smooth surface morphology, indicating good interfacial bonding, while PLA40TPS60 and PLA20TPS80 showed visible agglomeration spots of TPS. All samples displayed constant dripping of the melted material. However, PLA80TPS20 and PLA70TPS30 exhibited microcracks and voids, indicating poor interfacial bonding between PLA and TPS. When these two samples ignited, fragments of material detached instead of consistent dripping, suggesting that the minor proportion of TPS melted quickly compared to the less flammable PLA phase, causing detachment of PLA fragments aided by the presence of microcracks.

Based on these observations, it was confirmed that PLA60TPS40 and PLA40TPS60 had the optimal concentration of plasticizers to promote the flammability of PLA and improve the dispersion of TPS within PLA. PLA40TPS60 (19.2%) had a slightly higher limiting oxygen index (LOI) value compared to PLA60TPS40 (18.8%), indicating that a higher TPS content required a higher oxygen concentration for combustion despite the increased glycerol concentration. This might explain why PLA20TPS80 was not flammable like PLA40TPS60 and PLA60TPS40, despite having the highest glycerol concentration among all the blend bionanocomposite samples. A relatively high TPS content indicated a higher concentration of SPCNC within the samples, and it was hypothesized that the lignin content in cellulose generated char formation, which hindered flame propagation. However, the removal of lignin to achieve low water retention of SPCNC might have increased the flammability rate for samples with higher TPS content. Nevertheless, the effect of glycerol migration outweighed the reinforcement of SPCNC. In food packaging applications that prioritize biodegradability and cost-effective materials, PLA60TPS40 demonstrated acceptable physical stability but required improvement in flame-retardant properties.

9.9 CONCLUSIONS

To summarize, the abundance and non-toxic, degradable, and biocompatible properties of starch have attracted significant research interest, leading to the development of starch-based films. Simple laboratory methods like casting and extrusion/hot pressing can be used to create smart starch-based films. The emergence of biodegradable materials derived from renewable resources is inevitable, and they can replace synthetic polymers in important applications. Starch-based materials, particularly in industries like food packaging, can enhance the economic prospects of farmers while reducing the release of toxic substances and environmental pollution. However, there is room for improvement in the thermal resistance, mechanical properties, and barrier properties of starch-based films and composites without compromising their biodegradability. Challenges include starch's sensitivity to moisture and retrogradation processes, necessitating further research on starch, plasticizers, and nanoparticles to minimize water absorption and retrogradation and maintain mechanical strength and stiffness during storage. The development of these materials and their production is currently not cost effective, and addressing this issue is crucial to encourage the adoption of starch-based materials in packaging. Research should focus on creating packaging materials that are simpler and smarter and enable consumers to assess the quality, safety, shelf life, and nutritional value of the contents.

Sugar palm (*Arenga pinnata*) is a versatile tree with various traditional uses, and its fiber exhibits excellent mechanical properties that can rival other natural fibers like coir, oil palm fiber, kenaf, cotton, and jute. The combination of sugar palm fiber and starch from the same tree can create a "one-source" green composite. Utilizing sugar palm fiber and starch in green composites can contribute to reducing the environmental impact of synthetic polymers and fibers, decreasing reliance on petroleum products, and establishing sugar palm as a new industrial crop, particularly in tropical countries. This can lead to socio-economic empowerment of rural communities by generating more income and employment opportunities. However, the potential of sugar palm fiber and biopolymers in the composite industry for various industrial applications remains largely untapped. Further advanced characterization techniques and equipment, such as attenuated total reflection FTIR, AFM, x-ray photoelectron spectroscopy (XPS), FESEM, and thermal analysis instruments, should be employed to analyze and test sugar palm fibers, biopolymers, and their biocomposites. To enable effective packaging applications, it is crucial to determine the barrier properties, sealability, and moisture absorption of sugar palm-based films and biocomposites.

REFERENCES

Adawiyah, D. R., Sasaki, T., & Kohyama, K. (2013). Characterization of arenga starch in comparison with sago starch. *Carbohydrate Polymers, 92*(2), 2306–2313.

Aslam, M., Nadeem, H., Azeem, F., Zubair, M., Rasul, I., Muzammil, S., Zfzal, M., & Siddique, M. H. (2023). Applications of bioplastics in disposable products. In *Handbook of bioplastics and biocomposites engineering applications* (pp. 445–455). John Wiley & Sons.

Atikah, M. S. N., Ilyas, R. A., Sapuan, S. M., Ishak, M. R., Zainudin, E. S., Ibrahim, R., Atiqah, A., Ansari, M. N. M., & Jumaidin, R. (2019). Degradation and physical properties of sugar palm starch/sugar palm nanofibrillated cellulose bionanocomposite. *POLIMERY, 64*(10), 680–689.

Basiak, E., Lenart, A., & Debeaufort, F. (2017). Effect of starch type on the physico-chemical properties of edible films. *International Journal of Biological Macromolecules, 98*, 348–356.

Basiak, E., Lenart, A., & Debeaufort, F. (2018). How glycerol and water contents affect the structural and functional properties of starch-based edible films. *Polymers, 10*(4), 412.

Bibers, I., Tupureina, V., Dzene, A., & Kalnins, M. (1999). Improvement of the deformative characteristics of poly-β-hydroxybutyrate by plasticization. *Mechanics of Composite Materials, 35*, 357–364.

Chen, Z., Ma, Y., Gou, L., Zhang, S., & Wang, Z. (2023). Construction of caffeic acid modified porous starch as the dual-functional microcapsule for encapsulation and antioxidant property. *International Journal of Biological Macromolecules, 228*, 358–365.

Elbersen, H. W., & Oyen, L. P. A. (2010). *Sugar palm (Argena pinnata). Potential of sugar palm for bio-ethanol production.* FACT-Foundation.

Esmaeili, M., Pircheraghi, G., Bagheri, R., & Altstädt, V. (2019). Poly (lactic acid)/coplasticized thermoplastic starch blend: Effect of plasticizer migration on rheological and mechanical properties. *Polymers for Advanced Technologies, 30*(4), 839–851.

Fang, J., & Fowler, P. (2003). The use of starch and its derivatives as biopolymer sources of packaging materials. *Journal of Food Agriculture and Environment, 1*, 82–84.

Florencia, V., López, O. V., & García, M. A. (2020). Exploitation of by-products from cassava and ahipa starch extraction as filler of thermoplastic corn starch. *Composites Part B: Engineering, 182*, 107653.

Hasan, M., Rusman, R., Khaldun, I., Ardana, L., Mudatsir, M., & Fansuri, H. (2020). Active edible sugar palm starch-chitosan films carrying extra virgin olive oil: Barrier, thermomechanical, antioxidant, and antimicrobial properties. *International Journal of Biological Macromolecules, 163*, 766–775.

Hazrol, M. D., Sapuan, S. M., Ilyas, R. A., Othman, M. L., & Sherwani, S. F. K. (2020). Electrical properties of sugar palm nanocrystalline cellulose reinforced sugar palm starch nanocomposites. *POLIMERY, 65*(5), 363–372.

Hu, A., Li, L., Zheng, J., Lu, J., Meng, X., Liu, Y., & Rehman, R. (2014). Different-frequency ultrasonic effects on properties and structure of corn starch. *Journal of the Science of Food and Agriculture, 94*(14), 2929–2934.

Ilyas, R. A., Sapuan, S. M., Atiqah, A., Ibrahim, R., Abral, H., Ishak, M. R., Zainudin, E. S., Norizan, M. N., Mahamud, A., Ansari, M. N. M., Asyraf, M. R. M., Supian, A. B. M.,

& Ya, H. H. (2020). Sugar palm (Arenga pinnata [Wurmb.] Merr) starch films containing sugar palm nanofibrillated cellulose as reinforcement: Water barrier properties. *Polymer Composites*, *41*(2), 459–467.

Ilyas, R. A., Sapuan, S. M., Ishak, M. R., & Zainudin, E. S. (2018). Development and characterization of sugar palm nanocrystalline cellulose reinforced sugar palm starch bionanocomposites. *Carbohydrate Polymers*, *202*, 186–202.

Imre, B., & Pukánszky, B. (2013). Compatibilization in bio-based and biodegradable polymer blends. *European Polymer Journal*, *49*(6), 1215–1233.

Japar, S., & Sapaun Salit, M. (2012). The development and properties of biodegradable and sustainable polymers. *Journal of Polymer Materials*, *29*(1), 153.

Jiang, L., & Zhang, J. (2017). Biodegradable and biobased polymers. In *Applied plastics engineering handbook* (pp. 127–143). Elsevier.

Jumaidin, R., Sapuan, S. M., Jawaid, M., Ishak, M. R., & Sahari, J. (2016). Characteristics of thermoplastic sugar palm Starch/Agar blend: Thermal, tensile, and physical properties. *International Journal of Biological Macromolecules*, *89*, 575–581.

Jumaidin, R., Sapuan, S. M., Jawaid, M., Ishak, M. R., & Sahari, J. (2017). Effect of seaweed on mechanical, thermal, and biodegradation properties of thermoplastic sugar palm starch/agar composites. *International Journal of Biological Macromolecules*, *99*, 265–273.

Kalambur, S., & Rizvi, S. S. H. (2006). An overview of starch-based plastic blends from reactive extrusion. *Journal of Plastic Film & Sheeting*, *22*(1), 39–58.

Klähn, M., Krishnan, R., Phang, J. M., Lim, F. C. H., van Herk, A. M., & Jana, S. (2019). Effect of external and internal plasticization on the glass transition temperature of (Meth) acrylate polymers studied with molecular dynamics simulations and calorimetry. *Polymer*, *179*, 121635.

Li, C., Ju, B., & Zhang, S. (2023). Fully bio-based hydroxy ester vitrimer synthesized by crosslinking epoxidized soybean oil with doubly esterified starch. *Carbohydrate Polymers*, *302*, 120442.

Li, H., & Huneault, M. A. (2011). Comparison of sorbitol and glycerol as plasticizers for thermoplastic starch in TPS/PLA blends. *Journal of Applied Polymer Science*, *119*(4), 2439–2448.

Liu, H., Xie, F., Yu, L., Chen, L., & Li, L. (2009). Thermal processing of starch-based polymers. *Progress in Polymer Science*, *34*(12), 1348–1368.

Lu, D. R., Xiao, C. M., & Xu, S. J. (2009). Starch-based completely biodegradable polymer materials. *Express Polymer Letters*, *3*(6), 366–375.

Malik, M. K., Bhatt, P., Kumar, T., Singh, J., Kumar, V., Faruk, A., Fuloria, S., Fuloria, N. K., Subrimanyan, V., & Kumar, S. (2023). Significance of chemically derivatized starch as drug carrier in developing novel drug delivery devices. *The Natural Products Journal*, *13*(6), 40–53.

Mekonnen, T., Mussone, P., Khalil, H., & Bressler, D. (2013). Progress in bio-based plastics and plasticizing modifications. *Journal of Materials Chemistry A*, *1*(43), 13379–13398.

Mohd Nurazzi, N., Khalina, A., Chandrasekar, M., Aisyah, H. A., Ayu Rafiqah, S., Ilyas, R. A., & Hanafee, Z. M. (2020). Effect of fiber orientation and fiber loading on the mechanical and thermal properties of sugar palm yarn fiber reinforced unsaturated polyester resin composites. *POLIMERY*, *65*.

Mohamed, R., Mohd, N., Nurazzi, N., Aisyah, M. S., & Fauzi, F. M. (2017). Swelling and tensile properties of starch glycerol system with various crosslinking agents. In *IOP Conference Series: Materials Science and Engineering* (Vol. 223, No. 1, p. 012059). IOP Publishing.

Mouren, A., & Avérous, L. (2023). Sustainable cycloaliphatic polyurethanes: From synthesis to applications. *Chemical Society Reviews, 52*, 277–317

Nazrin, A., Sapuan, S. M., Zuhri, M. Y. M., Tawakkal, I. S. M. A., & Ilyas, R. A. (2021). Flammability and physical stability of sugar palm crystalline nanocellulose reinforced thermoplastic sugar palm starch/poly (lactic acid) blend bionanocomposites. *Nanotechnology Reviews, 11*(1), 86–95.

Norizan, M. N., Abdan, K., Salit, M. S., & Mohamed, R. (2017). Physical, mechanical and thermal properties of sugar palm yarn fibre loading on reinforced unsaturated polyester composites. *Journal of Physical Science, 28*(3).

Nurazzi, N. M., Khalina, A., Sapuan, S. M., Ilyas, R. A., Rafiqah, S. A., & Hanafee, Z. M. (2020). Thermal properties of treated sugar palm yarn/glass fiber reinforced unsaturated polyester hybrid composites. *Journal of Materials Research and Technology, 9*(2), 1606–1618.

Nurazzi, N. M., Khalina, A., Sapuan, S. M., & Rahmah, M. (2018). Development of sugar palm yarn/glass fibre reinforced unsaturated polyester hybrid composites. *Materials Research Express, 5*(4), 045308.

Ramakrishnan, R., Kulandhaivelu, S. V., Roy, S., & Viswanathan, V. P. (2023). Characterisation of ternary blend film of alginate/carboxymethyl cellulose/starch for packaging applications. *Industrial Crops and Products, 193*, 116114.

Rozilah, A., Jaafar, C. N. A., Sapuan, S. M., Zainol, I., & Ilyas, R. A. (2020). The effects of silver nanoparticles compositions on the mechanical, physiochemical, antibacterial, and morphology properties of sugar palm starch biocomposites for antibacterial coating. *Polymers, 12*(11), 2605.

Saha, T., Hoque, M. E., & Mahbub, T. (2020). Biopolymers for sustainable packaging in food, cosmetics, and pharmaceuticals. In *Advanced processing, properties, and applications of starch and other bio-based polymers* (pp. 197–214). Elsevier.

Sahari, J., Sapuan, S. M., Zainudin, E. S., & Maleque, M. A. (2012). Sugar palm tree: A versatile plant and novel source for biofibres, biomatrices, and biocomposites. *Polymers from Renewable Resources, 3*(2), 61–78.

Sahari, J., Sapuan, S. M., Zainudin, E. S., & Maleque, M. A. (2013). Thermo-mechanical behaviors of thermoplastic starch derived from sugar palm tree (Arenga pinnata). *Carbohydrate Polymers, 92*(2), 1711–1716.

Sahari, J., Sapuan, S. M., Zainudin, E. S., & Maleque, M. A. (2014). Physico-chemical and thermal properties of starch derived from sugar palm tree (Arenga pinnata). *Asian Journal of Chemistry, 26*(4), 955.

Sanyang, M. L., Ilyas, R. A., Sapuan, S. M., & Jumaidin, R. (2018). Sugar palm starch–based composites for packaging applications. In Mohammad Jawaid and Sarat Kumar Swain (Eds.), *Bionanocomposites for packaging applications* (pp. 125–147). Springer.

Sanyang, M. L., Sapuan, S. M., Jawaid, M., Ishak, M. R., & Sahari, J. (2016a). Development and characterization of sugar palm starch and poly (lactic acid) bilayer films. *Carbohydrate Polymers, 146*, 36–45.

Sanyang, M. L., Sapuan, S. M., Jawaid, M., Ishak, M. R., & Sahari, J. (2016b). Recent developments in sugar palm (Arenga pinnata) based biocomposites and their potential industrial applications: A review. *Renewable and Sustainable Energy Reviews, 54*, 533–549.

Shahi, N., Min, B., Sapkota, B., & Rangari, V. K. (2020). Eco-friendly cellulose nanofiber extraction from sugarcane bagasse and film fabrication. *Sustainability, 12*(15), 6015.

Shtarkman, B. P., & Razinskaya, I. N. (1983). Plasticization mechanism and structure of polymers. *Acta Polymerica, 34*(8), 514–520.

Soni, B., & Mahmoud, B. (2015). Chemical isolation and characterization of different cellulose nanofibers from cotton stalks. *Carbohydrate Polymers, 134*, 581–589.

Storz, H., & Vorlop, K. D. (2013). Bio-based plastics: Status, challenges and trends. *Applied Agricultural for Research*, *63*, 321–332.

Sulaiman, S., Cieh, N. L., Mokhtar, M. N., Naim, M. N., & Kamal, S. M. M. (2017). Covalent immobilization of cyclodextrin glucanotransferase on kenaf cellulose nanofiber and its application in ultrafiltration membrane system. *Process Biochemistry*, *55*, 85–95.

Supian, M. A. F., Amin, K. N. M., Jamari, S. S., & Mohamad, S. (2020). Production of cellulose nanofiber (CNF) from empty fruit bunch (EFB) via mechanical method. *Journal of Environmental Chemical Engineering*, *8*(1), 103024.

Syafiq, R. M. O., Sapuan, S. M., Zuhri, M. Y. M., Othman, S. H., & Ilyas, R. A. (2022). Effect of plasticizers on the properties of sugar palm nanocellulose/cinnamon essential oil reinforced starch bionanocomposite films. *Nanotechnology Reviews*, *11*(1), 423–437.

Tian, H., Li, Z., Lu, P., Wang, Y., Jia, C., Wang, H., Liu, Z., & Zhang, M. (2021). Starch and castor oil mutually modified, cross-linked polyurethane for improving the controlled release of urea. *Carbohydrate Polymers*, *251*, 117060.

Vieira, M. G. A., Da Silva, M. A., Dos Santos, L. O., & Beppu, M. M. (2011). Natural-based plasticizers and biopolymer films: A review. *European Polymer Journal*, *47*(3), 254–263.

Whistler, R. L., & Daniel, J. R. (1984). Molecular structure of starch. In *Starch: Chemistry and technology* (pp. 153–182). Elsevier.

Żołek-Tryznowska, Z., & Kałuża, A. (2021). The influence of starch origin on the properties of starch films: Packaging performance. *Materials*, *14*(5), 1146.

10 Biopolymers for Drug Delivery Applications
Modifications and Performance

10.1 INTRODUCTION

Biopolymers are compounds made of polymers that exist in living organisms like animals, plants, bacteria, and algae (Abdul Khalil et al., 2023). Biopolymers derived from renewable sources possess several advantageous features, including abundance in nature, biocompatibility, affordability, and complete degradation without leaving any harmful residues. These desirable qualities make biopolymers an excellent choice for biomedical applications. Biopolymers can be classified into polysaccharides, polypeptides, and polynucleotides based on the units they are composed of (Biswas et al., 2021). Typically, biopolymers are obtained from living organisms or waste materials through various chemical and modification processes. The physical and chemical properties of biopolymers, including their hierarchical assembly, are significantly influenced by the conditions under which they are processed (Russo et al., 2021). Hierarchical assembly of biopolymers contributes to their exceptional characteristics, such as high mechanical strength and super-hydrophobicity. Nevertheless, achieving materials that replicate hierarchical assembly using extracted biopolymers is a complex task (Zhang et al., 2021). Extracted biopolymers find extensive use in diverse applications like tissue engineering, delivery of therapeutic molecules, and food packaging. Regenerative tissue engineering aims to enhance the functionality of damaged tissues or malfunctioning organs by harnessing the body's inherent healing capabilities.

In the biomedical field, both synthetic and bio-polymers are utilized as scaffolds for cells to create an environment that promotes cell proliferation and the healing of damaged tissue. Biopolymers are particularly interesting due to their biocompatibility, biomimetic chemical entities, biodegradability, low toxicity, and release of non-toxic fragments during degradation (Veeman et al., 2021). Moreover, the structures of biopolymers resemble those of extracellular matrix macromolecules, allowing them to be compatible and function effectively within the host. Natural polymers derived from plant carbohydrates (e.g., chitosan, starch, carrageenan) and plant or animal proteins (e.g., whey protein, soy protein, collagen) are extensively employed in drug delivery applications, such as encapsulation, matrix systems, beads, scaffolds, nanoparticles, and liquid formulations (Pirsa & Aghbolagh Sharifi, 2020). These biopolymers serve as matrix materials, drug release modifiers, viscosity

modifiers, binding agents, film coating substances, disintegrating agents, solubilizing agents, emulsifying agents, suspending agents, gelling agents, and bioadhesives
(George & Suchithra, 2019). Starch-based biodegradable polymers are suitable for
drug delivery in the form of microcapsules, microspheres, implants, or hydrogels,
eliminating the need for surgical removal after drug depletion. Due to their biodegradability and nontoxicity, biopolymers have drawn much attention used in liver
and cardiac tissue engineering and wound healing as hydrogels, powders, and films
in tissue engineering (Sharma et al., 2021). Natural bioorganisms such as bacteria,
fungi, and algae help promote degradation or breakdown via aerobic or anaerobic
processes into small molecules, leaving behind organic by-products such as CO_2 and
H_2O (Garg, 2020).

Synthetic biopolymers like poly(l-lactic acid) (PLLA), poly(e-caprolactone)
(PCL), poly(glycolic acid) (PGA), polyvinyl alchohol (PVA), and poly(butylene
succinate) (PBS) have gained significant attention in the biomedical field (Elzoghby
et al., 2012). Additionally, biopolymers produced through microbial fermentation,
such as microbial polyesters (e.g., poly(3-hydroxybutyrate-co-3-hydroxyvalerate))
and microbial polysaccharides (e.g., curdlan, pullulan), find applications in biomedicine. Synthetic biodegradable polymers like poly(α-hydroxyesters), polyanhydrides,
and polyorthoesters are versatile and widely used for fabricating tissue engineering
matrices (Sultana et al., 2015). Biopolymer-based scaffolds can regenerate damaged
tissues and organs within the host body, promoting effective tissue regeneration
processes for applications in skin, cartilage, vascular systems, and bones, among
others. To be considered promising candidates for biomedical applications, these
scaffolds must meet essential criteria such as biocompatibility, biodegradability,
non-toxicity, non-inflammatory nature, non-immunogenicity, structural integrity,
optimal porosity, and favorable mechanical and physiological properties. During
the tissue regeneration process, scaffolds serve as three-dimensional structures that
mimic the extracellular matrix, providing an appropriate environment for tissue
regeneration. Utilizing biopolymer-based scaffolds can help mitigate the potential
side effects associated with non-biocompatible polymer implants, such as chronic
inflammation, immune reactions, and toxicity (Park et al., 2017; Yahya et al., 2021).
Recent advancements in biotechnology, particularly in cell regeneration with signal
molecules, offer promising techniques for biomedical applications. The development of smart biomaterials has prompted the need for smart biopolymer scaffolds
in biomedicine.

Proteins such as collagen, gelatin, keratin, sericin, and fibroin are utilized in the
fabrication of different formulations like films, Pickering emulsions, hydrogels,
nanogels, nanofibers, interconnected porous scaffolds, and 3D-printed scaffolds
(Figure 10.1) (Bealer et al., 2020). The electrospinning technique shows promise for
scaffold fabrication in tissue engineering applications (Mortimer & Wright, 2017).
These formulations have been explored for the regeneration of hard and soft tissues
such as wounds, bones, cartilage, and nerves. Plant-derived proteins like soy protein
and zein are also extensively studied for biomedical applications, as discussed in this
chapter. Extracted biopolymers often exhibit poor mechanical properties (toughness
and elasticity) and have limited film-forming capabilities. Consequently, the physical

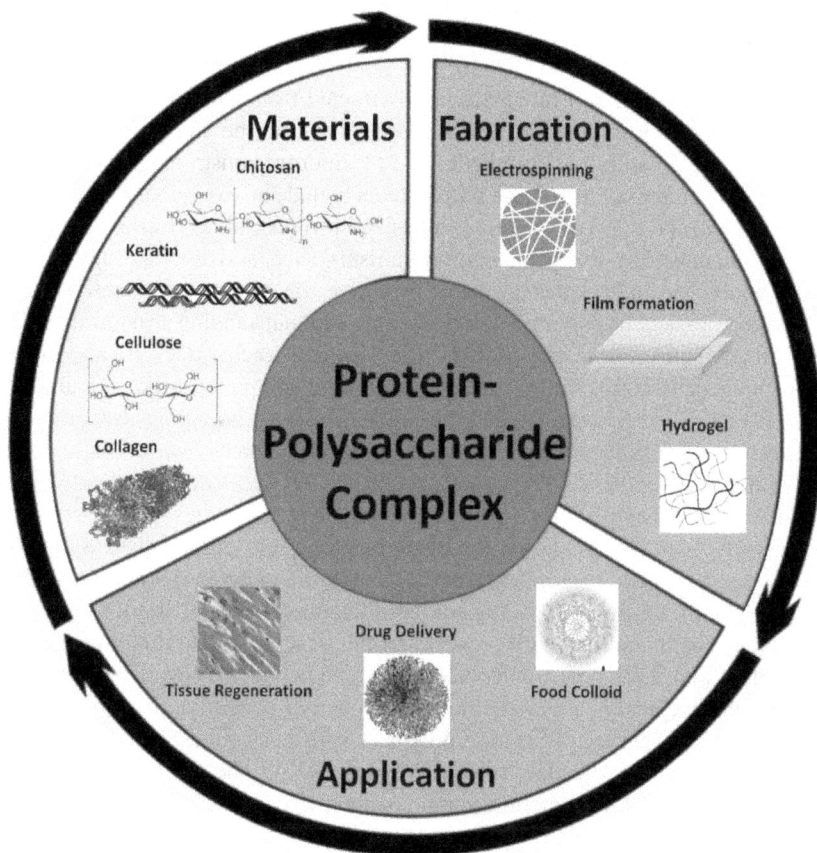

FIGURE 10.1 Biopolymers produced from different sources of proteins and polysaccharides, processed for a wide range of biomedical applications. Produced from Bealer et al. (2020).

properties of biopolymer scaffolds can be enhanced by reinforcing them with synthetic polymers and inorganic nanoparticles (Tabasum et al., 2019). Despite their numerous advantages, biopolymers have certain limitations, such as antigenicity and immunogenicity, which restrict their applications. However, careful optimization of the extraction procedure can eliminate antigenic residues and minimize these disadvantages. Several recent reviews have focused on protein-based polymers, discussing their use in drug delivery and tissue engineering applications. This chapter focuses on the background, crosslinking chemistry, and recent advances in the application of natural-based biopolymer scaffolds in drug delivery applications involving both natural and synthetic biopolymers. The chapter also addresses recent challenges that need to be considered to enhance the usefulness of existing scaffolds in future applications.

10.2 CLASSIFICATION AND CHARACTERISTICS OF BIOPOLYMERS

Depending on the origin of raw materials (extracted from natural resources such as starch, sugar, cellulose, and fossil oil), biopolymers can be categorized into three broad groups: (1) natural, (2) synthetic, and (3) microorganism-based biopolymers. The natural resources of producing biopolymers include microorganisms, plants, and animal tissues, and they can also be synthesized from aerial, terrestrial, and marine living organisms. For example, microorganisms such as bacteria, fungi, yeasts, molds, smuts, and many other forms of primitive life have great potential to offer an enormous variety of polymeric biomolecules with outstanding structural and biochemical attributes. These biopolymers comprise polysaccharides such as cellulose, chitin, chitosan, dextran, chitin, and hyaluronic acid and proteins like silk and keratin. Plants have been a precious and renewable source of both polysaccharides and proteins for a long time. Animals are also a potential source, whether highly developed or not and whether they live on the land, sea, or air. They can provide natural structures with strong potential in the biomedical field. Examples of these structures comprise glycosaminoglycans (chitin, hyaluronic acid, etc.), proteoglycans, and proteins (collagen, elastin, gelatin, etc.), with the addition of deoxyribonucleic acid, the genetic material present in all living sources (Thomas et al., 2020).

Based on the monomeric unit present and the structure, natural biopolymers can be classified into three types: polynucleotides, polypeptides (proteins), and polysaccharides. Nucleic acid is a generic biological material consisting of a large number of molecules in sequence. These are polymers with nucleotides as monomer units. There are two types of nucleic acids, DNA and RNA. These are linear polymers with smaller molecules (monomer units) attached in sequence. Several nucleotides are linearly linked by covalent bonds to form polynucleotides. A nucleotide molecule is composed of three distinct molecules: five-carbon sugar, a phosphate group, and a nitrogenous base. In DNA and RNA polynucleotide biopolymers, a nucleotide monomer has a phosphate group and is bonded to the sugar of the next nucleotide monomer, making a chain with a regular sugar-phosphate group (Biswas et al., 2021). In DNA, the sugar molecule is deoxyribose, but in RNA, it is ribose. Also, DNA biopolymers usually have a double-stranded chain, whereas RNA has linear chains of α-amino acids. A combination of polypeptide molecules results in the formation of proteins (Davidenko et al., 2019). The amino acids are covalently bonded with the help of peptide bonds. As shown in Figure 10.2, the individual amino acid linked in the protein chain is called a residue, and the linked chain of nitrogen, oxygen, and carbon is called the protein backbone. The picture shows how three amino acids are linked by peptide bonds, forming polypeptides. Proteins can be extracted from a wide variety of materials, including wool, leather, silk, gelatin, and collagen. Differences in the sequence of amino acids create varieties of proteins. Proteins have essential roles in living organisms, including immune responses, cell adhesion, and cell signaling.

Polysaccharides are long chains (linear or branched) of monosaccharide units bonded together by glycosidic linkages, but upon hydrolysis, they release the constituents of monosaccharides or oligosaccharides. Examples of polysaccharides include storage polysaccharides such as starch and glycogen and structural polysaccharides

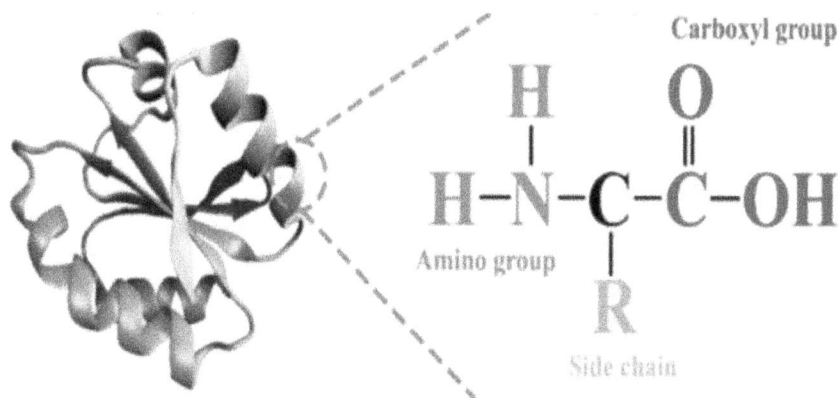

FIGURE 10.2 Structure of protein by the linked chain of nitrogen, oxygen, and carbon forming a protein backbone. Reproduced from Senthilkumaran et al. (2022).

TABLE 10.1
Three Classifications of Biopolymers

Classifications	Sources	Biopolymers
Natural	Extraction from biomass (plant, animal and plant protein, and lipids)	Starch, cellulose, alginate, carrageenan, chitosan, glycerides, waxes, gelatin, whey protein, soy protein, zein, wheat gluten
Synthetic	Biomass and petrochemical	Polylactic acid, polycaprolactone, polyvinyl alcohol, poly(glycolic) acid
Microbial	Microorganism fermentation	Bacterial cellulose, polyhydroxyalkanoates (PHA), polyhydroxybutyrate, poly(3-hydroxy-co-3-hydroxyvalerate) (PHBV)

such as cellulose and chitin (Pattanashetti et al., 2017). Pattanashetti et al. explained smart biopolymeric materials, which have the property to respond to significant changes due to small changes in the environment. They placed them in the category of environmentally sensitive polymers. They classified smart biopolymers into three categories: pH-sensitive smart biopolymers, thermosensitive smart biopolymers, and stimuli-responsive biopolymers (2017).

Riedel and Nickel (1999) classified biopolymers into three categories—polymer chain, polymer, and monomer. Polymer chain biopolymers are naturally synthesized polymers such as carbohydrates, proteins, and polyphenol resin. Polymer category biopolymers are biotechnologically synthesized polymers like polyhydroxybutyrate (PHB) and copolyesters. The third type is naturally synthesized monomer unit polymers such as plant oil and derivatives from sugar. Table 10.1 presents the classification of biopolymers into three categories. Recent technological advancements enable the vast application of biopolymers in the medical field, and these can be categorized into drug delivery applications, wound dressing and healing materials, tissue engineering, biomedical sensors, surgical scaffolds, and others. Controllable drug release in our bodies can be accomplished easily using biodegradable polymer encapsulation. In wound healing, highly biocompatible nonwovens can be used to replace human tissue, and simple sutures, staples, clips, or meshes are also available. Most biopolymers exhibit good film-forming behaviors, making them appropriate in high-performance applications and in traditional commodity uses. Biopolymer-based nonwovens can also be used in agriculture, filtration, hygiene, and protective clothing.

10.3 MODIFICATION TECHNIQUES FOR ENHANCING BIOPOLYMER PROPERTIES

Biopolymers composed of proteins, polysaccharides, and synthetic counterparts are preferred for medical applications. However, they lack the necessary mechanical properties and stability in aqueous environments. The incorporation of crosslinking techniques enhances the properties of these biomaterials; still, most available crosslinkers either negatively impact the functionality of the biopolymers or result in cytotoxic effects. Glutaraldehyde, the commonly used crosslinking agent, is challenging to handle, and conflicting views exist regarding its cytotoxicity when used for crosslinking materials. Recently, poly(carboxylic acids) have emerged as promising alternatives, capable of crosslinking in both dry and wet conditions. These acids have shown the potential to improve tensile properties, enhance stability under aqueous conditions, and facilitate cell attachment and proliferation. To achieve biopolymeric materials with the desired properties for medical applications, the development of green chemicals and innovative crosslinking approaches is imperative (Reddy et al., 2015).

Despite the recognized advantages and broad applicability of biomaterials, there are several limitations that hinder their use in biomedical applications (Butcher et al., 2014). Primarily, biopolymeric materials often lack sufficient mechanical properties and stability in aqueous and physiological environments, which are crucial for medical applications (Jiang et al., 2010). For instance, films and electrospun structures made from proteins tend to disintegrate under high humidity or in aqueous solutions (Hennink & van Nostrum, 2012). To overcome these limitations, crosslinking has emerged as a common approach (Hennink & van Nostrum, 2012). The primary purpose of crosslinking is to ameliorate the biomechanical properties of scaffolds by the formation of a firm network in a polymeric matrix. A crosslink is a physical or chemical bond that connects the functional groups of a polymer chain to another one through covalent bonding or supramolecular interactions such as ionic bonding or hydrogen bonding. Not only should an ideal crosslinker improve the mechanical performance of the polymer network, but it also should have no cytotoxic effects. As a result, crosslinking of polymer chains can affect physicochemical properties, including mechanical properties of the products such as tensile strength, stiffness, and strain, cell-matrix interactions, performance at higher temperatures, resistance to enzymatic and chemical degradation, gas permeation reduction, and shape memory retention. However, crosslinking also reduces degradability, diminishes the availability of functional groups in the crosslinked polymer, alters polymer rheology, and introduces potential processing difficulties and increased cytotoxicity (Madduma-Bandarage & Madihally, 2021).

Table 10.2 presents various types of crosslinkers and crosslinking techniques that are employed, depending on the specific biopolymer and desired property

TABLE 10.2

Assessment of the Cytotoxicity, Mechanical Properties, and Aqueous Stability for Biomedical Applications Based on the Crosslinkers, Crosslinking Methods, and Biomaterials That Have Been Developed

	Proteins						Carbohydrates				
	Gelatin	BSA	Collagen	Zein	Soy protein	Keratin	Cellulose	Starch	HA	Alginate	Chitosan
Glutaraldehyde	✓	✓	✓	✓	✓	✓	✓	✓	✓	✓	✓
EDC/NHS	✓	✓	✓	–	✓	✓	✓	–	✓	✓	✓
Epichlorohydrin	✓	✓	✓	–	–	–	✓	–	✓	–	✓
STMP	–	–	–	–	–	–	✓	✓	✓	–	✓
Citric acid	✓	–	✓	✓	–	–	✓	✓	–	–	✓
Dextran dialdehyde	✓	–	✓	–	✓	✓	–	✓	✓	–	✓
Genipin	–	–	✓	–	✓	–	–	–	✓	–	✓
PA	✓	–	✓	–	–	–	–	✓	–	✓	✓
Glyoxal	✓	✓	✓	✓	✓	✓	–	✓	✓	✓	✓
Chemical	✓	✓	✓	✓	✓	✓	✓	✓	✓	✓	✓
Physical (UV)	✓	–	✓	–	–	–	–	–	–	–	✓
Enzymatic	✓	✓	✓	✓	✓	✓	✓	✓	✓	✓	✓
Films	✓	✓	✓	✓	✓	✓	✓	✓	✓	✓	✓
Sponges, porous scaffolds	✓	✓	✓	✓	✓	✓	✓	✓	✓	✓	✓
Micro, nanoparticles	✓	✓	✓	✓	✓	✓	✓	✓	✓	✓	✓
Hydrogels	✓	✓	✓	–	✓	✓	✓	✓	✓	✓	✓
Electrospun fibers	✓	–	✓	✓	✓	✓	✓	–	✓	✓	✓
Micro fibers	✓	–	✓	–	✓	–	✓	–	–	✓	✓
In vitro cytotoxicity	✓	–	✓	–	–	✓	–	–	✓	✓	✓
In vivo cytotoxicity	–	–	✓	✓	–	✓	–	–	–	✓	✓
Mechanical properties	Weak	Weak	Weak	Weak	Acceptable	Acceptable	Good	Weak	Weak	Weak	Acceptable
Stability	Poor	Poor	Poor	Poor	Weak	Good	Good	Poor	Weak	Weak	Weak

Chemical crosslinking

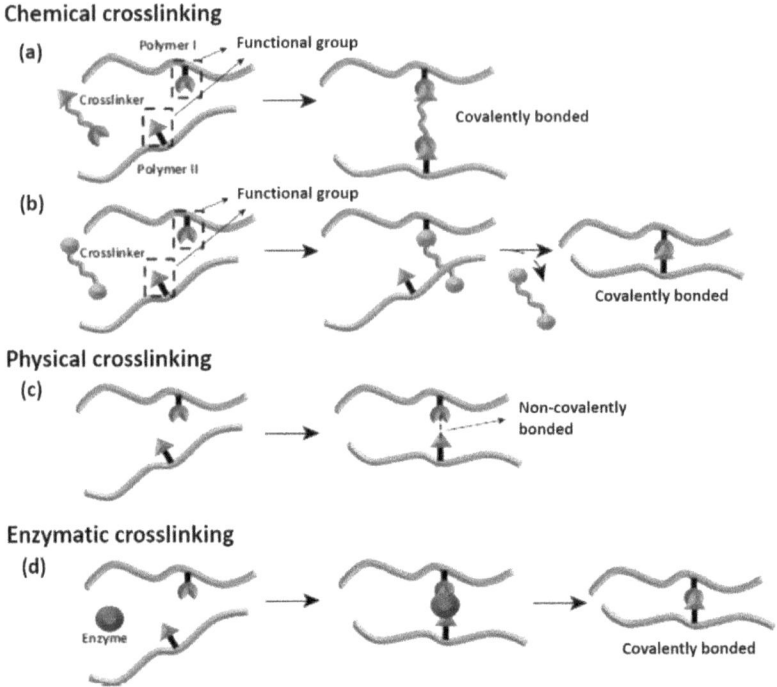

FIGURE 10.3 Illustration of three different approaches to crosslinking. (a) Chemical crosslinking occurs when the crosslinker is integrated into the bond, (b) chemical crosslinking takes place without the crosslinker being part of the bond, (c) physical crosslinking involves non-chemical means, and (d) enzymatic crosslinking is accomplished through enzymatic reactions. Reproduced from Reddy et al. (2015).

improvements, and Figure 10.3 shows the four different crosslinking approaches for biopolymers (Reddy et al., 2015). Among numerous chemical crosslinkers, glutaraldehyde is predominantly utilized because it can react with functional groups in both proteins and carbohydrates, leading to substantial improvements in mechanical performance (Marquié, 2001; Silva et al., 2004). Despite the improvements in mechanical performance, the cytotoxicity of glutaraldehyde-crosslinked materials has conflicting evidence (Lai, 2014; Umashankar et al., 2012). Nonetheless, the cytotoxicity of glutaraldehyde depends on its concentration, and concentrations up to 8% have been shown to be non-cytotoxic (Umashankar et al., 2012). Apart from glutaraldehyde, other chemicals such as carbodiimide, epichlorohydrin, and sodium metaphosphate have been used as crosslinking biopolymers, albeit with limited property enhancements due to their low crosslinking efficiency (Serdiuk et al., 2023). Recently, attempts have been made to employ carboxylic acids, like citric acid, for crosslinking biomaterials to improve mechanical properties and stability without compromising cytocompatibility. Crosslinking biomaterials with citric acid introduces pendant functionality, allowing the formation of ester bonds, resulting in improved hemocompatibility and increased availability of binding sites for bio-conjugation (Salihu et al., 2021; Zhao et al., 2015). This chapter provides an overview of the chemicals and techniques employed to crosslink biopolymeric materials intended for medical applications, with particular emphasis on protein-based biomaterials, which exhibit superior biocompatibility compared to synthetic polymer-based biomaterials but are inherently less stable in aqueous environments and therefore necessitate crosslinking.

10.3.1 GLUTARALDEHYDE

The utilization of glutaraldehyde as a crosslinker for biopolymers in medical applications has been extensive. Nevertheless, conflicting findings have been reported regarding the cytotoxicity of glutaraldehyde-crosslinked biopolymers. Additionally, the handling of glutaraldehyde during the crosslinking process is challenging due to its strong odor and low vapor pressure. The majority of research on glutaraldehyde crosslinking has been conducted through in vitro studies, while in vivo assessment of the crosslinked materials is essential to gain a meaningful understanding of their cytotoxicity and potential for medical applications (Oryan et al., 2018).

10.3.2 POLY(CARBOXYLIC ACIDS)

Poly(carboxylic acids) have the capability to react with hydroxyl and amine groups, enabling crosslinking of both polysaccharides and proteins (Figure 10.4) (Reddy et al., 2015). Crosslinking proteins with carboxylic acids has demonstrated biocompatibility and the desired property enhancements for biomaterials based on both proteins and carbohydrates. Previously, it was conventionally believed that crosslinking with carboxylic acids (with a minimum of three carboxylic groups) occurred only at high temperatures (150–175°C) and in the presence of catalysts. However, recent studies have shown that even carboxylic acids with two carboxylic groups can crosslink biopolymers in both wet and dry conditions, eliminating the need for potentially cytotoxic catalysts. In vitro investigations have confirmed the ability to crosslink fibers, films, electrospun structures, and phase-separated structures using citric acid. Nevertheless, further research employing in vivo approaches is required to establish the biocompatibility and suitability of biomaterials crosslinked with carboxylic acids for medical applications (Song et al., 2016). According to Oryan et al. (2018), despite its biocompatibility, widespread availability, low cost, high degree of crosslinking, and self-mineralization capabilities, citric acid-based crosslinkers typically require a catalyst during the crosslinking process and elevated temperatures for crosslinking, which may occasionally alter the characteristics of polymers.

(a)

Formic acid   Acetic acid   Propionic acid

Butyric acid   Valeric acid

Caproic acid   Benzoic acid

(b)

$$HC-\overset{O}{\underset{\|}{C}}-OH \quad \underset{-H_2O}{\longrightarrow} \quad HC-\overset{O}{\underset{\|}{C}} \diagdown O + HO-R_{cellulose} \longrightarrow HC-\overset{O}{\underset{\|}{C}}-O-R_{cellulose}$$

$$HC-\underset{\|}{C}-OH \qquad\qquad HC-\underset{\|}{C} \diagup \qquad\qquad HC-\underset{\|}{C}-O-H$$

| Polycarboxylic | Cyclic | Polycarboxylic acid bonded to cellulose |
| acid | anhydride | through an ester linkage |

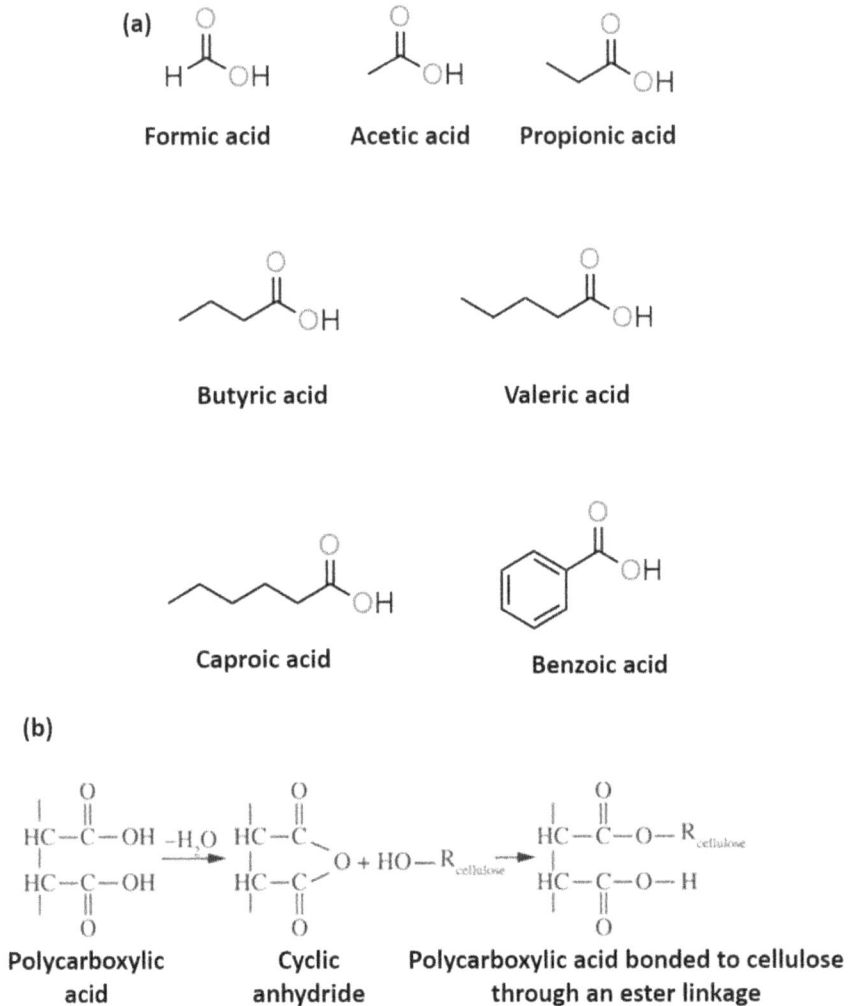

FIGURE 10.4   (a) Common names of polycarboxylic acid and (b) the mechanism of cross-linking of cellulose with polycarboxylic acids.

10.4   CROSSLINKING OF BIOPOLYMERS WITH POLYCARBOXYLIC ACID–BASED CROSSLINKERS

A study was conducted where aldehyde (glutaraldehyde) was combined with a car-boxylic acid (malic acid) and a dendrimer (EDC) to crosslink collagen gels. The results showed that crosslinking reduced enzymatic degradation and enhanced the adhesion and growth of L 929 cells (Saito et al., 2008). Instead of using a combi-nation of crosslinkers, collagen hydrogels crosslinked with dendrimers like EDC exhibited comparable resistance to collagenase degradation as glutaraldehyde (Duan & Sheardown, 2005). Another innovative method involved dual-crosslinking (using butanedioldiglycidyl ether as the crosslinking agent) and photopatterning to create hydrogels from hyaluronic acid with anisotropic swelling. These hydrogels demonstrated the ability to swell significantly while maintaining their structure, making them potentially valuable for applications in ophthalmology, wound healing, and other medical fields (Ahmed, 2015). Similarly, cellulose-based hydrogels exhibited controllable swelling (up to 720%) by adjusting the crosslinking conditions using 1,2,3,4-butanetetracarboxylic acid dianhydride, a carboxylic acid with four carboxyl groups (Kono & Fujita, 2012).

Plant proteins like wheat gluten and soy proteins have also been converted into fibers for potential use as scaffolds in tissue engineering and drug delivery (Samrot et al., 2023). While these fibers possess good mechanical properties in dry conditions, their stability in aqueous environments is inherently poor. Therefore, they have been crosslinked using carboxylic acids and other crosslinkers. By crosslinking the fibers with carboxylic acids, highly water-stable protein fibers were achieved. The crosslinking of proteins with carboxylic acids was found to follow a pseudo-first-order reaction, allowing control over the mechanical properties and degradation rates of the resulting fibers (Reddy et al., 2009a, 2009b).

To address the challenges associated with using aqueous solutions for crosslinking electrospun fibers, electrospun collagen scaffolds were crosslinked by exposure to glutaraldehyde vapor (Laha et al., 2016; Sisson et al., 2009). The crosslinked scaffolds exhibited increased tensile strength and resistance to collagenase degradation but had reduced porosity. However, contradictory findings regarding the cytotoxicity of glutaraldehyde-crosslinked fibers have been reported (Mekhail et al., 2011; Suesca et al., 2017; Yang et al., 2021). Citric acid has been used as a biocompatible alternative crosslinker for electrospun collagen fibers to overcome the cytotoxicity issue associated with glutaraldehyde (Jiang et al., 2010; Jiang et al., 2013). Due to limited available functional groups, a crosslinking extender was required to facilitate crosslinking with citric acid. The addition of glycerol, which possesses numerous hydroxyl groups, aided in crosslink formation, thereby improving the strength and stability of the electrospun fibers (Jiang et al., 2013). Similarly, electrospun matrices made from zein, a protein, exhibited weak tensile properties and rapid dissolution in aqueous solutions. Crosslinking these electrospun zein matrices with varying concentrations of citric acid resulted in matrices that maintained their structure even after 15 days of incubation in PBS at 37°C. Furthermore, the citric acid-crosslinked zein samples demonstrated higher attachment and proliferation rates of fibroblasts compared to electrospun PLA scaffolds (Jiang et al., 2010). Electrospun structures derived from chitosan, starch, and other polysaccharides have also undergone crosslinking to enhance their strength and stability. For instance, electrospun chitosan fibers (143–334 nm) were crosslinked with glycerol phosphate (GP), tripolyphosphate (TPP), and tannic acid (TA) (Kiechel & Schauer, 2013). These crosslinked fibers remained insoluble even in 1 M acetic acid after a 72-hour immersion period, although their cytocompatibility was not assessed. In contrast to conventional post-electrospinning crosslinking methods, in situ crosslinking of pullan/dextran mixtures with trisodium metaphosphite (STMP) was demonstrated by Shi et al. (2011).

This crosslinking approach reduced swelling and promoted the viability of human dermal fibroblasts, with observed actin stress fiber formation, suggesting potential applications in tissue engineering. However, in situ crosslinking is not feasible with most crosslinkers or polymers and may result in undesirable changes in material properties and decreased electrospinnability (Shi et al., 2011).

The citric acid crosslinked carboxymethyl cellulose hydrogels were successfully fabricated by a vacuum drying process through an esterification reaction (Aswathy et al., 2023). This study comparatively described the influence of polymer molecular weight and degree of substitution of the carboxymethyl group on crosslinking reaction for the first time. As compared to lower citric acid–concentration crosslinked hydrogels, the swelling ratio and moisture content are lower in a higher concentration of citric acid–crosslinked hydrogels. From these results, it is found that only 7%, 9%, and 10% citric acid concentration of 2carboxymethl cellulose and 7carboxymethyl cellulose samples showed good physicochemical and mechanical properties, and they were used for biological assays. Comparatively, 7carboxymethl cellulose hydrogel has more crosslinking density than 2carboxymethl cellulose hydrogel. All the hydrogels exhibited good cytocompatibility and hemocompatibility, but the 2carboxymethl cellulose hydrogel was not stable in cell culture media for further experiments. The data on the percentage of hemolysis for citric acid–crosslinked carboxymethyl cellulose hydrogels reveals that all the hydrogels exhibited a hemolysis value below 5%. According to the ISO standard (ISO–10,993–4), a material is regarded as safe if its hemolysis index is <5%. Hence, all hydrogels are hemocompatible and might be employed for tissue engineering applications based on the hemolysis value that was determined.

The vacuum drying process was utilized to successfully create hydrogels of carboxymethyl cellulose crosslinked with citric acid through an esterification reaction [32]. The study presents a comparative analysis of how the molecular weight of the polymer and the degree of substitution of the carboxymethyl group influence the crosslinking reaction. Hydrogels crosslinked with higher concentrations of citric acid exhibited lower swelling ratio and moisture content compared to those crosslinked with lower concentrations. Based on the findings, it was determined that hydrogels containing 7%, 9%, and 10% citric acid concentrations in 2-carboxymethyl cellulose and 7-carboxymethyl cellulose samples displayed favorable physicochemical and mechanical properties, making them suitable for biological assays. Among them, the 7-carboxymethyl cellulose hydrogel exhibited a higher crosslinking density than the 2-carboxymethyl cellulose hydrogel. All the hydrogels demonstrated good cytocompatibility and hemocompatibility, with the exception of the 2-carboxymethyl cellulose hydrogel, which was not stable in cell culture media for further experimentation. The hemolysis value of the citric acid crosslinked carboxymethyl cellulose hydrogels was found to be below 5%, indicating compliance with the hemolysis safety standard (ISO–10,993–4). Therefore, all the hydrogels are considered hemocompatible and have potential for use in tissue engineering applications based on their determined hemolysis value. Due to its possession of three carboxylic (COO-) groups and a hydroxyl (-OH) group, citric acid actively engages in hydrogen bonding interactions with other polymer networks, thereby enhancing their properties. Moreover, citric acid plays a significant role in cellular metabolism, as it serves as an intermediate product in the tricarboxylic acid cycle (Wang et al., 2023), which exhibits potential for various biomedical applications.

The hydrogels' *in vitro* cytocompatibility was evaluated using L929 mouse fibroblast cells. Extracts from the selected hydrogels were applied to the cell culture for a duration of 72 hours. The viability of the cells was determined using the MTT assay, which measures the metabolic activity of the cells. The results revealed that the percentage of viable cells was similar to the control group, with some hydrogels even exhibiting higher viability. Additionally, all the hydrogels stimulated cell viability and proliferation throughout the 72-hour culture period. Furthermore, investigations utilizing the Saos-2 osteoblast cell line demonstrated that the citric acid crosslinked carboxymethyl cellulose porous scaffold was cytocompatible and facilitated cell proliferation (Priya et al., 2021).

10.5   FABRICATION PROCESS OF BIOPOLYMERS

The choice of fabrication technique for designed biopolymer materials depends on their intended use. For instance, if a biocomposite is intended to deliver an anti-cancer component to the colon, it is crucial to develop components with a specific shape and structure that can withstand the mouth, stomach, and small intestine but degrade in the human system. Various methods are employed for preparing biopolymer-based drug delivery systems, including supercritical fluid extraction (Di Capua et al., 2017), desolvation (Pandey et al., 2020), electrospraying (Kurakula & Naveen, 2021), spray-drying (Oh et al., 2009), layer-by-layer self-assembly (Madrigal-Carballo et al., 2010), freeze-drying (Müller et al., 2013), and microemulsion (Chen et al., 2018).

Each method has its own advantages and disadvantages. Desolvation using solvents like ethanol or acetone is the simplest method for fabricating protein-based nanoparticles. In this method, nanoparticles are obtained in a turbid form by adding solvent to an aqueous mixture under constant stirring. The nanoparticle size can be controlled by adjusting the flow rate and volume of the desolvating agent. Electrospray technique is a gentle, single-step, and versatile method that utilizes electrostatic forces to break up a dielectric liquid into particles ranging from nanometers to micrometers in size from a macroscopic mass. Freeze-drying is suitable for temperature and pressure-sensitive compounds, where solvents are sublimed to yield dried porous particles. However, it is time-consuming and expensive and typically results in larger particle sizes. The layer-by-layer self-assembly method allows for the sequential formation of multilayer films through electrostatic, hydrophobic interactions and hydrogen bonding between layers. These interactions enable the deposition of alternating layers of biomaterials with opposite charges, providing precise control over nanoscopic features such as thickness, surface characteristics, and composition of the film. Microemulsions are prepared by dispersing biopolymers in two immiscible liquids in the presence of emulsifiers or surfactants. Nanomaterials obtained through this method are typically optically transparent, isotropic, and thermodynamically stable and have high drug solubilizing capacities. Despite challenges in their preparation, biopolymeric nanoparticles have the ability to transport bioactive compounds to target tissues, cells, and cell compartments. Although they exhibit limitations such as low drug-loading capacity and wide size distribution, biopolymer-based nanoparticles can be customized to specific sizes, making them a fascinating tool for researchers as long-lasting therapeutic agents (Bealer et al., 2020; Jones & McClements, 2010).

Another method widely used at the laboratory or industrial scale is the solvent casting method, also known as the solution function or wet processing method, which consists of creating an aqueous or hydroalcoholic mixture with the presence of the biopolymer. This method is based on the use of a solvent that allows the polymer to be suspended in a film-forming solution followed by solvent evaporation and polymer chain reformulation (Flórez et al., 2023). Alcohol, water, or other organic solvents are commonly used to dissolve the selected polymer. Sometimes, for the best results, the suspension polymer solution is heated or the pH adjusted. The polymer–solvent mixture is poured into a mold, drum, or flat surface, where it is left to dry for a specific time. Once the solvent is completely evaporated, the polymeric matrix is formed and can be peeled off from the mold. Casting is a simple method. However, there are a series of requirements to consider when applying the technique. One of the most important points is the solvent chosen to dissolve the polymer. If the polymer–polymer attraction forces in a solution are weaker than the polymer–solvent interactions, the chain segment will be stretched by solvent molecule diffusion. This results in swelling of the polymer matrix with the solvent. However, it should be noted that the dissolution capacity of a polymer varies depending on the solvent; hence the choice of solvent is an important point (Liu et al., 2015).

For a coating technique, the molecular weight of the polymer plays a significant role in this process and is an important consideration. It affects the rate at which the solvent can penetrate the polymer. Research has shown that higher molecular weight polymers dissolve more slowly compared to those with lower molecular weights (Zhang et al., 2001). This can be attributed to the fact that high molecular weight polymer chains have a slower rate of relaxation due to increased entanglement. As a result, they are unable to contract rapidly during cooling, creating a larger free volume. Consequently, higher molecular weight polymers exhibit a greater solvent penetration rate Additionally, these biopolymers possess sufficient cohesive strength and coalescence capacity. Another crucial requirement is that the polymer should be soluble in a volatile solvent or water. To achieve optimal results, it is necessary to create a stable solution with suitable viscosity (Olsen et al., 2003). The temperature and humidity of the environment are also critical factors for the successful development of the process (Abou Neel et al., 2013).

The solubility of biopolymers and the viscosity of the final solution are dependent on the molecular weight of the polymer. Higher molecular weight polymers result in more viscous solutions but also decrease the solubility of the polymer. Moreover, the creation of a coating that adheres well to the surface of the final product requires an estimation of the interfacial tension between the coating solution and the product surface. To ensure compatibility between the surface and the solution, it is beneficial to reduce the surface tension of the coating solution. This helps decrease the interfacial tension and enhances adhesion between the product and the coating (Jones & McClements, 2011).

One advantage of the coating method is that it does not require large amounts of substrate, since the layers created are usually only a few micrometers thick. It is also a simple and cost-effective technique compared to other polymer production methods. Additionally, it exhibits high adhesion capacity when coating complex geometries. However, despite its numerous benefits, the coating method has certain drawbacks that can compromise its reliability. For example, it may be susceptible to negative thermal effects such as cracking or delamination; lack of atmospheric protection leading to the penetration of contaminants into the substrate; and limitations related to the coating materials, such as different melting points and availability in various forms (sheets/powders), as well as biocompatibility between the components of the coating solution, among others (Chen et al., 2002).

10.6   BIOPOLYMERS IN DRUG DELIVERY APPLICATIONS

The primary requirement for any drug delivery system is the controlled release of a therapeutic agent in a customizable dosage, which can be achieved through successful conjugation between the therapeutic agent and a drug delivery vehicle (Nitta & Numata, 2013). However, the coupling of a drug with a delivery agent relies on the chemical structure of the drug molecules, which can be either hydrophilic or hydrophobic. Therefore, rational design of a drug delivery agent is crucial for successful conjugation (Larson & Ghandehari, 2012). Additionally, the drug carrier should prevent enzymatic degradation of the therapeutic agent prior to its controlled release. Following successful delivery, the drug carrier must possess non-toxicity and biodegradability and ideally be eliminated from the body without any adverse effects (Guadarrama-Escobar et al., 2023; Mustafai et al., 2023). Hybridizing biopolymers with nanoparticles enables the transportation of bioactive compounds to specific tissues, cells, and cell compartments. However, the preparation of such hybrids presents challenges, including limited drug-loading capacity and a wide size distribution. Despite these issues, biopolymer-based nanoparticles can be modified to achieve customized sizes, resulting in long-lasting sustainability. This characteristic makes them highly intriguing as therapeutic agents for researchers. Figure 10.5 illustrates a schematic diagram showcasing the process of drug loading in biopolymers and their mechanism of drug delivery.

In general, polymeric materials are excellent candidates for drug delivery agents. However, biopolymers and their derivatives exhibit superior properties such as ease of functionalization, water solubility, non-toxicity, biodegradability, and biocompatibility (Idrees et al., 2020; Janmohammadi et al., 2023). Moreover, these biopolymers

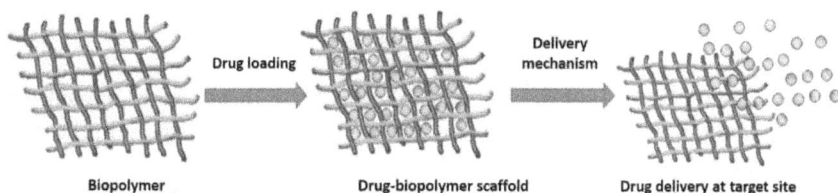

Biopolymer   Drug-biopolymer scaffold   Drug delivery at target site

FIGURE 10.5   Schematic diagram on the process of drug loading in biopolymers and their mechanism of drug delivery. Reproduced from Jacob et al. (2018).

can reduce drug toxicity through controlled release and prevent enzymatic degradation prior to releasing the therapeutic agent at the target site. The release of the therapeutic agent can be triggered by various stimuli such as physical (sonophoresis, temperature, light, magnetic or electric fields), chemical (biochemical, hypoxia, or pH), or environmental factors or a combination of multiple stimuli. In this context, biopolymers hold promise as drug delivery agents. Polysaccharides, polypeptides, and proteins are the three major classes of biopolymers extensively used in drug delivery systems.

Polysaccharides are carbohydrate molecules composed of long-chain monosaccharide or disaccharide units linked by glycosidic bonds and are high molecular weight derivatives of monosaccharides with repeating units. They are derived from plants (starch, hemicellulose, cellulose, agar, glucomannan, pectin, guar gum, and gum acacia), microbes (curdlan, gellan, dextran, xanthan), algae (alginate, carrageenan), and fungi (chitin, pullulan, scleroglucan) (Singh et al., 2019). Most polysaccharides are either neutral or possess a negative surface charge, with chitin being an exception, as it can have cationic polysaccharides. The functionality of polysaccharide molecules is a significant advantage, as it allows for control over the overall drug delivery systems. Common drug delivery systems include micro/nanocapsules, micro/nanospheres, polymer-drug conjugates, micelles, liposomes, and hydrogel formations, where therapeutic agents are dispersed or encapsulated within the biopolymer matrix (Barclay et al., 2019; Coelho et al., 2010). Grafting of smart polymers onto the biopolymer surface or forming biopolymer composites is also employed in drug delivery systems (Galaev & Mattiasson, 1999). Composite systems achieve drug diffusion rate control, inhibition of enzymatic degradation, alteration of pH-responsive range, or hydrogel formation. For instance, grafting bacterial nanocellulose with poly(acrylic acid) hydrogels demonstrated extended pH-responsive drug delivery compared to pristine bacterial nanocellulose due to changes in surface chemistry (Pötzinger et al., 2017). Additionally, composite hydrogel systems, such as alginate in bacterial nanocellulose scaffolds, exhibit a prolonged drug release rate due to intermolecular drug–drug interactions (Xue et al., 2017).

An instance of this is the fabrication of hydrogels by combining chitosan with polyethylene glycol (PEG) in the presence of a silane crosslinker. This hydrogel exhibits pH responsiveness and proves valuable in the realm of controlled drug delivery systems (Atta et al., 2015). Hydrogels composed of cellulose nanocrystals and gelatin also demonstrate remarkable sensitivity to pH, displaying their maximum swelling capacity at pH 3 (Ooi et al., 2016). Similarly, pH-responsive nanohydrogels are synthesized using tragacanth gum, a biopolymer, along with 3-aminopropyltriethoxysilane as a modifier and glycerol diglycidyl ether, PVA, and glutaraldehyde as crosslinkers (Hosseini et al., 2016). Calcium, ethylcellulose, and hydroxypropylmethyl cellulose can be employed to produce pectin-derived drug carriers, which hold considerable promise for targeted drug delivery in the colon (Liu et al., 2003).

Chitosan exhibits solubility in mildly acidic solutions while being insoluble in water and organic solvents. The solubility of chitosan in acidic conditions enables the preparation of natural hydrogels by utilizing modified chitin, such as partially deacetylated α-chitin nanofibers, which prove effective for drug delivery systems

(Xu et al., 2018). Chitosan is primarily derived from marine crustacean shells like crabs, shrimps, lobsters, and prawns in the form of chitin. Through alkaline deacetylation, chitin can be structurally modified into chitosan. Chitosan possesses high biocompatibility and biodegradability, making it highly valuable for pharmaceutical applications due to its non-toxicity, capacity for high drug loading, mucoadhesive properties, and high charge density. Its positively protonated amino groups facilitate interactions and enhance the bioavailability of drugs and their release in drug delivery systems. For instance, strawberry extract polyphenols, which carry a negative charge, can be effectively encapsulated within chitosan, and sustained release of polyphenol-loaded chitosan-tripolyphosphate complexes was observed *in vitro* at pH 7.4 (Pulicharla et al., 2016). Chitosan nanoparticles loaded with curcumin, synthesized through ionic gelation using sodium tripolyphosphate and barium chloride, exhibited a size below 500 nm. These curcumin-loaded nanoparticles demonstrated favorable drug release kinetics and sustained drug delivery against *Pseudomonas aeruginosa* in vitro (Samrot et al., 2018). In another study, a hybrid nanoparticle vehicle composed of vanillin-chitosan-calcium ferrite was prepared via ionic gelation, and the release of curcumin was investigated under various pH conditions and magnetic fields. The drug release mechanism was well-suited to the Higuchi model in most scenarios (Kamaraj et al., 2018).

In a recent study, thiolated silica nanoparticles coated with hydroxyethylcellulose (HEC) were synthesized through the self-assembly of 3-mercaptopropyltrimethoxysilane in an HEC solution. The size of the nanoparticles was found to be directly influenced by the amount of HEC present in the reaction mixture, consequently affecting the number of thiol groups on their surface. This information proves valuable for optimizing the size of the fabricated nanoparticles during synthesis, thereby aiding in the design of potential drug delivery systems (Mansfield et al., 2018). Another approach involved the copolymerization of bacterial cellulose obtained from coconut gel syrup with gelatin to form a hydrogel. Glutaraldehyde was utilized as a crosslinking agent in this process, resulting in a hydrogel nanocomposite that holds promise as a drug delivery system (Treesuppharat et al., 2017). Likewise, the drug loading capacity of bacterial cellulose was enhanced by incorporating graphene oxide (GO). This novel nanocomposite of bacterial cellulose implanted with GO demonstrated efficacy as a drug carrier in the treatment of drug delivery systems (Luo et al., 2017).

Dialdehyde nanocellulose (DACNC) and chitosan (CS) were used to fabricate cellulose nanocrystal/chitosan (CNC/CS) composites with varying weight ratios (Figure 10.6) (Xu et al., 2019). The DACNC were mixed with chitosan at the mass ratio of 1:1, 1:2, and 1:5, corresponding to CNC/CS1, CNC/CS2, and CNC/CS3. Fourier transform infrared spectroscopy (FTIR) analysis confirmed the successful reaction between the aldehyde groups on DACNC and the amino groups on chitosan. The isoelectric point and swelling ratio of the composites increased as the percentage of chitosan in the formulation increased. *In vitro* drug release studies were conducted using theophylline as a model drug. The cumulative release of the drug at pH 1.5 was significantly higher than at pH 7.4. Among the CNC/CS composites, CNC/CS3 exhibited the highest cumulative release, with approximately 85% at pH 1.5 and 23% at pH 7.4. The CNC/CS composites demonstrated excellent controlled drug release properties, making them potential candidates for gastric-specific

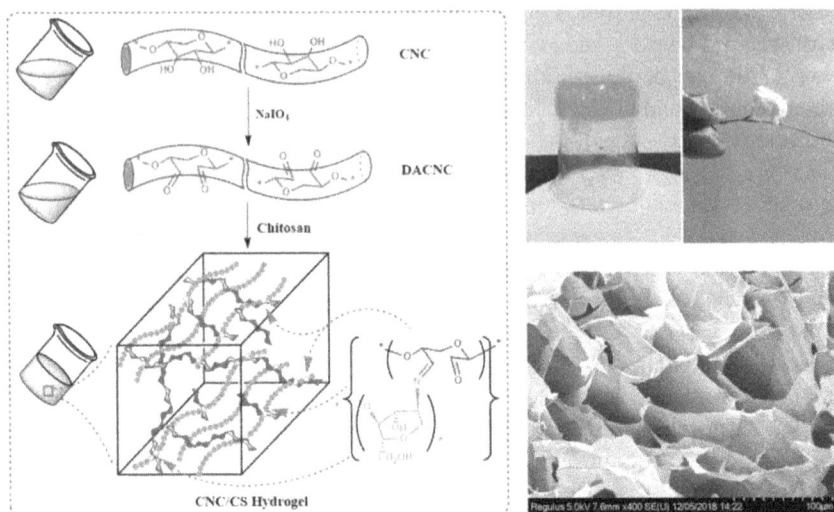

FIGURE 10.6 Cellulose nanocrystal/chitosan hydrogel for controlled drug release. Reproduced from Xu et al. (2019).

drug delivery. Further research could explore the development of new composites utilizing DACNC and other biocompatible natural polymers like gelatin, focusing on tailoring the porosity and drug release rates by adjusting the oxidation degree of DACNC and its loading in the composites. The encapsulation efficiency of theophylline in hydrogels was evaluated and is presented in Table 10.3, with CNC/CS3 showing the highest encapsulation efficiency, while CNC/CS1 exhibited the lowest. This variation in encapsulation efficiency may be attributed to the increasing swelling ratio associated with higher chitosan percentages in the composites.

A drug delivery system with minimal side effects was developed using starch-derived pullulan and a green chemistry approach. The system involved the incorporation of gold nanoparticles into pullulan, which were then loaded with 5-fluorouracil (5-Fu@AuNPs) and 5-Fu with folic acid (5-Fu@AuNPs-Fa). The cytotoxicity of these formulations was evaluated in vitro. Interestingly, the amount of 5-Fu required for 5-Fu@AuNPs-Fa to inhibit HepG2 cells by 50% was lower compared to free 5-Fu and 5-Fu@AuNPs. These findings suggest the potential of this system for cancer imaging and as a tool for drug delivery (Ganeshkumar et al., 2014). In another study, a hydroxyethyl starch-doxorubicin (HES-SS-DOX) conjugate was developed for anticancer drug delivery. This scaffold exhibited redox sensitivity and had an average diameter of approximately 20 nm. In vitro studies demonstrated the promising potential of this conjugate as a prodrug of doxorubicin (Hu et al., 2016). Nanoparticles composed of octenyl succinic anhydride maize starch were synthesized with an average diameter of 86.69 nm. These nanoparticles were crosslinked with epichlorohydrin using a microemulsion crosslinking reaction. The drug delivery mechanism was investigated using indomethacin as a model drug. Starch-based nanoparticles were prepared in a microemulsion system with [C3OHmim]Ac-in-oil, and their drug release behavior was studied (Qi et al., 2017). Starch hydrogel can serve as a vehicle for oral drug delivery of probiotic bacteria, specifically *Lactobacillus plantarum*. Encapsulation within starch hydrogel enables the probiotic bacteria to withstand the adverse conditions of the gastrointestinal tract and bile salt solution, providing enhanced resistance compared to non-encapsulated cells (Dafe et al., 2017).

10.7 RELEASE CHARACTERISTICS OF BIOPOLYMERS

A biopolymer particle can be specially designed to contain, shield, and release a specific functional element, such as a flavor, antimicrobial substance, antioxidant, or bioactive nutrient. Consequently, it may be necessary to tailor its design to release the active component at a specific location, such as the mouth, stomach, small intestine, or colon. To develop a mechanistic model for such processes, it is crucial to understand the underlying physicochemical mechanisms that govern the release. According to Jones and McClements (2010), four main mechanisms have been identified, differing primarily in the role played by the carrier particle in controlling the release of the biopolymer for drug delivery:

1. Diffusion: In this mechanism, the active component diffuses through the intact biopolymer particle matrix into the surrounding medium. The rate of mass transport depends on the substance's solubility in the particle matrix and its diffusion coefficient within the matrix. For biopolymer networks, the diffusion rate may also be influenced by the mesh size of the network relative to the size of the active component and any specific interactions between the network and the active component (such as electrostatic or hydrophobic attraction).
2. Erosion: The active component is released into the surrounding medium due to erosion processes occurring either at the outer layer or throughout the entire volume of the biopolymer matrix. Matrix erosion can result from physical, chemical, or enzymatic degradation processes, such as the dissociation of physical bonds (electrostatic, hydrophobic, or hydrogen bonds) or the chemical or enzymatic hydrolysis of covalent bonds.
3. Fragmentation: The active component is released into the medium as a result of physical disruption of the carrier, which may undergo fragmentation or fracturing under shear or compression forces. Although the bioactive substance still diffuses out of the particles, the release rate is accelerated due to the increased surface area and reduced diffusion path.
4. Swelling/Shrinking: Core release can be triggered by the absorption of solvent by the biopolymer particles, causing them to swell. For instance, an active component could be encapsulated within a solid biopolymer particle or a hydrogel biopolymer particle with a pore size small enough to prevent its leaching. Once the particle absorbs solvent molecules, it swells, enabling the active component to diffuse out. The active component could be loaded into the biopolymer particles by initially swelling them in its presence and then changing the solution conditions to induce shrinkage.

Mathematical theories have been developed to model various release mechanisms involving particulate systems (Pothakamury & Barbosa-Cánovas, 1995). Selecting an appropriate mathematical model requires an understanding of the physicochemical origins of the release mechanism, whether it involves diffusion, erosion, swelling, or fragmentation. To utilize these mathematical models, information about the structure and physicochemical properties of the biopolymer particles and the surrounding medium is essential. This includes details such as the initial particle size distribution, the concentrations of the active component within the particle and surrounding medium (equilibrium partition coefficients), and the rate at which the active component moves through the system (translational diffusion coefficients).

10.8 CONCLUSION

This chapter highlights the potential of biopolymers in biomedical applications, emphasizing their characteristic properties such as mechanical, thermal, compatibility, and optical properties. State-of-the-art processing methods for smart biopolymer fabrication are discussed, along with an overview of diverse biomedical applications. Various synthetic routes for biopolymer preparation and their significant parameters are compiled for multiple applications. It is suggested that the development and enhancement of biopolymer materials through functionalization or hybridization with different functionalities is crucial for targeted biomedical applications. For instance, blending biopolymers with other biodegradable polymers can yield tailor-made polymeric systems with desired physical properties and biodegradability. The preparation of bionanocomposites is identified as a promising approach to enhance polymeric properties while maintaining biodegradability. However, more research is needed to establish reliable drug delivery systems that address unresolved medical needs and consider the mechanism and time required for delivery to specific tissues or cellular compartments. Future advances in implementing biopolymer systems in therapeutic applications, including scalable production with controlled and targeted properties, hold great promise. Efficient processing methods for harvesting such biopolymer systems should be prioritized, particularly for biological systems. Integrating computational models to interpret the effects on material profiles, such as degradation and intra- and extra-cellular trafficking of polymeric nanoparticles, can further facilitate commercialization. In conclusion, intensive research on the surface and functionalization of biopolymers, such as nanocellulose, is essential to revolutionize biomedical materials. Modifying the chemical composition or surface characteristics can significantly impact the interaction between cells and materials, leading to the design of more efficient biopolymers for biomedical applications.

REFERENCES

Abdul Khalil, H. P. S., Yahya, E. B., Jummaat, F., Adnan, A. S., Olaiya, N. G., Rizal, S., Abdullah, C. K., Pasquini, D., & Thomas, S. (2023). Biopolymers based aerogels: A review on revolutionary solutions for smart therapeutics delivery. *Progress in Materials Science*, *131*, 101014.

Abou Neel, E. A., Bozec, L., Knowles, J. C., Syed, O., Mudera, V., Day, R., & Hyun, J. K. (2013). Collagen—emerging collagen based therapies hit the patient. *Advanced Drug Delivery Reviews*, *65*(4), 429–456.

Ahmed, E. M. (2015). Hydrogel: Preparation, characterization, and applications: A review. *Journal of Advanced Research*, *6*(2), 105–121.

Aswathy, S. H., NarendraKumar, U., & Manjubala, I. (2023). The influence of molecular weight of cellulose on the properties of carboxylic acid crosslinked cellulose hydrogels for biomedical and environmental applications. *International Journal of Biological Macromolecules*, *239*, 124282.

Atta, S., Khaliq, S., Islam, A., Javeria, I., Jamil, T., Athar, M. M., Shafiq, M. I., & Ghaffar, A. (2015). Injectable biopolymer based hydrogels for drug delivery applications. *International Journal of Biological Macromolecules*, *80*, 240–245.

Barclay, T. G., Day, C. M., Petrovsky, N., & Garg, S. (2019). Review of polysaccharide particle-based functional drug delivery. *Carbohydrate Polymers*, *221*, 94–112.

Bealer, E. J., Onissema-Karimu, S., Rivera-Galletti, A., Francis, M., Wilkowski, J., Salas-de la Cruz, D., & Hu, X. (2020). Protein—polysaccharide composite materials: Fabrication and applications. *Polymers*, *12*(2), 464.

Biswas, M. C., Jony, B., Nandy, P. K., Chowdhury, R. A., Halder, S., Kumar, D., Ramakrishna, S., Hassan, M., Ahsan, M. A., Hoque, M. E., & Imam, M. A. (2021). Recent advancement of biopolymers and their potential biomedical applications. *Journal of Polymers and the Environment*, 1–24.

Butcher, A. L., Offeddu, G. S., & Oyen, M. L. (2014). Nanofibrous hydrogel composites as mechanically robust tissue engineering scaffolds. *TRENDS in Biotechnology*, *32*(11), 564–570.

Chen, J., Xie, F., Li, X., & Chen, L. (2018). Ionic liquids for the preparation of biopolymer materials for drug/gene delivery: A review. *Green Chemistry*, *20*(18), 4169–4200.

Chen, T., Embree, H. D., Wu, L. Q., & Payne, G. F. (2002). In vitro protein—polysaccharide conjugation: Tyrosinase-catalyzed conjugation of gelatin and chitosan. *Biopolymers: Original Research on Biomolecules*, *64*(6), 292–302.

Coelho, J. F., Ferreira, P. C., Alves, P., Cordeiro, R., Fonseca, A. C., Góis, J. R., & Gil, M. H. (2010). Drug delivery systems: Advanced technologies potentially applicable in personalized treatments. *EPMA Journal*, *1*, 164–209.

Dafe, A., Etemadi, H., Dilmaghani, A., & Mahdavinia, G. R. (2017). Investigation of pectin/starch hydrogel as a carrier for oral delivery of probiotic bacteria. *International Journal of Biological Macromolecules*, *97*, 536–543.

Davidenko, N., Cameron, R., & Best, S. (2019). Natural biopolymers for biomedical applications. In Narayan, R., Wang, M., Laurencin, C., Yu, X. J., (Eds.), *Encyclopedia of Biomedical Engineering, 1*, 162–176. Elsevier: Cambridge, MA, USA.

Di Capua, A., Adami, R., & Reverchon, E. (2017). Production of luteolin/biopolymer microspheres by supercritical assisted atomization. *Industrial & Engineering Chemistry Research*, *56*(15), 4334–4340.

Duan, X., & Sheardown, H. (2005). Crosslinking of collagen with dendrimers. *Journal of Biomedical Materials Research Part A: An Official Journal of the Society for Biomaterials, The Japanese Society for Biomaterials, and the Australian Society for Biomaterials and the Korean Society for Biomaterials*, *75*(3), 510–518.

Elzoghby, A. O., Samy, W. M., & Elgindy, N. A. (2012). Protein-based nanocarriers as promising drug and gene delivery systems. *Journal of Controlled Release*, *161*(1), 38–49.

Flórez, M., Cazón, P., & Vázquez, M. (2023). Selected biopolymers' processing and their applications: A review. *Polymers*, *15*(3), 641.

Galaev, I. Y., & Mattiasson, B. (1999). 'Smart' polymers and what they could do in biotechnology and medicine. *TRENDS in Biotechnology*, *17*(8), 335–340.

Ganeshkumar, M., Ponrasu, T., Raja, M. D., Subamekala, M. K., & Suguna, L. (2014). Green synthesis of pullulan stabilized gold nanoparticles for cancer targeted drug delivery. *Spectrochimica Acta Part A: Molecular and Biomolecular Spectroscopy, 130,* 64–71.

Garg, S. (2020). Bioremediation of agricultural, municipal, and industrial wastes. In *Waste management: Concepts, methodologies, tools, and applications* (pp. 948–970). IGI Global.

George, B., & Suchithra, T. V. (2019). Plant-derived bioadhesives for wound dressing and drug delivery system. *Fitoterapia, 137,* 104241.

Guadarrama-Escobar, O. R., Serrano-Castañeda, P., Anguiano-Almazán, E., Vázquez-Durán, A., Peña-Juárez, M. C., Vera-Graziano, R., Morales-Florido, M. I., Rodriguez-Perez, B., Rodriguez-Cruz, I. M., Miranda-Calderón, J. E., & Miranda-Calderón, J. E. (2023). Chitosan nanoparticles as oral drug carriers. *International Journal of Molecular Sciences, 24*(5), 4289.

Hennink, W. E., & van Nostrum, C. F. (2012). Novel crosslinking methods to design hydrogels. *Advanced Drug Delivery Reviews, 64,* 223–236.

Hosseini, M. S., Hemmati, K., & Ghaemy, M. (2016). Synthesis of nanohydrogels based on tragacanth gum biopolymer and investigation of swelling and drug delivery. *International Journal of Biological Macromolecules, 82,* 806–815.

Hu, H., Li, Y., Zhou, Q., Ao, Y., Yu, C., Wan, Y., Xu, H., Li, Z., & Yang, X. (2016). Redox-sensitive hydroxyethyl starch—doxorubicin conjugate for tumor targeted drug delivery. *ACS Applied Materials & Interfaces, 8*(45), 30833–30844.

Idrees, H., Zaidi, S. Z. J., Sabir, A., Khan, R. U., Zhang, X., & Hassan, S. (2020). A review of biodegradable natural polymer-based nanoparticles for drug delivery applications. *Nanomaterials, 10*(10), 1970.

Jacob, J., Haponiuk, J. T., Thomas, S., & Gopi, S. (2018). Biopolymer based nanomaterials in drug delivery systems: A review. *Materials Today Chemistry, 9,* 43–55.

Janmohammadi, M., Nazemi, Z., Salehi, A. O. M., Seyfoori, A., John, J. V., Nourbakhsh, M. S., & Akbari, M. (2023). Cellulose-based composite scaffolds for bone tissue engineering and localized drug delivery. *Bioactive Materials, 20,* 137–163.

Jiang, Q., Reddy, N., & Yang, Y. (2010). Cytocompatible cross-linking of electrospun zein fibers for the development of water-stable tissue engineering scaffolds. *Acta Biomaterialia, 6*(10), 4042–4051.

Jiang, Q., Reddy, N., Zhang, S., Roscioli, N., & Yang, Y. (2013). Water-stable electrospun collagen fibers from a non-toxic solvent and crosslinking system. *Journal of Biomedical Materials Research Part A, 101*(5), 1237–1247.

Jones, O. G., & McClements, D. J. (2010). Functional biopolymer particles: Design, fabrication, and applications. *Comprehensive Reviews in Food Science and Food Safety, 9*(4), 374–397.

Jones, O. G., & McClements, D. J. (2011). Recent progress in biopolymer nanoparticle and microparticle formation by heat-treating electrostatic protein—polysaccharide complexes. *Advances in Colloid and Interface Science, 167*(1–2), 49–62.

Kamaraj, S., Palanisamy, U. M., Mohamed, M. S. B. K., Gangasalam, A., Maria, G. A., & Kandasamy, R. (2018). Curcumin drug delivery by vanillin-chitosan coated with calcium ferrite hybrid nanoparticles as carrier. *European Journal of Pharmaceutical Sciences, 116,* 48–60.

Kiechel, M. A., & Schauer, C. L. (2013). Non-covalent crosslinkers for electrospun chitosan fibers. *Carbohydrate Polymers, 95*(1), 123–133.

Kono, H., & Fujita, S. (2012). Biodegradable superabsorbent hydrogels derived from cellulose by esterification crosslinking with 1, 2, 3, 4-butanetetracarboxylic dianhydride. *Carbohydrate Polymers, 87*(4), 2582–2588.

Kurakula, M., & Naveen, N. R. (2021). Electrospraying: A facile technology unfolding the chitosan based drug delivery and biomedical applications. *European Polymer Journal,* *147*, 110326.

Laha, A., Yadav, S., Majumdar, S., & Sharma, C. S. (2016). In-vitro release study of hydrophobic drug using electrospun cross-linked gelatin nanofibers. *Biochemical Engineering Journal, 105*, 481–488.

Lai, J. Y. (2014). Interrelationship between cross-linking structure, molecular stability, and cytocompatibility of amniotic membranes cross-linked with glutaraldehyde of varying concentrations. *RSC Advances, 4*(36), 18871–18880.

Larson, N., & Ghandehari, H. (2012). Polymeric conjugates for drug delivery. *Chemistry of Materials, 24*(5), 840–853.

Liu, D., Nikoo, M., Boran, G., Zhou, P., & Regenstein, J. M. (2015). Collagen and gelatin. *Annual Review of Food Science and Technology, 6*, 527–557.

Liu, L. S., Fishman, M. L., Kost, J., & Hicks, K. B. (2003). Pectin-based systems for colon-specific drug delivery via oral route. *Biomaterials, 24*(19), 3333–3343.

Luo, H., Ao, H., Li, G., Li, W., Xiong, G., Zhu, Y., & Wan, Y. (2017). Bacterial cellulose/graphene oxide nanocomposite as a novel drug delivery system. *Current Applied Physics, 17*(2), 249–254.

Madduma-Bandarage, U. S. K., & Madihally, S. V. (2021). Synthetic hydrogels: Synthesis, novel trends, and applications. *Journal of Applied Polymer Science, 138*(19), 50376.

Madrigal-Carballo, S., Lim, S., Rodriguez, G., Vila, A. O., Krueger, C. G., Gunasekaran, S., & Reed, J. D. (2010). Biopolymer coating of soybean lecithin liposomes via layer-by-layer self-assembly as novel delivery system for ellagic acid. *Journal of Functional Foods, 2*(2), 99–106.

Mansfield, E. D. H., Pandya, Y., Mun, E. A., Rogers, S. E., Abutbul-Ionita, I., Danino, D., Williams, A. C., & Khutoryanskiy, V. V. (2018). Structure and characterisation of hydroxyethylcellulose–silica nanoparticles. *RSC Advances, 8*(12), 6471–6478.

Marquié, C. (2001). Chemical reactions in cottonseed protein cross-linking by formaldehyde, glutaraldehyde, and glyoxal for the formation of protein films with enhanced mechanical properties. *Journal of Agricultural and Food Chemistry, 49*(10), 4676–4681.

Mekhail, M., Wong, K. K. H., Padavan, D. T., Wu, Y., O'Gorman, D. B., & Wan, W. (2011). Genipin-cross-linked electrospun collagen fibers. *Journal of Biomaterials Science, Polymer Edition, 22*(17), 2241–2259.

Mortimer, C. J., & Wright, C. J. (2017). The fabrication of iron oxide nanoparticle-nanofiber composites by electrospinning and their applications in tissue engineering. *Biotechnology Journal, 12*(7), 1600693.

Müller, A., Ni, Z., Hessler, N., Wesarg, F., Müller, F. A., Kralisch, D., & Fischer, D. (2013). The biopolymer bacterial nanocellulose as drug delivery system: Investigation of drug loading and release using the model protein albumin. *Journal of Pharmaceutical Sciences, 102*(2), 579–592.

Mustafai, A., Zubair, M., Hussain, A., & Ullah, A. (2023). Recent progress in proteins-based micelles as drug delivery carriers. *Polymers, 15*(4), 836.

Nitta, S. K., & Numata, K. (2013). Biopolymer-based nanoparticles for drug/gene delivery and tissue engineering. *International Journal of Molecular Sciences, 14*(1), 1629–1654.

Oh, J. K., Lee, D. I., & Park, J. M. (2009). Biopolymer-based microgels/nanogels for drug delivery applications. *Progress in Polymer Science, 34*(12), 1261–1282.

Olsen, D., Yang, C., Bodo, M., Chang, R., Leigh, S., Baez, J., Carmichael, D., Perälä, M., Hämäläinen, E.-R., Polarek, J., & Jarvinen, M. (2003). Recombinant collagen and gelatin for drug delivery. *Advanced Drug Delivery Reviews, 55*(12), 1547–1567.

Ooi, S. Y., Ahmad, I., & Amin, M. C. I. M. (2016). Cellulose nanocrystals extracted from rice husks as a reinforcing material in gelatin hydrogels for use in controlled drug delivery systems. *Industrial Crops and Products, 93*, 227–234.

Oryan, A., Kamali, A., Moshiri, A., Baharvand, H., & Daemi, H. (2018). Chemical crosslinking of biopolymeric scaffolds: Current knowledge and future directions of crosslinked engineered bone scaffolds. *International Journal of Biological Macromolecules, 107*, 678–688.

Pandey, V., Haider, T., Jain, P., Gupta, P. N., & Soni, V. (2020). Silk as a leading-edge biological macromolecule for improved drug delivery. *Journal of Drug Delivery Science and Technology, 55*, 101294.

Park, S. B., Lih, E., Park, K. S., Joung, Y. K., & Han, D. K. (2017). Biopolymer-based functional composites for medical applications. *Progress in Polymer Science, 68*, 77–105.

Pattanashetti, N. A., Heggannavar, G. B., & Kariduraganavar, M. Y. (2017). Smart biopolymers and their biomedical applications. *Procedia Manufacturing, 12*, 263–279.

Pirsa, S., & Aghbolagh Sharifi, K. (2020). A review of the applications of bioproteins in the preparation of biodegradable films and polymers. *Journal of Chemistry Letters, 1*(2), 47–58.

Pothakamury, U. R., & Barbosa-Cánovas, G. V. (1995). Fundamental aspects of controlled release in foods. *Trends in Food Science & Technology, 6*(12), 397–406.

Pötzinger, Y., Kralisch, D., & Fischer, D. (2017). Bacterial nanocellulose: The future of controlled drug delivery? *Therapeutic Delivery, 8*(9), 753–761.

Priya, G., Madhan, B., Narendrakumar, U., Suresh Kumar, R. V., & Manjubala, I. (2021). In vitro and in vivo evaluation of carboxymethyl cellulose scaffolds for bone tissue engineering applications. *ACS Omega, 6*(2), 1246–1253.

Pulicharla, R., Marques, C., Das, R. K., Rouissi, T., & Brar, S. K. (2016). Encapsulation and release studies of strawberry polyphenols in biodegradable chitosan nanoformulation. *International Journal of Biological Macromolecules, 88*, 171–178.

Qi, L., Ji, G., Luo, Z., Xiao, Z., & Yang, Q. (2017). Characterization and drug delivery properties of OSA starch-based nanoparticles prepared in [C3OHmim] Ac-in-oil microemulsions system. *ACS Sustainable Chemistry & Engineering, 5*(10), 9517–9526.

Reddy, N., Li, Y., & Yang, Y. (2009a). Alkali-catalyzed low temperature wet crosslinking of plant proteins using carboxylic acids. *Biotechnology Progress, 25*(1), 139–146.

Reddy, N., Li, Y., & Yang, Y. (2009b). Wet cross-linking gliadin fibers with citric acid and a quantitative relationship between cross-linking conditions and mechanical properties. *Journal of Agricultural and Food Chemistry, 57*(1), 90–98.

Reddy, N., Reddy, R., & Jiang, Q. (2015). Crosslinking biopolymers for biomedical applications. *TRENDS in Biotechnology, 33*(6), 362–369.

Riedel, U., & Nickel, J. (1999). Natural fibre-reinforced biopolymers as construction materials—new discoveries. *Die Angewandte Makromolekulare Chemie, 272*(1), 34–40.

Russo, T., Fucile, P., Giacometti, R., & Sannino, F. (2021). Sustainable removal of contaminants by biopolymers: A novel approach for wastewater treatment. Current state and future perspectives. *Processes, 9*(4), 719.

Saito, H., Murabayashi, S., Mitamura, Y., & Taguchi, T. (2008). Characterization of alkali-treated collagen gels prepared by different crosslinkers. *Journal of Materials Science: Materials in Medicine, 19*, 1297–1305.

Salihu, R., Abd Razak, S. I., Zawawi, N. A., Kadir, M. R. A., Ismail, N. I., Jusoh, N., Mohamad, M. R., & Nayan, N. H. M. (2021). Citric acid: A green cross-linker of biomaterials for biomedical applications. *European Polymer Journal, 146*, 110271.

Samrot, A. V., Burman, U., Philip, S. A., Shobana, N., & Chandrasekaran, K. (2018). Synthesis of curcumin loaded polymeric nanoparticles from crab shell derived chitosan for drug delivery. *Informatics in Medicine Unlocked, 10*, 159–182.

Samrot, A. V., Sathiyasree, M., Rahim, S. B. A., Renitta, R. E., Kasipandian, K., Krithika Shree, S., Rajalakshmi, D., Shobana, N., Dhiva, S., Abirami, S., Visvanathan, S., Mohanty, B. K., Sabesan, G. S., & Chinni, S. V. (2023). Scaffold using chitosan, agarose, cellulose, dextran and protein for tissue engineering: A review. *Polymers, 15*(6), 1525.

Senthilkumaran, A., Babaei-Ghazvini, A., Nickerson, M. T., & Acharya, B. (2022). Comparison of protein content, availability, and different properties of plant protein sources with their application in packaging. *Polymers, 14*(5), 1065.

Serdiuk, V., Shevchuk, O., Tetiana, K., Bukartyk, N., & Tokarev, V. (2023). Synthesis of reactive copolymers with peroxide functionality for cross-linking water-soluble polymers. *Journal of Applied Polymer Science, 140*(1), e53254.

Sharma, S., Sudhakara, P., Singh, J., Ilyas, R. A., Asyraf, M. R. M., & Razman, M. R. (2021). Critical review of biodegradable and bioactive polymer composites for bone tissue engineering and drug delivery applications. *Polymers, 13*(16), 2623.

Shi, L., Le Visage, C., & Chew, S. Y. (2011). Long-term stabilization of polysaccharide electrospun fibres by in situ cross-linking. *Journal of Biomaterials Science, Polymer Edition, 22*(11), 1459–1472.

Silva, R. M., Silva, G. A., Coutinho, O. P., Mano, J. F., & Reis, R. L. (2004). Preparation and characterisation in simulated body conditions of glutaraldehyde crosslinked chitosan membranes. *Journal of Materials Science: Materials in Medicine, 15*(10), 1105–1112.

Singh, A. K., Bhadauria, A. S., Kumar, P., Bera, H., & Saha, S. (2019). Bioactive and drug-delivery potentials of polysaccharides and their derivatives. In *Polysaccharide carriers for drug delivery* (pp. 19–48). Elsevier.

Sisson, K., Zhang, C., Farach-Carson, M. C., Chase, D. B., & Rabolt, J. F. (2009). Evaluation of cross-linking methods for electrospun gelatin on cell growth and viability. *Biomacromolecules, 10*(7), 1675–1680.

Song, K., Xu, H., Xie, K., & Yang, Y. (2016). Effects of chemical structures of polycarboxylic acids on molecular and performance manipulation of hair keratin. *RSC Advances, 6*(63), 58594–58603.

Suesca, E., Dias, A. M. A., Braga, M. E. M., De Sousa, H. C., & Fontanilla, M. R. (2017). Multifactor analysis on the effect of collagen concentration, cross-linking and fiber/pore orientation on chemical, microstructural, mechanical and biological properties of collagen type I scaffolds. *Materials Science and Engineering: C, 77*, 333–341.

Sultana, N., Hassan, M. I., & Lim, M. M. (2015). *Composite synthetic scaffolds for tissue engineering and regenerative medicine.* Springer.

Tabasum, S., Younas, M., Zaeem, M. A., Majeed, I., Majeed, M., Noreen, A., Iqbal, M. N., & Zia, K. M. (2019). A review on blending of corn starch with natural and synthetic polymers, and inorganic nanoparticles with mathematical modeling. *International Journal of Biological Macromolecules, 122*, 969–996.

Thomas, S., Gopi, S., & Amalraj, A. (2020). *Biopolymers and their industrial applications: From plant, animal, and marine sources, to functional products.* Elsevier.

Treesuppharat, W., Rojanapanthu, P., Siangsanoh, C., Manuspiya, H., & Ummartyotin, S. (2017). Synthesis and characterization of bacterial cellulose and gelatin-based hydrogel composites for drug-delivery systems. *Biotechnology Reports, 15*, 84–91.

Umashankar, P. R., Mohanan, P. V., & Kumari, T. V. (2012). Glutaraldehyde treatment elicits toxic response compared to decellularization in bovine pericardium. *Toxicology International, 19*(1), 51.

Veeman, D., Sai, M. S., Sureshkumar, P., Jagadeesha, T., Natrayan, L., Ravichandran, M., & Mammo, W. D. (2021). Additive manufacturing of biopolymers for tissue engineering and regenerative medicine: An overview, potential applications, advancements, and trends. *International Journal of Polymer Science, 2021*, 1–20.

Wang, M., Xu, P., & Lei, B. (2023). Engineering multifunctional bioactive citrate-based biomaterials for tissue engineering. *Bioactive Materials*, *19*, 511–537.

Xu, J., Liu, S., Chen, G., Chen, T., Song, T., Wu, J., Shi, C., He, M., & Tian, J. (2018). Engineering biocompatible hydrogels from bicomponent natural nanofibers for anticancer drug delivery. *Journal of Agricultural and Food Chemistry*, *66*(4), 935–942.

Xu, Q., Ji, Y., Sun, Q., Fu, Y., Xu, Y., & Jin, L. (2019). Fabrication of cellulose nanocrystal/chitosan hydrogel for controlled drug release. *Nanomaterials*, *9*(2), 253.

Xue, Y., Mou, Z., & Xiao, H. (2017). Nanocellulose as a sustainable biomass material: Structure, properties, present status and future prospects in biomedical applications. *Nanoscale*, *9*(39), 14758–14781.

Yahya, E. B., Amirul, A. A., HPS, A. K., Olaiya, N. G., Iqbal, M. O., Jummaat, F., A. K., A. S., & Adnan, A. S. (2021). Insights into the role of biopolymer aerogel scaffolds in tissue engineering and regenerative medicine. *Polymers*, *13*(10), 1612.

Yang, L., Xie, S., Ding, K., Lei, Y., & Wang, Y. (2021). The study of dry biological valve crosslinked with a combination of carbodiimide and polyphenol. *Regenerative Biomaterials*, *8*(1), rbaa049.

Zhang, H., Yoshimura, M., Nishinari, K., Williams, M. A. K., Foster, T. J., & Norton, I. T. (2001). Gelation behaviour of konjac glucomannan with different molecular weights. *Biopolymers: Original Research on Biomolecules*, *59*(1), 38–50.

Zhang, J., Cheng, Y., Xu, C., Gao, M., Zhu, M., & Jiang, L. (2021). Hierarchical interface engineering for advanced nanocellulosic hybrid aerogels with high compressibility and multifunctionality. *Advanced Functional Materials*, *31*(19), 2009349.

Zhao, X., Liu, Y., Li, W., Long, K., Wang, L., Liu, S., Wang, Y., & Ren, L. (2015). Collagen based film with well epithelial and stromal regeneration as corneal repair materials: Improving mechanical property by crosslinking with citric acid. *Materials Science and Engineering: C*, *55*, 201–208.

11 Crosslinking Networks of Functional Biopolymer Hydrogels

11.1 INTRODUCTION

Hydrogels are three-dimensional polymer networks with a strong attraction to water, but their ability to dissolve is prevented by their chemically or physically interconnected structure (Ahmed, 2015). Hydrogels are created from macromolecules that contain hydrophilic groups, such as hydroxyl (-OH), carboxylic acid (-COOH), sulfonic acid (-SO$_3$H), and amide groups (-CONH- and -CONH$_2$-), either embedded within or grafted onto the polymeric backbone. Due to their hydrophilic components, hydrogels can absorb anywhere from a fraction to thousands of times their dry weight in water or physiological fluids (Abdul Khalil et al., 2022). Because of their hydrophilicity, hydrogels have become highly valuable in biomedical applications, offering adjustable properties and the ability to mimic certain aspects of natural tissues. The term "network" indicates that crosslinks are necessary to prevent the hydrophilic polymer chains or segments from dissolving into the aqueous phase. Hydrogels can also be described rheologically. Aqueous solutions of hydrophilic polymers at low or moderate concentrations, without significant chain entanglement, exhibit Newtonian behavior. However, the introduction of crosslinks between different polymer chains results in viscoelastic networks that can even display purely elastic behavior. The water-absorbing capacity of hydrogels has made them a subject of interest not only for researchers investigating the fundamental aspects of swollen polymeric networks but also for their widespread application in various technological areas, such as contact lenses, protein separation, cell encapsulation matrices, and controlled drug and protein release devices. When fully hydrated, hydrogels exhibit physical properties similar to living tissue and natural rubber. They are soft, smooth, and capable of rapidly returning to their original shape after undergoing minor deformation. Moreover, hydrogels possess low surface energy, promoting biocompatibility and minimizing protein and cell adhesion from surrounding tissues upon implantation. The use of hydrogels has become ubiquitous across a wide range of biomedical, pharmaceutical, and functional food industries (Abdul Khalil et al., 2023).

Hydrogels can be produced from synthetic polymers like poly(N-isopropyl acrylamide) (pNiPAAM) and poly(hydroxyethyl methacrylate) (pHEMA) (Jain et al., 2020). Alternatively, they can be derived naturally from tissues (e.g., hyaluronic acid, chondroitin sulfate, collagen, gelatin) or from substances like chitosan, alginate, and cellulose, which have gained attention in hydrogel design (Luo

DOI: 10.1201/9781003416043-11

et al., 2023). Synthetic polymer hydrogels are appealing because they offer precise chemical structures and can be designed at the molecular level, resulting in a wide range of environmentally responsive hydrogels. However, many synthetic hydrogels are not biodegradable and may induce local inflammation and toxicity due to trace chemicals. Hydrogels can also be prepared by combining a synthetic polymer with a biopolymer, two different biopolymers, or two different synthetic polymers. The integration of different macromolecules can lead to structures that better mimic living tissue, offer enhanced performance through added functionality, exhibit synergistic effects, or improve stability and biodegradability. One advantage of using macromolecules derived from plants or animals is their inherent properties, such as bioactivity, degradability, and biocompatibility, as they are susceptible to human enzymes. Among the many carbohydrate-based biopolymers, chitosan, hyaluronic acid, pectin, heparin sulfate, and alginate have a well-documented history of safe use, biocompatibility, biodegradability, and low toxicity. For example, hyaluronic acid, a non-sulfated glycosaminoglycan, demonstrates unique viscoelastic behavior and plays a crucial role in regulating cell adhesion and tissue morphogenesis through specific chemical receptors. However, biopolymers also have potential drawbacks, such as weaker mechanical properties, wider molecular weight distributions, undefined chemical compositions, and possible immune responses depending on their sources. Hydrogels composed of decellularized extracellular matrix (ECM) show promise due to their native fiber arrangement and bioactivity, but they often lack mechanical strength, have heterogeneous structures, and offer limited control over physicochemical signal presentation (Dhand et al., 2021).

The formation of hydrogels involves linking biopolymers together either through their natural intermolecular interactions or via chemical modifications that allow for crosslinking. The introduction of various hydrogel crosslinking chemistries to biopolymers has expanded the range of achievable hydrogel properties. However, some properties necessary for many biomedical applications may still be lacking (Zhang & Khademhosseini, 2017). Polymer blends and composite hydrogel formulations have enabled a wider range of physical properties for hydrogels (Neves et al., 2020). Nevertheless, these materials can negatively affect encapsulated cells, undergo phase separation, and experience degradation of their properties over time. To enhance the desired properties of biopolymer hydrogels, there is growing interest in the design and incorporation of interpenetrating secondary networks. Interpenetrating polymer network hydrogels consist of independent yet interwoven polymer networks at the molecular level. Unlike conventional composite hydrogels, these networks cannot be separated without breaking crosslinks. Interpenetrating polymer network hydrogels offer improved mechanical strength, efficient drug loading capacity, and controlled swelling behavior. Various physical and covalent chemistries can be employed to crosslink individual networks within interpenetrating polymer network hydrogels, either simultaneously or sequentially. These hydrogels are characterized by their structure (topography, porosity, swelling), mechanical properties (tensile/compressive modulus, toughness, fracture energy), and functionality (degradation, in vitro and in vivo biocompatibility) (Dhand et al., 2021).

Double network hydrogels are a specific type of interpenetrating polymer network hydrogels characterized by two interpenetrating yet independent networks with contrasting properties. The molar concentration of the secondary network is 20–30 times higher than that of the primary network. The primary network, which is rigid and stiff, acts as a sacrificial network, while the ductile secondary network can undergo significant deformations, resulting in exceptional strength and toughness. Ionic-covalent entanglement hydrogels, on the other hand, are dual network hydrogels formed by an ionically crosslinked rigid network and a covalently crosslinked elastic network, enabling self-healing of the primary network upon rupture (Chimene et al., 2018; Sears et al., 2020). Figure 11.1 shows a wide range of physical or covalent crosslinking methods have been employed towards the formation of hydrogel interpenetrating polymer network, with synthesis occurring primarily through either sequential crosslinking or simultaneous crosslinking that has been proposed by Dhand et al. (2021).

In contrast to interpenetrating polymer network hydrogels, hybrid or composite hydrogels are single network hydrogels that incorporate nano- or microstructures, usually through physical attachment or chemical conjugation strategies. Some of

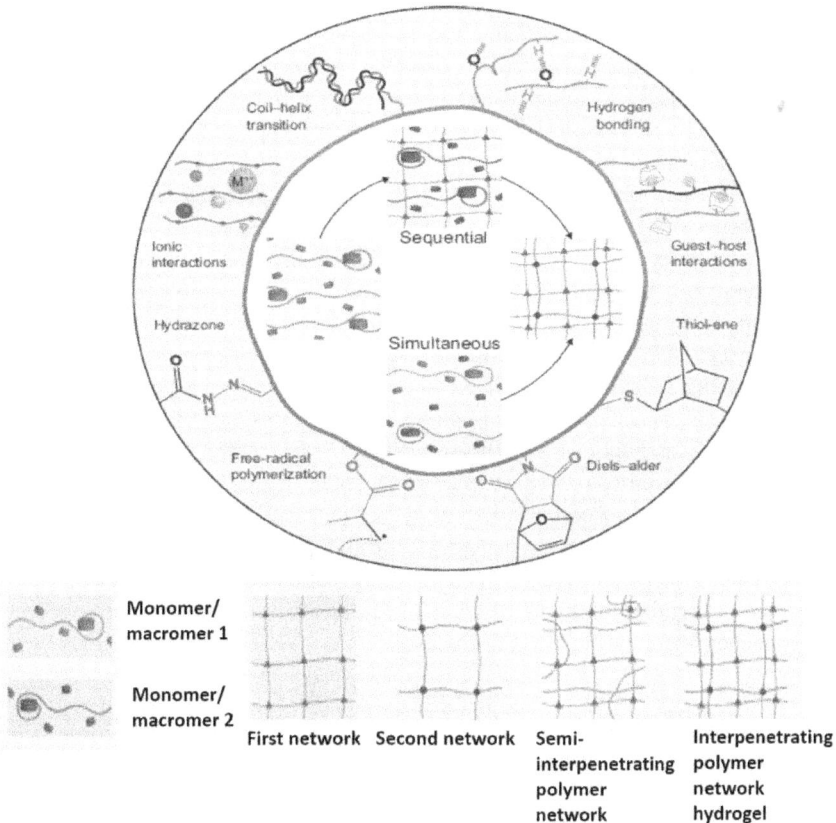

FIGURE 11.1 The type of crosslinking chemistry utilized for each network influences the interactions between networks within interpenetrating polymer network hydrogels. Reproduced from Dhand et al. (2021).

these hydrogels have shown remarkable improvements in mechanical properties, often combining the strengths of the individual components (Pei et al., 2021). However, they may require substantial amounts of reinforcing agents to achieve desirable mechanical properties. Common fillers like gold nanorods, hydroxyapatite, carbon nanotubes (CNTs), and graphene may aggregate over time and undergo irreversible failure, limiting their self-healing ability (Nurazzi et al., 2021). Therefore, this chapter describes and discusses novel surface functionalization networks achieved through chemical and physical crosslinking methods to design functional hydrogels. It also outlines the characteristics and potential applications of these hydrogels in relation to their preparation methods.

11.2 OVERVIEW OF BIOPOLYMER HYDROGELS

Hydrogels have the ability to imitate characteristics found in numerous tissues, and significant progress has been made in customizing hydrogel properties, such as mechanics and degradation, for broad biomedical applications. The utilization of synthetic polymers in constructing hydrogels has led to noteworthy advancements due to the precise control over chemical structures, minimal batch variability, and ease of sourcing (Janoušková, 2018). However, recent developments have involved creating hydrogels from biological substances, such as biopolymers, in order to introduce specific inherent biofunctionality (Van Vlierberghe et al., 2011). Biopolymers are natural polymers sourced from animals and plants, encompassing a diverse range of polysaccharides (e.g., sugars) and polypeptides (e.g., proteins). Notable examples of polysaccharides include hyaluronic acid, chondroitin sulfate, heparin, dextran, alginate, cellulose, chitin, and chitosan. Representative examples of polypeptides include gelatin, silk fibroin, albumin, elastin, keratin, and engineered polypeptides with unique functionalities. The selection of specific polysaccharides or polypeptides can confer desirable properties on hydrogels, such as cell adhesion and degradability.

Biopolymers used for hydrogel formation can generally be classified into two groups: polysaccharides and polypeptides. These molecules consist of repeating units of sugars or peptides, which govern the diverse properties of biopolymers. Biopolymers inherently possess features that make them appealing as biomaterials, including chemical compositions that facilitate cell interactions and degradation. Biopolymer hydrogels can be formed through polymer entanglement, achieved through high molecular weight or concentration, assembly based on specific functionalities of certain biopolymers (e.g., charge), or interpolymer crosslinking resulting from chemical modifications of the biopolymer (Muir & Burdick, 2020).

11.3 IMPORTANCE OF SURFACE FUNCTIONALIZATION

Modifying biopolymers through chemical means is often necessary to facilitate the formation of hydrogels. These modifications target different chemical groups present in the repeat units of the biopolymers, such as amines, hydroxyl groups, and carboxylic acids. This enables the utilization of various crosslinking methods, including mixing, light, redox, and thermal techniques. The mechanical properties of the resulting hydrogels are primarily influenced by the degree of biopolymer modification, the extent of crosslinking, the concentration of biopolymers, and the specific crosslinking chemistry employed (Klein & Poverenov, 2020). For highly stable hydrogels, chemical groups that allow for covalent crosslinking, such as free radical chain polymerization and click reactions, are commonly used (Figure 11.2). However, dynamic covalent crosslinking, such as Schiff base and disulfide bonds, can also be employed to achieve a balance between hydrogel stability and desirable features like self-healing behavior (Figure 11.2). On the other hand, if a less stable hydrogel is desired, physical crosslinking methods such as hydrogen bonding and metal-ligand coordination are typically utilized. These physical crosslinks exhibit properties like shear thinning and gradual disassembly over time (Figure 11.2). Moreover, diverse biopolymer networks can be combined, such as interpenetrating networks, to further modify hydrogel properties according to specific application requirements. Figure 11.3 illustrates common chemical reactions employed for biopolymer modification, including esterification, amidation, etherification, and carbamate formation.

As previously stated, the inclusion of crosslinks in a hydrogel is necessary to prevent the dissolution of hydrophilic polymer chains when exposed to an aqueous

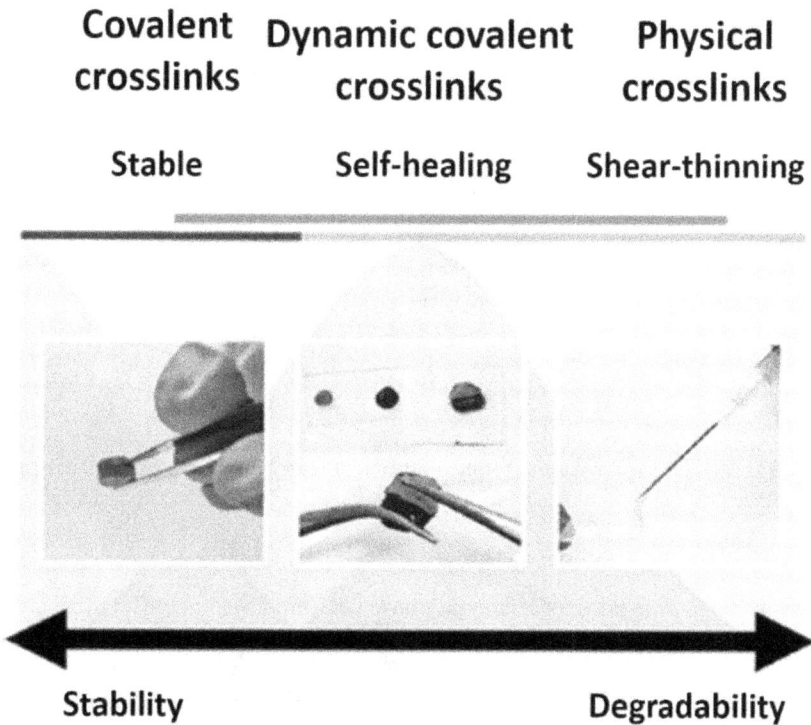

FIGURE 11.2 General hydrogel properties as a function of crosslink type. Reproduced from Muir & Burdick (2020).

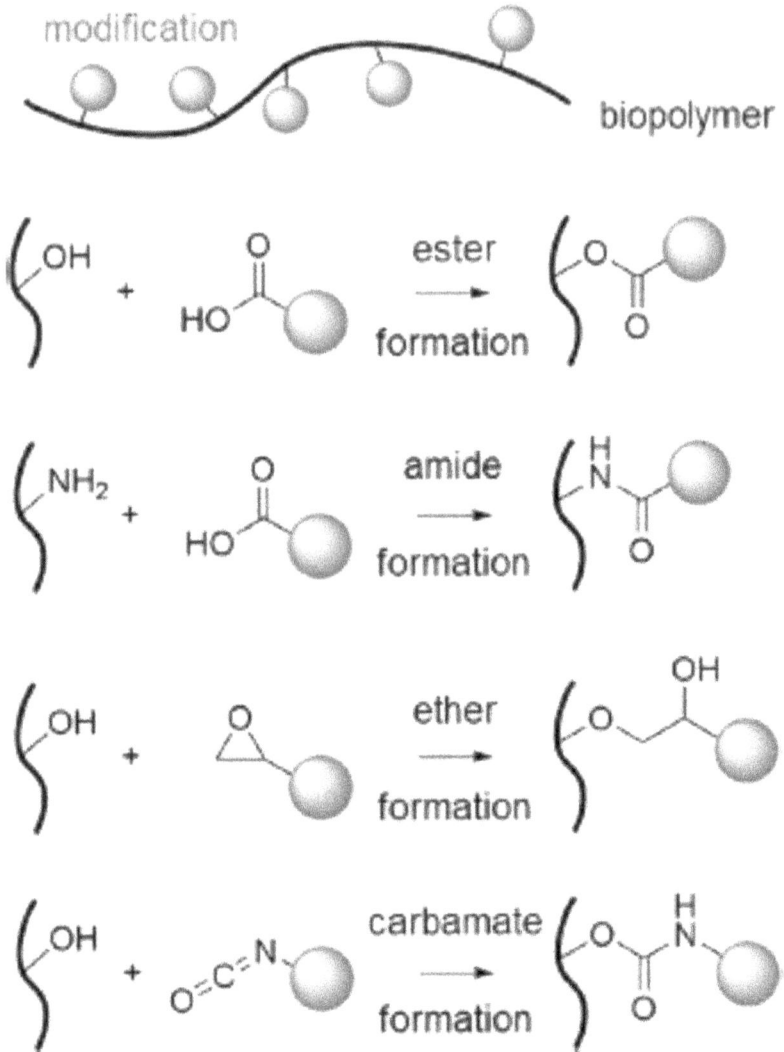

FIGURE 11.3 Chemical reactions for modification of biopolymers like esterification, amidation, etherification, and carbamate formation. Reproduced from Muir & Burdick (2020).

environment. A wide range of techniques have been employed to create these crosslinks and prepare hydrogels. To enhance the biodegradability of hydrogels, labile bonds are often incorporated into the gels. These bonds can exist in either the polymer backbone or the crosslinks utilized during gel formation. Under physiological conditions, such labile bonds can be broken enzymatically or chemically, primarily through hydrolysis (Park et al., 1993). It is essential to exert control over the degradation kinetics, meaning that the parameters influencing the degradation properties need to be adjustable. However, degradability alone is not the ultimate solution. Once hydrogels are implanted, it is crucial that they exhibit excellent biocompatibility and that the degradation byproducts generated are minimally toxic (Chiong et al., 2021). These byproducts should either be metabolized into harmless substances or excreted through renal filtration. In general, hydrogels demonstrate favorable biocompatibility. Their hydrophilic surface possesses a low interfacial free energy when in contact with bodily fluids, resulting in minimal protein and cell adhesion to these surfaces (Elbert & Hubbell, 1996).

11.4 CROSSLINKING NETWORKS FOR FUNCTIONALITY OF BIOPOLYMER HYDROGELS

In chemically crosslinked hydrogels, covalent bonds are present between different polymer chains, while in physically crosslinked hydrogels, dissolution is prevented by physical interactions, which exist between different polymer chains. The chemically crosslinked hydrogels can be created by radical polymerization, chemical reaction of complementary groups, high-energy irradiation, and crosslinking by enzymes. Physically crosslinked hydrogels can be created through ionic interactions, crystallization, crosslinked hydrogels from amphiphilic block and graft copolymers, and hydrogen bonds and protein interactions (Hennink & van Nostrum, 2012).

11.5 CONCLUSIONS AND FUTURE PERSPECTIVES

Chemical crosslinking is a widely used technique for enhancing the mechanical properties of hydrogels. However, the crosslinking agents traditionally employed often pose toxicity and environmental concerns, leading to undesirable reactions with bioactive substances in the hydrogel matrix. To overcome these issues, physical crosslinking methods such as radiation or electron beams have emerged as alternative approaches. Radiation crosslinking offers several advantages, including controlled crosslinking by adjusting the dosage, energy efficiency, cleanliness, and absence of unwanted residuals in the final products. In recent years, significant progress has been made in the development of novel hydrogel systems. Fundamental research has greatly contributed to our understanding of this unique material class. Hydrogels have demonstrated great promise in various applications, such as encapsulating living cells and facilitating controlled release of pharmaceutically active proteins. Numerous crosslinking techniques have been developed and are currently available for hydrogel preparation. Physically crosslinked hydrogels, in particular, are intriguing for entrapping labile bioactive substances and living cells due to their mild formation conditions and lack of organic solvents. Despite the availability of physical crosslinking methods, there is still a need for new approaches in hydrogel research. It is anticipated that principles from the expanding field of supramolecular chemistry will be utilized to design novel hydrogel types with tailored properties, preferably prepared in an aqueous environment. Supramolecular chemistry offers the potential to create dynamic, reversible, and stimuli-responsive hydrogels with enhanced functionality. Additionally, protein engineering techniques hold promise for developing hydrogel systems with precise control over their microstructure and properties, enabling customization for specific applications. Furthermore, it is foreseeable that hydrogel systems triggering gel formation through specific stimuli such as temperature, pH, or particular compounds will be further explored and applied in pharmaceutical, biomedical, and agricultural applications. These stimulus-responsive hydrogels can enable targeted drug delivery, tissue engineering, and controlled release of agrochemicals, among other applications. By incorporating stimuli-sensitive elements into hydrogel design, it becomes possible to achieve on-demand release and precise spatiotemporal control over the gel properties. Overall, ongoing research in hydrogel development aims to overcome the limitations of traditional crosslinking methods and harness the advantages of physical crosslinking, supramolecular chemistry, protein engineering, and stimulus-responsive systems. These advancements will pave the way for the creation of safer, more efficient, and highly functional hydrogels, with a wide range of potential applications in various fields.

REFERENCES

Abdul Khalil, H. P. S., Muhammad, S., Yahya, E. B., Amanda, L. K. M., Abu Bakar, S., Abdullah, C. K., Aiman, A. R., Marwan, M., & Rizal, S. (2022). Synthesis and characterization of novel patchouli essential oil loaded starch-based hydrogel. *Gels*, *8*(9), 536.

Abdul Khalil, H. P. S., Yahya, E. B., Jummaat, F., Adnan, A. S., Olaiya, N. G., Rizal, S., Abdullah, C. K., Pasquini, D., & Thomas, S. (2023). Biopolymers based aerogels: A review on revolutionary solutions for smart therapeutics delivery. *Progress in Materials Science*, *131*, 101014.

Ahmed, E. M. (2015). Hydrogel: Preparation, characterization, and applications: A review. *Journal of Advanced Research*, *6*(2), 105–121.

Chimene, D., Peak, C. W., Gentry, J. L., Carrow, J. K., Cross, L. M., Mondragon, E., Cardoso, G. B., Kaunas, R., & Gaharwar, A. K. (2018). Nanoengineered ionic—covalent entanglement (NICE) bioinks for 3D bioprinting. *ACS Applied Materials & Interfaces*, *10*(12), 9957–9968.

Chiong, J. A., Tran, H., Lin, Y., Zheng, Y., & Bao, Z. (2021). Integrating emerging polymer chemistries for the advancement of recyclable, biodegradable, and biocompatible electronics. *Advanced Science*, *8*(14), 2101233.

Dhand, A. P., Galarraga, J. H., & Burdick, J. A. (2021). Enhancing biopolymer hydrogel functionality through interpenetrating networks. *TRENDS in Biotechnology*, *39*(5), 519–538.

Elbert, D. L., & Hubbell, J. A. (1996). Surface treatments of polymers for biocompatibility. *Annual Review of Materials Science*, *26*(1), 365–394.

Hennink, W. E., & van Nostrum, C. F. (2012). Novel crosslinking methods to design hydrogels. *Advanced Drug Delivery Reviews*, *64*, 223–236.

Jain, A., Bajpai, J., Bajpai, A. K., & Mishra, A. (2020). Thermoresponsive cryogels of poly (2-hydroxyethyl methacrylate-co-N-isopropyl acrylamide)(P (HEMA-co-NIPAM)): Fabrication, characterization and water sorption study. *Polymer Bulletin*, *77*, 4417–4443.

Janoušková, O. (2018). Synthetic polymer scaffolds for soft tissue engineering. *Physiological Research*, *67*.

Klein, M., & Poverenov, E. (2020). Natural biopolymer-based hydrogels for use in food and agriculture. *Journal of the Science of Food and Agriculture*, *100*(6), 2337–2347.

Luo, Y., Tan, J., Zhou, Y., Guo, Y., Liao, X., He, L., Li, D., Li, X., & Liu, Y. (2023). From crosslinking strategies to biomedical applications of hyaluronic acid-based hydrogels: A review. *International Journal of Biological Macromolecules*, 123308.

Muir, V. G., & Burdick, J. A. (2020). Chemically modified biopolymers for the formation of biomedical hydrogels. *Chemical Reviews*, *121*(18), 10908–10949.

Neves, S. C., Moroni, L., Barrias, C. C., & Granja, P. L. (2020). Leveling up hydrogels: Hybrid systems in tissue engineering. *TRENDS in Biotechnology*, *38*(3), 292–315.

Nurazzi, N. M., Sabaruddin, F. A., Harussani, M. M., Kamarudin, S. H., Rayung, M., Asyraf, M. R. M., Aisyah, H. A., Norrrahim, M. N. F., Ilyas, R. A., Zainudin, E. S., Sapuan, S. M., Khalina, A., & Abdullah, N. (2021). Mechanical performance and applications of CNTs reinforced polymer composites: A review. *Nanomaterials*, *11*(9), 2186.

Park, H., Park, K., & Shalaby, W. S. (1993). *Biodegradable hydrogels for drug delivery*. CRC Press.

Pei, X., Wang, J., Cong, Y., & Fu, J. (2021). Recent progress in polymer hydrogel bioadhesives. *Journal of Polymer Science*, *59*(13), 1312–1337.

Sears, C., Mondragon, E., Richards, Z. I., Sears, N., Chimene, D., McNeill, E. P., Gregory, C. A., Gaharwar, A. K., & Kaunas, R. (2020). Conditioning of 3D printed nanoengineered ionic—covalent entanglement scaffolds with iP-hMSCs derived matrix. *Advanced Healthcare Materials*, *9*(15), 1901580.

Van Vlierberghe, S., Dubruel, P., & Schacht, E. (2011). Biopolymer-based hydrogels as scaffolds for tissue engineering applications: A review. *Biomacromolecules*, *12*(5), 1387–1408.

Zhang, Y. S., & Khademhosseini, A. (2017). Advances in engineering hydrogels. *Science*, *356*(6337), eaaf3627.

12 Current Challenges and Future Prospects of Biopolymer Blends and Biopolymer-Based Nanocomposites

12.1 INTRODUCTION

The plastic revolution began with Bakelite's synthesis in 1907, marking the beginning of the polymer age. Since then, thousands of synthetic polymers have been developed (Jothimani et al., 2019). As of 2019, 8,300 million metric tons (Mt) of virgin polymers have been produced. Of that amount, only 9% of produced synthetic polymers were recycled, and the rest remained in the natural environment. By 2050, it is estimated that the production of synthetic polymers will increase to 12,000 million metric tons. This increase in non-biodegradable plastic production could have a disastrous effect on nature. A proactive approach to developing fast-degrading polymers and utilizing natural polymers could be one solution to reduce this impact (Jothimani et al., 2019). Also, the increasing environmental impacts of pollution due to fossil-based polymers have made the production of new materials with sustainable properties urgent. Biodegradable polymers derived from renewable materials have opened up a new opportunity to replace the utilization of synthetic polymers with the capacity to degrade naturally by the action of microorganisms (Rosseto et al., 2019).

Environmentally friendly systems are the appropriate solution to reduce non-biodegradable plastic waste and replace it with green and eco-friendly polymer products. Biopolymers, considered green polymeric matrices or plastic natural feedstock by synthetic routes, frequently have poorer characteristics and performances than traditional polymeric matrices. The present chapter underlines how hybrid blends containing renewable polymers, in concurrence with synthetic polymers and additives, have great potential in enhancing the moisture and gas barrier properties of biobased materials and how they are not economically practicable to be utilized without polymeric blends with low-priced plastics of comparable necessary characteristics. However, green biobased polymers can be used in specific applications due to their limited characteristics, including high cost and poor mechanical and thermomechanical properties compared to traditional commodity polymers. Developing green and eco-friendly polymeric blends with acceptable characteristics cannot overcome the limitations of biopolymers alone. However, biodegradable plastic blends need accurate cautions, postconsumer organization, and additional design to

DOI: 10.1201/9781003416043-12

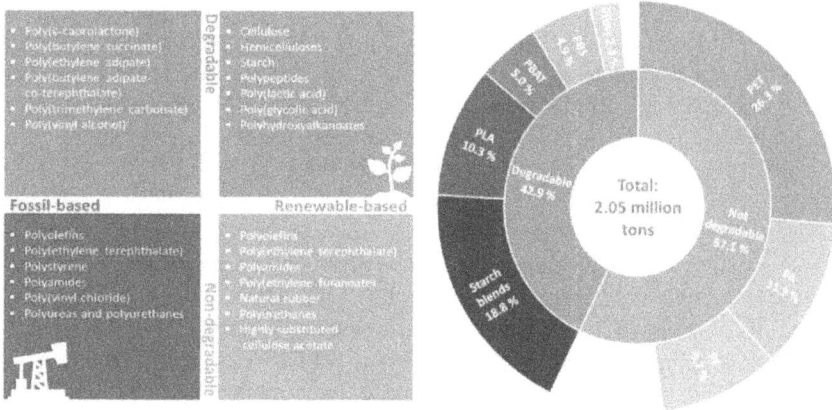

FIGURE 12.1 Biopolymer classification and global production capacities of biopolymers in 2017 according to European Bioplastics. Reproduced from Kumar et al. (2023).

achieve fast biodegradation in various environmental conditions (Luzi et al., 2019). Figure 12.1 presents the biopolymer classification and global production capacities of biopolymers in 2017 according to European Bioplastics.

12.2 BIOPOLYMER DERIVATIVES AND PROCESSING METHODS

By definition, a bioplastic can be defined as a polymer partly or wholly derived from biological sources such as sugarcane; potato starch; or the cellulose from trees, straw, and cotton. According to European Bioplastics, a material can be considered a bioplastic if it is biobased, biodegradable, or both. Also, biobased material might not be biodegradable because the property of biodegradation does not rely on the resource basis of a material but is rather linked to its chemical structure. In other words, 100% biobased plastics may be non-biodegradable, and 100% fossil-based plastics can biodegrade. Preparation of biopolymeric blends with synthetic polymers is one practice to enhance the characteristics of biopolymers, changing their degradation rates and modulating the cost of the materials. For example, polymer blends of olefins with biodegradable polymers are the most popular, especially for degradable packaging applications that impart a sufficient decrease of volume in the landfill by partial and bulk disintegration (Luzi et al., 2019).

12.3 PRODUCTION AT MASS SCALE

Biopolymers or natural-based polymer advancement has led to versatile applications useful in day-to-day life in every possible area in the last century. The replacement of synthetic polymers with biodegradable and renewable materials like biobased polymers has become important because of their greater availability (Rosseto et al., 2019). In addition, limitations of reserves of fuel resources has created growing interest in bioplastics. The scarcity of fossil fuels has led to growing costs for plastic products; therefore finding alternative raw materials to replace plastics has become increasingly urgent (Chen, 2014). In 2015, the production capacities of biobased and biodegradable plastics accounted for only 1% of the total global plastic production. This percentage is anticipated to witness substantial growth in the upcoming years. By 2020, the share of biobased and biodegradable plastics had increased to 2.5% of fossil-based production. It's worth noting that, on a weight basis, most biobased and biodegradable plastics are generally more expensive than their fossil-based counterparts. Nevertheless, this cost disparity can be offset by specific material properties inherent to biopolymers, particularly in terms of cost reduction, especially during the end-of-life phase. To date, there are various kind of biobased plastics products that are cost competitive. Plus, the price of fossil-based plastics depends on oil prices, which are likely to fluctuate from time to time, whereas biobased plastics that depend on biomass are more likely to be stable. Improved properties of biopolymers either alone or in composition with other material produce technologically advanced applications. However, higher cost production and difficult preparative methods limit their production (Jothimani et al., 2019).

The higher price of biobased and biodegradable plastics generally contribute to their higher density.

12.4 ENVIRONMENTAL AND SUSTAINABILITY FACTORS

12.4.1 GREENHOUSE GAS EMISSIONS

The upsurge of biobased polymer utilization is largely due to the environmental dilemma of climate change. The production of carbon-based sources in plastic manufacturing has a major effect on greenhouse gases in the atmosphere and increases CO_2 emissions (Chen, 2014). On December 11, 1997, an environmental treaty was signed and became law on the February 16, 2005. Environmental issues related to greenhouse gas emission (GHG) and climate change were definitely raised and drew global attention, particularly in the face of toxic waste, pollution, contamination, exhaustion of natural resources, and environmental deterioration (Luzi et al., 2019). The Intergovernmental Panel on Climate Change (IPCC) trajectory to 2050 was focusing on stabilizing atmospheric GHG concentrations at 450 ppm CO_2, which requires emission reduction of 80% compared to that in 1990 (Chen, 2014).

12.4.2 BIODEGRADABLE AND RECYCLABLE PLASTICS

Biobased polymers are plastics derived from renewable resources. However, the term does not necessarily imply biodegradability. Despite becoming alternatives to fossil-based plastics and offering greenhouse gas emission savings, another main purpose of the production of biobased polymers is their durability, which depends on end-of-life energy recovery during incineration or recycling (Chen, 2014). The application of plastics in packaging sectors alone contributes to 26% of manufacturing volume, and the quantity of plastic waste from packaging products could grow by two-fold and three-fold by 2030 and 2050, respectively. The fact that plastic materials are extremely useful in various sectors of applications and the positive and synergic combination of main characteristics make them appropriate materials to be utilized. According to Plastic Recyclers Europe, the European association representing over 115 members involved in plastic recycling, their data from 2014 indicates that a mere 35.9% of post-consumer plastic waste was repurposed, with an additional 38.6% of such waste being earmarked for energy recovery (Luzi et al., 2019). The balance of the percentage mostly ends up in landfills. With the intention of reducing the effects of plastic waste effect on the environment, biobased plastics have been selected based on their short lifespan.

In addition, to expand the spectrum of sustainability, incorporating resources and practices that move a step closer toward sustainability, increasing the renewable amount or lowering the overall weight of petroleum-based plastics have been considered suitable options. Sustainability is required for our generation to manage resources based on the average quality of life that can potentially be shared by future generations. Today's sustainable plastics are not automatically biodegradable and even contain polyolefin made from renewable feedstock. Plastic quality is also defined via its durability. Durability might be the opposite of biodegradability. It refers to the ability of a material to offer long-term performance without significant deterioration and resist the effects of use and ageing. The durability of plastic depends on the chemical structure of the polymer and the conditions the plastic is subjected to. For example, PLA can be durable under indoor conditions with a lifespan in the vicinity of 10 to 20 years but at the same time be compostable under industrial conditions (Verma & Goh, 2018).

Biobased materials are defined in European standard EN 16755 as materials produced from biomass. They can be produced wholly or partly from biomass. Renewable materials are produced from materials collected from resources which are naturally replenished on a human timescale. Biobased material can be considered renewable as long as new crop cultivation balances harvesting. However, biobased and biodegradable are not equivalent. Biodegradable materials are materials that can be broken down by microorganisms (e.g. bacteria or fungi) into water and naturally occurring gases, including carbon dioxide (CO_2) and methane (CH_4), and biomass. Biodegradable materials can be also produced from petrochemical-based polymers, and not all biobased materials are biodegradable (Verma & Fortunati, 2017). The properties of types of polymers and their biodegradability properties are presented in Figure 12.2. The biodegradability of materials depends strongly on the environmental conditions, including temperature, presence of microorganisms, oxygen, and water. The properties of the degradation of the polymer might be different based on the soil conditions; humid or dry climate; surface or marine water; or human-made systems like home composting, industrial composting, or anaerobic digestion (Verma & Fortunati, 2017).

Biodegradable materials can also be categorized as composted materials. However, not all biodegradable materials are compostable. Similar to biodegradable materials, compostable materials can break down under similar conditions but require higher temperatures (55 to 60°C). According to the EN 13432 standard,

	Petrochemical	Partly bio-based	Bio-based
Non-biodegradable	PE, PP, PET, PS, PVC	Bio-PET, PTT	Bio-PE
Biodegradable	PBAT, PBS(A), PCL	Starch blends	PLA, PHA, Cellophane

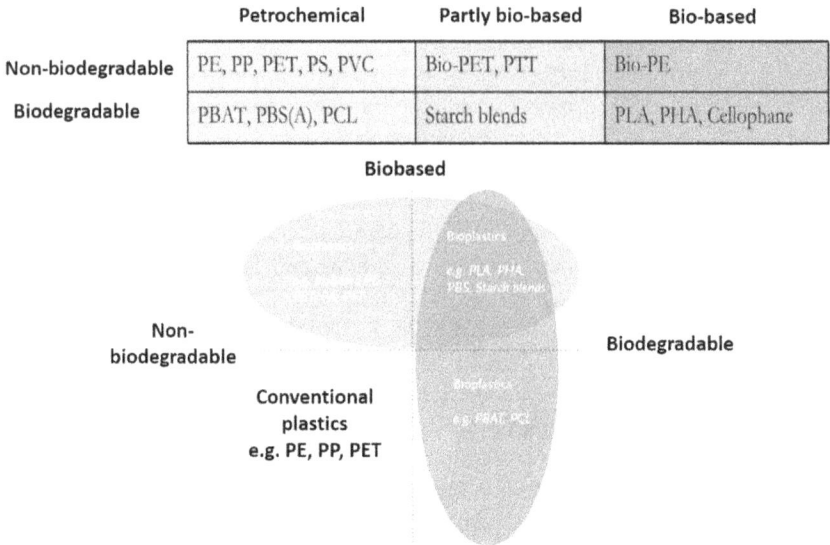

FIGURE 12.2 Biobased vs. petrochemical-based plastics and biodegradable vs. non-biodegradable plastics.

plastic products, especially from the packaging industries, can be only considered compostable if they meets these conditions (Verma & Fortunati, 2017):

- The packaging material and its relevant organic components (> 1 wt.%) can biodegrade naturally
- Disintegration of the packaging must take place within a certain period
- The plastic products must not have any negative effect on the composting process
- The quality of the compost must not be negatively affected by the derived materials

Recycling samples, on the other hand, are defined as material that can be reprocessed to become a new product. Recycling can be divided into two types, mechanical and chemical. Used plastic products, for example, are collected, sorted, and reprocessed into new plastic products, known as mechanical recycling. In chemical recycling, the product is usually broken down to monomers and applied again in the production of a new polymer.

12.5   FEASIBILITY FOR INDUSTRIAL PURPOSES IN THE NEAR FUTURE

Biopolymer materials have garnered significant attention due to their sustainable origins. These biopolymers are primarily derived from natural materials such as cellulose, chitosan, starch, and proteins (collagen, soy, casein, etc.). Developing new biopolymers should prioritize simplicity, cost-effectiveness, and recyclability, offering substantial employment opportunities. Moreover, new hybrid systems are expected to emerge, as these biodegradable polymers hold promise for practical and essential properties.

One particularly anticipated area of research involves the development of structurally refined biodegradable shape memory polymers. This entails identifying new polymer classes, characterizing material properties, and assessing their in vivo performance. Despite the existence of known biopolymers like polylactic acid (PLA), polyhydroxybutyrate (PHB), and polyhydroxyalkanoate (PHA), properties such as shape memory have not been demonstrated. Ongoing research is focused on discovering finely tuned and ideal biopolymers that enable temporal control of shape; possess the required mechanical properties; exhibit desired biodegradation rates; and ensure important biocompatibility with tissues, blood, and non-inflammatory responses. Many studies on biodegradable shape memory polymers are still in the early stages, necessitating further in vivo investigations to validate their efficacy in new treatment modalities, drug delivery applications, and developing retractable and drug-eluting stents.

One of the most challenging aspects is the utilization of biopolymers in biodegradable polymer drug-eluting stents. Recent advancements in these biopolymers have proved their effectiveness and safety as long-lasting polymer drug-eluting stents. Techniques for creating biodegradable micro-particles for protein delivery, including botulinum toxins, have been developed. These strategies involve precipitating and washing proteins with natural solvents to remove water, followed by dispersion in polymer-dissolved natural solvents to prevent exposure to water/solvent interfaces and preserve the bioactivity of the protein tablets. Micro-particles can be fabricated through template or emulsion methods. These micro-particles, composed of one or more biodegradable polymers with entrapped protein agents such as botulinum toxin, can also be formulated into thermals or crosslinked hydrogels. The stability of the protein within the micro-particles, along with the controlled release of the agents, ensures sustained efficacy. Second-generation drug-eluting stents have raised concerns about the risk of late and very late stent thrombosis compared to bare-metal stents, necessitating prolonged dual antiplatelet therapy. Despite extensive investigations, the exact mechanisms behind these late events still need to be elucidated. Stent polymers have been implicated as potential factors. The persistence of durable polymers after complete drug release has been associated with allergic reactions and inflammatory responses. Third-generation drug-eluting stents with more biocompatible or biodegradable polymers have been developed to address this issue.

However, the strength and durability of biodegradable materials may be limited, reducing their value in specific critical medical applications. Inconsistent rates of biodegradation can pose challenges in predicting and controlling degradation within the body, as degradation rates can vary widely depending on the specific material and its use. Toxic byproducts released during the degradation process of certain biodegradable materials can harm the body and limit their applicability. Compatibility with implants and existing medical equipment may also constrain using biodegradable materials in specific applications. Addressing these issues will require ongoing research and development efforts to enhance biodegradable materials' strength, uniformity, durability, and compatibility. Additionally, streamlining regulatory approval processes and promoting the utilization of biodegradable materials in the medical industry will be essential for their widespread adoption.

12.6 CURRENT PROSPECTS AND MARKET POTENTIAL OF BIOPOLYMER-BASED PRODUCTS

12.6.1 PRODUCTION CAPACITIES

As demand grows and more advanced biopolymers, technologies, and devices emerge, the bioplastics industry continues to expand and diversified. According to the data collected by European Bioplastics in collaboration with the Nova Institute (2020), biopolymer-based products currently account for around 1% of the 368 million tons of plastic manufactured per year. However, global bioplastic production capacities are expected to grow from around 2.23 million tons in 2021 to around 2.87 million tons in 2025.

According to European Bioplastics (2020), approximately 2.21 million tons of biopolymer-based plastics were produced in 2020. These include biodegradable biopolymers such as PLA, PBS, PBAT, PHA, starch blends, and other biodegradable polymers, which account for about 58.1%, and non-biodegradable biobased plastics such as biobased polyethylene (PE) and polyethylene terephthalate (PET), as well as biobased PA and PEF, which make up just over 40% of global bioplastics manufacturing. However, the total production of biodegradable plastics is projected to increase by 62.7% by year 2025. This shows that the industry and the market focus have shifted to emphasize biodegradable products, mainly affected by environmental concerns. Biopolymers are being used in a growing range of applications, including packaging, catering, consumer electronics, automobiles, agriculture/horticulture, toys, textiles, and others. In 2020, the packaging sector appeared to be the highest for biopolymer-based production and attained about 555 tons, followed by textiles, agriculture, automotive, building, coatings, electronics, and others. With the rising capacities of practical polymers, segments such as packaging, agriculture, automotive and transportation, building and manufacturing, and electric and electronics are all on the increase. Apart from that, the biomedical sector could be another that would create demands in biopolymer-based products, especially during the pandemic (COVID-19) season.

12.6.2 GEOGRAPHICAL REGIONS

In terms of business geography, the Asia-Pacific region is projected to expand at the fastest pace during the forecast period. Globally, Asia-Pacific is the leading manufacturer and user of biopolymer-based products. According to figures published by European Bioplastics (2020), the Asia-Pacific region accounts for about 46% of global bioplastic production, which was the highest among other regions such as Europe, North America, South America, and Australia, which only accounted for 26%, 17%, 10%, and 1%, respectively. Although Europe currently houses one-fourth of the world's bioplastics processing potential, Asia continues to lead when it comes to actual bioplastics production and regional capability growth. Several key factors that contribute to the growth of bioplastic productions in Asia Pacific region could be:

- Emphasis on pollution-free economic growth, prompting manufacturers to seek out biodegradable materials by the Chinese government.
- The Indian food and grocery industry is the sixth largest in the world, with retail accounting for about 70% of total revenue. Furthermore, food and food goods are one of the most important segments in the country's retail industry, with a market size of USD 490 billion in 2013 and a forecast of INR 61 lakh crore (USD 894.98 billion) by 2020.
- Increasing interest in items such as fish, meat, and vegetables, which require packaging to keep them shelf stable.
- The increasing trend of ready-packed meals, which needed plastic packaging.
- The increasing trend of online shopping.

China and India are now the world's leading producers of food and beverages. Organized retail, supermarkets, and online shopping centers are growing exponentially. With that, the use of biopolymers for packaging positively affects the industry. Furthermore, according to a recent report by European Bioplastics, the majority of new biobased polymer capacity investment is projected to occur in the Asia-Pacific region, owing to better access to feedstock and a favorable political climate.

12.6.3 INDUSTRY EXPANSION

The industry is focused on raising biopolymer processing capabilities. Increasing anxiety about the sustainability of packaging has resulted in a significant increase in demand for biopolymer-based products. Biopolymers are being used in making bioplastics for packaging by a number of consumer product companies, including Samsung. Samsung's electronics division unveiled its latest sustainability strategy in January 2019, which includes updating the packaging of its phones, laptops, and appliances. Plastic will no longer be used in products such as televisions, refrigerators, and washing machines. Instead, they'll be packaged in bags made of recycled plastic or biopolymer-based plastics, which materials are derived from vegetable fats, corn starch, and sugar cane (Samsung Electronics, 2019).

Aside from that, rising global industry expansion and stricter environmental legislation are likely to fuel biopolymer demand in the near future. For example, Novozymes, a Danish enzyme producer, formed a joint venture with Carbios, which is a prominent biotechnology company based in France, to commercialize the manufacturing of biodegradable plastics, mainly using plastic-degrading enzymes, in January 2019 (Pohjakallio, 2020). In the same year in October, Swiss food company Nestlé opened a new testing facility aimed at addressing the issue of packaging waste (Boz et al., 2020). In addition, starting in January 2020, the Fraunhofer Institute for Microstructure of Materials and Systems and the scientific film manufacturer Polifilm Extrusion GmbH collaborated on developing biopolymer-based films for food packaging as part of a joint research project. The experts in this project are using chitosan coatings to improve the susceptibility of food to microbes (Mihalca et al., 2021). Over the projected period, the biopolymer packaging market is estimated to grow at a CAGR of 12.6%. (2021–2026) (European Bioplastics, 2020). There are a few key reasons for the advancement of biopolymer-based products in the market:

- Industry is concentrating on improving biopolymer processing capacities.
- Vendors in industry now have greater control of macromolecular structures thanks to advanced biopolymer manufacturing technology, allowing new generations of commodity polymers to compete with costly specialty polymers.
- The production of newly researched and developed materials, especially advanced biopolymer formulations, is assisting in meeting a variety of requirements for other packaging functions such as food safety and consumer safety.
- Technological advancements can propel the industry to new heights. During the forecast era, the market is projected to be driven by biopolymer innovations.
- Scientists have produced biopolymer films from seaweeds, potato peels, and fruit pomace, which can be used to store perishable food items like bread, dried fruits, and spices.
- Between 2025 and 2040, the European Union (EU) and the American Chemical Council's Plastics Division have set a goal of using 100% recyclable, recycled, and compostable plastics. These rules are likely to have a major impact on the plastic packaging industry in the future.

All these key reasons would increase the demand for mono-material plastics and biodegradable polymers in a variety of industries. Currently, the high manufacturing prices of all types of packaging are a source of concern. Production is expected to rise as a result of the regulations, and prices are expected to fall as a result of economies of scale. Moreover, the imminent effect of public fears about coronavirus viral exposure and worries about the virus's potential to live on surfaces is likely to drive a much longer-term market for new products and products with antiviral and antibacterial properties. Enhanced active drug products with reduced toxicity and high efficacy are being researched. Consumer demand for such materials in everyday consumer goods is expected to increase dramatically following the COVID-19 season, as consumers are likely to retain their interest in and preferences for products enhanced with antiviral and antibacterial properties during this difficult period.

12.7 CONCLUSIONS

The production of biopolymers is experiencing steady growth, and there are ongoing advancements in the processes used to acquire the necessary raw materials from renewable sources and enhance polymerization techniques. One of the most captivating areas of research revolves around biopolymers due to their significant role in ensuring environmental safety, particularly as packaging materials in the food industry. Moreover, they have demonstrated remarkable applications in the medical and agricultural fields, such as wound healing and drug delivery. These biopolymers are potential substitutes for toxic and undesirable synthetic single-use plastics. Biopolymers, or biopolymers derived from renewable resources, encompass various substances like carboxymethyl cellulose, pectins, hemicellulose, starch, pullulan, xanthan gum, agar, guar gum, alginate, gum karaya, and gellan. Several copolymerization methods (e.g., block, random, or grafting) have been proposed and implemented to improve the properties of biopolymers. Physical blending, another commonly employed method, creates biopolymer products with different physical and morphological attributes. These methods result in biopolymers that exhibit both an accelerated rate of biodegradation and the necessary mechanical strength for the intended application. There is a growing demand for sustainable development in the field of biopolymers, aiming to achieve properties comparable to traditional polymers for industrial applications. However, one major obstacle in this progress is the high cost associated with industrial processes. Research in biopolymers based on PLA and chitosan has increased due to their biocompatibility and suitability for implantable medical applications. Despite these advancements, the utilization of biodegradable polymers remains less than 1% when compared to synthetic plastics.

Various industries, including consumer goods, packaging, and automobile manufacturing, highly demand biopolymer materials. Companies are investing in the research and development of biopolymers and expanding their production facilities to meet this demand. Initially, the biopolymer industry focused on developing biodegradable materials to address the issue of plastic waste. However, the current emphasis is on producing durable materials using sustainable processes involving biomass-derived substrates, reducing greenhouse gas emissions, and decreasing reliance on fossil resources. Additionally, the production of biobased polymers is not limited to substituting petrochemical counterparts but also involves the development of new applications with improved or novel properties.

The future of the biopolymer market is expected to witness further growth and development, both in terms of the variety and quantity of biobased materials. In particular, advancements in processes to utilize lignocellulosic biomass, a waste product, for biopolymer production are anticipated to significantly reduce production costs and make renewable-derived plastics even more attractive. Considering prospects and expectations, several essential factors have been taken into account. These include the cost-effectiveness of manufacturing food packaging materials, as it often involves expensive equipment; the recommendation to incorporate eco-friendly additives and agents in green packaging materials; and the need to comply with regulations and legislation related to food packaging materials in the EU and other regions. These measures aim to ensure high food safety and transparency for consumers.

Furthermore, biopolymers are characterized by their excellent mechanical performance, affordability, low density, and low CO_2 emissions. They are also non-toxic when in contact with food and completely biodegradable. Therefore, the torrefaction process is highly recommended for biopolymer production. These features open up opportunities for a wide range of applications. In addition to emerging applications, scientists are now focusing on utilizing biopolymers in automotive manufacturing. This is due to their low density and low CO_2 emissions when burned. By assembling vehicle components from low-density biocomposites, it is possible to reduce CO_2 emissions, save fuel, and lower costs. Consequently, the relationship between vehicle weight and CO_2 emissions is expected to become a significant trend by 2020–2030.

REFERENCES

Boz, Z., Korhonen, V., & Sand, C. K. (2020). Consumer considerations for the implementation of sustainable packaging: A review. *Sustainability (Switzerland)*, *12*(6). https://doi.org/10.3390/su12062192

Chen, Y. J. (2014). Bioplastics and their role in achieving global sustainability. *Journal of Chemical and Pharmaceutical Research*, *6*(1), 226–231.

Jothimani, B., Venkatachalapathy, B., Karthikeyan, N. S., & Ravichandran, C. (2019). A review on versatile applications of degradable polymers. In D. Gnanasekaran (Ed.), *Material horizons: From nature to nanomaterials* (pp. 403–422). Springer Nature Ltd. https://doi.org/10.1007/978-981-13-8063-1_17

Kumar, A., Mishra, R. K., Verma, K., Aldosari, S. M., Maity, C. K., Verma, S., Patel, R., & Thakur, V. K. (2023). A comprehensive review of various biopolymer composites and their applications: From biocompatibility to self-healing. *Materials Today Sustainability*, 100431. https://doi.org/10.1016/j.mtsust.2023.100431

Luzi, F., Torre, L., Kenny, J. M., & Puglia, D. (2019). Bio- and fossil-based polymeric blends and nanocomposites for packaging: Structure-property relationship. *Materials*, *12*(3). https://doi.org/10.3390/ma12030471

Mihalca, V., Kerezsi, A. D., Weber, A., Gruber-Traub, C., Schmucker, J., Vodnar, D. C., Dulf, F. V., Socaci, S. A., Fărcaş, A., Mureşan, C. I., Suharoschi, R., & Pop, O. L. (2021). Protein-based films and coatings for food industry applications. *Polymers*, *13*(5). https://doi.org/10.3390/polym13050769

Pohjakallio, M. (2020). Secondary plastic products—examples and market trends. In *Plastic waste and recycling*. https://doi.org/10.1016/b978-0-12-817880-5.00018-9

Rosseto, M., Rigueto, C. V. T., Krein, D. D. C., Balbé, N. P., Massuda, L. A., & Dettmer, A. (2019). Biodegradable polymers: Opportunities and challenges. In A. Sand & E. Zaki (Eds.), *Biodegradable polymers: Opportunities and challenges, organic polymers* (pp. 481–505). IntechOpen. https://doi.org/10.5772/intechopen.88146

Samsung Electronics. (2019). *A sustainable future*. Samsung Electronics Sustainability Report. https://www.samsung.com/us/aboutsamsung/sustainability/strategy

Verma, D., & Fortunati, E. (2017). Biobased and biodegradable plastics-facts and figures. *Food & Biobased Reseacrh*, *4*(1722). https://doi.org/10.1007/978-3-319-68255-6_103

Verma, D., & Goh, K. L. (2018). Functionalized graphene-based nanocomposites for energy applications. In *Functionalized graphene nanocomposites and their derivatives: Synthesis, processing and applications* (pp. 219–243). Elsevier Inc. https://doi.org/10.1016/B978-0-12-814548-7.00011-8.

What are bioplastics? (2020). European Bioplastics. https://www.european-bioplastics.org/bioplastics/

Index

Note: Page numbers in *italics* indicate a figure and page numbers in **bold** indicate a table on the corresponding page.

For Product Safety Concerns and Information please contact our EU
representative GPSR@taylorandfrancis.com
Taylor & Francis Verlag GmbH, Kaufingerstraße 24, 80331 München, Germany

9 7 8 1 0 3 2 5 4 2 6 5 2